Ionic
Liquids
in
Chemical
Analysis

Analytical Chemistry Series

Charles H. Lochmüller, Series Editor
Duke University

Quality and Reliability in Analytical Chemistry
George-Emil Bailescu, Raluca-Ioana Stefan, Hassan Y. Aboul-Enein

HPLC: Practical and Industrial Applications, Second Edition
Joel K. Swadesh

Ionic Liquids
in
Chemical Analysis

Edited by
Mihkel Koel

CRC Press
Taylor & Francis Group
Boca Raton London New York

CRC Press is an imprint of the
Taylor & Francis Group, an **Informa** business

CRC Press
Taylor & Francis Group
6000 Broken Sound Parkway NW, Suite 300
Boca Raton, FL 33487-2742

First issued in paperback 2020

© 2009 by Taylor & Francis Group, LLC
CRC Press is an imprint of Taylor & Francis Group, an Informa business

No claim to original U.S. Government works

ISBN-13: 978-0-367-57747-6 (pbk)
ISBN-13: 978-1-4200-4646-5 (hbk)

Library of Congress Cataloging-in-Publication Data

Ionic liquids in chemical analysis / edited by Mihkel Koel.
    p. cm. -- (Analytical chemistry ; 3)
    Includes bibliographical references and index.
    ISBN-13: 978-1-4200-4646-5
    ISBN-10: 1-4200-4646-2
    1. Ionic solutions. 2. Chemistry, Analytic. 3. Solution (Chemistry) I. Koel', Mikhkel' Nikolaevich. II. Title. III. Series.

QD561.I635 2009
541'.372--dc22   2008013341

Visit the Taylor & Francis Web site at
http://www.taylorandfrancis.com

and the CRC Press Web site at
http://www.crcpress.com

Contents

Foreword

This excellent book is a well-timed event in a rapidly expanding area within chemistry and within the analytical/measurement community in chemistry. It comes at a time when there is growing pressure for many of us to become familiar with this *new* area and to speculate on the implications for our particular niche in chemical measurement science.

Much of the popular reference to ionic fluids seems to be focused on the *green* chemical use of these remarkable substances as more ecologically and environmentally acceptable solvents in analytical procedures. We are a long way from any evidence that every possible use of organic solvents in analytical procedures has an ionic fluid analog. Only time will tell.

I am led to reflect on my early ventures into the chemical literature of the late nineteenth century driven by curiosity and the need in those days of old to meet the foreign language requirements of the typical American doctoral program. One can find articles based purely on the discovery that, for example, acetone is a very useful solvent in organic synthesis! Recall, dear reader, that it took decades to adopt standards for what purity meant for many of the common organic solvents. As late as the 1970s, there was enough variance on the definition of purity that high-performance liquid chromatography methods, which were successful in one area of the world failed in others entirely on the basis of impurities in the components of the mobile phases used. Even pure methanol and pure acetonitrile were not readily available everywhere. My point is that before ionic fluid substitutes can be found for the use of compendial methods, there may be a good deal of work to be done.

Dense ionic fluids are not all that new if one examines the many applications of molten salt use in chemistry to date. A good deal of the work is in electrochemistry where the relatively high temperatures are less of a limitation but the relation between low-temperature molten salts and ionic fluids certainly exists. It would be wise neither to completely depend on nor to completely ignore all that has been learned with molten salts and molten salt chemistry. Some highly reactive, easily oxidized metals are readily purified in molten salt solvent systems without the problems with oxygen or the decomposition of water with release of hydrogen.

How will the microscopic properties of ionic fluids complicate the application of these remarkable substances to new kinds of analysis? If they are used for stationary phases (or mobile phases) in, for example, chromatography, it is likely that the familiar dominant factors in retention—volatility or solubility—will remain. Which of the unique properties of ionic fluids will have significant secondary effects on selectivity may be of concern. Will the *double layer* at the surface favor some sort of Gibbs isotherm with surface sorption competing with bulk dissolution? It is fine to speak of a dense ionic fluid as being electroneutral but how will that dense ionic atmosphere influence solubility for ionic species? Will mixed ionic fluids make analytical use of the common mechanistic organic strategy possible, where different solvents are used to favor by-product production, but, in the case of measurement, favor a more easily measured analyte? One could double the size of this book by speculating on what we will know 30 years hence. Only time will tell and thus there is a lot of work to be done.

Charles H. Lochmüller
Duke University

Acknowledgments

First of all, my sincere thanks to all authors who kindly agreed to participate in this project. They have done a very good job in preparing their chapters and have been cooperative in further finalizing them in time. I am truly grateful for that.

I would like to thank the Taylor & Francis production team, particularly our first project coordinator Amber Donley.

The support of Tallinn University of Technology must be acknowledged in preparation of this book, especially Chair of Analytical Chemistry Prof. Mihkel Kaljurand, whose constant interest and friendly comments helped very much in the development of this project.

Mihkel Koel
Tallinn University of Technology Tallinn, Estonia

Editor

Mihkel Koel defended his dissertation (Candidate of Chemical Sciences) in analytical chemistry at Leningrad State University (now St Petersburg, Russia) in 1989. He undertook long-term scientific visits to Duke University (United States), Helsinki University (Finland), and USIS Fulbright Fellowship (6 months in Los Alamos National Laboratory, 1999).

Since 1990, he has been a senior research scientist in the Department of Analytical Chemistry in the Institute of Chemistry of Estonian Academy of Sciences. Since 2002 he has been affiliated with the Faculty of Science, Institute of Chemistry, and Chair of Analytical Chemistry at Tallinn University of Technology.

His interests have included such areas of analytical chemistry as separation science, supercritical fluid extraction, chromatography, thermal analysis, and mathematical analysis of the chemical data (chemometrics).

Contributors

Jared L. Anderson
Department of Chemistry
University of Toledo
Toledo, Ohio

Gary A. Baker
Oak Ridge National Laboratory
Oak Ridge, Tennessee

Sheila N. Baker
Neutron Scattering Science Division
Oak Ridge National Laboratory
Oak Ridge, Tennessee

Rolf W. Berg
Technical University of Denmark
Lyngby, Denmark

Alain Berthod
Laboratory of Analytical
  Sciences
University of Lyon
Villeurbanne, France

Joan Frances Brennecke
Department of Chemical and
  Biomolecular Engineering
University of Notre Dame
Notre Dame, Indiana

Frank V. Bright
Oak Ridge National Laboratory
Oak Ridge, Tennessee

Samuel Carda-Broch
Department of Analytical
  Chemistry
University of Jaume I
Castellon de la Plana, Spain

Sheng Dai
Chemical Sciences Division
Oak Ridge National Laboratory
Oak Ridge, Tennessee

Urszula Domańska
Warsaw University of Technology
Warsaw, Poland

Vladimir M. Egorov
Lomonosov Moscow State University
Moscow, Russia

Ralf Giernoth
University of Cologne
Cologne, Germany

Christopher Hardacre
School of Chemistry and
  Chemical Engineering
Queen's University
Belfast, Northern Ireland,
  United Kingdom

William T. Heller
Oak Ridge National Laboratory
Oak Ridge, Tennessee

Mihkel Kaljurand
Tallinn University of Technology
Tallinn, Estonia

Mihkel Koel
Tallinn University of Technology
Tallinn, Estonia

Zulema K. Lopez-Castillo
Department of Chemical and
  Biomolecular Engineering
University of Notre Dame
Notre Dame, Indiana

Huimin Luo
Nuclear Science and Technology
  Division
Oak Ridge National Laboratory
Oak Ridge, Tennessee

Taylor A. McCarty
Oak Ridge National Laboratory
Oak Ridge, Tennessee

Berlyn Rose Mellein
Department of Chemical and
  Biomolecular Engineering
University of Notre Dame
Notre Dame, Indiana

Claire Lisa Mullan
School of Chemistry and Chemical
  Engineering
Queen's University
Belfast, Northern Ireland,
  United Kingdom

Igor V. Pletnev
Lomonosov Moscow State
  University
Moscow, Russia

Maria-Jose Ruiz-Angel
Department of Analytical
  Chemistry
University of Valencia
Burjassot, Spain

Svetlana V. Smirnova
Lomonosov Moscow State
  University
Moscow, Russia

Apryll M. Stalcup
University of Cincinnati
Cincinnati, Ohio

Andreas Tholey
University of Saarland
Saarbrücken, Germany

Merike Vaher
Tallinn University
  of Technology
Tallinn, Estonia

Tristan Gerard Alfred Youngs
School of Mathematics
  and Physics
Queen's University
Belfast, Northern Ireland,
  United Kingdom

Common ionic liquids— structure, name, and abbreviation

Common ILs include ammonium, phosphonium, sulfonium, guanidinium, pyridinium, imidazolium, and pyrrolidinium cations. However, use of nomenclature vary among researchers. It is prefered to abbreviate throughout this book as $[C_nC_mC_zIm]$, $[C_npy]$, and $[C_nC_mpyr]$ for the alkylimidazolium, pyridinium, and pyrrolidinium cations, respectively, where Im stands for imidazolium, py for pyridinium, and pyr for pyrrolidinium. The number of carbons in the N-alkyl chains are expressed by n, m, and z.

The most common anions are Cl⁻, Br⁻, $[BF_4]^-$, and $[PF_6]^-$. For simplicity, trifluoromethanesulfonyl $[CF_3SO_2]^-$ anion is abbreviated as [TfO], *bis*(trifluoromethanesulfonyl)imide $[(CF_3SO_2)_2N]^-$ anion as $[Tf_2N]^-$ (Tf is a short-hand notation for triflate), and dicyanamide $[N(CN)_2]^-$ anion as [dca]. There are several examples of alkylsulfate anions, which we will abbreviate $[C_nSO_4]^-$, where n is the carbon number of alkyl chain.

Some Examples of Names, Structural Formulas, and Abbreviations Used

Ionic Liquid	Structural Formula	Abbreviation
1-Ethyl-3-methylimidazolium hexafluorophosphate		$[C_2C_1Im][PF_6]$
1-Butyl-3-methylimidazolium hexafluorophosphate		$[C_4C_1Im][PF_6]$
1-Butyl-3-methylimidazolium chloride		$[C_4C_1Im]Cl$

(continued)

[C₆C₁Im][Tf₂N]

[C₁C₁Im][C₁SO₄]

[C₄C₁Im][C₁SO₄]

[C₂C₁Im][C₂SO₄]

1-Hexyl-3-methylimidazolium
bis(trifluoromethylsulfonyl)imide

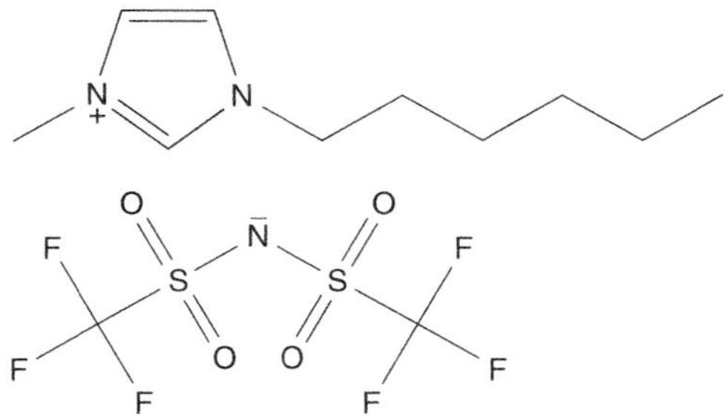

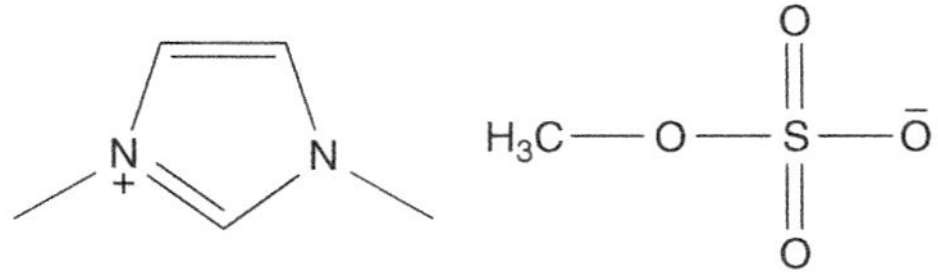

1,3-Dimethylimidazolium methylsulfate

1-Butyl-3-methylimidazolium
methylsulfate

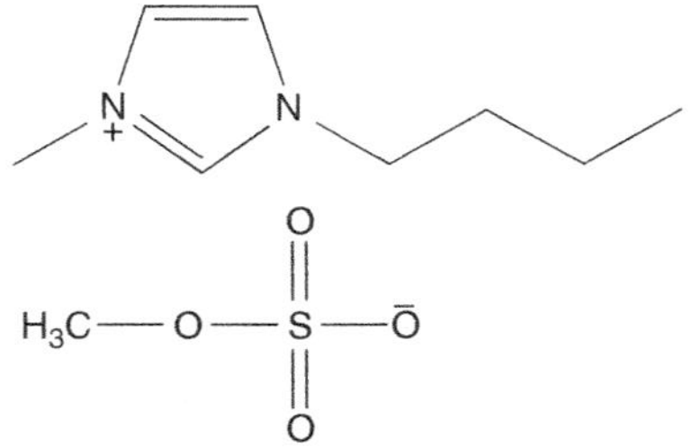

1-Ethyl-3-methylimidazolium
ethylsulfate

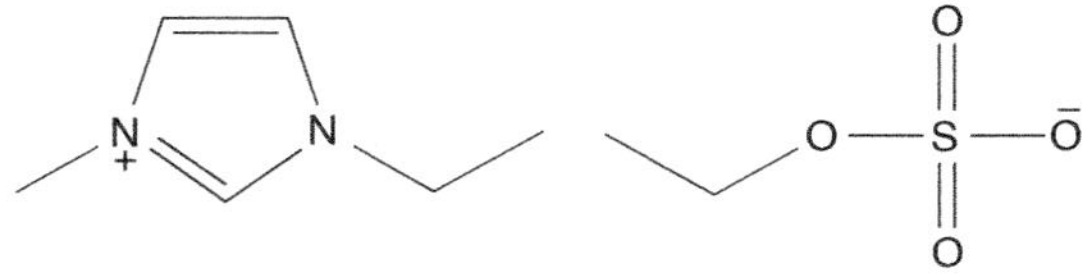

(Continued)

Ionic Liquid	Structural Formula	Abbreviation
1-Butyl-3-methylimidazolium octylsulfate		$[C_4C_1Im][C_8SO_4]$
1-Ethyl-3-methylimidazolium tosylate		$[C_2C_1Im][TOS]$
1-Hexyloxymethyl-3-methylimidazolium tetrafluoroborate		$[C_6H_{13}OCH_2-C_1Im][BF_4]$

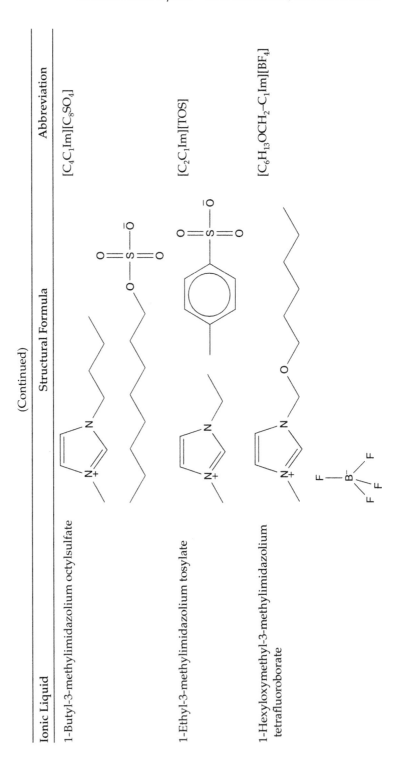

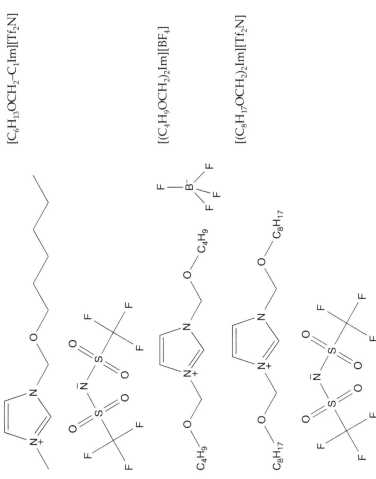

[C₆H₁₃OCH₂–C₁Im][Tf₂N]

1-Hexyloxymethyl-3-methylimidazolium
bis(trifluoromethylsulfonyl)imide

[(C₄H₉OCH₂)₂Im][BF₄]

1,3-Dibutyloxymethylimidazolium
tetrafluoroborate

[(C₈H₁₇OCH₂)₂Im][Tf₂N]

1,3-Dioctyloxymethylimidazolium
bis(trifluoromethylsulfonyl)imide

(continued)

(Continued)

Ionic Liquid	Structural Formula	Abbreviation
1,3-Didecyloxymethylimidazolium *bis*(trifluoro-methylsulfonyl)imide		$[(C_{10}H_{21}OCH_2)_2Im][Tf_2N]$
1-Butyl-3-methylimidazolium 2-(2-methoxyethoxy)-ethysulfate		$[C_4C_1Im][MDEGSO_4]$
4-Butyl-pyridinium tetrafluoroborate		$[4-C_4py][BF_4]$
Ethyl-(2-hydroxyethyl)–dimethyl-ammonium tetrafluoroborate		$[(C_1)_2C_2\ HOC_2N][BF_4]$

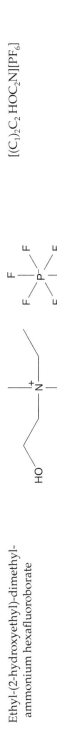

Name	Structure	Abbreviation
Ethyl-(2-hydroxyethyl)-dimethyl-ammonium hexafluoroborate		$[(C_1)_2C_2\ HOC_2N][PF_6]$
Butyl-(2-hydroxyethyl)-dimethyl-ammonium bromide		$[(C_1)_2C_4\ HOC_2N]Br$
Hexyl(2-hydroxyethyl)dimethyl-ammonium bromide		$[(C_1)_2C_6\ HOC_2N]Br$
Hexyl(2-hydroxyethyl)dimethyl-ammonium tetrafluoroborate		$[(C_1)_2C_6\ HOC_2N][BF_4]$
Dideyldimethylammonium nitrate		$[(C_{10})_2(C_1)_2N][NO_3]$
(Benzyl)dimethylalkylammonium nitrate	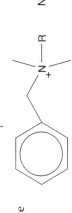	$[Be(C_1)_2C_nN][NO_3]$

(continued)

(Continued)

Ionic Liquid	Structural Formula	Abbreviation
Tetrabutylphosphonium methanesulfonate		$[(C_4)_4P][C_1SO_3]$
1-(4-Methoxyphenyl)imidazolium trifluoromethanesulfonate		$[CH_3O–PhC_1Im][TfO]$
1-Benzyl-3-methylimidazolium trifluoromethanesulfonate		$[BeC_1Im][TfO]$

Note: The less common ionic liquids not seen in this list are described in the text.

List of abbreviations

2,5-Dihydroxybenzoic acid	DHB
α-Cyano-4-hydroxycinnamic acid	HCCA
β-Cyclodextrin	β-CD or HP-β-CD
Acetonitrile	ACN
Aqueous two-phase liquid system	ATPS
Atmospheric pressure chemical ionization	APCI
Background electrolyte	BGE
Calcium binding proteins	CALB
Calix[4]arene-*bis*(*tert*-octylbenzo-crown-6)	BOBCalixC6
Capillary electrochromatography	CEC
Capillary electrophoresis	CE
Capillary gel electrophoresis	CGE
Capillary zone electrophoresis	CZE
Charge transfer	CT
Chiral stationary phases	CSP
Circular dichroism	CD
Collision-induced decomposition	CID
Conductor-like screening model for real solvents	COSMO-RS
Countercurrent chromatography	CCC
Critical micelle concentration	CMC
Crown ether dicyclohexano-15-crown-5	DC15C5
Crown ether dicyclohexano-18-crown-6	DC18C6
Cytochrome *c*	Cyt-c
Dibenzothiophene	DBT
Differential scanning calorimetry	DSC
Diffusion-ordered spectroscopy	DOSY
Dimethylated β-cyclodextrin	β-DM
Dimethyldinonylammonium bromide	DMDNAB
Dimethylsulfoxide	DMSO
Dioctylsulfosuccinate	docSS
Diode array detector	DAD
Double-stranded DNA	dsDNA
Dual spin probe	DSP
Electrochemiluminescence	ECL

Electron impact	EI
Electroosmotic flow	EOF
Electrospray ionization	ESI
Empirical potential structure refinement	EPSR
Ethanol	EtOH
Ethyl *tert*-butyl ether	ETBE
Extended x-ray absorption fine structure	EXAFS
Fast atom bombardment	FAB
Flory–Benson–Treszczanowicz model	FBT
Fourier transform infared	FTIR
Fourier transform MS	FTMS
Gas chromatography	GC
Gas–liquid chromatography	GLC
Group contribution method of predicting activity coefficients	UNIFAC
Heteronuclear multiple-quantum correlation	HMQC
High-pressure liquid chromatography	HPLC
High-resolution magic angle spinning NMR	HR-MAS NMR
Indium tin oxide	ITO
Inductively coupled plasma	ICP
Ionic liquid	IL
Ionic liquid matrices	ILM
Laser desorption/ionization	LDI
Linear solvation free energy relationship	LSFER
Liquid chromatography	LC
Liquid phase microextraction	LPME
Liquid–liquid equilibrium	LLE
Mass spectrometry	MS
Matrix-assisted laser desorption/ionization	MALDI
Mercury film electrode	MFE
Methanol	MeOH
Micellar electrokinetic capillary chromatography	MEKC
Multiple linear regression analysis	MLRA
Near infrared	NIR
Nicotinamide adenine dinucleotide	NADH
N-methyl-2-pyrrolidinone	NMP
Nonaqueous capillary electrophoresis	NACE
Nuclear magnetic resonance	NMR
Nuclear Overhauser effect	NOE
Nuclear Overhauser effect spectroscopy	NOESY
Peak asymmetry factor	PAF
Permethylated β-cyclodextrin	β-PM
Phosphorous oxychloride	$POCl_3$
Phosphorous trichloride	PCl_3
Polydimethylsiloxane	PDMS

Poly(ethylene oxide)	PEO
Polyethylene glycol	PEG
Polyoxyethylene-100-stearylether	Brij 700
Polyoxyethylene-23-laurylether	Brij 35
Polytetrafluoroethylene	PTFE
Quality assessment and quality control	QA and QC
Radial distribution functions	RDFs
Reversed phase LC	RPLC
Rotating frame Overhauser effect spectroscopy	ROESY
Secondary ion mass spectrometry	SIMS
Self-assembled monolayer	SAM
Sinapinic acid	SA
Sodium dodecylsulfate	SDS
Solid phase microextraction	SPME
Solid–liquid phase equilibria	SLE
Static structure factor	S(Q)
Surface-confined ionic liquids	SCIL
Surface-enhanced Raman scattering	SERS
Task-specific ionic liquids	TSILs
Tetradecyltrimethylammonium bromide	TTAB
Tetrahydrofuran	THF
Thermogravimietric analysis	TGA
Time of flight	ToF
Total correlation spectroscopy	TOCSY
Ultraviolet–visible	UV–Vis
Upper critical solution temperature	UCST
Vapor–liquid equilibrium	VLE
Vogel–Tammann–Fulcher equation	VTF
Wall-coated open tubular	WCOT
X-ray photoelectron spectroscopy	XPS

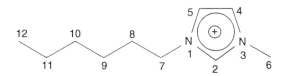

Scheme 1 Numbering scheme on the 1-hexyl-3-methylimidazolium cation, $[C_6C_1Im]^+$, showing the three ring protons H2, H4, and H5.

Introduction

In sale et sole existunt omnia
(Life depends on salts and sunshine)

The development of analytical chemistry continues at a steady rate and every new discovery in chemistry, physics, molecular biology, and materials science finds a place in analytical chemistry as well. The place can either be a new tool for existing measurement challenges or a new challenge to develop stable and reliable methods. Two examples are the advent of nano-structure materials and alternative solvents, both of which saw their main development in the past decade. Nanostructural materials pose a new scale of measurement challenge in size and number. New solvents with their environmentally benign properties offer a possibility for wasteless operation.

What happens when salt is melting and can this melt be used in chemical processes? This question has a long history, and usually organic chemists are not interested in finding solutions because of high temperatures related to melts. It was true up to the end of the twentieth century when room temperature molten salts became available. Their difference from common liquids is emphasized by calling them *ionic liquids* (ILs). The term is used loosely to describe organic salts with their melting point close to or below room temperature. ILs define a *class* of fluids rather than a small group of individual examples. And this was a successful choice of the name. They form liquids composed in the majority of ions. This gives these materials the potential to behave very differently in contrast to conventional molecular liquids when used as solvents. ILs promise entirely new ways to do solution chemistry, which could improve both measurement and the impact of the amount of waste into the environment.

Historically, the following four main steps must be mentioned: the preparation of ethylammonium nitrate $[C_2H_5NH_3][NO_3]$ by Paul Walden in 1914 is recognized by many as the first IL. This compound has a melting point of 12°C but owing to its high reactivity has not really found a use [1]. This was the outcome of his studies of conductivity and electrical properties of salt solutions, especially nonaqueous solutions of organic salts. He conducted very systematic studies with different solvents and salts, and his special interest was in ammonium salts. But Walden himself pointed on the work of C. Schall on alkyl-quinoline triiodide's as low-melting organic salts [2], which are nowadays under the study as tailor-made ILs [3].

At that time, Paul Walden (1863–1957) (because of his Latvian origin, he is known in Latvia as Pauls Valdens) was working in Riga Polytechnicum. Later, Riga Polytechnicum was restructured and became an official university of Russian Empire—Riga Polytechnic Institute. In 1910, Walden became a member of the Russian Academy of Science. Between 1911 and 1915 in St Petersburg, Walden was head of the Laboratory of Chemistry at the Academy of Science. Later, he joined Riga Polytechnic Institute again, but during World War I, the Institute was evacuated to Moscow where Walden worked until his return to Riga in 1918. Although Latvia had gained independence, at that time free and international contacts were not supported, and Walden left for Germany and served as Head of the Chemistry Department of the University of Rostock from 1919 to 1934. Paul Walden made a series of specific discoveries in chemistry (e.g., *Walden Inversion*) and can be regarded as the founder of two new scientific fields—dynamic stereochemistry and the electrochemistry of nonaqueous solutions.

The main consensus seems to be that the first major studies of room temperature molten salts were made in the 1940s by a group led by Frank Hurley and Tom Weir at Rice University. When they mixed and gently warmed powdered pyridinium halides with aluminum chloride, the powders reacted, giving a clear, colorless liquid [4–7]. These mixtures were meant to be used in electrochemistry, particularly in electroplating with aluminum.

The third step was the introduction of alkylimidazolium salts in the early 1980s [8]. It was the discovery of 1-ethyl-3-methylimidazolium-based chloroaluminate ILs in 1982 that accelerated activities in the area of room temperature ILs (RTILs). The named salt is one of the most widely studied room temperature melt systems, which is liquid at room temperature for compositions between 33 and 67 mol% $AlCl_3$. The exciting property of halogenoaluminate ILs is their ability for acid–base chemistry, which can be varied by controlling the molar ratio of the two components. This kind of tuning makes these ILs attractive as nonaqueous reaction media. Aluminum chloride, however, reacts readily with water, which has limited the use of these types of ILs to the electrodeposition of metals and some synthetic reactions requiring very strong Lewis acid catalysts.

Almost that time, in the 1980s, the term *ionic liquid* became more popular to describe organic salts that melt below ~100°C and have an appreciable liquid range.

The fourth step is related to the search for air- and water-stable ILs, which followed 10 years later, and this gave a real push for further developments in this area. Air- and water-stable molten salts can be obtained using the weakly complexing anion in the imidazolium compound [9].

With these studies, the alkylimidazolium-based salts became almost synonyms of ILs. Possibly, the most widely studied one is 1-butyl-3-methylimidazolium hexafluorophosphate, which is liquid at room temperature and melts at −8°C. This particular IL has been studied for a wide range of applications.

An excellent short history of ILs, which covers the crucial moments of this area up to 1994, is presented by eyewitness to and participant in crucial developments, Professor John S. Wilkes [10].

ILs have proved to be as media not only for potentially *green* synthesis, but also for novel applications in the analysis, where the unique properties of these liquid materials provide new options based on different chemical and physical properties. ILs can be applied not only in the existing methods where it is always needed to improve sensitivity and selectivity of the analysis, but their different behavior and properties can offer original solutions in chemical analysis; and the search for new applications of ILs is growing in every area of chemistry including analytical chemistry.

However, introduction of the dimension of green chemistry into the assessment of analytical methods should be a natural trend of development in chemistry and it should coincide with its general policy. Some of the principles of green chemistry, such as prevention of waste generation, safer solvents and auxiliaries, design for energy efficiency, safer chemistry to minimize the potential of chemical accidents, and the development of instrumental methods are directly related to analytical chemistry. Analytical methods are developing fast, whereas concern about the safety of environment, water, and food is a strong driving force. Also, the public needs confirmation that chemical products and processes are safe.

Several reviews have been published about ILs and analytical chemistry, fortunately now we have main players in this field in one place who kindly agreed to provide their contributions. This book is an attempt to collect experience and knowledge about the use of ILs in different areas of analytical chemistry such as separation science, spectroscopy, and mass spectrometry that could lead others to new ideas and discoveries. In addition, there are chapters providing information of studies on determination of physicochemical properties, thermophysical properties and activity coefficients, phase equilibrium with other liquids, and discussion about modeling, which are essential to know beforehand, also for wider applications in analytical chemistry.

There are very promising examples of the use of these unique materials in the primary literature. ILs have good solvating properties together with broad spectral transparency, making them suitable solvents for spectroscopic measurements. It was tried to cover interest in different extraction techniques also, starting with separation of gases and ending with metals. It has been demonstrated that task-specific ILs (TSILs) have advantages compared to common solvents used as separation media in the liquid–liquid extraction process achieving both high efficiency and selectivity of separation. The main advantage of ILs for other applications in analytical chemistry lies in their low volatility, which makes ILs useful as solvents for working in both high temperature (GC stationary phases) and high vacuum (MALDI matrixes) environments. The reader will find changes in style and emphasis in different chapters that have been prepared by different authors; hopefully this variety is the strength of this book.

Using an IL as a solvent or an electrolyte medium, it is possible to achieve a broader range of operational temperatures and conditions, relative to other conventional electrolytic media, and make them promising materials in various electrochemical devices for analytical purposes, such as sensors and electrochromic windows also. However, this book is not intended to give a detailed coverage of analytical electrochemistry with ILs. To justify this, we use the words of Prof. Keith Johnson, University of Regina, Canada: "There is a vast literature on electrochemistry of ILs, both the recent organic and semiorganic salts and the inorganic 'molten' salts. This consists of papers in which the liquids are characterized by their electrochemistry or the properties of solutes in them are studied electrochemically. That does indeed give a pool of information on electrochemistry in these systems just as is available for aqueous or organic solutions (see Ohno's book [11] or Baizer's book [12] for the latter). However, if one has a problem of determining X in Y, treating the sample with an ionic system may be a means to making a solution of X, but electroanalysis of said solution to measure X may not be straightforward. It appears not to have been tried and would likely be prohibitively expensive through IL consumption. The picture may change in a few years but at present there is little to say—certainly not enough for a book chapter."

The approach in this book will be that of a tutorial providing an aid to the novice to enter the area that will include both new and senior scientists. Therefore, it is not expected to give a complete coverage of the literature in the area. Also time sets the limit and we had to stop on the middle of 2007. We do not expect that ILs will solve every problem in chemistry (in our case analytical chemistry), but our hope is to help find a proper area where the use of these materials could be the most advantageous. Thus, for ILs the future is bright and the future has to be green.

References

1. Walden, P., Ueber die Molekulargrösse und electrische Leitfähigkeit einiger geschmolzenen Salze. *Bulletin de l'Academie Imperiale des Sciences de St. Petersburg,* 405–422, 1914.
2. Schall, C., Conductivity measurements of fused organic salts. *Zeitschr. f. Elektrochem.,* 14, 397–405, 1908.
3. Jork, C., Kristen, C., Pieraccini, D., Stark, A., Chiappe, C., Beste, Y.A., and Arlt, W., Tailor-made ionic liquids, *J. Chem. Thermodyn.,* 37, 537–558, 2005.
4. Hurley, F.H., Electrodeposition of aluminum. U.S. Patent No. 2 446 331, 1948.
5. Wier, T.P. and Hurley, F.H. Electrodeposition of aluminum. U.S. Patent No. 2 446 349, 1948.
6. Wier, T.P., Electrodeposition of aluminum. U.S. Patent No. 2 446 350, 1948.
7. Hurley, F.H. and Wier, T.P., The electrodeposition of aluminium from nonaqueous solutions at room temperature. *J. Electrochem. Soc.,* 98, 207–212, 1951.

8. Wilkes, J.S., Levisky, J.A., Wilson, R.A., and Hussey, C.L., Dialkylimidazolium chloroaluminate melts: A new class of room-temperature ionic liquids for electrochemistry, spectroscopy, and synthesis. *Inorg. Chem.*, 21, 1263–1264, 1982.

9. Wilkes, J.S. and Zaworodtko, M.J., Air and water stable 1-ethyl-3-methylimidazolium based ionic liquids. *J. Chem. Soc. Chem. Comm.*, 13, 965–967, 1992.

10. Wilkes, J.S., A short history of ionic liquids—from molten salts to neoteric solvents, *Green Chem.*, 4, 73–80, 2002.

11. Hiroyuki Ohno (Ed.), *Electrochemical Aspects of Ionic Liquids*, Wiley, Hoboken, NJ, p. 408, 2005.

12. Lund, H. and Baizer, M.M. (Eds), *Organic Electrochemistry: An Introduction and Guide* (3rd Edition), Dekker, New York, p. 639, 1991.

chapter one

General review of ionic liquids and their properties

Urszula Domańska

Contents

1.1 Introduction

Ionic liquids (ILs) can be composed from a large number of cations and anions, with an estimated number of possible ILs on order 10^{18}, which makes this class of compounds as one of the largest known in chemistry. Knowledge of the physical properties of ILs and the phase behavior with gases, liquids, and solids (including inorganic salts) is important for evaluating and selecting ILs for each application as well as process design. ILs have recently

1

become very popular as potential solvents for industrial applications in many different disciplines of science and environment. Enormous progress was made during the recent 10 years to synthesize new low-melting ILs that can be handled under ambient conditions, and nowadays more than 350 ILs are already commercially available. Although the main interest is still focused on the synthesis in room temperature, IL as solvent/catalyst, on electrochemistry, physical chemistry in ILs, various aspects of thermodynamics, and ILs in analytical chemistry are discussed in this chapter. The focus is placed on air- and water-stable ILs as they will presumably dominate various fields of chemistry in the future. Indeed, because each IL has its own unique properties, it should be possible to design a compound, a solvent, an additive, nanostructure particles, sensors, gels, or mixtures to suite particular applications.

Relevant literature available up to February 2007 has been covered. The presentation is restricted to systems which seem to have the best prospects for successful use in different fields of chemistry and chemical engineering.

International organizations, for example, IUPAC Thermodynamics, started to collect the physical–chemical and thermodynamic properties of ILs about 10 years ago. Now, for the first time we can find massive data in two data banks: Dortmund Data Bank, Germany [1] and NIST Boulder Colorado, USA [2].

It is generally known that the examined properties and phase behavior of ILs vary on cation and anion structures changing. Some typical trends will be presented in this chapter on the basis of the structural effect on the interactions between counterpart ions (see, for example Ref. 3, the spoon-shaped structure of the unit cell of the 1-dodecyl-3-methylimidazolium hexafluorophosphate, $[C_{12}C_1Im][PF_6]$), and between the IL and the solvent, or the coexisting compound. The structure of IL and its interaction with the environment is extremely important in applications in analytical chemistry [4].

This chapter reviews developments in physical–chemical properties, thermophysical properties, phase equilibria, activity coefficients, modeling, and electrochemistry.

1.2   Effect of the structure on physical–chemical properties

The physical–chemical properties of ILs depend on the nature and size of both their cation and anion constituents. Their application in science and industries is merited because ILs have some unique properties, such as a negligible vapor pressure, good thermal stability, tunable viscosity and miscibility with water, inorganic and organic substances, a wide electrochemical window, high conductivity, high heat capacity and solvents available to control reactions. Despite their wide range of polarity and hydrogen-bonding ability, these new solvents are liquid from 180 K (glass transition) to 600 K. Possible choices of cation and anion that will result in the formation of ILs are numerous. The most popular five different well-known classes of ILs are as follows: imidazolium, pyridinium, pyrrolidinium quaternary ammonium, and tetra alkylphosphonium ILs. Of these, the most popular in experimental laboratory work worldwide are undoubtedly 1,3-dialkylimidazolium salts, primarily due to the attractive and easy-tailored physical properties. To do this, however, it is necessary to assume that ILs are the substances of which the local structural (i.e., electronic and steric) features may be correlated with their physical–chemical properties. The effect of the cation and anion structures has been studied in detail. The influence on physical–chemical properties and phase behavior is the subject of discussion in this chapter.

From early research it is known that ILs are hydrogen-bonded substances with strong interionic interactions. For example, the isomorphous crystals of 1-ethyl-3-methylimidazole iodide and bromide structures consist of layers of anions and cations interconnected by an extended network of hydrogen bonds. Each cation is hydrogen-bonded to three anions and each anion is hydrogen-bonded to three cations [5]. On the other hand, 1-ethyl-3-methylimidazolium chloride has a more complicated crystallographic structure—chloride anion is hydrogen-bonded to the three cations, but to different ring protons than iodide and bromide salts. Focusing on the other anions, there is a clear anion effect seen in the properties with special place for *bis*(trifluoromethylsulfony)imide, $[Tf_2N]^-$ ion, which appears not to fit in the simple trend, as this anion becomes more basic, the hydrogen bond donor ability of the IL decreases. Usually, there is an important Coulombic contribution to the hydrogen bonds formed between the IL and the other solvents especially for $[Tf_2N]^-$ and $[NO_3]^-$ anions [6]. Interactions between two molecular solvents are usually described by their polarity, as expressed through their dielectric constants. Since this scale is unable to provide adequate correlations with many experimental data with ILs, the hydrogen-bond acidity, hydrogen-bond basicity and dipolarity and polarizability effects have mainly been used. Changing the cation or anion, their solvent properties can differ considerably from one another as well as from traditional molecular solvents. Two different ILs that have essentially identical *polarity* ratings or descriptors can produce very different results when used as solvents for organic reactions, gas–liquid chromatography (GLC), or extraction. ILs with additional functional groups are capable of having additional interactions with other solvents or dissolved molecules. By demonstrating their structure and diversity of functionality, they are capable of most types of interactions as dispersive, π–π, n–π, hydrogen bonding, dipolar, and ionic/charge–charge interactions already mentioned. In every solution, there can be a number of different (in terms of type and strength) and often simultaneous solute–solvent interactions. Several approaches have been proposed that allow one to examine and categorize different IL–molecular solvent interactions. The Rohrschneider–McReynolds constants were originally developed to characterize liquid stationary phases for gas chromatography on the basis of several different interaction parameters [7]. The solvation parameter model developed by Abraham has been used to characterize either liquid- or gas-phase interactions between solute molecules and liquid phases [8–10]. The classification of the 17 ILs based on dipolarity and hydrogen-bond basicity characteristics has been presented to provide a model that can be used to pick ILs for specific organic reactions, liquid extractions, or GLC stationary phases [11].

The structural factors of the cation are focused on symmetry, polarity (charge density), number of carbon atoms in the alkane substituent and its flexibility, the rotational symmetry of the head ring, the cyclic and branched structures, and the functional tail group. Similar structural factors have an influence on the properties of anions, including charge delocalization either by a large volume of the central atom, or by the presence of the perfluoroalkyl chain. It was recently very punctiliously discussed by Hu and Xu [12], and also the discussion on many other ILs, including pyrazinium, piperazinium, and chiral ILs, has been presented.

Chemical structures, names, and abbreviations of modern typical ILs, most of which will be discussed in this chapter, are listed in Table 1 of the Introduction.

The structure of the extremely popular anion $[Tf_2N]^-$ is specified—the charge from the central nitrogen is delocalized onto the neighboring sulfur atoms, but not to any great extent onto the four sulfonyl oxygen atoms. Thus, the delocalized charge is confined to the molecule and shielded by the oxygens and terminal $-CF_3$ groups from Coulombic interactions with neighboring cations [13]. The cations or anions containing oxygen, such as alkoxy groups, for example $[C_6H_{13}OCH_2-C_1Im]^+$ or $[MDEGSO_4]^-$, tend to have specific interactions due to hydrogen bonding, or the dipole–dipole interactions. The acidic C(2)-H group of the imidazolium ring is bonded to a carbon located between two electronegative nitrogen atoms and contrary to ammonium, phosphonium, and sulfonium, ILs could engage in H-bonding. C(2)-H has greater ability to hydrogen bonding than the D(4), or C(5)-H groups, proved by the phase equilibria measurements and the interaction with alcohols [14], spectroscopic nuclear magnetic resonance (NMR) measurements, and *ab initio* calculations [15–17]. The H-bonding ability of the anion strongly depends, on its effective charge density and its symmetry.

1.3   Densities, viscosities, and transport properties

The effect of the cation/anion structure on density was perfectly discussed for more than 300 ILs in a recent publication [12]. ILs are mostly denser than water with values ranging from 1 for typical ILs to 2.3 g cm^{-3} for fluorinated ILs. For example, the densities of three salts are: $[C_8C_1Im][BF_4]$ 1.08 g cm^{-3} or $[C_{10}C_1Im][BF_4]$ 1.04 g cm^{-3}, whilst trifluoromethylethylpyridazinium *bis*(trifluoromethylsulfonyl)imide, 2.13 g cm^{-3} at $T = 298.15$ K [18,19]. Density depends strongly on the size of the ring in the cation, on the length of the alkyl chain in the cation, on the symmetry of ions and on the interaction forces between the cation and the anion. The ILs with aromatic head ring, in general, present greater densities than pyridinium head ring ILs and than do imidazolium ring ILs. Density increases with increasing symmetry of their cations. The increases of an alkyl chain diminish the densities in a systematic manner (sometimes only slightly, as was shown above) [12,20]. ILs with functional groups reveal higher densities than those of alkyl chains. The densities of ILs based on 1,3-dialkylimidazolium cations increase for typical anions in the order: $[Cl]^- < [BF_4]^- < [C_2SO_4]^- < [PF_6]^- < [Tf_2N]^-$. Very popular in many recent studies, especially in the liquid–liquid extraction research, $[C_6C_1Im][Tf_2N]$ displayed density equal to 1.372 g cm^{-3} at $T = 298.15$ K [12]. For ammonium salts, an increase in an alkyl chain at cation increases densities. The influence of popular anions and cations of imidazolium salts is presented in Figure 1.1.

It is clearly shown that salts with $[Tf_2N]^-$ anion have higher density than salts with $[PF_6]^-$ anion. The influence of the alkyl chain on the density of specific alkoxy- and acetoxy-ammonium salts is shown in Figure 1.2.

Densities are higher for the alkoxymethyl(2-hydroxyethyl)-dimethylammonium salts than for (2-acetoxyethyl)alkoxymethyl-dimethylammonium salts [21]. The increases of the alkyl chain at the cation from ethyl- to dodecyl- decrease the density from 1.45 to 1.21 g cm^{-3}. The densities of imidazolium salts with popular cation $[C_2C_1Im]^+$ are presented in Table 1.1.

It is easy to notice that the lowest value of density is with dicyanamide anion, $[dca]^-$, and the highest with the *bis*(trifluoromethylsulfonyl)imide anion, $[Tf_2N]^-$.

There is little data available covering on density as a function of pressure and temperature. Typical imidazolium salts, $[C_4C_1Im][PF_6]$, $[C_4C_1Im][BF_4]$, $[C_4C_1Im][Tf_2N]$, $[C_6C_1Im][Tf_2N]$, and $[C_8C_1Im][BF_4]$, as well as $[C_8C_1Im][PF_6]$, were measured from 298 to 333/343 K and up to 60, or 200 MPa [28–30]. Recently, the densities of phosphonium salts (trihexyltetradecylphosphonium

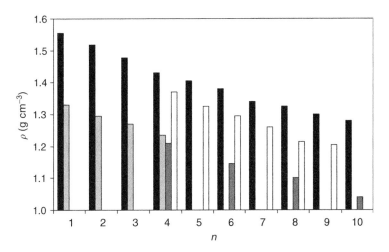

Figure 1.1   The influence of cation and anion on the density of imidazolium $[C_nC_1Im][X]$ ionic liquids at $T = 298.15$ K, where $n = 1–10$ and anions $[X]^-$ are $[Tf_2N]^-$ (black), $[AlCl_4]^-$ (light grey), $[BF_4]^-$ (dark grey), and $[PF_6]^-$ (white).

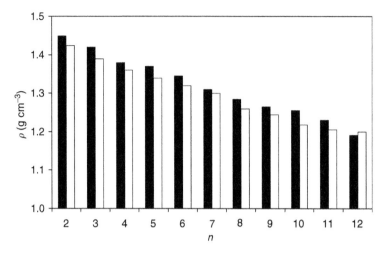

Figure 1.2 The influence of anion on the density of ammonium ionic liquids with alkoxy and acetoxy cations and [Tf₂N]⁻ anion at $T = 298.15$ K: alkoxymethyl(2-hydroxyethyl)-dimethyl cation (black) and (2-acetoxyethyl)alkoxymethyl-dimethyl cation (white). (Adapted from Pernak, J., Chwała, P., and Syguda, A., *Polish. J. Chem.*, 78, 539, 2004.)

Table 1.1   Densities, ρ, at 298.15 K of Several Ionic Liquids with $[C_2C_1Im]^+$ Cation

Anion	ρ (g cm^{-3})	Reference
$[BF_4]^-$	1.240	22
	1.279	23
	1.280	24
$[Tf_2N]^-$	1.520[a]	25
	1.518	23
$[CF_3SO_3]^-$	1.390[a]	25
$[CF_3CO_2]^-$	1.390	26
	1.285[a]	25
$[C_2SO_4]^-$	1.236	Our value
[dca]$^-$	1.060	27

[a] At $T = 295.15$ K.

chloride, or acetate, or *bis*(trifluoromethylsulfonyl)imide) at 298–333 K and up to 65 MPa were published [31]. The densities of these ILs are significantly lower than that of commonly used imidazolium-based ILs. With an increasing pressure the density increases.

Transport properties play an important role in chemical reactions, electrochemistry, and liquid–liquid extraction. This concerns mainly the viscosity of ILs and their solutions with molecular solvents. Viscosity of ILs, typically at the level of 10–500 cP at room temperature, is much higher than that characteristic of water ($\eta(H_2O) = 0.89$ cP at 298.15 K) and aqueous solutions. The high dynamic viscosity (viscosity coefficient) of ILs causes difficulties

Table 1.2   Viscosity, η, at 298.15 K of Several
Ionic Liquids with $[C_4C_1Im]^+$ Cation

Anion	η (cP)	Reference
$[I]^-$	963	37
$[NO_3]^-$	266	38
$[BF_4]^-$	233[a]	32
$[PF_6]^-$	312[a]	32
$[NbF_6]^{-b}$	95	39
$[CF_3SO_3]^-$	90[c]	32
$[CF_3CO_2]^-$	73[c]	32
$[Tf_2N]^-$	52[c]	32

[a] At T = 303.15 K.
[b] Hexafluoroniobate anion.
[c] At T = 293.15 K.

in practice. It affects the diffusion of solutes and practical issues, such as stirring and pumping. Data of viscosity at different temperatures were published by many authors [12,32–36]. The influence of the anion for a few salts with the $[C_4C_1Im]^+$ cation is presented in Table 1.2.

The viscosity of ILs is determined by van der Waals forces and hydrogen-bonded structures. Electrostatic forces and the shift of charge at the anion may also play an important role. For the same $[C_4C_1Im]^+$ cation, the viscosities for typical anions decrease in the order: $[I]^- > [PF_6]^- > [BF_4]^- > [TfO]^- > [CF_3CO_2]^- > [Tf_2N]^-$. For a series of 1-alkyl-3-methylimidazolium cations, increasing the alkyl chain length from butyl to octyl increases the hydrophobicity and the viscosity of the IL (for $[PF_6]^-$ and $[BF_4]^-$ anions), whereas densities and surface tension values decrease. This is due to stronger van der Waals forces between cations, leading to an increase in the energy required for molecular motion [40]. It is expected that the replacement of the alkyl chain by a hydroxyl functional group would increase the viscosity by increasing the H-bonding. Pernak et al. [21] found that for alkyl chain lengths from 2 to 12 on alkoxymethyl(2-hydroxymethyl)dimethylammon ium, (A), or (2-acetoxyethyl)-alkoxymethyldimethyl-ammonium, (B), salts with $[Tf_2N]^-$ anion, the viscosity monotonously increases (75–225 cP for A and 160–290 cP for B). The same effect was observed for the 1-[(1R,2S,5R)-(-)-menthoxymethyl]-3-alkylimidazolium *bis*(trifluoromethylsulfonyl)imides [41]; and new, chiral ammonium-based ILs containing the same (1R,2S,5R)-(-)-menthyl group and $[Tf_2N]^-$ anion [42]. Unfortunately, the values of viscosity of chiral ammonium-based ILs, 710–880 cP (303.15 K), as well as chiral pyridinium-based ILs, 550–1003 cP (323.15 K), were very high [42,43].

Viscosity temperature dependence in ILs is more complicated than in most molecular solvents, because most of them do not follow the typical Arrhenius behavior. Most temperature studies fit the viscosity values into the Vogel–Tammann–Fulcher (VTF) equation, which adds an additional adjustable parameter (glass transition temperature) to the exponential term.

In general, all ILs show a significant decrease in viscosity as the temperature increases. A systematic study of a possible description covering ILs by the Arrhenius or VTF equations was made by Okoturo and Van der Noot [44].

The viscosity decreases and the transport properties will improve after adding one or two organic solvents or water [45]. Pure ILs with [dca]$^-$ have much lower viscosities [27]; for [C$_2$C$_1$Im][dca] the viscosity value is 21 cP at 298.15 K. Small contamination of [Cl]$^-$ or [Br]$^-$ anions in ILs from the synthesis increases the viscosity.

New lithium salts used in electrochemistry (e.g., LiPF$_6$, LiCF$_3$SO$_3$, LiAsF$_6$, and so on) have much lower melting points and ion transport properties than conventional lithium salts, and they could be considered as ILs. The viscosity values for lithium salts, LiTFA-n, depending on the number of oxyethylene groups in the oligo-ether substituents, are from 370 to 790 cP (303.15 K) [46].

Because the properties (melting point, density, viscosity) and behavior of the ILs can be adjusted to suit an individual synthesis type, extraction, heat transfer, or electrochemistry problems, they can truly be described as designer solvents.

1.3.1   Molar volume

Values of molar volumes can be calculated from densities measured for the liquid salt, or can be calculated as for hypothetical subcooled liquid at 298.15 K using the group contribution method [47]. As expected, the molar volumes of 1,3-dialkylimidazolium salts and quaternary ammonium salts increase progressively as the length of alkyl chain of the substituent increases. Some molar volumes values at 298.15 K are listed in Table 1.3.

The molar volumes at higher temperatures and the influence of the increasing temperature can be shown for N-butylpyridinium tetrafluoroborate,

Table 1.3   Molar Volume, V_m, at 298.15 K of Several Ionic Liquids

Ionic Liquid	V_m(cm^3 mol^{-1})	Reference
[C$_6$C$_1$Im][BF$_4$]	221.3	48
[C$_6$C$_1$Im][PF$_6$]	244.3	48
[C$_6$C$_1$Im][Tf$_2$N]	325.9	49
[C$_1$C$_1$Im][C$_1$SO$_4$]	156.7	50
[C$_4$C$_1$Im][C$_1$SO$_4$]	206.5	50
[C$_4$C$_1$Im][C$_8$SO$_4$]	326.2	50
[C$_4$C$_1$Im]Cl	186.7	51
[C$_{12}$C$_1$Im]Cl	325.8	51
[(C$_1$)$_2$C$_2$HOC$_2$N]Br	179.8	52
[(C$_8$H$_{17}$OCH$_2$)$_2$Im][Tf$_2$N]	542.9	53
[(C$_4$)$_4$P][C$_1$SO$_3$]	387.3	54

[C$_4$py][BF$_4$]. Three molar volume values at three temperatures are 185.4 cm^3 mol^{-1} (313.15 K), 186.3 cm^3 mol^{-1} (323.15 K), and 187.4 cm^3 mol^{-1} (333.15 K) [55]. A detailed discussion of the influence of the nature of cations and anions on the molar volume has to follow from densities.

1.3.2 Excess molar volume

To design any process involving ILs for use on an industrial scale, it is necessary to know the range of densities in binary mixtures and the intermolecular IL—solvent interactions. From the first literature data [56] for {[C$_2$C$_1$Im][BF$_4$] + H$_2$O} at temperatures 315.15 and 353.15 K, positive values of the excess molar volumes were observed. For the similar system [57] of {[C$_2$C$_1$Im][BF$_4$] + H$_2$O} at 286.15 K and the equimolar composition, $V^E_{m(max)}$, was 0.21 cm^3 mol^{-1}. For [C$_1$C$_1$Im][C$_1$SO$_4$] with water, the V^E_m values were observed as a sinusoidal curve with $V^E_{m(min)} = -0.18$ cm^3 mol^{-1} at $x_{IL} = 0.42$ [50]. With longer the alkyl chain on the cation from methyl to butyl, V^E_m decreased; for [C$_4$C$_1$Im][C$_1$SO$_4$] with water, $V^E_{m(min)} = -0.26$ cm^3 mol^{-1} at equimolar composition [50]. For the same cation and [PF$_6$]$^-$ anion measured with acetonitrile and methanol (298.15–318.15 K) [58] and with acetone, 2-butanone, 3-pentanone, cyclopentanone, and ethyl acetate at 298.15 K [59], negative values of V^E_m were found. Recently, new data were presented for [C$_4$C$_1$Im][PF$_6$], [C$_6$C$_1$Im][PF$_6$], [C$_8$C$_1$Im][PF$_6$], and [C$_1$C$_1$Im][C$_1$SO$_4$] with 2-butanone, ethyl acetate, and 2-propanol (293.15–303.15 K) [60], and negative excess molar volumes were also observed ($V^E_{m(min)} = -2.18$ cm^3 mol^{-1} at $x_{2-but} = 0.3$ at 303.15 K). The excess molar volumes, V^E_m, have been determined for [C$_1$C$_1$Im][C$_1$SO$_4$] with an alcohol (methanol, or ethanol, or 1-butanol), and for [C$_4$C$_1$Im][C$_1$SO$_4$] with an alcohol (methanol, or ethanol, or 1-butanol, or 1-hexanol, or 1-octanol, or 1-decanol) and for [C$_4$C$_1$Im][C$_8$SO$_4$] with an alcohol (methanol, or 1-butanol, or 1-hexanol, or 1-octanol, or 1-decanol) at 298.15 K and atmospheric pressure [50]. These systems exhibit very negative or positive molar excess volumes, V^E_m, and negative, or positive excess molar enthalpies, H^E_m, predicted by the Flory–Benson–Treszczanowicz (FBT) model [61]. Negative molar excess volumes, V^E_m, are attributed to hydrogen bonding between the short chain alcohols and the ILs, and to efficient packing effects. The FBT model overestimates the self-association of an alcohol in the solutions and shifts the calculated curves to the higher alcohol mole fraction [50].

V^E_m was negative for all mixtures of {[C$_4$C$_1$Im][C$_1$SO$_4$] (1) + an alcohol (methanol, or ethanol, or 1-butanol), or water (2)} and positive for mixtures of {[C$_4$C$_1$Im][C$_1$SO$_4$] (1) + an alcohol (1-hexanol, or 1-octanol, or 1-decanol) (2)} over the entire composition range (see Figure 1.3 for [C$_4$C$_1$Im][C$_1$SO$_4$] + 1-hexanol) [50].

Less negative values of the excess molar volumes were obtained for the IL with the longer alkyl chain in the cation for the same alcohol. The structure of [C$_4$C$_1$Im][C$_1$SO$_4$] is less H-bonded than [C$_1$C$_1$Im][C$_1$SO$_4$] in the pure state.

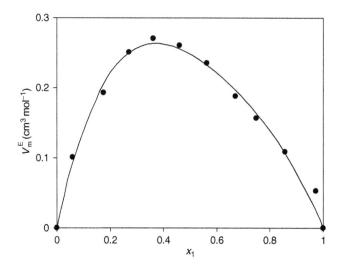

Figure 1.3 V_m^E for $\{[C_4C_1Im][C_1SO_4]$ (1) + 1-hexanol (2)$\}$ mixtures at 298.15 K. Points, experimental results from Domańska, U. et al. Solid line, Redlich–Kister correlation. (Adapted from Domańska, U., Pobudkowska, A., and Wiśniewska, A., *J. Solution Chem.*, 35, 311, 2006.)

This is probably due to the fact that the molecule of $[C_4C_1Im][C_1SO_4]$ is more flat and that is why this salt is liquid at room temperature in contrast to $[C_1C_1Im][C_1SO_4]$. There is no doubt that the corresponding values for the self-association of ILs and alcohols are responsible for these results. V_m^E was positive for all mixtures of $\{[C_4C_1Im][C_8SO_4]$ (1) + an alcohol (2)$\}$ over the entire composition range with the exception of methanol. The excess molar volume data become more positive in the following order: methanol < ethanol < 1-butanol < 1-hexanol < 1-octanol < 1-decanol. Comparison between $[C_4C_1Im]$ $[C_1SO_4]$ and $[C_4C_1Im][C_8SO_4]$ shows that by increasing the alkyl chain length in the anion, the interaction between unlike molecules decreases and the packing effect is the worst. V_m^E is more positive for $[C_4C_1Im][C_8SO_4]$ in every alcohol.

The excess molar volumes for $\{[C_8C_1Im][BF_4]$ + 1-butanol, or 1-pentanol$\}$ were found very small and negative in the alcohol-rich range of the mixture composition and positive in the alcohol-poor range [62]. More positive values were observed for 1-pentanol ($V_{m(max)}^E = 0.92$ cm^3 mol^{-1} at equimolar composition and 298.15 K).

The influence of temperature and pressure on the excess molar volume is not very well known. For ILs the V_m^E values were observed more negative at higher temperature [60,63]. Increasing the pressure from 0.1 to 20 MPa at the same temperature, less negative values of V_m^E were observed [63]. The influence of temperature on the V_m^E values at the pressure 15 MPa for $\{[C_1C_1Im][C_1SO_4]$ + methanol$\}$ is presented in Figure 1.4.

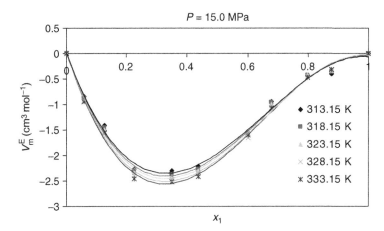

Figure 1.4 V_m^E for $\{[C_1C_1Im][C_1SO_4]$ (1) + methanol (2)$\}$ under the high pressure 15.0 MPa for five temperatures [63]. Solid lines, Redlich–Kister correlation. (Adapted from Heintz, A. et al., *J. Solution Chem.*, 34, 1135, 2005.)

Strong intermolecular interactions between the hydroxyl group and the IL lead to the negative values of excess molar volumes, V_m^E, and excess molar enthalpies H_m^E. The strongly negative H_m^E curve for $\{[C_1C_1Im][C_1SO_4]$ + water$\}$, predicted by the FBT model, was dominated by the chemical contribution, and the curve was strongly shifted to the higher IL mole fraction range. The negative excess molar enthalpies were predicted at the equimolar composition for $\{[C_1C_1Im][C_1SO_4]$ + methanol, or ethanol, or 1-butanol$\}$. For the negative values of H_m^E, the strongest IL-alcohol hydrogen bond exceeds that for the O—H···O between the alcohol molecules ($-21.8/-21.9$ kJ mol^{-1}) and between the IL molecules themselves. The values of H_m^E were found mainly positive for IL mixtures with hydrocarbons [64,65] and only for ($[C_2C_1Im][Tf_2N]$ + benzene) value of H_m^E have appeared negative. Using the Progogine–Defay expression, an estimation of the excess molar volume and enthalpy for the system $\{[C_4C_1Im][BF_4]$ + water$\}$ was done [66] and the results at 278 K were: $V_{m(max)}^E = 0.92$ cm^3 mol^{-1} and $H_{m(max)}^E = 3000$ J mol^{-1}. The results indicate that IL interactions with water are weaker than in many studied alcoholic solutions, especially in the presence of alcohols with the short alkyl chain. The observed inverse dependence on the temperature for aqueous or alcoholic mixtures refers to the special trend of the packing effect of ILs into hydroxylic solvents and its strong dependence on the steric hindrance of aliphatic residues in the cation, or anion, or an alcohol (e.g., $[C_4C_1Im][C_8SO_4]$).

1.3.3   Isobaric expansivity and isothermal compressibility

Knowledge of temperature and pressure dependence of physical–chemical properties is very useful to estimate the values of derived parameters, such as the thermal expansion coefficient, α_p, and the isothermal compressibility, κ_T.

The ILs do not expand appreciably at the commonly measured temperature range from 298.15 to 343.15 K. This small expansion with the temperature is best quantified by the volume expansivity, defined as

$$\alpha_p \equiv -\left(\frac{\partial \ln \rho}{\partial T}\right)_p \equiv \left(\frac{\partial \ln V}{\partial T}\right)_p \tag{1.1}$$

Because the molar volumes and densities of ILs are usually linear functions of temperature, the isobaric expansivities are easy to obtain from linear fits of the density data. The α_p values of ILs are in the range of $4-6 \times 10^{-4}$ K^{-1}, whereas the values of α_p for most molecular organic liquids are significantly higher ($8-12 \times 10^{-4}$ K^{-1}). The isobaric expansivities of ILs are similar to those of water [28], which range from 2.57×10^{-4} K^{-1} (298.2 K) to 5.84×10^{-4} K^{-1} (343.2 K). For example the isobaric expansivities are: $\alpha_p = 8.63 \times 10^{-4}$ K^{-1} (298.2 K) for 1-methylimidazole [28] and $\alpha_p = 8.47 \times 10^{-4}$ K^{-1} (298.2 K) for N-methyl-2-pyrrolidinone [67]. Some values of the isobaric expansivities at ambient pressure for typical ILs are presented in Table 1.4.

Systematic density measurements at a wide range of temperature and pressure [28,31,63,68,69] were helpful to obtain isothermal compressibility, which is calculated using the isothermal pressure derivative of density according to Equation 1.2

$$\kappa_T \equiv \left(\frac{\partial \ln \rho}{\partial p}\right)_T \equiv \frac{1}{\rho}\left(\frac{\partial \rho}{\partial p}\right)_T \tag{1.2}$$

The isothermal compressibility, κ_T, for some ILs at ambient pressure is presented in Table 1.5.

For the long alkyl chain phosphonium ILs, the isothermal compressibilities were found at ambient pressure about 20% higher than those for the

Table 1.4 The Isobaric Expansivities, α_p, at 298.2 K of Several Ionic Liquids

Ionic Liquid	$\alpha_p \cdot (10^4/K^{-1})$	Reference
$[C_4C_1Im][PF_6]$	6.11	28
$[C_8C_1Im][BF_4]$	6.24	28
$[4C_4py][BF_4]$	5.43	28
$[C_4C_1Im][Tf_2N]$	6.36	68
$[C_1C_1Im][C_1SO_4]$	5.25	50
$[C_2C_1Im][C_2SO_4]$	5.65	Our value
$[(C_6H_{13})_3P(C_{14}H_{29})][Tf_2N]$	7.43[a]	31

[a] At 0.21 MPa.

Table 1.5   The Isothermal Compressibility, κ_T at Different Temperatures
of Several Ionic Liquids

Ionic Liquid	$\kappa_T \cdot (10^4 / \mathrm{MPa^{-1}})$	T/K	Reference
$[C_4C_1Im][BF_4]$	3.96	298.2	29
$[C_4C_1Im][PF_6]$	4.17	298.2	29
$[C_4C_1Im][Tf_2N]$	5.26	298.2	68
$[C_6C_1Im][Tf_2N]$	5.61	298.2	68
$[C_1C_1Im][C_1SO_4]$	3.63	313.15	63
$[C_2C_1Im][C_2SO_4]$	3.41	298.15	Our value
$[(C_6H_{13})_3P(C_{14}H_{29})][Tf_2N]$	6.12	298.2/0.21 MPa	31

imidazolium-based ILs [31]. At higher pressures this effect was diminished. This was suggested as the typical influence of long alkyl chains on the packing effect and free-volume effect.

1.4   Surface tension and micellization

Without doubt, a complete picture of the surface tension of pure ILs and their solution and the parameters that govern the mechanism of adsorption connected with ILs would be incredibly useful in the study and improvement of industrially relevant catalysis and surface reaction processes. This information will be necessary for chemical engineering of larger scale reactions. Surface tension can reveal some fundamental features of a liquid, but few studies of this property have been reported [12]. A single compilation of surface tension values, including eight variously substituted imidazolium liquids, has shown [33] that the values of surface tension range from 33.8 mNm^{-1} for $[C_8C_1Im]Cl$ through 46.6 mNm^{-1} for $[C_4C_1Im][BF_4]$ and 48.8 mNm^{-1} for $[C_4C_1Im][PF_6]$ to 54.7 mNm^{-1} for $[C_4C_1Im]I$. The values of surface tension for $[C_1C_1Im][C_1SO_4]$ and $[C_4C_1Im][C_1SO_4]$ at 298.15 K are 56.5 and 41.4 mNm^{-1} (our results), respectively.

The surface tension of ILs are lower than that for water (72.7 mNm^{-1} at 293.15 K, 0.1 MPa) but higher than that for alkane (23.39 mNm^{-1} for decane at 298.15 K, 0.1 MPa). The aromatic substituent at the imidazolium ring in place of the alkyl chain lowers the surface tension [70], for example, Ph(CH$_2$)$_n$, where $n = 1$–3 in 1-alkyl(aralkyl)-3-methylimidazolium salts, $[Ph(CH_2)_nC_1Im][Tf_2N]$ salts show values 40.8–43.5 mNm^{-1}. The only lower values were observed for the tetralakylammonium salts with perchlorate anion, $[ClO_4]^-$; from 9.6 to 9.7 mNm^{-1} for different alkyl substituents [71]. The surface tension is observed to increase as the number of methylene units at the imidazolium or ammonium cation is increased. The value of surface tension of $[C_4C_1Im][PF_6]$ is close to that of imidazole, whereas the surface tension of the $[C_{12}C_1Im][PF_6]$ salt approaches those of an alkane. In general, the surface tension of 1,3-dialkylimidazolium ILs decreases in the order: $[I]^- > [PF_6]^- > [BF_4]^- > [C_1SO_4]^- > [Tf_2N]^-$.

Less viscous [Tf$_2$N]$^-$ containing salts have higher densities and lower surface tension than their [PF$_6$]$^-$ containing analogs.

The surface tension of the aqueous solutions of 1,2-dimethyl-3-*N*-hexadecyl-imidazolium tetrafluoroborate, [C$_{16}$C$_1$C$_1$Im][BF$_4$], the IL surfactant, monotonously decreases with an increasing surfactant concentration and forms a hydrophobic surface on the polymer–clay nanocomposites surface [72]. ILs have been found to be very useful solvents in micelle formation (impressive solvation ability) with many different surfactants. Best-known surfactant aggregates are normal micelles that spontaneously form surfactant aggregates (with the hydrophobic tails toward the center and the hydrophilic head groups at the outer surface) in water. Reverse or inverse micelles also can be formed in nonpolar organic solvents. The dissolution of surfactants in ILs also depresses the surface tension in a manner analogous to aqueous solutions. The two main differences are that the initial surface tension is lower than in water and that the critical micelle concentration (CMC) is generally higher than that for the same surfactant in water [73]. On the other hand, the surfactant/ionic liquid [C$_{10}$C$_1$Im]Br is a strongly amphiphilic salt and forms aggregates/micelles in aqueous solutions at very low concentrations 0.038 mol dm^{-3} [74]. It has been shown that the CMC values of surfactants may be significantly modified with appropriate ILs of different hydrophobic–hydrophilic properties. It strongly depends also on the length of the alkyl chain of imidazolium cation [75]. It was proved that ammonium ILs, [(C$_1$)$_2$C$_3$HOC$_2$N]Br and [(C$_1$)$_2$C$_4$HOC$_2$N]Br, also significantly lower the CMC of surfactant *n*-hexadecyl-trimethylammonium bromide in aqueous solutions. The lower CMC was either due to the increasing concentration of the Br$^-$ anion or to the formation of mixed micelles (cation of surfactant + cation of the IL) [76]. The influence of temperature on the CMC is very low. The new data in this field can be interpreted as resulting from the strong interaction of the IL with the surfactant.

1.5   Melting point, glass transition, and thermal stability

As a type of substances, ILs have been defined to have melting points below 373 K and most of them are liquid at room temperature. Salts with a halogen anion, revealing a higher melting temperature are known as the precursors of ILs. For example, the melting temperatures of [C$_4$C$_1$Im]Cl, [C$_8$C$_1$Im]Cl, [C$_{10}$C$_1$Im]Cl, and [C$_{12}$C$_1$Im]Cl are 341.9, 285.4, 311.2, and 369.8 K, respectively [51], whereas the melting temperatures of [C$_4$C$_1$Im][Tf$_2$N], [C$_8$C$_1$Im][Tf$_2$N], [C$_{10}$C$_1$Im][Tf$_2$N], and [C$_{12}$C$_1$Im][Tf$_2$N] are 267.0 K [70], liquid (189 K, glass transition), 244.0 K [70], and 256.3 K [75], respectively. For the longer alkane chain as C$_{10}$, the melting point increases. Both, cations and anions have influence on the low melting points of ILs. Usually, the increase in anion size and its asymmetric substitution leads to a decrease in the melting point. As we can see from the preceding examples, the size and symmetry of the cation have an important impact on the melting points of ILs. For the short chain alkyl substituents in 1,3-dialkylimidazolium salts, an increase in the alkyl chain decreases the melting temperature. The flexibility of the cation and anion is also an important factor. An increase of alkyl chain length enhances the molar volume and chain flexibility of the cation. The U-shaped plots of melting points (melting temperature versus chain length, for $n = 0$–20) were presented for many imidazolium and pyridinium salts in Ref. 12. Increasing the alkyl chain length from 0 to 8 the melting temperature decreases and increasing from 8 to 20 it causes a monotonous increasing of the melting point. The melting points of imidazolium salts increase with the degree of chains branching [12]. The same influence of the alkyl chain length was observed for different substituents for chiral ammonium-based ILs; namely, trialkyl[(1R,2S,5S)-(-)-menthoxymethyl]ammonium tetrafluoroborate [42]. Changing one alkyl substituent from methyl to ethyl decreases the melting temperature 26 K; from ethyl to n-decyl the melting temperature decreases 107 K. For chiral imidazolium chlorides the melting point decreases monotonously from 411 K (methyl group) to 318 K (n-hexyl group) and to the liquid phase (oil for n-heptyl and higher n-alkanes) [41]. Different relations were observed for the symmetrical dialkoxymethyl-substitued imidazolium salts with [Cl]$^-$, [BF$_4$]$^-$, [PF$_6$]$^-$, and [Tf$_2$N]$^-$ anions [77]. Melting temperatures were 305 K for didecyloxyimidazolium bis(trifluoromethylsulfonyl)imide or 318 K for didodecyloxyimidazolium bis(trifluoromethylsulfonyl)imide [41]. The melting points of the ILs based on [PF$_6$]$^-$, or [BF$_4$]$^-$, or [Tf$_2$N]$^-$ decreased in the order mentioned earlier (see Table 1.6).

The glass transition temperatures for the most popular ILs are between 213 and 183 K. It is very interesting that also for the glass transition temperatures versus cation chain length the U-shape plot was observed for most of imidazolium hexafluorophosphate, or tetrafluoroborate salts, or imidazolium bis(trifluoromethylsulfonyl)imides, or ammonium bis(trifluoromethylsulfonyl)imides [12]. The minimum of the curve is for $n = 5$ or 6. The changes of glass transition temperatures with changing length of the alkyl chain are much smaller than the melting temperatures—see the values for chiral imidazolium and ammonium-based salts [41,42]. It was also observed that the melting temperatures were not so low (between 318 and 328 K) for these ILs.

Popular ILs are thermally stable up to 700 K. Thermal stability is limited by the same factors that contribute to the melting temperature. For many salts the [Tf$_2$N]$^-$ anion presents the highest decomposition temperature and a lower melting temperature. For example [81], for the [C$_4$C$_1$Im]$^+$ cation the temperature of decomposition increases in the order: [Br]$^-$ < [BF$_4$]$^-$ < [CF$_3$SO$_3$]$^-$ < [Tf$_2$N]$^-$. Low decomposition temperature was observed [80] for benzalkonium nitrite, [Be(C$_1$)$_2$C$_n$N][NO$_2$], quaternary ammonium-based IL with $T_{onset} = 436$ K. A very high decomposition temperature was observed [82] for 1,2,3,4,5-methylimidazolium bis(trifluoromethylsulfonyl)imide, [(C$_1$)$_5$Im][Tf$_2$N], $T_{onset} = 743$ K. Usually, the difference in the decomposition temperature between particular salts with different cations is rather insignificant. For example [12], for

Table 1.6   Melting Temperature, T_{fus}/K, and Decomposition
Temperature, T_{dec}/K

Ionic Liquid	T_{fus} (K)	T_{dec} (K)	Reference
$[C_6H_{13}OCH_2-C_1Im][BF_4]$	190.7 (glass)	522[a]	78
$[C_6H_{13}OCH_2-C_1Im][Tf_2N]$	194.0 (glass)	530[a]	78
$[(C_1)_2C_2HOC_2N]Br$	541.4	559[b]	79
$[(C_1)_2C_2HOC_2N][BF_4]$	426.8	583[c]	53
$[(C_1)_2C_2HOC_2N][PF_6]$	272.0	556[c]	53
$[(C_1)_2C_4HOC_2N]Br$	359.3	525[c]	53
$[4C_4py][BF_4]$	361.9	575[b]	53
$[Be(C_1)_2C_n][NO_3]$	309.4	215[c]	80
$[(C_{10})_2(C_1)_2N][NO_3]$	291.9	234[c]	80
$[C_4C_1Im][Tf_2N]$	271.0	695[c]	81
$[C_3C_1C_1Im][Tf_2N]$	288.0	735[c]	81
$[C_1C_1Im][C_1SO_4]$	308.9	649[c]	82

[a] T_{dec}, decomposition temperature determined from onset to 5 wt% mass loss.
[b] T_{dec}, decomposition temperature determined from onset to 80 wt% mass loss.
[c] T_{dec}, decomposition temperature determined from onset to 50 wt% mass loss.

alkoxymethylimidazolium   *bis*(trifluoromethylsulfonyl)imides,   the   decomposition temperature is ~493 K, for dialkylimidazolium salts, from butyl to
1-hexyl-3-methylimidazolium *bis*(trifluoromethylsulfonyl)imides, the decomposition temperature is ~700 K. Branching the alkyl chain decreases the thermal stability of imidazolium ILs. Thermal stability increases with increasing
anion charge density if the cationic charge density is also high [12].

Finally, Table 1.6 depicts summarized data for some ILs observed in the
differential scanning calorimetry (DSC) and thermogravimetric analysis
(TGA) experiment.

1.6   Solid–liquid phase equilibria

Experimental diagrams of solid–liquid phase equilibria (SLE) have been
shown to play a dominant role in the development of the crystallization process. Experience in process synthesis and development, which covers petrochemical, fine chemicals, pharmaceutical, polymer, and IL processes that
projects on crystallization and solids processing need not only SLE diagrams
but also crystallization kinetics and mass-transfer limitations to optimize
the final process design. The most important is, however, knowledge of the
solubility of the desired product in the selected solvent or binary solvent
mixture. In every process, a wide knowledge of the pressure, temperature,
and the overall composition of the mixture is needed. Previously, vapor–
liquid equilibrium (VLE) played a key role in the design of separation (distillation) processes. The production of high-molecular-weight chemicals needs

the thermodynamic basis of SLE in binary and ternary mixtures at normal and high pressure to design and synthesize crystallization processes. In this case, phase diagrams help to visualize the regions in the composition space where the system exists as a single phase or a mixture of multiple phases, providing a better understanding of the thermodynamic limitations imposed by the phase behavior. Solubility is strongly dependent on both, intramolecular forces (solute–solute and solvent–solvent) and intermolecular forces (solute–solvent).

Systematic investigations into the physical–chemical properties and phase equilibria of binary systems with imidazoles, benzimidazoles, and their derivatives with organic solvents and water have been presented [83–89]. The purpose of these measurements was to obtain basic information on the interaction of the imidazole or benzimidazole rings with organic solvents (alcohols, ethers, ketones, aromatic hydrocarbons, chlorohydrocarbons, and alkanes) and water, having in mind the imidazolium ILs. The simple imidazolium molecules with two hydrophilic nitrogen groups (1 and 3) could imply specific interactions between them and with the solvent. Benzimidazoles or phenylimidazoles have a large aromatic group substituted at the imidazole ring, causing new interaction effects, manifested in an enhancement of the intermolecular interaction and in changing the solution structure. The molecular rearrangement in the solution depends on the possibility of hydrogen bond formation between the imidazole molecules themselves and between the imidazole and the polar solvents. In the solutions with benzene or toluene, the n–π interaction was expected. The hydrogen bonds were responsible for the new crystal structure, described as blocks of imidazole salt the mono-, di-, or tetracarboxylic acids [90]. The results of SLE measurements have confirmed that binary mixtures of imidazoles with organic solvents or water present simple eutectic mixtures. Solubility depends on the kind of the substituent at the imidazole ring, the melting temperature, and the melting enthalpy of the solute. For example, the solubility in 1-octanol, or in dipentyl ether, or in 2-pentanone decreases in the order: 1,2-dimethylimidazole > 1H-imidazole > 2-methyl-1H-imidazole > 2-methylbenzimidazole > benzimidazole > phenylimidazole [83–89]. The solubilities of imidazoles and benzimidazoles in alcohols, ethers, and ketones decrease with an increase in the number of carbon atoms of the solvent molecule. Their solubilities were always lower than the ideal solubility, and activity coefficients in the saturated solutions were higher than one. Branching of alcohols, ethers, and ketones decreases the solubility of imidazoles and benzimidazoles. In some mixtures the miscibility gap with the upper critical solution temperature (UCST) was observed, for example, 1,2-dimethylimidazole and 1H-imidazole in dibutyl ether or dipentyl ether [84]. The solubility of imidazoles in alcohols, ethers, and ketones shows different solute–solvent interactions; the substitution of the methyl group in position 2 of the imidazole ring increases the solubility. Solubility in cyclic ether or ketone was higher than in linear ethers or ketones because of the similarity of the structure [84,85]. The solubility

of a certain imidazole in different solvents decreases in the order: water >
alcohol > ketone > ether. The solubilities of some benzimidazoles in dichloromethane and nitrobenzene were much lower than those in alcohols and
much higher than those in ethers. The solubilities of benzimidazoles were
higher than those of phenylimidazoles [88]. Thus, imidazoles, benzimidazoles, and phenylimidazoles were chosen as the precursors of ILs in the
solubility measurements not only because of the large hydrophobic aromatic
groups but also for the known specific interactions of nitrogen atoms, or the
hydrogen atom with the solvent. It was noted that the interaction with oxygen from the nitro group of the solvent and with the oxygen from the carbonyl group in ketones was similar.

The results obtained with imidazoles were useful for predicting the
physical properties and phase equilibria properties of ILs synthesized on
the base of an imidazole molecule. The systematic study of the solubilities
of 1-alkyl-3-methylimidazolium chloride, namely, $[C_4C_1Im]Cl$, $[C_8C_1Im]Cl$,
$[C_{10}C_1Im]Cl$, and $[C_{12}C_1Im]Cl$ in alcohols (C_2–C_{12}) has confirmed that complete phase diagrams are eutectic mixtures [91–93]. The complete phase diagrams $[C_4C_1Im]Cl$, $[C_{10}C_1Im]Cl$, and $[C_{12}C_1Im]Cl$ were found to show eutectic
behavior in *tert*-butyl alcohol, 1-decanol, and 1-dodecanol [92,93]. The liquidus curves of these salts in primary, secondary, and tertiary alcohols exhibit
similar shapes. The solubility increases in the order: 1-butanol > 2-butanol >
tert-butyl alcohol. The solubility was lower than the ideal one, and the activity
coefficients in the saturation solution were higher than one. The solubilities
of 1,3-dialkylimidazolium chlorides in alcohols decrease with an increase
of the alkyl chain of an alcohol from ethanol to 1-dodecanol with the exception of $[C_4C_1Im]Cl$ in 1-butanol. This can be explained by the best packing
effect in the solution for the same number of carbon atoms of the solvent and
butyl substituent at the imidazole ring [93]. The SLE are presented [92,93]
as an example for three systems {$[C_4C_1Im]Cl$, or $[C_{10}C_1Im]Cl$, or $[C_{12}C_1Im]Cl$ +
tert-butyl alcohol} in Figure 1.5. Figures 1.5 and 1.6 show the influence of the
melting temperature of a solute and the alkyl chain substituent at the imidazolium ring on the SLE.

The solubility of IL is strictly dependent on melting temperature of the
solute, which is generally the result of the alkane or phenyl substituents
at imidazolium ring. The conclusions can be taken from the solubilities of
benzimidazole and 2-phenylimidazole in water [94] and of every measured
imidazoles in organic solvents [83–89].

The immiscibility in the liquid phase was observed for $[C_{10}C_1Im]Cl$ with
water and for $[C_8C_1Im]Cl$ with water and 1-octanol [51]. For both salts the
solubility in 1-octanol was higher than that in water. Only $[C_8C_1Im]Cl$ was
liquid at room temperature (melting point, T_{fus} = 285.4 K) [51]. The binary
mixtures of $[C_{12}C_1Im]Cl$ with *n*-alkanes and ethers have shown a very flat
liquidus curve, but only in {$[C_{12}C_1Im]Cl$ + *n*-dodecane, or methyl 1,1-dimeth-
ylether} the immiscibility in the liquid phase was observed for the very low
solvent mole fraction [95].

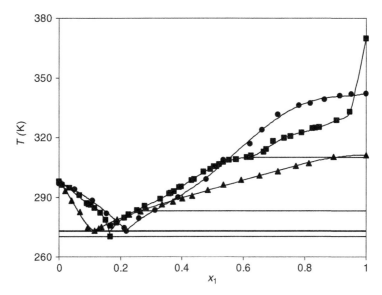

Figure 1.5 Solid–liquid equilibria for [C₄C₁Im]Cl (●), or [C₁₀C₁Im]Cl (▲), or [C₁₂C₁Im]Cl (■) with *tert*-butyl alcohol. (Adapted from Domańska, U., Bogel-Łukasik, E., and Bogel-Łukasik, R., *J. Phys. Chem. B*, 107, 1858, 2003; Domańska, U. and Bogel-Łukasik, E., *Fluid Phase Equilib.*, 218, 123, 2004.)

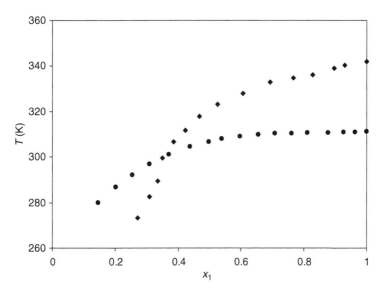

Figure 1.6 Solid–liquid equilibria diagrams for different ionic liquids (1) in 1-octanol are [C₄C₁Im]Cl (◆) and [C₁₀C₁Im]Cl (●). (Adapted from Domańska, U. and Bogel-Łukasik, E., *Fluid Phase Equilib.*, 218, 123, 2004; Domańska, U. and Bogel-Łukasik, E., *Ind. Eng. Chem. Res.*, 42, 6986, 2003.)

An important advantage of ILs is that the anion, the cation, and the composition can be arranged based on the required properties. Generally, anions contribute to the overall characteristics of ILs and determine the air and water stability. One of the most important properties of ILs is that the melting point and the solubility can be easily changed by the structural variation of the cation or the anion. For example, the IL with the $[C_4C_1Im]^+$ cation with different anions will show the SLE diagram with immiscibility in the liquid phase from $x_{IL} = 10^{-4}$ up to 0.58 for $[Cl]^-$ anion, liquid–liquid equilibria (LLE) up to 0.9 ($T = 310$ K) for $[C_1SO_4]^-$ anion and LLE up to 0.97 ($T = 305$ K) for $[PF_6]^-$ anion [95]. Thus, for the ILs with $[C_4C_1Im]^+$ cation, solubility in water decreases in the order: $[Cl]^- > [C_1SO_4]^- > [PF_6]^-$.

After changing the anion from chloride (ILs described earlier) to hexafluorophosphate, it is easy to obtain [96] salts with very low melting temperatures, for example, $T_{fus} = 332.8$ K for $[C_2C_1Im][PF_6]$ and $T_{fus} = 276.4$ K for $[C_4C_1Im][PF_6]$. The SLE diagrams with immiscibility in the liquid phase with the UCST were observed in the mixtures of $\{[C_2C_1Im][PF_6]$ + benzene, or toluene, or ethylbenzene, or o-xylene, or m-xylene, or p-xylene$\}$ [96]. The liquidus curves exhibited a similar shape for different solvents and differences between solvents were small. The evident differences were observed in the liquid phase—the mutual solubility of $[C_2C_1Im][PF_6]$ in benzene and its alkyl derivatives decreased with an increase of the alkyl substituent at the benzene ring. In alcohols, the only system with complete miscibility in the liquid phase was observed with methanol $\{[C_2C_1Im][PF_6]$ + methanol$\}$ [97]. For longer chain alcohols, the immiscibility with UCST was noted. The mutual solubility of $[C_2C_1Im][PF_6]$ with alcohols decreases with an increase in the molecular weight of an alcohol. Solubility in methanol and ethanol is higher than in aromatic hydrocarbons; in 1-propanol it is comparable with that in toluene [97].

SLE were measured [53,98] also for the dialkoxyimidazolium ILs, $[(C_4H_9OCH_2)_2Im][BF_4]$, $[(C_6H_{13}OCH_2)_2Im][BF_4]$, $[(C_6H_{13}OCH_2)_2Im][Tf_2N]$, $[(C_8H_{17}OCH_2)_2Im][Tf_2N]$, $[(C_{10}H_{21}OCH_2)_2Im][Tf_2N]$, and for pyridinium IL, $[py][BF_4]$. The phase equilibria of these salts were measured with common popular solvents: water, or alcohols or n-alkanes, or aromatic hydrocarbons. These salts mainly exhibit simple eutectic systems with immiscibility in the liquid phase with UCSTs, not only with aromatic hydrocarbons, cycloalkanes, and n-alkanes but also with longer chain alcohols. The higher interaction may be expected between two alkoxy groups with polar solvent such as water or alcohol. The better solubility of the IL in a chosen solvent means the possible hydrogen bonding between the IL and the solvent. The second alkoxy group in the molecule of the IL causes a higher melting temperature and a solid compound in the room temperature, but also stronger interaction with the solvent. Only the $[(C_4H_9OCH_2)_2Im][BF_4]$ salt with tetrafluoroborate anion exhibits a small miscibility gap in alcohols (ethanol, 1-octanol) in the IL low mole fraction (the area of immiscibility shifts toward solvent-rich region). A much higher binary liquid area was observed in benzene, where the n–π interaction rather than hydrogen bonding exists. Greater mutual solubilities were observed for $[Tf_2N]^-$ anion even for longer alkoxy chains on the cation (C_6, or C_8 versus C_4). Complete miscibility in the liquid phase was observed for $[(C_6H_{13}OCH_2)_2Im][Tf_2N]$, or $[(C_6H_{13}OCH_2)_2Im][BF_4]$ in 1-butanol, or 1-hexanol, or 1-octanol, or 1-decanol, or benzene contrary to mixtures with hexane, cyclohexane, or water [98]. Decreasing interaction with polar solvent was observed by replacing the anion from $[Tf_2N]^-$ to $[BF_4]^-$ in dihexyloxy salts as in many other ILs. The systems of $[(C_6H_{13}OCH_2)_2Im][BF_4]$ and alcohols (1-hexanol, 1-octanol, or 1-decanol) have shown SLE diagrams with complete miscibility in the liquid phase, whereas salt with one alkoxy group $[C_6H_{13}OCH_2–C_1Im][BF_4]$ exhibited liquid–liquid immiscibility with the UCST, that increasing with the increasing length of the alkyl chain of alcohol [98]. Similar interaction and complete miscibility was observed for $[(C_8H_{17}OCH_2)_2Im][Tf_2N]$ or $[(C_{10}H_{21}OCH_2)_2Im][Tf_2N]$ in alcohols (ethanol, 1-octanol) and in benzene. The solubility decreases as the molecular weight of the alcohol increases. The eutectic point in the binary system of $\{[(C_4H_9OCH_2)_2Im][BF_4]$ (1) + benzene (2)$\}$ was $T_{1,e} = 260.3$ K, $x_{1,e} = 0.450$ and for the binary system $\{[(C_8H_{17}OCH_2)_2Im][Tf_2N]$ (1) + benzene (2)$\}$ was $T_{1,e} = 271.8$ K, $x_{1,e} = 0.331$. In every system with $[(C_4H_9OCH_2)_2Im][BF_4]$, the solid–solid phase transition was noted in a phase diagram.

The longer alkoxy chain of the cation for two ILs with the same anion causes higher melting temperature of the compound and lower solubility [98]. The influence of the pyridinium ring, or the imidazolium ring on the solubility was discussed using the solubility measurements of N-decyloxy-methyl-3-amido-pyridinium tetrafluoroborate in alcohols (ethanol, 1-butanol, 1-hexanol, 1-dodecanol) and in benzene. Unfortunately, this compound has only one alkoxy group and one new amido-group which will cause additional interaction with the solvent. Anyway, the melting temperature was higher than it was observed for the alkoxyimidazolium salts and solubility was lower. Complete miscibility was observed for alcohols with the exception of 1-dodecanol, for which a small area of immiscibility was observed in the low-IL mole fraction [53].

SLE of quaternary ammonium ILs in alcohols, hydrocarbons, and water have been measured for many salts. Systematic studies of SLE phase diagrams for quaternary ammonium salts $[(C_1)_2C_2HOC_2N]Br$, $[(C_1)_2C_3HOC_2N]Br$, $[(C_1)_2C_4HOC_2N]Br$, and $[(C_1)_2C_6HOC_2N]Br$ in water and alcohols have been published [52,79]. Other anions including $[BF_4]^-$, $[PF_6]^-$, $[dca]^-$, and $[Tf_2N]^-$ have also been investigated [53].

Quaternary ammonium salts are well-known cationic surfactants and popular phase-transfer (PT) catalysts. In addition, these salts exhibit both antimicrobial activities and antielectrostatic effects. Another useful compound which belongs to the ammonium salt group is chinoline chloride, also known as vitamin B_4; it is an essential component that ensures proper functioning of the nervous system and is widely used as a feed additive for livestock.

The solubility of [(C$_1$)$_2$C$_n$HOC$_2$N]Br in 1-octanol increases when the alkyl chain of cation increases in almost the whole concentration range. In water, the solubility of [(C$_1$)$_2$C$_2$HOC$_2$N]Br is much lower than that of the other three salts and the best solubility was observed in Ref. 79 for [(C$_1$)$_2$C$_6$HOC$_2$N]Br. The results mainly depend on the temperature of melting and the enthalpy of melting of the solute. It can be stated that the short alkane substituent at the ammonium cation causes the highest melting temperature and the worst solubility of this salt in 1-octanol and water. These results are opposite to those presented earlier for 1-alkyl-3-methylimidazolium chlorides, observed for much longer alkane chains (C$_4$ versus C$_{12}$) in 1-octanol and water [51]. The solid–liquid phase diagrams have shown simple eutectic mixtures. SLE and LLE diagrams for [(C$_1$)$_2$C$_4$HOC$_2$N][BF$_4$] (1) in 1-octanol are presented in Figure 1.7. The immiscibility in this system was observed in the IL mole fraction range from $x_1 = 0.45$ to 0.78 [53].

The effect of interaction with alcohols is different for [(C$_1$)$_2$C$_2$HOC$_2$N]Br and [(C$_1$)$_2$C$_2$HOC$_2$N] [BF4]. The solubility of [(C$_1$)$_2$C$_2$HOC$_2$N]Br in alcohols from ethanol to 1-dodecanol did not present the miscibility gap [52], whereas [(C$_1$)$_2$C$_2$HOC$_2$N] [BF4] showed immiscibility in alcohols at high temperatures. These systems exhibit UCST higher than 435 K for every system measured [53].

The solubility of [(C$_1$)$_2$C$_2$HOC$_2$N]Br in primary alcohols (C$_2$–C$_{12}$) decreases with an increase of the alkyl chain of an alcohol [52]. Differences in solubilities between the primary and the secondary alcohols are not

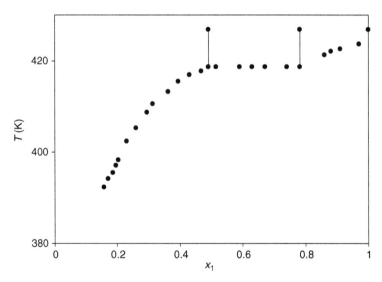

Figure 1.7 Solid–liquid and liquid–liquid equilibria (SLE/LLE) diagrams for [(C$_1$)$_2$C$_4$HOC$_2$N][BF$_4$] (1) in 1-octanol: experimental points (●); solid line designated LLE. (Adapted from Domańska, U., *Thermochim. Acta*, 448, 19, 2006.)

significant. Branching of alcohols decreases the solubility of ammonium salts, as it was observed for imidazolium salts. It was noted that the solubility of $[(C_1)_2C_2HOC_2N]Br$ in ethanol was higher than the ideal solubility; the activity coefficients were lower than one. This strong interaction with the solvent can be explained by the similarity of an ethyl group connected with an OH group on the cation and the same chain of ethanol.

In polar solvents such as alcohols one can expect stronger interaction between unlike molecules than in alcohol or IL itself. Clearly, hydrogen bonding, or $n-\pi$, or other interactions of the cation or anion of the IL with the solvent play an important role in controlling liquid–liquid and solid–liquid phase behavior of ammonium and imidazolium-based ILs. However, the existence of the LLE in these mixtures is the evidence that the interaction between some IL and the solvent is not significant.

The SLE diagram for a longer alkyl chain IL (2-hydroxy-ethyl)dimethyl undecyloxymethylammonium dicyanamide, $[C_{11}OC_1EtOH(C_1)_2N][dca]$ (1) in 1-octanol presents a typical SLE/LLE phase diagram—a simple eutectic system with immiscibility in the liquid phase with the UCST. The influence of the $[dca]^-$ anion in spite of the long alkyl chain makes this salt liquid at room temperature ($T_{fus} = 283.5$ K). Therefore, the choice of the anion can have a huge effect on the phase behavior of ammonium and imidazolium ILs.

Similar to the imidazolium ILs, very low solubility of the ammonium ILs was observed in alkanes. For example, the solubility of butyl(2-hydroxyethyl) dimethylammonium bromide, $[(C_1)_2C_4HOC_2N]Br$, exhibited a simple eutectic system with immiscibility in the liquid phase in the IL mole fraction range from $x_{IL} = 0.02$ to 0.70 [53]; the other ammonium salt, (benzyl)dimethyl-alkylammonium nitrate, $[Be(C_1)_2C_nN][NO_3]$ showed very small solubility in the liquid phase in hexadecane and it was slightly better in hexane (immiscibility from $x_{IL} = 10^{-4}$ to 0.90) [99]; for the ammonium salt with an alkane substituent only, didecyldimethylammonium nitrate, $[(C_{10})_2(C_1)_2N][NO_3]$, the solubility in the liquid phase was better in hexane (immiscibility from $x_{IL} = 10^{-4}$ to 0.50) than in hexadecane, where the immiscibility was observed in the whole IL mole fraction [100]. In all systems with imidazolium and ammonium salts, an increase in the alkyl chain length of the alkane (solvent) resulted in a decrease of solubility.

Generally in cycloalkanes, solubility is very similar to that in alkanes. The solubility of imidazolium IL, 1-ethyl-3-methylimidazolium tosylate, $[C_2C_1Im][TOS]$ in cyclohexane and cycloheptane displays a very large miscibility gap in the liquid phase (close to its melting temperature) [101,102]. Ammonium IL, $[(C_1)_2C_4HOC_2N]Br$, has shown lower solubility in cyclohexane than in heptane and the immiscibility gap was from $x_{IL} = 0.02$ to 0.90 [53].

Changing the solvent from an aliphatic hydrocarbon to an aromatic hydrocarbon demonstrates that the interaction is most likely due to $n-\pi$ interactions between the oxygen atom of the IL and the benzene ring. Solid at room temperature ILs present usually much lower immiscibility gap in benzene than in alkanes or cycloalkanes. For example, $[(C_4H_9OCH_2)_2Im][BF_4]$ shows immiscibility with UCST, $T_c = 278$ K in the low mole fraction of the IL [53]. In the phase diagrams of an ammonium IL, (benzyl)dimethylalkylammonium nitrate [Be(C_1)$_2C_nN$] [NO_3] with benzene and toluene reported, small immiscibility was observed in the liquid phase with the UCST at high-IL mole fraction [99].

The solubility of quaternary phosphonium salt, tetrabutylphosphonium methanesulfonate, [(C_4)$_4$P][C_1SO_3] in alcohols, or alkylbenzenes was investigated [54]. The systems with alcohols (1-butanol, or 1-hexanol, or 1-octanol, or 1-decanol, or 1-dodecanol) and aromatic hydrocarbons (benzene, or toluene, or ethylbenzene, or propylbenzene) have shown simple eutectic systems; the immiscibility in the liquid phase was detected in binary mixtures with the aromatic hydrocarbons, ethylbenzene and propylbenzene [54]. The solubility of [(C_4)$_4$P][C_1SO_3] in alcohols increases in the order from 1-dodecanol to 1-butanol. The results mainly depend on the melting temperature of the solvent. It can be stated that a better solubility is observed at the lowest melting temperature of the solvent. Increasing the alkyl chain length of the alcohol causes an increase in the eutectic temperature and shifts the composition of the eutectic point to the higher IL mole fraction.

The eutectic point of the system ([(C_4)$_4$P][C_1SO_3] + benzene) is shifted toward a much lower IL mole fraction in comparison to that of ([(C_4)$_4$P][C_1SO_3] + 1-alcohol). Experimental phase diagrams of SLE and LLE with alkylbenzenes were characterized mainly by the following: (1) the mutual solubility of [(C_4)$_4$P][C_1SO_3] in ethylbenzene and propylbenzene decreases with an increase in the alkyl substituent at the benzene ring; (2) the UCST of the system increases when the alkyl chain length at benzene ring increases; (3) the immiscibility gap was found at low concentration of the IL [54].

As compared to conventional organic solvents, ILs are much more complex solvents, capable of undergoing many types of interactions. Therefore, a single *polarity* term fails to describe the type and magnitude of individual interactions that they make.

The SLE in a mixture of a solid (1) in a liquid may be expressed in a very general manner by Equation 1.3

$$-\ln x_1 = \frac{\Delta_{fus}H_1}{R}\left(\frac{1}{T} - \frac{1}{T_{fus,1}}\right) - \frac{\Delta_{fus}C_{p,1}}{R}\left(\ln\frac{T}{T_{fus,1}} + \frac{T_{fus,1}}{T} - 1\right) + \ln\gamma_1 \quad (1.3)$$

where x_1, γ_1, $\Delta_{fus}H_1$, $\Delta_{fus}C_{p,1}$, $T_{fus,1}$, and T stand for the mole fraction, the solute activity coefficient, the enthalpy of fusion, the difference in the solute heat capacity between the liquid and the solid at the melting temperature, the melting temperature of the solute (1), and the measured equilibrium temperature, respectively.

In many studies of the SLE of ILs, three methods have been used to derive the solute activity coefficients, γ_1, from the so-called correlation equations that describe the Gibbs free energy of mixing (GE), the Wilson [103], UNIQUAC ASM [104], and NRTL1 [105] models. Historically, the UNIQUAC associated–solution theory has been proposed to reproduce the VLE, LLE, and SLE of binary (alcohol + unassociated component, or alcohol) mixtures as well as for ternary mixtures of two alcohols with one nonpolar component. The model postulates the formation of linear multisolvated complexes from the i-mers of one or two alcohols. The modified form of the NRTL equation proposed by Renon was presented by Nagata and co-workers [105] by substituting the local surface fraction for the local mole fraction and further by including Guggenheim's combinatorial entropy for athermal mixtures whose molecules differ in size and shape. The resulting equations involve three adjustable parameters and are extended to multicomponent systems without adding ternary parameters.

Two adjustable parameters of the equations can be found by an optimization technique using Marquardt's or Rosenbrock's maximum likelihood method of minimization

$$\Omega = \sum_{i=1}^{n} \left[T_i^{exp} - T_i^{cal}(x_{1i}, P_1, P_2) \right]^2 \tag{1.4}$$

where Ω is the objective function (OF), n is the number of experimental points, T_i^{exp} and T_i^{cal} denote the experimental and calculated equilibrium temperatures, respectively, corresponding to the concentration x_{1i}. P_1 and P_2 are model parameters resulting from the minimization procedure. The root-mean-square deviation of temperature is defined as follows:

$$\sigma_T = \left(\sum_{i=1}^{n} \frac{\left(T_i^{exp} - T_i^{cal} \right)^2}{n-2} \right)^{1/2} \tag{1.5}$$

Parameters r_i and q_i of the UNIQUAC ASM model are calculated with the following relationships:

$$r_i = 0.029281 \, V_m \tag{1.6}$$

$$q_i = \frac{(Z-2)r_i}{Z} + \frac{2(1-l_i)}{Z} \tag{1.7}$$

where V_m is the molar volume of pure compound i at 298.15 K. The molar volume of IL V_{m1} was measured (298.15 K), or was calculated as for a hypothetical subcooled liquid by the group contribution method [47]. The coordination number Z was assumed to be 10 and the bulk factor l_i was assumed to be zero for the linear molecule or one for the cyclic molecule. The calculations were carried out by the use of the literature data set of association for alcohols or for water. The temperature dependence of association constants was calculated from the van't Hoff relation, assuming the enthalpy of hydrogen-bond formation to be temperature independent. Some values of model parameters obtained by fitting solubility curves together with the corresponding standard deviations are given in Table 1.7.

For most of the systems with alcohols, the description of SLE was given by the average standard mean deviation $(\sigma_T) < 2$ K for UNIQUAC ASM and NRTL 1 equations. The procedure of correlation has been described in many articles [52–54,79,84–88,91–94]. Using GE models the solute activity coefficients in the saturated solution, γ_1, were described.

For the systems with alcohols, the description of SLE given by the UNIQUAC ASM equation (assuming the association of alcohols) did not provide any better results. It can be understood as a picture of a very complicated interaction between the molecules in the solution: it means that it exists not only in the association of alcohol molecules but also between alcohol and IL molecules and between IL molecules themselves. Parameters shown in Table 1.7 may be helpful to describe activity coefficients for any concentration, temperature, and to describe ternary mixtures. They are also useful for the complete thermodynamic description of the solution.

1.6.1   Miscibility with water

At room temperature, all popular imidazolium salts such as $[C_nC_1Im][PF_6]$ or $[C_nC_1Im][Tf_2N]$ are insoluble in water, but at the same time they are very hygroscopic. The miscibility gap was observed for water with $[C_2C_1Im][Tf_2N]$ and $[C_4C_1Im][Tf_2N]$ at 353.15 K [106]. LLE were measured for $[C_2C_1Im][PF_6]$ and $[C_4C_1Im][PF_6]$ with water at the temperature range 315–323 K [107]. It was found that the mutual solubility increases with increasing temperature. The thermodynamic analysis of phase behavior and physical–chemical properties was made for $[C_4C_1Im][PF_6]$, $[C_8C_1Im][PF_6]$, and $[C_8C_1Im][BF_4]$ with water at temperature 295 K [108]. The mutual solubilities of water and $[C_8C_1Im][PF_6]$ are lower (the binary liquid phases were observed in the range of the IL mole fraction from $x_{IL} = 3.5 \times 10^{-4}$ to 0.2×10^{-4}) than those seen for the IL with a shorter alkyl chain, $[C_4C_1Im][PF_6]$, (the binary liquid phases were observed in the range of the IL mole fraction from $x_{IL} = 1.29 \times 10^{-3}$ to 0.26×10^{-3}). As observed in the other properties, changing the anion from $[PF_6]^-$ to $[BF_4]^-$ increases the mutual solubilities substantially [108].

The solubility of the IL in water is mainly determined by the anion of the IL. Typical imidazolium ILs based on halide-, ethanoate-, nitrate-, and trifluoroacetate-based anions are fully water-soluble [109]. Salts with $[BF_4]^-$ and $[CF_3SO_3]^-$ anions are partly soluble in water. The influence of the anion on the solubility of $[C_4C_1Im]Cl$ and $[C_4C_1Im][BF_4]$ in water is shown in Figure 1.8.

For chloride anion, complete miscibility in the liquid phase was observed and for tetrafluoroborate anion, the LLE area was observed in the low mole fraction of the IL and at low temperature. It is known that the $[C_4C_1Im][BF_4]$ is water miscible at 298.15 K; the UCST is at 278 K [110]. The immiscibility with water was measured for $[C_4C_1Im][PF_6]$, $[C_8C_1Im][PF_6]$, $[C_{10}C_1Im][PF_6]$,

Table 1.7 Correlation of Solubility Data (Solid–Liquid Phase Equilibria) of {IL (1) + Solvent (2)} by Means of the Wilson, UNIQUAC ASM, and NRTL1 Equations: Values of Parameters and Measures of Deviations

Solvent	Parameters			Deviations			Reference
	Wilson $g_{12}-g_{11}$ $g_{12}-g_{22}$ (J mol⁻¹)	UNIQUAC ASM Δu_{12} Δu_{21} (J mol⁻¹)	NRTL1[a] Δu_{12} Δu_{21} (J mol⁻¹)	Wilson σ_T (K)	UNIQUAC ASM σ_T (K)	NRTL1 σ_T (K)	
Benzimidazole + 1-propanol	−3301 7088	3117 −3341	−502 −2990	0.93	1.01	0.92	87
2-Methylbenzimidazole + 1-hexanol	8365 −3788	−1622 642	−54435 −98.32	2.75	1.64	1.19	87
[C₄C₁Im]Cl + 1-hexanol	−2922 12690	−3284 44272	−5234 4539	2.93	1.59	1.70	93
[C₁₂C₁Im]Cl + n-decane	5173 6631	1243 −32	3318 4790	2.72	4.72		95
[(C₈H₁₇-OCH₂)₂Im][Tf₂N] + 1-octanol	3057 3331	377 −422	3032 1353	2.25	4.66	2.75	53
[(C₁₀H₂₁OCH₂)₂Im][Tf₂N] + benzene	−556 505	—	505 −578	0.84	—	0.84	53
[py][BF₄] + 1-hexanol	29526 1563	—	1360 11835	2.50	—	2.34	53
[(C₁)₂C₂HOC₂N]Br + 1-butanol	—	−39 −1922	233[b] −9057	—	4.00	4.00	52

[a] Calculated with the third nonrandomness parameter α = 0.3.
[b] Calculated with the third nonrandomness parameter α = 0.1.

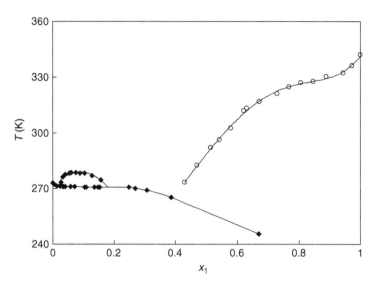

Figure 1.8 The influence of anion on the solubility of imidazolium ILs in water: [C$_4$C$_1$Im]Cl (O) and [C$_4$C$_1$Im][BF$_4$] (◆). (Adapted from Domańska, U. and Bogel-Łukasik, E., *Fluid Phase Equilib.*, 218, 123, 2004; Lin, H. et al., *Fluid Phase Equilib.*, 253, 130, 2007.)

and [C$_{10}$C$_1$Im][BF$_4$] at 298.15 K [111]. When more polar anion is taken, like in the [C$_1$C$_1$Im][C$_1$SO$_4$] or [C$_4$C$_1$Im][C$_1$SO$_4$], complete miscibility with water is observed [50].

There is no doubt that also the length of the alkyl chain of the cation has an influence on solubility in water, as in other solvents. For a longer chain $n \geq 4$, the LLE is observed. For example, [C$_4$C$_1$Im][BF$_4$] forms a biphasic system at low temperatures and at low mole fraction of the IL (see Figure 1.8). Comparing the [BF$_4$]$^-$ anion with the [PF$_6$]$^-$ anion, it can be concluded that the [PF$_6$]$^-$-based ILs dissolve less water than those of the [BF$_4$]$^-$. The solubility of water in the IL decreases with the increasing alkyl chain length on the cation [109].

The immiscibility in the liquid phase was measured for [C$_8$C$_1$Im]Cl and [C$_{10}$C$_1$Im]Cl with water, and the UCST ~322 and ~323 K, respectively, was observed [61]. The influence of the polar alkoxy group in the cation or anion is presented in Figure 1.9.

As for many other solvents, two alkoxy groups in the cation increase solubility; the area of two liquid phases is lower for [(C$_6$H$_{13}$OCH$_2$)$_2$Im][Tf$_2$N] than for [C$_6$H$_{13}$OCH$_2$–C$_1$Im][Tf$_2$N]; the UCST is >370 K for both salts [98]. Much higher interaction with water is presented by [C$_4$C$_1$Im][MDEGSO$_4$], where two alkoxy groups are in the anion and four extra oxygen atoms in the sulfate group [112]. The immiscibility was observed only at very narrow and low IL mole fraction with UCST >375 K.

The influence of the alkyl chain of the cation and the type of the anion on the SLE diagram {IL + water} is shown for ammonium ILs in Figure 1.10.

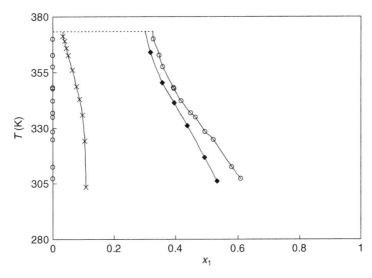

Figure 1.9 The influence of the alkoxy group in the cation or anion on solubility in water: $[C_6H_{13}OCH_2–C_1Im][Tf_2N]$ (○), $[(C_6H_{13}OCH_2)_2Im][Tf_2N]$ (◆), and $[C_4C_1Im]$ $[MDEGSO_4]$ (×). (Adapted from Domańska, U. and Marciniak A., *Fluid Phase Equilib.*, 260, 9, 2007; Domańska, U. and Marciniak, A., *Green Chem.*, 9, 262, 2007.)

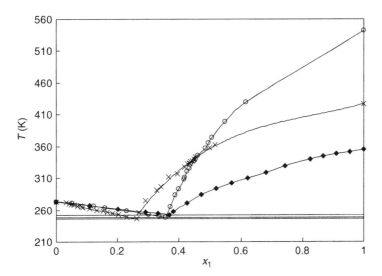

Figure 1.10 The influence of the alkyl chain of cation and the type of anion on the solid–liquid phase equilibria diagram {IL + water}: $[(C_1)_2C_2HOC_2N]Br$ (○), $[(C_1)_2C_6HOC_2N]Br$ (◆), and $[(C_1)_2C_2HOC_2N] [BF_4]$ (×). (Adapted from Domańska, U. and Bogel-Łukasik, R., *J. Phys. Chem. B*, 109, 12124, 2005.)

The difference between the liquidus curve of [(C$_1$)$_2$C$_2$HOC$_2$N]Br and that of [(C$_1$)$_2$C$_6$HOC$_2$N]Br is the result of the IL melting temperature. The results of measurements of the solubility of [(C$_1$)$_2$C$_2$HOC$_2$N]Br, [(C$_1$)$_2$C$_3$HOC$_2$N]Br, [(C$_1$)$_2$C$_4$HOC$_2$N]Br, and [(C$_1$)$_2$C$_6$HOC$_2$N]Br in water did not show the immiscibility in the liquid phase [79]. It was stated before that increasing the alkyl chain at the cation of ammonium salt decreases the melting temperature of the IL. Changing the anion [BF$_4$]$^-$ versus [Br]$^-$, the same effect was observed— lowering the melting temperature of the IL [53,79]. Substitution of the bromide anion for a tetrafluoroborate anion increases the solubility in water. As can be seen from Figure 1.10, three simple eutectic mixtures with complete miscibility in the liquid phase were detected for these binary systems.

The interesting influence of the cation on the SLE diagram {IL + water} can be observed [99,100] from the diagrams of ammonium salts [(C$_{10}$)$_2$(C$_1$)$_2$N][NO$_3$] and [Be(C$_1$)$_2$C$_n$N][NO$_3$]. Simple liquidus curve and no immiscibility in the liquid phase for the didecyldimethylammonium cation with the eutectic point shifted strongly to the solvent-rich side was noted (see Figure 1.11).

Water, being a polar solvent, can interact with the anion [NO$_3$]$^-$ but the long alkane chains (C$_{10}$) of the cation are hydrophobic; hence, there is little interaction of water with the cation, and thus the anion entity is favored by the (IL + water) mixtures. When one alkyl chain (C$_{10}$) was changed by the benzyl group and the other one by the longer alkane chain (C$_{12}$–C$_{14}$) as the substituents at the cation, the decrease of solubility was reported. These new substituents had strong influence on packing effects in the liquid phase,

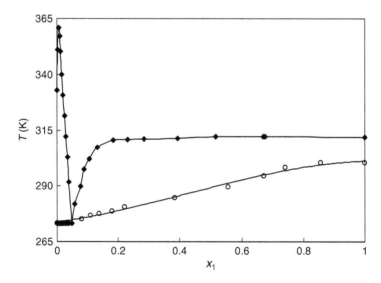

Figure 1.11 The influence of cation on the solid–liquid phase equilibria diagram {IL + water}: [(C$_{10}$)$_2$(C$_1$)$_2$N][NO$_3$] (O) and [Be(C$_1$)$_2$C$_n$][NO$_3$] (◆). (Adapted from Domańska, U., Bąkała, I., and Pernak, J., *J. Chem. Eng. Data*, 52, 309, 2007; Domańska, U., Ługowska, K., and Pernak, J., *J. Chem. Thermodyn.*, 39, 729, 2007.)

resulting in immiscibility at low weight fraction of the IL. The UCST for the system $\{[\text{Be}(C_1)_2C_nN][NO_3] + H_2O\}$ was observed at a temperature, $T = 360.8$ K, and a mass fraction, $w_{IL} = 0.11$. Inversely to water, a miscibility gap in the liquid phase was not observed for $[\text{Be}(C_1)_2C_nN][NO_3]$ with short chain alcohols [99].

It is clear that ILs are capable of absorbing significant amounts of water, whether in contact with vapor or liquid. Using ILs as a solvent in the synthesis in reactions with moisture-sensitive compounds, as it is the case with metal-organic synthesis, it is necessary to take into account the hygroscopicity of ILs. Hydrophobic ILs are, in fact, hygroscopic. It was proved that chlorides and nitrate ILs absorb much more water from atmospheric air at ambient temperature than the corresponding hexafluorophosphate or tetrafluoroborate ILs [109]. It is widely known that pure ILs based on the $[\text{PF}_6]^-$ anion can readily decompose to produce hydrogen fluoride (HF) in the presence of water. The use of ILs based on the $[\text{PF}_6]^-$ anion in industrial applications will be very limited.

It has been shown by Cammarata et al. [113] that water molecules absorbed by ILs from air are bound via hydrogen bonding with typical anions of imidazolium ILs with the concentration of dissolved water in the range 0.2–1.0 mol dm^{-3}. The conclusion was that most of the water molecules at these concentrations exist in symmetric 1:2 type associating complexes: anion–HOH–anion. The associating enthalpy of this H-bonding was estimated from spectral shifts, with the values in the range 8–13 kJ mol^{-1}. The strength of H-bonding between water molecules and anions was found to increase in the order: $[\text{PF}_6]^- < [\text{SbF}_6]^- < [\text{BF}_4]^- < [\text{Tf}_2N]^- < [\text{ClO}_4]^- < [\text{TfO}]^- < [\text{NO}_3]^- < [\text{CF}_3\text{CO}_2]^-$. The presence of water in the IL may affect many of their solvent properties such as density, surface tension, polarity, viscosity, and conductivity. It has been stated as well that substitution of the C2 proton in the imidazolium ring with a CH_3 group has no effect on the molecular state of water [113].

1.6.2   1-Octanol/water partition coefficient

The partition coefficient K_{OW} of an organic compound in the 1-octanol/water system is used to assess the bioaccumulation potential and the distribution pattern of drugs and pollutants. The partition coefficient of imidazole and ILs strongly depends on the hydrogen bond formed by these molecules and is less than one due to the high solubility in water. The low value of the 1-octanol/water partition coefficient is required for new substances, solvents, insecticides to avoid bioaccumulation. K_{OW} is an extremely important quantity because it is the basis of correlations to calculate bioaccumulation, toxicity, and sorption to soils and sediments. Computing the activity of a chemical in human, fish, or animal lipid, which is where pollutants that are hydrophobic will appear, is a difficult task. Thus, it is simpler to measure the 1-octanol/water partition coefficient. This parameter is used as the primary parameter characterizing hydrophobisity.

The thermodynamic data needed to calculate the 1-octanol/water partitioning is the LLE in the ternary system, in binary phases:

$$f_i^o \left(x_i^o, T, P \right) = f_i^w \left(x_i^w, T, P \right) \tag{1.8}$$

where f_i and x_i are the fugacity and mole fraction of species i, and the superscripts o and w denote the 1-octanol and water phases, respectively.

$$K_{OW} = \frac{c_i^o}{c_i^w} \tag{1.9}$$

where c is the molar concentration of species i in the 1-octanol and water phase.

For very hydrophobic compounds, when K_{OW} is very high (10^5), there is evidence from field studies involving fish, birds, and animals that there is a bioaccumulation or bioconcentration up to food chain [114]. The pesticide DDT is an example of bioaccumulating chemical with a K_{OW} of $\sim 10^6$. The thermodynamic model for air/water and 1-octanol/water activity coefficients and the importance of the activity coefficient at infinite dilution measurements have been presented as well [114].

Twelve imidazolium-based ILs were studied experimentally by Ropel et al. [115] (mainly the *bis*(trifluoromethylsulfonyl)imides), and some of these results are compared with other literature data of imidazoles and ILs in Table 1.8.

As we can see, the log K_{OW} is negative for benzimidazole or 2-methylbenzimidazole and for ILs with a longer alkyl chain at the cation, for example, $[C_6C_1Im]^+$ and $[C_8C_1Im]^+$. It was shown that for salts with the $[Tf_2N]^-$ anion K_{OW} regularly increases as the alkyl chain length on the cation increases [115]. It was also shown that the tested ILs were not dangerous to the atmosphere and water. The bioaccumulation of $[C_6C_1Im][Tf_2N]$ was calculated on level 0.08–3.16, which is much less than 250, a moderate border of bioaccumulation [115]. The values of K_{OW} are very low for ILs that are totally miscible with water, for example, $[C_4C_1Im][NO_3]$. K_{OW} is an extremely important quantity to describe the hydrophobisity or hydrophilisity of a compound and can be correlated with bioaccumulation and toxicity in fish and other aquatic organisms. Since all of K_{OW} values of popular ILs are very small, the conclusion is that these ILs will not accumulate or concentrate in the environment. Nevertheless, the new analytical methods are tested to measure the concentration of ILs in aqueous environmental samples. The method based on the cation exchange solid-phase extraction, followed by the selective elution, was proposed by using HPLC or electrophoretic methods [117,118]. The overall procedure was verified by using standard spiked samples of tap water, seawater, and freshwater. The structural variability of commercially available ILs as well as the abundance of theoretically accessible ILs was

Table 1.8 1-Octanol/Water Partition Coefficients, K_{OW}, at 289.15 K

Imidazole or Ionic Liquid	K_{OW}	Reference
1*H*-imidazole	0.49[a]	83
2-Methyl-1*H*-imidazole	0.67[a]	83
1,2-Dimethylimidazole	0.83[a]	83
Benzimidazole	16.90[a]	83
2-Methylbenzimidazole	28.51[a]	83
[C$_{10}$C$_1$Im]Cl	0.52[a]	51
[C$_{12}$C$_1$Im]Cl	0.73[a]	51
[C$_4$C$_1$Im][PF$_6$]	0.022, 0.020[b]	115,116
[C$_4$C$_1$Im][BF$_4$]	0.003	115
[C$_4$C$_1$Im][NO$_3$]	0.0038	115
[C$_4$C$_1$Im][Tf$_2$N]	0.11–0.62	115
[C$_6$C$_1$Im][Tf$_2$N]	1.42–1.66	115
[C$_8$C$_1$Im][Tf$_2$N]	6.3–11.1	115
[C$_2$C$_1$Im][PF$_6$]	0.015[b]	115
[C$_2$C$_1$Im][Tf$_2$N]	0.09–0.11	115
[C$_6$H$_{13}$OCH$_2$–C$_1$Im][Tf$_2$N]	0.39[a]	98
[C$_6$H$_{13}$OCH$_2$–C$_1$Im][BF$_4$]	0.64[a]	98
[(C$_6$H$_{13}$OCH$_2$)$_2$Im][BF$_4$]	0.44[a]	98

[a] Values calculated from solubilities in 1-octanol and water with a correction for mutually saturated solvents.
[b] At temperature 303 K.

illustrated and the consequences for an integrated risk assessment accompanying the development process were presented by Jastorff et al. [119]. The side-chain effect of the cation on toxicity for imidazolium type ILs was confounded by more complex biological testing. The influence of an anion on cytotoxicity was also shown for the first time [119].

The experimental data on the adsorption of ILs onto natural soils have shown that the desorption of ILs from agricultural soils decreases with an increase of the alkyl chain length [120]. It was shown that only 2% of an initially sorbed [C$_6$C$_1$Im]$^+$ cation as compared with 10% of the [C$_4$C$_1$Im]$^+$ cation was desorbed from the soils investigated. The effect of the different size of the cationic head group on IL sorption can be assessed by comparing the larger *N*-butyl-4-methylpyridinium with the smaller 1-butyl-3-methylimidazolium. For clay agricultural soil, their sorption coefficients do not differ very much, and for fluvial meadow soils, they are even identical. But for forest- and fluvial agricultural soils, alkylimidazolium sorption is twice as high as that of alkylpyridinium. It was stated that for these two soils the additional steric availability of the active sorption site may play an important role in the sorption process before molecular interactions occur [120].

Since ILs are mostly nonvolatile, they cannot contribute to atmospheric pollution. However, as it was shown that they show some solubility in water, studies of their effect on aquatic organisms or adsorption on soils are just starting.

1.6.3   High-pressure solid–liquid phase equilibria

High-pressure phase equilibrium data of mixtures including ILs are of importance for safety and efficient operation of chemical plants and oil transportation and refineries. The SLE in binary mixtures that contains IL and the molecular solvent: {[C_2C_1Im][TOS] (1) + cyclohexane, or benzene (2)} and {[C_1C_1Im][C_1SO_4] (1) + 1-hexanol (2)} have been measured under very high pressures up to about 900 MPa at the temperature range from T = 328 to 363 K for the technological use [101]. The thermostatted apparatus was used to measurements of transition pressures from the liquid to the solid state. The pressure–temperature-composition relation of the high-pressure SLE, polynomial based on the Yang model was satisfactorily used [121]. The freezing and melting temperatures at a constant composition increase monotonously with pressure. The high-pressure experimental results obtained at isothermal conditions (p–x) were interpolated to the well-known T–x diagrams. The experimental results at high pressures were compared for every system to those at normal pressure. The influence of the interaction between the IL and the solvent on the solubility at 0.1 MPa and high pressure up to 900 MPa was discussed.

The system {[C_2C_1Im][TOS] (1) + cyclohexane (2)}, measured at ambient pressure, was found to exhibit immiscibility in the liquid phase. At high pressure only the characteristic flat liquidus curve in a broad range of IL's mole fraction was observed. The effect of pressure on the SLE of the binary mixtures has been measured at various constant compositions and temperatures. The liquid–solid transitions were determined for the whole IL concentration range from x_1 = 0 up to 1. The solubility of [C_2C_1Im][TOS] was lower than that of [C_1C_1Im][C_1SO_4], not only because of the higher melting temperature, but also because of the polar solvent (1-hexanol). At 0.1 MPa and at high pressure, the solubility of [C_2C_1Im][TOS] in benzene was higher than in cyclohexane. Figure 1.12 presents the liquidus curves of [C_1C_1Im][C_1SO_4] in 1-hexanol at different temperatures and pressures. For many ILs the eutectic point was at very low mole fraction of the IL and was not measured, especially under high pressure.

The influence of the high pressure on the liquidus curve in the tested systems was typical: with increasing pressure the freezing curves shifted monotonously to higher temperatures; the difference of the pressure from 0.1 to 160 MPa increased the melting temperature of the [C_2C_1Im][TOS] ΔT = ~40.1 K. The melting temperature $T_{fus,1}$ of [C_2C_1Im][TOS] changed from $T_{fus,1}$ = 322.9 K (0.1 MPa) to 363.1 K (160 MPa) and that of [C_1C_1Im][C_1SO_4]

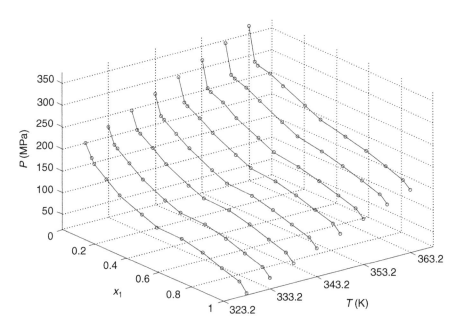

Figure 1.12 Influence of pressure on the liquidus curve of {[C_1C_1Im][C_1SO_4] (1) + 1-hexanol (2)} up to 360 MPa (Adapted from Domańska, U. and Morawski, P., *Green Chem.*, 9, 361, 2007.); experimental points (○), solid line, description with the Yang equation. (Adapted from Yang, M. et al., *Fluid Phase Equilib.*, 204, 55, 2003.)

changed from $T_{fus,1}$ = 308.9 K (0.1 MPa) to 343 K (160 MPa). These results show that high pressure has much stronger influence on the thermophysical properties of [C_2C_1Im][TOS] than on [C_1C_1Im][C_1SO_4]. The slope of dp/dT is totally different for the ILs and the popular molecular solvents [101].

The experimental high pressure data (P, x_1) at constant temperature can be correlated by the equation proposed by Yang et al. [121]. They showed that the activity coefficient γ_1 at high pressure can be expressed as

$$\ln \gamma_1 = \sum_{i=0}^{3} a_i \left(\frac{1}{T} - \frac{1}{T_{fus,1}} \right)^i + a' \left(\ln \frac{T}{T_{fus,1}} + \frac{T_{fus,1}}{T} - 1 \right) \qquad (1.10)$$

where a_i and a' are two adjustable parameters. After substituting Equation 1.10 into Equation 1.3, some simplifications can be obtained.

$$\ln x_1 = \sum_{i=0}^{3} b_i \left(\frac{1}{T} - \frac{1}{T_{fus,1}} \right)^i + b' \left(\ln \frac{T}{T_{fus,1}} + \frac{T_{fus,1}}{T} - 1 \right) \qquad (1.11)$$

where $b_0 = -a_0$, $b_1 = -a_1 - \Delta_{fus}H_1/R$, $b_2 = -a_2$, $b_3 = -a_3$, $b' = -a' + \Delta_{fus}C_{p,1}/R$.

The value of the second term on the right-hand side of Equation 1.11 is small and this term can be neglected. Thus, Equation 1.11 may be rewritten in a simple form as

$$\ln x_1 = \sum_{i=0}^{3} b_i \left(\frac{1}{T} - \frac{1}{T_{\text{fus},1}} \right)^i \tag{1.12}$$

SLE curves are dependent on pressure. With increasing pressure, the SLE curves shift to a higher temperature. The b_i terms in Equation 1.12 were found to be pressure-dependent and this dependence can be expressed as follows:

$$b_i = \sum_{j=0}^{2} D_{ji} P^j \tag{1.13}$$

The OF used in the fit of the parameters of Equations 1.12 and 1.13 was as follows:

$$\text{OF} = \sum_{i=1}^{n} w_i^{-2} \left(\ln x_{1i}^{\text{cal}} \left(T_i, P_i, D_{ji} \right) - \ln x_{1i} \right) \tag{1.14}$$

where $\ln x_{1i}^{\text{cal}}$ denotes the values of the logarithm of the solute mole fraction calculated from Equation 1.11, $\ln x_{1i}$ denotes the logarithm of the experimental solute mole fraction and T_i, P_i, D_{ji} express temperature, pressure, and coefficients from Equation 1.13, respectively; w_i is the weight of the calculated values described by means of the error propagation formula.

Direct experimental points and the results of the correlation are shown in 3D diagram in Figure 1.12. Solid lines represent the results of the fit with the Yang equation. The description of the experimental data is quite good not only for ILs but also for many systems. Greater standard deviations are observed usually for higher pressures or close to the eutectic point.

1.7 Liquid–liquid phase equilibria

A major reason for the interest in ILs is their negligible vapor pressure which decreases the risk of technological exposure and solvent loss to the atmosphere. The solid–liquid and liquid–liquid measurements of ILs systems based on N,N'-dialkyl-substituted imidazolium cations or ammonium cations or phosphonium cations are attracting increasing attention for applications in liquid–liquid extraction [122–126]. Knowledge of the SLE is of paramount importance for the design of separation processes, especially cooling, evaporation, and antisolvent crystallization. There is a pressing need to develop better solvents for separation, especially in the case of nonideal complexing mixtures. First technological investigations have been conducted about the suitability of ILs as entrainers for the extractive distillation and as extraction solvents for the liquid–liquid extraction. ILs were found to be capable of breaking a multitude of azeotropic systems. The nonvolatility of ILs in combination with their remarkable separation efficiency and selectivity enable new processes for the separation of azeotropic mixtures which, in comparison to conventional separation processes, might offer a potential for cost-savings [122,126]. Until now, it cannot be predicted which ILs are the best for certain applications. With the increasing comprehension on how the structure of an IL affects its physical properties one will be able to use the advantages of ILs over volatile organic solvents. Owing to their unique structures and properties they may be compared with dendrimers or hyperbranched polymers. For example, it is possible to extract out a tetrahydrofuran (THF)–water mixture using $[C_2C_1Im][BF_4]$. This IL easily breaks the azeotropic ethanol–water phase behavior by interacting selectively with water (The conventional entrainer is 1,2-ethanediol). The different interactions IL–water and IL–THF result not only in an increasing relative volatility of THF but also in a phase split of the liquid mixture [126]. It was shown also that ILs may give much higher selectivities for the separation of aliphatics from aromatics (cyclohexane/benzene) than N-methyl-2-pyrrolidinone (NMP) or NMP + water mixture [122]. Systematic investigations on the activity coefficient at infinite dilution for different solvent/ionic liquid systems have proved unbelievably attractive effects (will be discussed later in this chapter).

A large miscibility gap exists in the mixtures of $[C_2C_1Im][Tf_2N]$ or $[C_2C_1Im][C_2SO_4]$ with aliphatic (cyclohexane) and aromatic (benzene) hydrocarbons [64,65]. The excess molar enthalpy for the {benzene + $[C_2C_1Im][Tf_2N]$} was highly negative (approximately $-750\,J\,mol^{-1}$) and for the {cyclohexane + $[C_2C_1Im][Tf_2N]$} was positive ($\sim 450\,J\,mol^{-1}$). This means that ILs are excellent entrainers for the separation of aliphatic from aromatic hydrocarbons by extractive distillation or extraction.

In early measurements the solubility of $[C_2C_1Im][PF_6]$ in aromatic hydrocarbons (benzene, toluene, ethylbenzene, *o*-xylene, *m*-xylene, and *p*-xylene) and that of $[C_4C_1Im][PF_6]$ in the same aromatic hydrocarbons, and in *n*-alkanes (pentane, hexane, heptane, and octane), and in cyclohydrocarbons (cyclopentane and cyclohexane) has been presented [96].

Systematic studies of IL solubility measurements have been presented in the literature [14,48,50,53,62,66,78,89,94–100,112,127–134]. A systematic investigation on the mixtures of $[C_4C_1Im][PF_6]$ with alcohols (ethanol, 1-propanol, and 1-butanol) was presented very early [129,130]. Methanol was found completely miscible with $[C_4C_1Im][PF_6]$ at temperatures >273 K. A systematic decrease in solubility was observed with an increase of the alkyl chain of an alcohol. The measurements were repeated with the analytical UV spectroscopy method used to determine very small solubilities of the IL in the alcohols [131].

The same conclusions resulted from the solubility measurements of
[C_2C_1Im][Tf_2N] in alcohols (1-propanol, 1-butanol, and 1-pentanol) and from
those of [C_8C_1Im][BF_4] with alcohols (1-butanol and 1-pentanol) [62,132].
An interesting analysis of LLE in the mixtures of {IL ([C_4C_1Im][BF_4], [C_4C_1Im]
[Tf_2N], [$C_6C_1C_1$Im][Tf_2N], [C_6C_1Im][BF_4], [C_4C_1Im][TfO], [C_2C_1Im][Tf_2N],
[$C_4C_1C_1$Im] [Tf_2N]) + 1-alcohol (from 1-propanol to 1-dodecanol, 2-propanol)
or water} has revealed the most important factors that govern the phase
behavior of ILs with alcohol and water [14]. That discussion was continued
for {IL ([C_8C_1Im][BF_4], [C_6C_1Im][Tf_2N], [C_8C_1Im][Tf_2N], [$C_6C_1C_1$Im][Tf_2N]) +
1-alcohol (1-hexanol, 1-octanol)} [133]. As in many published results, the alkyl
chain length increasing on the alcohol was found to cause an increase in the
UCST of the system, and that increasing the alkyl chain length on the cation
results in a decrease in the UCST of the system (see Figure 1.13) [133].

At the same time, the solubilities of [C_2C_1Im][PF_6] in alcohols (methanol,
ethanol, 1-propanol, 2-propanol, 1-butanol, 2-butanol, *tert*-butyl alcohol,
and 3-methyl-1-butanol) were measured [97]. The LLE measurements of
[C_2C_1Im][PF_6] or [C_4C_1Im][PF_6] have shown mutual solubilities with aliphatic,
cyclic, and aromatic hydrocarbons being the function of the chain length of the
alkyl substituent at the imidazole ring [96,97]. The solubility of [C_2C_1Im][PF_6]
in alcohols decreases with an increase of the molecular weight of the solvent
and is higher in secondary alcohols than in primary alcohols. In every case,
with the exception of methanol, the mutual LLE were observed. The shape

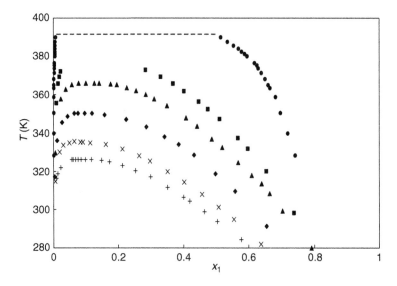

Figure 1.13 Liquid–liquid equilibria in the systems of {[C_nC_1Im][PF_6] (1) + 1-butanol
(2)}; Experimental points: [C_2C_1Im][PF_6] (●), [C_4C_1Im][PF_6] (■), [C_3C_1Im][PF_6] (▲),
[C_6C_1Im][PF_6] (◆), [C_7C_1Im][PF_6] (×), [C_8C_1Im][PF_6] (+), and boiling temperature of the
solvent (dashed line). (Adapted from Domańska, U. and Marciniak A., *J. Phys. Chem. B*,
108, 2376, 2004; Wu, C-T. et al., *J. Chem. Eng. Data*, 48, 486, 2003.)

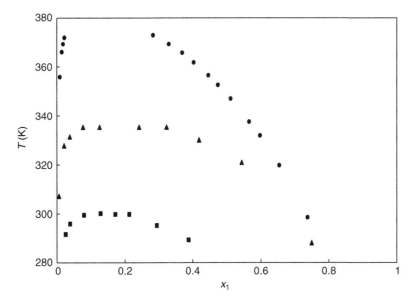

Figure 1.14   Liquid–liquid equilibria in the systems of {[C₄C₁Im][X] (1) + 1-butanol (2)}; Experimental points: [C₄C₁Im][PF₆] (●), [C₄C₁Im][BF₄] (▲), and [C₄C₁Im][Tf₂N] (■). (Adapted from Wu, C.-T. et al., *J. Chem. Eng. Data*, 48, 486, 2003; Crosthwaite, J.M. et al., *J. Phys. Chem. B*, 108, 5113, 2004.)

of the equilibrium curve was similar for [C₂C₁Im][PF₆] in every alcohol and for different ILs in the same alcohol. LLE in the systems of {[Cₙ₁C₁Im][PF₆] + 1-butanol} are presented in Figure 1.13. The influence of alkyl chain on the solubility in 1-butanol is shown in Ref. 97 for [C₂C₁Im][PF₆], and in Ref. 130 for [C₄C₁Im][PF₆], [C₃C₁Im][PF₆], [C₆C₁Im][PF₆], [C₇C₁Im][PF₆], and [C₈C₁Im][PF₆]. The influence of the anion is shown in Figure 1.14.

The LLE in the systems of {[C₄C₁Im][X] + 1-butanol} are presented in Refs 14 and 130 for [PF₆]⁺, [BF₄]⁺, and [Tf₂N]⁺.

The observation of the UCST was limited very often by the boiling temperature of the solvent. The solubility of [C₂C₁Im][PF₆] and [C₄C₁Im][PF₆] in aromatic hydrocarbons and in alcohols decreases with an increase of the molecular weight of the solvent. The difference on the solubility in *o-*, *m-*, and *p*-xylene was not significant. Solubility was better in secondary alcohols than in primary alcohols. For example, for [C₂C₁Im][PF₆] solubilities in methanol and ethanol were higher than those in aromatic hydrocarbons. The miscibility gap in C₃ alcohols was bigger than that in benzene, but comparable with the solubility in toluene; solubility in 3-methyl-1-butanol was very low and similar to ethylbenzene and *o-*, *m-*, and *p*-xylene [96,97].

The LLE of twenty binary systems containing [C₆H₁₃OCH₂–C₁Im][BF₄] with aliphatic hydrocarbons (*n*-pentane, *n*-hexane, *n*-heptane, or *n*-octane) and aromatic hydrocarbons (benzene, toluene, ethylbenzene, *o*-xylene, *m*-xylene, or *p*-xylene) were presented [78]. Also, the mixtures of [C₆H₁₃OCH₂–C₁Im][Tf₂N] and cyclohexane, or with aliphatic hydrocarbons (*n*-hexane, or *n*-heptane), or with aromatic hydrocarbons (benzene, toluene, or ethylbenzene, *o*-xylene, *m*-xylene, or *p*-xylene) have been measured [78]. The LLE or SLE of binary systems comprising $[C_6H_{13}OCH_2–C_1Im][Tf_2N]$ with an alcohol (1-butanol, 1-hexanol, or 1-octanol), a ketone (3-pentanone or cyclopentanone), or water were measured as well as binary systems comprising $[C_6H_{13}OCH_2–C_1Im][BF_4]$ with an alcohol (methanol, ethanol, 1-butanol, 1-hexanol, or 1-octanol), a ketone (3-pentanone or cyclopentanone), or water [98].

Many factors that control the phase behavior of ILs with other liquids, especially with alcohols have been discussed in many published articles focusing on pyridinium ILs as opposed to imidazolium ILs [134], or imidazolium ILs as opposed to ammonium ILs [52] or phosphonium IL [54]. The impact of different alcohol and IL characteristics including alcohol chain length, cation (alkyl or alkoxy) chain length, and cations with different substituent groups on the imidazolium ring or nitrogen in ammonium ILs or different substituent groups on the pyridynium cation have been discussed [14,52,54,66,79,91–93,97,98,127,129–134].

For a better understanding of the IL behavior and with a view to the application in chemical engineering or the development of thermodynamic models, reliable experimental data are required. Basically, an IL can act as both a hydrogen-bond acceptor (anion) and donor (cation) and would be expected to interact with solvents with both accepting and donating sites. On the other hand, polar solvents as alcohols are very well-known to form hydrogen-bonded net with both high enthalpies and constants of association. Hence, they would be expected to stabilize with hydrogen bond donor sites; it was interesting to increase the polarity of the cation or anion of the IL and increasing hydrogen bonding opportunities with the anion by the replacement of $[Cl]^-$ or $[PF_6]^-$ or $[BF_4]^-$ anions by $[C_1SO_4]^-$ with four atoms of oxygen, or with the $[MDEGSO_4]^-$ anion by adding the alkoxy group to the alkane chain at anion or at cation at the imidazole ring, $[C_6H_{13}OCH_2–C_1Im]^+$ [98,112,127].

ILs based on the 1-alkyl-3-methylimidazolium cation, $[C_nC_1Im]^-$, or tetraalkylammonium cation, $[(C_n)_4N]^-$, are among the most popular and commonly studied or used in technological improvements. As for the anions, *bis*(trifluoromethylsulfonyl)imide, $[Tf_2N]^-$, and alkylsulfate, $[C_nSO_4]^-$ (n = 1, 2, 8), are much superior compared to the more commonly investigated hexafluorophosphate, $[PF_6]^-$, and tetrafluoroborate, $[BF_4]^-$, being hydrolytically stable and less viscous. Comparing the results of the solubility of ILs in typical solvents from different publications, it can be concluded that the miscibility gap in the liquid phase increases in the order: alcohol < aromatic hydrocarbon < cycloalkane < *n*-alkane [50–54,66,78,79,95–100,127–136].

The $[C_2C_1Im][PF_6]$ or $[C_4C_1Im][PF_6]$ has shown immiscibility in alkanes (*n*-hexane, *n*-heptane, and *n*-octane) in the whole IL mole fraction range from $x_{IL} = 10^{-4}$ to 0.98 [96,135]; the salt with longer alkane chain, $[C_{12}C_1Im]Cl$, has shown a simple eutectic system with *n*-octane, and *n*-decane and eutectic system with a small miscibility gap in the low IL mole fraction in n-dodecane [95]; $[C_1C_1Im][C_1SO_4]$ and $[C_4C_1Im][C_1SO_4]$ have shown immiscibility in alkanes (n-pentane, n-hexane, n-heptane, n-octane, and n-decane) in the whole IL mole fraction range from $x_{IL} = 10^{-4}$ to 0.98 [128]; $[C_4C_1Im][C_8SO_4]$ was better soluble in n-alkanes (n-hexane, n-heptane, n-octane, and n-decane)— the immiscibility gap for n-heptane was observed in the IL mole fraction range from $x_{IL} = 10^{-4}$ to 0.5 at room temperature [50]; for the $[C_6H_{13}OCH_2-C_1Im][BF_4]$ with n-alkanes (n-pentane, n-hexane, n-heptane, and n-octane) the immiscibility gap was observed in the IL mole fraction range from $x_{IL} = 10^{-4}$ to 0.85–0.95 and slightly better solubility was observed for $[C_6H_{13}OCH_2-C_1Im][Tf_2N]$—the equilibrium curve was at $x_{IL} = 0.80$–0.83 for n-hexane and n-heptane [78]; much better solubility was observed for dihexyloxy-imidazolium salt $[(C_6H_{13}OCH_2)_2Im][Tf_2N]$ in n-hexane—the equilibrium curve was at $x_{IL} = 0.52$ [98]; the IL with alkoxy group in anion $[C_4C_1Im]$ $[MDEGSO_4]$ has shown immiscibility in n-hexane in the whole IL mole fraction range from $x_{IL} = 10^{-4}$ to 0.98 at room temperature [112]. In these systems the small free volume effects and the van der Waals interactions between the alkanes chains of cation and/or anion and n-alkane can only explain the differences in solubilities.

Similar small solubility in n-alkanes was observed for the ammonium ILs. For example, the solubility of butyl(2-hydroxyethyl)dimethylammonium bromide, $[(C_1)_2C_4HOC_2]Br$, exhibited a simple eutectic system with immiscibility in the liquid phase in the IL mole fraction range from $x_{IL} = 0.02$ to 0.70 [53]; the other ammonium salt: (benzyl)dimethylalkylammonium nitrate, $[Be(C_1)_2C_nN][NO_3]$, showed very small solubility in the liquid phase in n-hexadecane and slightly better in hexane (immiscibility in the range from $x_{IL} = 10^{-4}$ to 0.90) [99]; for the ammonium salt with an n-alkane substituents only: didecyldimethylammonium nitrate, $[(C_{10})_2(C_1)_2N][NO_3]$, the solubility in the liquid phase was better in n-hexane (immiscibility in the range from $x_{IL} = 10^{-4}$ to 0.50), but in n-hexadecane the immiscibility was observed in the whole IL mole fraction [100].

In all systems with imidazolium and ammonium salts, an increase in the alkyl chain length of the n-alkane (solvent) resulted in a decrease of solubility.

Generally, in cycloalkanes the solubility is very similar to that in n-alkanes. For example, the $[C_2C_1Im][PF_6]$ or $[C_4C_1Im][PF_6]$ has shown immiscibility in cycloalkanes (cyclopentane, cyclohexane) in the whole IL mole fraction range from $x_{IL} = 10^{-4}$ to 0.90 at room temperature [96,135]; the solubility of $[C_6H_{13}OCH_2-C_1Im][BF_4]$ and of $[C_6H_{13}OCH_2-C_1Im][Tf_2N]$ in cyclohexane was much higher—the immiscibility gap was observed in the IL mole fraction range from $x_{IL} = 10^{-4}$ to 0.72 and to 0.65. Usually, the solubility in cycloalkanes is better than in n-alkanes [78,112].

The influence of the length of the alkyl chain of the imidazolium IL's cation on the solubility in n-alkanes or cycloalkanes, and the interaction with a solvent has already been discussed for different ILs by the comparison of different thermodynamic properties [45,53,136]. For example, the solubility of two salts $[C_1C_1Im][C_1SO_4]$ and $[C_4C_1Im][C_1SO_4]$ in cycloalkanes, aromatic hydrocarbons, and other solvents (alcohols, ethers, and ketones) increases by increasing the alkyl chain on the cation [34,127]. Sometimes similarity of the molecule, imidazolium ring and the cycloalkane ring, results in better solubility in cycloalkane than in *n*-alkanes. An opposite reaction was observed for ammonium salts; for example, $[(C_1)_2C_4HOC_2N]Br$ is much less soluble in the liquid phase in cyclohexane than in *n*-hexane [53].

The phase diagrams of ILs with benzene and benzene derivatives differ from those of aliphatic hydrocarbons. The solubility of ILs is much higher in aromatic hydrocarbons than in aliphatic and cyclohydrocarbons, which is the result of interaction with the benzene ring. The interaction can be as $n-\pi$ and $\pi-\pi$ or even forming well-organized phases such as clatrate-type structures in the liquid phase [96,112,128,137]. In a very recent work, it was found that in the system $([C_2C_1Im][Tf_2N]$ + benzene) an equimolar congruently melting inclusion compound, $[C_2C_1Im][Tf_2N]\cdot C_6H_6$, exists [137]. The immiscibility in the liquid phase was observed for $[C_2C_1Im][PF_6]$ and $[C_4C_1Im][PF_6]$ in benzene in the IL mole fraction range from $x_{IL} = 10^{-4}$ to 0.45 at room temperature [96]; simple eutectic system without immiscibility in the liquid phase was observed for $[C_{12}C_1Im]Cl$ in benzene [95]; much better solubility in benzene than in *n*-alkanes and cycloalkanes was observed in [128] for $[C_2C_1Im][C_1SO_4]$ and $[C_4C_1Im][C_1SO_4]$; very strong interaction was observed for $[C_6H_{13}OCH_2-C_1Im][BF_4]$ and for $[C_6H_{13}OCH_2C_1Im][Tf_2N]$ with benzene, where the immiscibility gap was shown only in a narrow area, in the IL mole fraction range from $x_{IL} = 10^{-4}$ to 0.20 and to 0.10, respectively [78]; the $[C_4C_1Im][MDEGSO_4]$ (the IL with alkoxy group in anion) has shown immiscibility in benzene in the twice high IL mole fraction range from $x_{IL} = 10^{-4}$ to 0.40 at room temperature [112]. The area of two liquid phases decreases with an increase in the chain length of the alkyl substituent at the imidazole ring (e.g., $[C_2C_1Im][PF_6]$ versus $[C_4C_1Im][PF_6]$, or $[C_2C_1Im][C_1SO_4]$ versus $[C_4C_1Im][C_1SO_4]$). This trend is perfectly shown for $[C_nC_1Im][Tf_2N]$, where n = 1–10 with benzene, toluene, and α-methylstyrene. Untypical shape of the LLE curves was observed for $[C_4C_1Im][Tf_2N]$ and $[C_6C_1Im][Tf_2N]$ at high temperatures at ~450 K [138]. The interaction between these two ILs and benzene derivatives has revealed that unusually high-temperature demixing may occur. The UCST increases (solubility decreases) also with an increase of the alkyl chain at the benzene ring (solvent) [96,128].

The phase diagrams of an ammonium IL, $[Be(C_1)_2C_nN][NO_3]$, with benzene and toluene have shown low immiscibility in the liquid phase with UCST at high IL mole fraction [99].

For the phosphonium salt, $[(C_4)_4P][C_1SO_3]$, in toluene and ethylbenzene, the immiscibility was observed only in a very narrow area of the hydrocarbon high mole fraction [54].

Usually, the UCST increases as the length of the alkyl chain of hydrocarbon or cyclohydrocarbon increases. This trend was observed for systems presented earlier and in binary mixtures with 1,3-dialkylimidazolium salts as [C_2C_1Im][PF_6] and [C_4C_1Im][PF_6], or of [C_1C_1Im][C_1SO_4] and [C_4C_1Im][C_1SO_4] [127,128] or of [$C_6H_{13}OCH_2$–C_1Im][BF_4] [78,98].

The LLE for ILs and common solvents such as alcohols is very important for developing ILs for liquid–liquid extraction processes. Previous studies in many laboratories have shown this potential. Most of the measured mixtures were of {IL + short chain alcohol} binary systems. It is well known that an increase in the alkyl chain length of the alcohol resulted in an increase in the UCST. Nevertheless, the solubilities of many ILs were measured in 1-octanol (important value for the description of bioaccumulation) [14,50–54, 79,98–100,112,127,133]. The other short chain alcohols were pointed out earlier.

The ability of the anion of the IL to accept hydrogen bonding with an alcohol was observed to be inversely related to the effect that the anion had on the UCST. Although the anion was observed to have a significant effect on the solubility in alcohols, the importance of the anion decreased as the length of the alkyl chain on the cation increased [133]. The longer alkyl chain cases the stronger the interaction with an alcohol through dispersion forces [133]. The substitution of the acidic C2 hydrogen on the imidazolium ring with a methyl group increases UCST due to a reduction in hydrogen bonding with an alcohol [133].

The hydrogen bonding between the IL and the $CHCl_3$ causes extremely interesting phase behavior of a binary system. The interaction between hydrogen of $CHCl_3$ and the π system of the imidazolium ring together with that between the chlorine atoms and the acidic hydrogen atoms of the same ring, C2, C4, and C5 is responsible for the entropy/enthalpy balance of mixing as the temperature increases [139]. The phase diagram of [C_4C_1Im][Tf_2N] and [C_3C_1Im][Tf_2N] and their mixtures with $CHCl_3$ were encountered for the first time in binary and quasibinary liquid solutions of ILs with UCST and lower critical solution temperature (LCST) together. This type of behavior is usually restricted to aqueous or polymer solutions. It was also indicated via electrospray mass spectrometric studies that in the mixture the cation–anion contact pairs 1:2 and 2:1 exist together with simple cations and anions [139].

In addition to the experimental results of phase equilibria, the correlation with the widely known GE models was assigned to. It was indicated by many authors that SLE, LLE, and VLE data of ILs can be correlated by Wilson, NRTL, or UNIQUAC models [52,54,64,79,91–101,106,112,131,134]. For the LLE experimental data, the NRTL model is very convenient, especially for the SLE/LLE correlation with same binary parameters of nonrandom two-liquid equation for mixtures of two components. For the binary systems with alcohols the UNIQUAC equation is more adequate [131]. For simplicity, the IL is treated as a single neutral component in these calculations. The results may be used for prediction in ternary systems or for interpolation purposes. In many systems it is difficult to obtain experimentally the equilibrium curve at very low solubilities of the IL in the solvent. Because this solubility is on the level of mole fraction $x_{IL} \sim 10^{-3}$ or 10^{-4}, sometimes only the spectroscopic method has to be used [131]. From many published results [65,127–130] and COSMO-RS predictions the mole fraction at equilibrium temperatures of the dilute IL solutions can be assumed as $x_{IL} = 10^{-4}$.

The solute activity coefficients, γ_1, of the saturated solutions were correlated for many mixtures by the NRTL model describing the excess Gibbs energy [140]

$$\frac{G^E}{RT} = x_1 x_2 \left[\frac{\tau_{21} G_{21}}{x_1 + x_2 G_{21}} + \frac{\tau_{12} G_{12}}{G_{12} x_1 + x_2} \right] \tag{1.15}$$

$$\tau_{12} = (g_{12} - g_{22})/RT \tag{1.16}$$

$$\tau_{21} = (g_{21} - g_{11})/RT \tag{1.17}$$

$$G_{12} = \exp(-\alpha_{12} \tau_{12}) \tag{1.18}$$

$$G_{21} = \exp(-\alpha_{12} \tau_{21}) \tag{1.19}$$

$$\ln(\gamma_1) = x_2^2 \left[\tau_{21} \left(\frac{G_{21}}{x_1 + x_2 G_{21}} \right)^2 + \frac{\tau_{12} G_{12}}{\left(x_2 + x_1 G_{12} \right)^2} \right] \tag{1.20}$$

Model adjustable parameters $(g_{12} - g_{22})$ and $(g_{21} - g_{11})$ were found by the minimization of the OF:

$$\text{OF} = \sum_{i=1}^{n} \left[(\Delta x_1)_i^2 + (\Delta x_1^*)_i^2 \right] \tag{1.21}$$

where n is the number of experimental points and Δx defined as

$$\Delta x = x_{cal} - x_{exp} \tag{1.22}$$

The root-mean-square deviation of the mole fraction was defined as follows:

$$\sigma_x = \left(\frac{\sum\limits_{i=1}^{n} (\Delta x_1)_i^2 + \sum\limits_{i=1}^{n} (\Delta x_1^*)_i^2}{2n - 2} \right)^{1/2} \tag{1.23}$$

Usually, the modeled binodal curves reproduced the experimentally observed data (with different constant nonrandomness factor, α) to a satisfactory degree [96–98,112,131,134].

The conductor-like screening model for real solvents (COSMO-RS) is a unique method for predicting the thermodynamic properties of mixtures on the basis of unimolecular quantum chemical calculations for individual molecules [141,142]. In this approach, all information about solutes and solvents is extracted from initial quantum chemical COSMO calculations, and only very few parameters are adjusted to the experimental values of partition coefficients and vapor pressures of a wide range of neutral organic compounds. COSMO-RS is capable of predicting activity coefficients, partition coefficients, vapor pressures, and solvation free energies of neutral compounds. In addition, successful applications of COSMO-RS to predict the thermodynamic properties of ions in solution and ILs have been reported [65,130,131,143,144]. The LLE properties have been calculated from the equity

$$x_i^I \gamma_i^I = x_i^{II} \gamma_i^{II}$$
(1.24)

where indices I and II denote the liquid phases, x_i is the mole fractions of the two solvents and γ_i the activity coefficients of the solvents as computed by COSMO-RS.

The COSMO-RS predictions correspond better to experimental results for less polar ILs such as $[C_4C_1Im][PF_6]$ [107,131] than for $[C_4C_1Im][C_1SO_4]$ or $[C_1C_1Im][C_1SO_4]$ [127,128]. This can be explained partly by the stronger polarity of especially $[C_1C_1Im][C_1SO_4]$, which results in possibly larger errors in COSMO-RS predictions. COSMO-RS overestimates the polarity difference of the compounds and thus the miscibility gap prediction is too wide and the UCST is too high for $[C_4C_1Im][C_1SO_4]$ or $[C_1C_1Im][C_1SO_4]$ [127,128]. This is especially true for the binary mixtures with alcohols. As one would expect this overestimation is a result of strong A–B interaction, IL–an alcohol for the investigated mixtures. It can be noticed as well that for the {$[C_4C_1Im][C_1SO_4]$ or $[C_1C_1Im][C_1SO_4]$ + ketone} binary mixtures the COSMO-RS prediction was the best. Generally, it was found that the $[C_1SO_4]^-$ anion is extremely polar and that prediction by COSMO-RS is very difficult.

At the end of this section the effect of inorganic salts on the LLE of the IL's aqueous solutions is worth underlining. It has been proved that the addition of NaCl, Na_2SO_4, and Na_3PO_4 salts to the binary mixture of {$[C_4C_1Im][BF_4]$ + H_2O} causes an increasing salting-out effect being the function of concentration of an inorganic salt. All three inorganic salts reveal significant upward shifts of the UCST of the systems [145]. The magnitude of the LL demixing temperature shifts depends on both the water-structuring nature of the salt and its concentration. The interaction effects are correlated with the ionic strength of the solution and the Gibbs free energy of the hydration of the inorganic salt [145].

1.7.1   Liquid–liquid equilibria in ternary systems

ILs exhibit many properties, which allow for their application as selective solvents in solvent extraction. Large selectivities, remarkable loading capacities for certain components, low vapor pressures, comparatively low melt, and solution viscosities as well as a good thermal and chemical stability are only some of their characteristics described in this chapter.

Separation technology evolved during the twentieth century, driven primarily by advances in the petrochemical industry. Several technologies, such as distillation, extraction, and adsorption, have been known and used for quite some time, and other technologies, such as membranes, have evolved and are being used in new applications. Recent materials, such as ILs or hyperbranched polymers, and other scientific and engineering advances provide the potential for wider opportunities in almost all separation technologies.

The potential of ILs in the field of extractive distillation was shown in Refs 126 and 146 for the separation of THF–water and ethanol–water azeotropes using $[C_2C_1Im][BF_4]$. This IL easily breaks the azeotropic ethanol–water phase behavior by interacting selectively with water.

The LLE in the similar ternary systems {$[C_4C_1Im][PF_6]$ + water + ethanol} was measured in three temperatures (290.15, 298.15, and 313.25 K) [66]. From these results it can be concluded that with decreasing temperature three liquid phases can be expected.

Some ILs were tested as extracting agents to separate ethanol from its mixtures with *tert*-amyl ethyl ether [147–149]. Industrially, ether purification is carried out using water as the extraction solvent. The $[C_8C_1Im]Cl$ IL has shown better solute-distribution ratio than in the system with water [148]. Also, higher values of selectivity were obtained in comparison with $[C_4C_1Im][TfO]$ and water [147]. Better results were only observed [149] with $[C_2C_1Im][C_2SO_4]$. The purification of the other ether, ethyl *tert*-butyl ether, from its mixtures with ethanol was tested with four ILs—$[C_4C_1Im][TfO]$ [150], $[C_2C_1Im][C_2SO_4]$ [151], $[C_2C_1Im][C_1SO_3]$ [150], and $[C_2C_1Im][TfO]$ [152]. It has been found that 1-ethyl-3-methylimidazolium methanesulfonate leads to very high values of solute distribution ratios and selectivities (>2000) comparing with three other ILs.

These two ethers are used as blending agents (antiknock effect) in the petroleum industry. Especially, ethyl *tert*-butyl ether is expected to become the most widely used octane booster on gasoline blending. Ethyl *tert*-butyl ether is mainly produced on an industrial scale by the reaction of isobutene with an excess of ethanol. Thus, the separation from ethanol is very important.

A successful extraction of heptane from the azeotropic mixture with ethanol by means of a liquid–liquid operation with the IL, $[C_6C_1Im][PF_6]$, was recently published [153].

The separation of aromatic hydrocarbons (benzene, toluene, ethyl benzene, and xylenes) from methane to *n*-decane aliphatic hydrocarbon mixtures is challenging since the boiling points of these hydrocarbons are in a close range and they have several combinations from azeotropes. For years, the sulfolane, NMP or NMP with water as cosolvent were used in the industry. Several ILs were tested for this extraction by Meindersma et al. [123] in ternary LLE systems and by many scientists via the activity coefficients at infinite dilution measurements. It has been found that many ILs are suitable for the extraction of benzene/n-alkane or toluene/n-alkane (n-hexane, n-heptane) mixtures. This discussion will be presented later in this chapter.

The requirements of a suitable IL for the separation of aromatic and aliphatic hydrocarbons are the high solubility of aromatic hydrocarbons in the IL and low solubility of aliphatic hydrocarbons in the IL. The latter is easy to obtain, but for the former, the strong interaction, n–π or π–π , has to be assumed. There are quite many publications [154–158] concerning the LLE in ternary systems {IL + aromatic hydrocarbon + aliphatic hydrocarbon} using $[C_2C_1Im][I_3]$, $[C_4C_1Im][I_3]$, $[C_2C_1Im][Tf_2N]$, $[C_8C_1Im]Cl$, $[C_6C_1Im][BF_4]$, $[C_6C_1Im][PF_6]$, $[C_2C_1Im][C_2SO_4]$, $[C_1C_1Im][C_1SO_4]$, $[C_4C_1Im][C_1SO_4]$, and 4-methyl-N-butylpyridinium tetrafluoroborate, $[4C_1$–$C_4py][BF_4]$. For the mixtures of C_7, C_{12}, and C_{16} n-alkanes, it was found that the selectivity given by $[C_8C_1Im][Cl]$ increases with increasing carbon number of the n-alkane [156]. The best solubility of toluene was observed [158] for $[4C_1$–$C_4py][BF_4]$ in comparison with three imidazolium salts: $[C_2C_1Im][C_2SO_4]$, $[C_1C_1Im][C_1SO_4]$, and $[C_4C_1Im][C_1SO_4]$. It was also stated that $[4C_1$–$C_4py][BF_4]$ was the best solvent for the aromatic/aliphatic separation from these four ILs [158]. From this experiment and from LLE of binary systems {IL + benzene, or toluene} it can be assumed that alkoxyimidazolium ILs will demonstrate high selectivities for the aromatic/aliphatic separation.

Some interesting research has focused on the influence of the alkyl-substituent length on ternary LLE and separation of systems benzene/n-hexane [159]. For the benzene/n-hexane separation the $[C_nC_1Im][Tf_2N]$ ILs were chosen, with n ranging from 2 to 12. The mutual solubility IL + benzene increases when the alkyl chain of the cation increases. For $[C_{12}C_1Im][Tf_2N]$ complete miscibility was presented. Unfortunately, the increasing n-hexane concentration in the IL as the chain length increases lowers the distribution ratio. It has an influence on the ternary diagram. In conclusion, it was indicated that ILs with longer alkyl chains demonstrated lower selectivities. Conversely, an optimum solute distribution ratio is achieved for the ILs with 8–10 carbon atoms in their alkyl chain. Because of the high viscosity of longer chain ILs, the $[C_4C_1Im][Tf_2N]$ was chosen from these mixtures [155,159].

Similar good results of the separation of aromatic and aliphatic hydrocarbons were recently obtained with ethyl(2-hydroxyethyl)dimethylammonium *bis*(trifluoromethylsulfonyl)imide, $[(C_1)_2C_2HOC_2N][Tf_2N]$, at 298.15 K [160]. The separation of *m*-xylene from *n*-octane by extraction with $[(C_1)_2C_2HOC_2N]$ $[Tf_2N]$ was observed with the distribution ration of 0.3 and selectivities of range 22–31. The other ammonium salt as $[(C_1)_2C_4HOC_2N][BF_4]$ or 1,3-dihexyl-oxymethyl-imidazolium tetrafluoroborate was not so successful in this separation [161].

1.7.2   Extraction of metal ions

In separation, ILs are mainly used to extract metal ions with popular ILs [C_4C_1Im][PF_6], [C_6C_1Im][PF_6], or other ILs with [PF_6]$^-$ or [Tf_2N]$^-$ anions [162–167]. ILs have usually hydrophobic character that allows them to extract hydrophobic compounds in biphasic separation. Metal ions tend to stay in the aqueous solution being hydrated. Therefore, to remove metal ions from the aqueous phase into the hydrophobic IL, extractants (ligands) are normally needed to form complexes to increase the hydrophobicity of the metal. In more details this area will be covered in chapter 10 of the book.

1.8   Vapor–liquid phase equilibria

Since the evaporation of the molecule involves stretching of the hydrogen-bonded structure of the IL, it is natural to examine the change of vapor pressure for binary mixtures with molecular solvents. The vapor pressure of mixtures {[C_4C_1Im][BF_4] or [C_4C_1Im]Br + 2,2,2-trifluoroethanol} were measured in the concentration range from 40 to 90 mass percent of the IL and in the temperature range from 293.2 to 573.2 K [168]. The highest vapor pressure was found for 40 mass percent of IL at 370.2 K: $p = 100.5$ kPa.

VLE of binary mixtures of alcohol (methanol or ethanol or 1-propanol) or benzene with [C_4C_1Im][Tf_2N] were studied by using a static method [169]. The activity coefficients of solvents in these mixtures were higher than one except the system with methanol, where the values were lower than one for the mole fraction of alcohol $x = 0$ to 0.35. The same method was used for the VLE measurements of {1,2-ethanediol or 1,4-butanediol + [C_2C_1Im][Tf_2N]} mixtures and {metoxybenzene or hydroxymethylbenzene + [C_2C_1Im][Tf_2N]} mixtures [170]. Pressures were very low and activity coefficients were higher than one, except the high concentrated mixtures of hydroxymethylbenzene. The same static method based on the transpiration technique used for the low pressure substances was investigated in the mixtures of {dimethyl adipate or ethyl benzoate or benzylamine + [C_2C_1Im][Tf_2N]}. The measured pressure for binary mixtures was lower than 360 Pa and calculated from the VLE activity coefficients was higher than one [170].

The VLE was measured for three ILs: [C_4C_1Im][PF_6], [C_8C_1Im][PF_6], and [C_8C_1Im][BF_4] with absorbed water using a gravimetric microbalance at three temperatures, 283.15, 298.15, and 308.15 K. The solubility of water vapor was the highest in [C_8C_1Im][BF_4]. It was suggested that water is more soluble in the system where the counteranion is [BF_4]$^-$ rather than [PF_6]$^-$, because of the higher charge density for [BF_4]$^-$ anion [171].

Important information for the separation of aromatics from aliphatic hydrocarbons was derived from the VLE data at three temperatures 303.15, 333.15, and 353.15 K for [C_1C_1Im][Tf_2N], [C_2C_1Im][Tf_2N], and [C_4C_1Im][Tf_2N] with benzene and cyclohexane [64]. Based on the observed larger miscibility gaps for cyclohexane (higher activity coefficients compared to benzene), suggestions can be made about the suitability of these ILs for the separation process. It was also confirmed that the miscibility gap increases with decreasing alkyl chain from butyl to methyl at the imidazolium ring [64].

Low vapor pressure was also measured for the binary system of {THF + [C_2C_1Im][BF_4]} at temperature 337.15 K: for 0.3 mole fraction of THF it was 84.2 kPa [172]. Also, ternary VLE was measured for four ILs, [C_2C_1Im][BF_4], [C_4C_1Im]Cl, [C_4C_1Im][BF_4], and [C_8C_1Im][BF_4] with water and THF [144]. Tailoring of ILs as entrainers in extractive distillation was discussed.

The VLE of many systems {cyclohexane or cyclohexene or methanol or ethanol + [C_6C_1Im][Tf_2N] or [C_2C_1Im][Tf_2N]} at 353.10 K were measured together with the activity coefficients at infinite dilution, and the prediction of thermodynamic behavior was made using original UNIFAC, mod. UNIFAC, and COSMO-RS(01) models [65]. A homogenous behavior was observed for methanol and ethanol, and a miscibility gap was obtained for cyclohexane and cyclohexene. The VLE correlations with UNIQUAC show excellent agreement with the experimental results. For the first time also the group contribution method UNIFAC was used for the prediction of thermodynamic properties of ILs with perspective results [65]. Important measurements of the VLE with polar solvents, methanol, ethanol, 2-propanol, acetone, THF, water with three ILs ([C_2C_1Im][Tf_2N], [C_4C_1Im][Tf_2N], and [C_1C_1Im][$(CH_3)_2PO_4$]) have been presented at 353.15 K [106]. The complete miscibility was observed for [C_2C_1Im][Tf_2N] and [C_4C_1Im][Tf_2N] with acetone, alcohols, and THF, and the miscibility gap was found for water. Also strong negative deviation from Raoult's law was observed for the systems of water, or methanol, or ethanol with [C_1C_1Im][$(CH_3)_2PO_4$]). The high selectivity of [C_1C_1Im][$(CH_3)_2PO_4$] was found for the separation of THF/water by extractive distillation [106].

The VLE was also measured for binary and ternary systems of {ethanol + [C_2C_1Im][C_2SO_4]} and {ethanol + ethyl *tert*-butyl ether + [C_2C_1Im][C_2SO_4]} at 101.3 kPa [151]. This ternary system does not exhibit a ternary azeotrope. The possibility of [C_2C_1Im][C_2SO_4] use as a solvent in liquid–liquid extraction or as an entrainer in extractive distillation for the separation of the mixture ethanol/ethyl *tert*-butyl ether was discussed [151].

The estimated critical temperatures and normal boiling temperatures using the Eötvos and Guggenheim equation, from the pure substances data of surface tension and density, were presented by Rebelo et al. [172]. Normal boiling temperatures of ILs were found inherently high because the strong, long-range Coulomb interactions prevent the particles separation into the gas phase. When the ion size decreases and the charge density increases the normal boiling temperature increases. For example, the values are 725 and 646 K for the [C_2C_1Im][Tf_2N] and [C_4C_1Im][Tf_2N], respectively [172].

In every VLE measurement cited earlier, the vapor pressure of pure IL was zero. It is widely believed that a fundamental characteristic of ILs is that they reveal no measurable vapor pressure. In 2006, Earle et al., demonstrated that a range of pure, aprotic ILs such as [C_4C_1Im][Tf_2N], [C_4C_1Im][PF_6], [C_2C_1Im][C_2SO_4], and many others can be vaporized under vacuum at

473–573 K and then recondensed at lower temperatures [173]. It was also demonstrated that there is a possibility of separating two ILs by distillation without their thermal decomposition. On the other hand, for example, BASF used ILs in the extractive distillation process at temperatures of 448 K and pressure of 5 kPa for more than three months without loss of the IL. The assumed nonvolatility of ILs had been a basis of their common reputation as "green" solvents.

This chapter contains many interesting sections that cover the development of the subject from a wide range of perspectives. There is no doubt that the VLE data that exist today will enable studies of separation by more polar or hydrophobic ILs and the first of such studies can only be a short way off.

1.9   Activity coefficients in ionic liquids

Thermodynamic activity coefficients can be determined from the phase equilibrium measurements, and they are a measure of deviations from Raoult's law. Data of the activity coefficients covering the whole range of liquid composition of {IL + molecular solvent} mixtures have been reported in the literature and discussed in sections 1.6, 1.7, and 1.8 as the values obtained from the SLE, LLE, and VLE data. When a strong interaction between the IL and the solvent exists, negative deviations from ideality should be expected with the activity coefficients lower than one.

Activity coefficients at infinite dilution, γ_{13}^{∞}, of organic solutes in ILs have been reported in the literature during the last years very often [1,2,12,45,64, 65,106,123,144,174–189]. In most cases, a special technique based on the gas chromatographic determination of the solute retention time in a packed column filled with the IL as a stationary phase has been used [45,123,174–176,179,181–187]. An alternative method is the "dilutor technique" [64,65,106, 178,180]. A lot of γ_{13}^{∞} (where 1 refers to the solute, i.e., the organic solvent, and 3 to the solvent, i.e., the IL) provide a useful tool for solvent selection in extractive distillation or solvent extraction processes. It is sufficient to know the separation factor of the components to be separated at infinite dilution to determine the applicability of a compound (a new IL) as a selective solvent.

In many publications, the activity coefficients at infinite dilution γ_{13}^{∞} have been determined for alkanes, alk-1-enes, alk-1-ynes, cycloalkanes, aromatic hydrocarbons, carbon tetrachloride, and methanol in the IL at different temperatures. Figure 1.15 presents a plot of the activity coefficients, γ_{13}^{∞}, against chain length, n of hydrocarbons: n-alkanes, alk-1-enes, alk-1-ynes, and cycloalkanes for [C$_6$C$_1$Im][Tf$_2$N] [183].

For every hydrocarbon the activity coefficient increases with an increase of the carbon chain length. Usually, the value of activity coefficient decreases when the interaction between the solvent and the IL increases. It can be seen in Figure 1.16, where the plot of the activity coefficients, γ_{13}^{∞}, for different C$_6$ hydrocarbons in [C$_6$C$_1$Im][Tf$_2$N] is presented [183]. The value of the activity

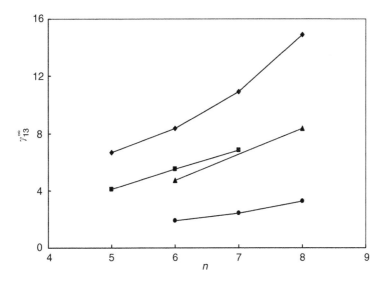

Figure 1.15 Plot of activity coefficients, γ_{13}^{∞}, against chain length, n, of hydrocarbons for [C$_6$C$_1$Im][Tf$_2$N]: n-alkanes (◆), alk-1-enes (▲), alk-1-ynes (●), and cycloalkanes (■). (Adapted from Letcher, T.M. et al., *J. Chem. Thermodyn.*, 37, 1327, 2005.)

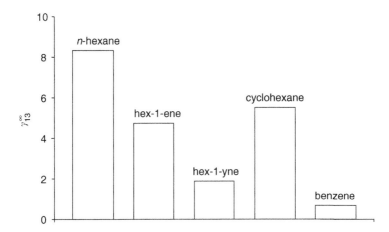

Figure 1.16 Plot of activity coefficients, γ_{13}^{∞}, for different C$_6$ hydrocarbons in [C$_6$C$_1$Im][Tf$_2$N]. (Adapted from Letcher, T.M. et al., *J. Chem. Thermodyn.*, 37, 1327, 2005.)

coefficient in benzene is the lowest. The influence of the alkyl chain in the cation is presented in Figure 1.17 for the [Tf$_2$N]$^-$ anion.

The plot of the activity coefficients, γ_{13}^{∞}, against alkyl chain length, n, in the cation [C$_n$C$_1$Im][Tf$_2$N] is shown for n-hexane, hex-1-ene, benzene, and cyclohexane. The increase of the alkyl chain at the cation decreases the activity coefficient for every kind of hydrocarbon in the IL.

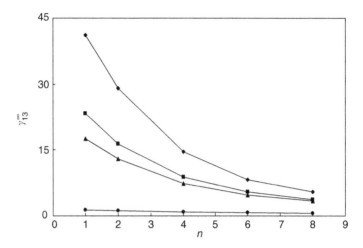

Figure 1.17 Plot of activity coefficients, γ_{13}^∞, against alkyl chain length, n, in cation [C_nC_1Im][Tf_2N]: n-hexane ($\blacklozenge$), hex-1-ene ($\blacktriangle$), benzene ($\bullet$), and cyclohexane ($\blacksquare$).

The selectivity S_{ij}^∞ is the ratio of the activity coefficients at the infinite dilution and is given by

$$S_{ij}^\infty = \frac{\gamma_{i3}^\infty}{\gamma_{j3}^\infty} \tag{1.25}$$

where i and j refer to the liquids to be separated and it usually refers to hexane and benzene, or n-heptane and toluene, respectively. The selectivity value is used to determine the potential of the ionic solvent for extractive distillation in the separation of aromatic compounds from aliphatic compounds. The capacity

$$k^\infty = \frac{1}{\gamma_{j3}^\infty} \tag{1.26}$$

determines the ratio of the feed flow; that is, the capacity has an influence on the operating costs. Often an increase of selectivity leads to a decrease in capacity. It is shown in Figure 1.18, where the selectivities and capacities are plotted against the alkyl chain length of the cation of [C_nC_1Im][Tf_2N].

It can be seen that the best IL for the economic realization of this process is [C_4C_1Im][Tf_2N].

GLC is a well-established and accurate method used to obtain γ_{13}^∞ and the partial molar excess enthalpies at infinite dilution values $\Delta H_1^{E\infty}$, which is determined from the Gibbs–Helmholtz equation:

$$\left[\frac{\partial \ln \gamma_i^\infty}{\partial(1/T)}\right] = \frac{\Delta H_1^{E\infty}}{R} \tag{1.27}$$

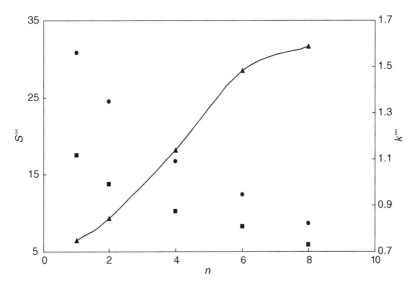

Figure 1.18   Selectivities at infinite dilution, S^∞, and capacities, k^∞, against the n-alkyl chain length, n in the cation of ionic liquid $[C_nC_1Im][Tf_2N]$: S^∞ (n-hexane/benzene) (●), S^∞ (cyclohexane/benzene) (■), and k^∞ (benzene) (—▲—).

In general, within each of the series, n-alkane, alk-1-ene, alk-1-yne, and cycloalkane, the $\Delta H_1^{E\infty}$ values increase with the increasing carbon number of solute (molecular solvents) [184–186]. Opposite effects were observed for $[C_8C_1Im]Cl$ and $[C_4C_1Im][MDEGSO_4]$ [178,182]. The values of $\Delta H_1^{E\infty}$ for hex-1-yne, hept-1-yne, benzene, and carbon tetrachloride are negative for some ILs [182,183]. The negative value (-480 J mol^{-1}) was also observed for benzene in $[C_2C_1Im][Tf_2N]$ [177]. The value for benzene in $[C_6C_1Im][Tf_2N]$, $\Delta H_1^{E\infty} = -900$ J mol^{-1} is more negative, indicating that the intermolecular π–π interactions are higher for longer alkyl substituent at the imidazole ring [183].

The selectivity S_{ij}^∞ values for the separation of hexane (i)/benzene (j) mixture at $T = 298.15$ K using different IL compounds and some very polar solvents obtained from the literature are presented in Table 1.9. The results demonstrate a significant influence of the cation on the S_{ij}^∞ values. For a given anion $[Tf_2N]^-$ the values of S_{ij}^∞ are much higher for shorter alkyl chain at the imidazole ring (see also Figure 1.18).

The results at $T = 298.15$ K for $[C_1C_1Im]^+$, $[C_2C_1Im]^+$, $[C_4C_1Im]^+$, and $[C_6C_1Im]^+$ are 29.0, 24.4, 22.9, 16.7, and 12.4, respectively. It was shown earlier in many investigations that the selectivities of ILs in separating organic liquids are often higher than those for commonly used solvents, such as NMP, or NMP + water. For the imidazolium compounds with the $[Tf_2N]^-$ anion, the selectivity increases with the decreasing alkyl chain length at the imidazole ring for different separation mixtures. Furthermore, in Table 1.9, it was shown that by changing an anion from $[Tf_2N]^-$ to $[PF_6]^-$, or $[BF_4]^-$, an

Table 1.9   Selectivities, S_{ij}^{∞}, at Infinite Dilution of Various
Solvents for the *n*-Hexane/Benzene Separation
$S_{ij}^{\infty} = \gamma_{i3}^{\infty}/\gamma_{j3}^{\infty}$ at $T = 298.15$ K

Solvent	S_{ij}^{∞}	Reference
Sulfolane	30.5	188
Dimethylsulfoxide	22.7	188
Diethylene glycol	15.4	188
N-methyl-2-pyrrolidinone	12.5	188
[4C$_4$C$_1$pyr][Tf$_2$N]	16.7[a]	65
[C$_2$C$_1$Im][C$_2$SO$_4$]	41.4[a]	177
[C$_6$C$_1$Im][PF$_6$]	21.6	179
[C$_6$C$_1$Im][BF$_4$]	23.1	179
[C$_2$C$_1$Im] [Tf$_2$N]	24.4[b]	65
[C$_2$C$_1$Im][Tf$_2$N]	22.9	176
[C$_2$C$_1$Im][Tf$_2$N]	37.5	184
[C$_2$C$_1$C$_1$Im][Tf$_2$N]	25.3	176
[C$_4$C$_1$Im][Tf$_2$N]	16.7[b]	177
[C$_6$C$_1$Im][Tf$_2$N]	12.4	183

[a] Extrapolated value.
[b] Interpolated value.

increase in selectivity is observed. Comparing different results for many
ILs, the selectivity S_{ij}^{∞} and capacity values for the separation of *n*-hexane (i)/
benzene (j) mixture have to be calculated; also the density, viscosity, toxicity,
and cost for the investigated IL have to be discussed. The results presented
show that the activity coefficients and intermolecular interactions of different solutes with the IL are very much dependent on the chemical structure
of the IL.

   The effect of the alkyl chain length, the structure of the cation and anion,
and the final selection of the IL for aromatic/aliphatic extraction (mainly
toluene/*n*-heptane) have been presented by Meindersma et al. [123]. The
ILs [C$_1$C$_1$Im][C$_1$SO$_4$], [C$_2$C$_1$Im][C$_2$SO$_4$], and 4-methyl-*N*-butylpyridinium salt
[4C$_1$C$_4$py][BF$_4$] have been chosen as the best for the extraction experiment.
In conclusion, [4C$_1$C$_4$py] [BF$_4$] showed the best combination of capacity
and selectivity. Also, some other ILs appeared suitable for this extraction:
[C$_4$C$_1$Im][BF$_4$], [3C$_1$C$_4$py][C$_1$SO$_4$], and [C$_2$C$_1$Im][C$_2$SO$_4$]. The ammonium based
[(C$_1$)$_2$C$_2$HOC$_2$N][Tf$_2$N] can also be not neglected [160].

   Recently, ILs similar to those presented in this section have been under
intense investigation. The quantitative structure-property relationship
(QSPR) method to the analysis of γ_{13}^{∞} values obtained in different laboratories
was used for the correlation and prediction. These studies have shown that
QSPR is a powerful tool for extending experimental activity coefficient data,

especially for new solutes [189]. Excellent results of modeling of the activity coefficients of tetramethylammonium, tetraethylammonium, tetrapropylammonium, and methylammonium salts with different anions in water at 298.15 K have been obtained with the electrolyte–nonrandom two-liquid (NRTL) equation [190].

In conclusion, the examination of selectivity has been done together with capacity in a variety of solutes in ILs to provide a picture of interactions and costs.

1.10   Heat capacity of ionic liquids and heat of mixing

Over the past decades, heat capacities have constructed an impressive database in the ILs, with respect to both the solid and the liquid phases. Heat capacities are needed to estimate heating and cooling requirements as well as the heat-storage capacity. From the beginning, heat capacities of popular ILs were measured in Refs 191–193 for $[C_2C_1Im][PF_6]$, $[C_4C_1Im][PF_6]$, $[C_6C_1Im][PF_6]$, $[C_4C_1Im]Cl$, $[C_4C_1Im][Tf_2N]$, $[C_2C_1Im][BF_4]$, $[C_4C_1Im][BF_4]$, and $[C_4C_1C_1Im][Tf_2N]$. Systematic measurements of imidazolium ILs at two temperatures 298.15 and 323.15 K, especially of *bis*(trifluoromethylsulfonyl)imides were performed by Fredlake et al. [81]. For example, the value for $[C_2C_1Im][Tf_2N]$ was 524.3 J mol^{-1} K^{-1}; the lowest value was measured for $[C_4C_1Im]Br$, 316.7 J mol^{-1} K^{-1}. Heat capacity increases with temperature and with the increasing number of atoms in the IL. Some of these salts ($[C_4C_1Im][PF_6]$ and $[C_4C_1Im][Tf_2N]$) were measured again by Troncoso et al. [193] at the temperature range between 283 and 328 K. The deviations from the data of Fredlake et al. [81] were observed and a slight influence of the *effect of Cl*$^-$ content was stated. Again, the same ILs and *N,N*-methyl propyl pyrrolidinium salt with $[Tf_2N]^-$ anion were measured at the temperature range from 283.15 to 358.15 K [194].

The values of heat capacities of 24 ILs, pyridinium-based, imidazolium, and ammonium ILs were presented by the same laboratory a year later [195]. The high value (766 J mol^{-1} K^{-1} at 298 K) was observed for 1-hexyl-2-propyl-3,5-diethylpyridinium salt, $[1C_6–2C_3–3,5C_2py][Tf_2N]$. It was found that heat capacity increases linearly with increasing molar mass for these compounds that are comprised of a limited number of different atoms. A series of pyridinium and imidazolium-based ILs mainly with trifluoromethanesulfonate anion, $[TfO]^-$, have been measured by Diedrichs and Gmehling [124]. The values between 300 and 800 J mol^{-1} K^{-1} were observed for different ILs.

The heat capacities of series of ammonium bromides ($[(C_1)_2C_3HOC_2N]Br$, $[(C_1)_2C_4HOC_2N]Br$, and $[(C_1)_2C_6HOC_2N]Br$) were measured for the solid and liquid phase, and the difference in the solute heat capacity between the liquid and solid phase at the melting temperature, $\Delta_{fus}C_{p,1}$ has been determined by the DSC analysis [79]. Negative values of $\Delta_{fus}C_{p,1}$ were observed for these compounds.

The excess molar enthalpies at 323 K and 1.3 MPa were measured in [64] for hydrocarbons (hex-1-ene, cyclohexane, benzene, and cyclohexene) in the $[C_2C_1Im][Tf_2N]$. The negative excess enthalpies were observed (-730 J mol^{-1} at $x_H = 0.63$) only in the mixtures with benzene, expected from the discussion about the interactions in the solution. Much more data can be found in the two existing data banks [1,2]; for example, in Dortmund Data Bank, 37 systems are accessible.

The heats of the solution, $\Delta_{sol}H$, of ILs $[C_2C_1Im][BF_4]$ and $[C_4C_1Im][BF_4]$ in acetonitrile as a function of concentration were measured at 298.15 K [196]. The values were approximately -4.5 J mol^{-1} for small concentrations of the IL. The heat of the solution at the infinite dilution as well as the enthalpy of transfer from water to methanol and acetonitrile for four ILs was detected [196]. Enthalpies of the solution of six organic solutes (methanol, *tert*-butanol, 1-hexanol, chloroform, toluene, and ethylene glycol) in $[C_2C_1Im][Tf_2N]$ have been measured, and the enthalpies of mixing have been calculated [196]. Excess molar enthalpies were negative for chloroform and toluene.

Calorimetric measurements are time consuming and very expensive because of the amount of the IL taken to the experiment. The results usually show the same type of interaction as in other experiments as the activity coefficients at the infinite dilution or solubility measurements. From calorimetric measurements it can be observed that the molar heat capacities depend linearly on the temperature and increase proportionally to the alkyl chain length of the cation.

1.11   Electrochemical properties

In recent years, the 1,3-dialkylimidazolium salts have been vigorously studied, because they are expected to be a new electrolyte medium. Especially when the counter anion of such imidazolium salt is exchanged from halogens to popular anions such as $[PF_6]^-$, $[BF_4]^-$, $[Tf_2N]^-$, $[TfO]^-$, and others, which are known to be water stable and less viscous liquids at ambient temperatures. This characteristic (low viscosity and correspondingly high ionic conductivity, chemical and thermal stability, nonflammability and a wide electrochemical potential window) is preferable for electrolyte solution for lithium battery. Various trials have been made to have a fundamental understanding of these ILs and to introduce them in practical applications, such as batteries, capacitors, and electrochemical solar cells. The formation of polymer electrolytes based on these ILs has also been proposed. However, the fundamental understanding of the ionic transport properties is not sufficient. The problems of ionic diffusion coefficient, the degree of ionic association, and the interaction between ions have not been clearly revealed. Recently, the critical review of ILs as electrolytes was presented from the point of view of their possible application in electrochemical processes and devices. The viscosity, melting and freezing points, conductivity, temperature dependence of conductivity, electrochemical stability, and possible application in aluminum electroplating, lithium batteries, and in electrochemical capacitors were reviewed [36]. The introduction to novel ILs in electrochemistry was presented in the book edited by Ohno [197].

For many years scientists have been searching a new class of highly conductive polymer electrolytes containing ILs based on pyridinium, or imidazolium halide, and aluminum chloride [198–207]. Poly(1-butyl-4-vinylpyridinium halides) (chloride or bromide), pyCl or pyBr with AlCl$_3$ have revealed an eutectic system with 1:1 congruently low melting compound (~303 K) with conductivity ~12.6 mScm^{-1} [198]. Mixtures of three salts, triethylammonium benzoate, lithium acetate, and Li[Tf$_2$N] in 7/2/1 molar ratio form a stable molten salt with the glass transition temperature and melting temperature 221 and 328 K, respectively [199]. The ionic conductivity at 333 K is 1 mScm^{-1}. The compatibility of the ionic melts with polymers has been explored by using several commercially available polymers (polyacrylonitrile, poly(vinyl butyral)). Polymer-based solid electrolytes, especially poly(ethylene oxide) (PEO), are of growing importance in solid-state electrochemistry [200–204]. PEO dissolves inorganic salts and transports the dissociated ions along the segmental motion of the polyether chain. PEO has been studied as an ion-conductive matrix which is useful in many applications, such as secondary batteries, sensors, electrochromic displays, and other electrochemical devices. The PEO oligomers having sulfonamide groups or ethylimidazolium groups on the chain end have shown drastic decrease of glass transition temperature and improvement in ionic conductivity [201]. For 10 mol% of addition of Li$^+$ anion to PEO unit, the small maximum of ionic conductivity at 323 K was observed. Generally, the ionic conductivity is dependent on the number of carrier anions and on the segmental motion of the polymer matrix. Derivatives of PEO having both vinyl group and imidazolium salt structure on their ends with changing chloride anion to [Tf$_2$N]$^-$ anion have revealed very high conductivity, 12 mScm^{-1} at 303 K. The same effects were obtained by changing bromide counter anion to [Tf$_2$N]$^-$ anion in dimethylimidazolium-alkylene salts. The high conductivity of these salts was kept even after the addition of lithium salt, Li[Tf$_2$N] [204]. A number of novel salts with ammonium and pyrrolidinium-based cations and the [Tf$_2$N]$^-$ anion were tested in its lithium salts. The conductivities for some derivatives were ~1.4 or ~2.2 mScm^{-1} at 298.15 K [205]. The zwitterionic-type molten salts having a sulfonamide group instead of sulfonate group mixed with Li[Tf$_2$N] had an ionic conductivity of 10 mScm^{-1} [205]. Different polymer gel electrolytes with alkali metal ILs consisted of sulfate anion, imidazolium cation, and alkali metal cation displayed excellent ionic conductivity—10^{-5} mScm^{-1} [206].

Ionic conductivity for the chosen pure ILs was determined in a wide range of temperatures (263–373 K) [208]. For example, the ionic conductivities for [C$_2$C$_1$Im][BF$_4$] and [C$_2$C$_1$Im][Tf$_2$N] at 294 K were about 2.2 and 0.8 mScm^{-1}, respectively [199]. Much higher values for these two ILs were collected by Galiński et al. [36]. It was also shown in Ref. 208 that at 303 K the ionic conductivity decreases from 17.2 to 10.6 mScm^{-1} after adding 1 M of Li[BF$_4$].

The ionic conductivity of $[C_2C_1Im][BF_4]$ was higher than of $Li[C_2C_1Im][BF_4]$. The high influence of water and ethanol on ionic conductivity was observed for 1-allyl-3-methylimidazolium salt. The conductivity increases with an increase of temperature and the addition of water and ethanol. For example, the conductivity was changed from 3.82 mScm^{-1} for pure IL at 293.15 K to 8.17 mScm^{-1} for the mixture (IL + 0.3 water + 0.1 ethanol in mole fraction) [209].

The influence of alkyl chain, anion, and water on conductivity was tested in Ref. 210 for $[C_2C_1Im][Tf_2N]$, $[C_4C_1Im][Tf_2N]$, $[C_6C_1Im][Tf_2N]$, and $[C_4C_1Im][PF_6]$. The conductivities at 293.15 K for these ILs were from 7.73 to 1.09 mScm^{-1}. The addition of water always increases conductivity. It means that for the entire temperature the conductivity decreases when the alkyl chain increases. The $[PF_6]^-$ anion decreases the conductivity in comparison with the $[Tf_2N]^-$ anion.

The effect of the addition of water and molecular solvents such as propylene carbonate, N-methylformamide, and 1-methylimidazole on the conductivity of $[C_4C_1Im][Br]$ and $[C_2C_1Im][BF_4]$ was measured at 298 K [211]. The mixture of the IL and the molecular solvent or water showed a maximum on the conductivity/mole fraction IL curves. The maximum for nonaqueous solvents was at the level of approximately 18–30 mScm^{-1} at low mole fraction of the IL and the maximum for water was at level approximately 92–98 mScm^{-1} [211]. The conductivity of a mixture of these two ILs depends monotonically on the composition. The temperature dependence of the conductivity obeys the Arrhenius law. Activation energies, determined from the Arrhenius plot, are usually in the range of 10–40 kJ mol^{-1}. The mixtures of two ILs or of an IL with molecular solvents may find practical applications in electrochemical capacitors [212].

1.12   Conclusions and perspectives

In recent years, physical–chemical measurements on supported ILs have been performed by using many techniques that have appeared to be a performing tool to derive physical–chemical properties and the reaction mechanism of catalytic reaction on ILs. These studies have been motivated by the necessity to bridge the material gap in many fields of sciences with novel compounds.

Before using the ILs, it must be remembered that they can be dramatically altered by the presence of impurities. Impurities can change the nature of these compounds. The main contaminants are halide anions and organic bases, arising from unreacted starting material and water. The influence of water and chloride anion on the viscosity and density of ILs has already been extensively discussed by many authors [56]. The hydrophilic/hydrophobic behavior is important for the solvation properties of ILs as it is necessary to dissolve reactants, but it is also relevant for the separation and extraction processes and in electrochemical processes. Furthermore, the water content of ILs can affect the rates and selectivity of reaction (this problem was not discussed in this chapter) and can be taken as a cosolvent in extraction process as it is with NMP and water. The solubility of water in ILs is, moreover, an important factor for industrial applications of these solvents. Unfortunately, the toxicity of ILs discussed in this chapter and possible pathway into the environment is not known. The phase equilibria of ILs with water were discussed in Section 1.6.1. The interaction with water has also been investigated through theoretical calculations. Molecular dynamics simulation of mixtures of 1,3-dimethylimidazolium ILs with water have been performed as a function of composition [213].

It is impossible to measure all possible combinations of the anions and cations, including the possibility of using mixtures of ILs, because the number of these novel substances is extremely high. To have the possibility to tailor the best IL for a certain application, more theoretical work has to be done, for example, the prediction of the melting temperatures of quaternary ammonium ILs.

Recently, Rebelo and coworkers [172] presented a method to estimate the critical temperatures of some ILs based on the temperature dependence of their surface tension and liquid densities. The molar enthalpies of vaporization of a series of commonly used ILs were also determined for the first time. The molar enthalpies of vaporization of $[C_nC_1Im][Tf_2N]$ ILs in the function of the alkyl chain length have been presented [214]. The critical properties (T_c, P_c, V_c), the normal boiling temperatures, and the acentric factors of 50 ILs were determined as well for the first time [215].

Studies on physical, chemical, and thermodynamic properties of ILs, high-quality data on reference systems, the creation of the comprehensive database, the review efforts, the theoretical modeling of physical–chemical properties, also, the development of acceptable thermodynamic models such as COSMO-RS (01), or mod. UNIFAC is a strategy that will help to make progress in the field of ILs.

ILs have many fascinating properties which make them of fundamental interest to all chemists and physicists, because both thermodynamics and kinetics of reactions carried out in ILs are different to those in conventional molecular solvents. It was shown by many reviewers that by choosing the correct IL, high product yields can be obtained, and a reduced amount of waste can be produced in a given reaction. Often the IL can be recycled, and this leads to a reduction of the costs of the processes. It must be emphasized that reactions in ILs are not difficult to perform and usually require no special apparatus or methodologies.

Recent data and other scientific and engineering advances provide the potential for expanded opportunities in almost all separation technologies. Future separation needs are related to the pharmaceutical, biomedical, and other biotech industries, microelectronics, aerospace, and alternative fuels (i.e., hydrogen) segments of the economy. In addition, nanotechnology will impact separations in general with respect to scale and materials. Environmental concerns, such as CO_2 levels in the atmosphere, will continue to provide the impetus for improved separation technologies.

Very little research has been conducted on the potential application of ILs as lubricants or heat transfer fluids. ILs have the potential to compete with even the most successful synthetic organic and silicone-based compounds in the market.

The unique properties of ILs and the ability to tailor properties by the choice of the cation, the anion, and substituents open the door to many new technologies. ILs offer the opportunity for chemical engineers to revolutionize the industry. However, there are a variety of challenges that are associated with turning ILs from scientific concepts into practical fluids for industrial use. These have to do with cost, lack of corrosion data, and absence of toxicity data.

References

1. Gmehling, J. *Data Bank of Ionic Liquids*, DECHEMA, DETHERM on the WEB, Thermophysical Properties of Pure Substances & Mixtures, http://www.ddbst.de/new/Default.htm
2. *Ionic Liquids Database (IL Thermo)*, Thermodynamics Research Center, National Institute of Standards and Technology, Boulder, CO, USA, http://ilthermo.boulder.nist.gov/ILThermo/mainmenu.uix
3. Gordon, C.M. et al., Ionic liquid crystals: hexafluorophosphate salts, *J. Mater. Chem.*, 8, 2627, 1998.
4. Koel, M., Ionic liquids in chemical analysis, *Cr. Rev. Anal. Chem.*, 35, 177, 2005.
5. Elaiwi, A. et al., Hydrogen bonding in imidazolium salts and its implications for ambient-temperature halogenoaluminate (III) ionic liquids, *J. Chem. Soc. Dalton Trans.*, 3467, 1995.
6. Crowhurst, L. et al., Solvent-solute interaction in ionic liquids, *Phys. Chem. Chem. Phys.*, 5, 2790, 2003.
7. Armstrong, D.W., He, L., and Liu, Y.-S., Examination of ionic liquids and their interaction with molecules, when used as stationary phases in gas chromatography, *Anal. Chem.*, 71, 3873, 1999.
8. Abraham, M.H., Scales of solute hydrogen-bonding: their construction and application to physicochemical and biochemical processes, *Chem. Soc. Rev.*, 22, 73, 1993.
9. Abraham, M.H. and Acree, Jr. W.E., Comparative analysis of solvation and selectivity in room temperature ionic liquids using the Abraham linear free energy relationship, *Green. Chem.*, 8, 906, 2006.
10. Sprunger, L. et al., Characterization of room temperature ionic liquids by the Abraham model cation-specific and anion-specific equation coefficients, *J. Chem. Info. Model.*, 47, 1123, 2007.
11. Anderson, J.L. et al., Characterizing ionic liquids on the basis of multiple solvation interactions, *J. Am. Chem. Soc.*, 124, 14247, 2002.
12. Hu, Y-F. and Xu, Ch-M., Effect of the structures of ionic liquids on their physical-chemical properties and the phase behavior of mixtures involving ionic liquids, *Chem. Rev.*, 2006, Doi: 10.1021/cr0502044.
13. Golding, J. et al., Weak intermolecular interactions in sulfonamide salts: structure of 1-ethyl-2-methyl-3-benzyl imidazolium bis[(trifluoromethyl)sulfonyl] amide, *Chem. Commun.*, 1593, 1998.

14. Crosthwaite, J.M. et al., Liquid phase behavior of imidazolium-based ionic liquids with alcohols, *J. Phys. Chem. B*, 108, 5113, 2004.

15. Dupont, J. and Spencer, J., On the noninnocent nature of 1,3-dialkylimidazolium ionic liquids, *Angew. Chem. Int. Ed.*, 43, 5296, 2004.

16. Cole, A.C. et al., Novel Brønsted acid ionic liquids and their use as dual solvent-catalysts, *J. Am. Chem. Soc.*, 124, 5962, 2002.

17. Morrow, T.I. and Maginn, E.J., Molecular dynamics study of the ionic liquid 1-n-butyl-3-methylimidazolium hexafluorophosphate, *J. Phys. Chem. B*, 106, 12807, 2002.

18. Branco, L.C. et al., Preparation and characterization of new room temperature ionic liquids, *Chem. Eur. J.*, 8, 3671, 2002.

19. Singh, R.P. et al., Quaternary salts containing the pentafluorosulfanyl (SF$_5$) group, *Inorg. Chem.*, 42, 6142, 2003.

20. Marsh, K.N., Boxall, J.A., and Lichtenthaler, R., Room temperature ionic liquids and their mixtures—a review, *Fluid Phase Equilib.*, 219, 93, 2004.

21. Pernak, J., Chwała, P., and Syguda, A., Room temperature ionic liquids—new choline derivatives, *Polish. J. Chem.*, 78, 539, 2004.

22. Earle, M.J. et al., The distillation and volatility of ionic liquids, *Nature*, 439, 831, 2006.

23. Noda, A., Hayamizu, K., and Watanabe, M., Pulsed-gradient spin-echo ^{1}H and ^{19}F NMR ionic diffusion coefficient, viscosity and ionic conductivity of non-chloroaluminate room-temperature ionic liquids, *J. Phys. Chem. B*, 105, 4603, 2001.

24. Nishida, T., Tashiro, Y., and Yamamoto, M., Physical and electrochemical properties of 1-alkyl-3-methylimidazolium tetrafluoroborate for electrolyte, *Fluorine Chem.*, 120, 135, 2003.

25. Xu, W. and Angell, C.A., Solvent free electrolytes with aqueous solution—like conductivities, *Science*, 302(5644), 422, 2003.

26. Yanes, E.G. et al., Capillary electrophoretic application of 1-alkyl-3-methyl-imidazolium-based ionic liquids, *Anal. Chem.*, 73, 3838, 2001.

27. MacFarlane, D. et al., Low viscosity ionic liquids based on organic salts of the dicyanamide anion, *Chem. Commun.*, 1430, 2001.

28. Gu, Z. and Brennecke, J.F., Volume expansivities and isothermal compressibilities of imidazolium and pyridynium based ionic liquids, *J. Chem. Eng. Data*, 47, 339, 2002.

29. Gomes de Azevedo, R. et al., Thermophysical and thermodynamic properties of 1-butyl-3-methylimidazolium tetrafluoroborate and 1-butyl-3-methylimidazolium hexafluorophosphate over an extended pressure range, *J. Chem. Eng. Data*, 50, 997, 2005.

30. Gomes de Azevedo, R. et al., Thermophysical and thermodynamic properties of ionic liquids over an extended pressure range: [bmim][NTf$_2$] and [hmim][NTf$_2$], *J. Chem. Thermodyn.*, 37, 671, 2005.

31. Esperanca, J.M.S.S. et al., Densities and derived thermodynamic properties of ionic liquids. 3. Phosphonium-based ionic liquids over an extended pressure range, *J. Chem. Eng. Data*, 51, 237, 2006.

32. Bourbigou, O.H. and Magna L., Ionic liquids: perspectives for organic and catalytic reactions, *J. Mol. Cat.*, 182–183, 419, 2002.

33. Huddleston, J.G. et al., Characterization and comparison of hydrophilic and hydrophobic room temperature ionic liquids incorporating the imidazolium cation, *Green Chem.*, 3, 156, 2001.

34. Trulove, P. and Mantz, R. in Wassercheid, P. and Welton, T. (Eds), *Ionic Liquids in Synthesis*, Wiley-VCH, Weinheim, 2003, 112.
35. Rooney, D.W. and Seddon, K.R. in Wypych, G. (Ed.), *Handbook of Solvents*, Chem. Tec. Publishing, Toronto, 2001, 1463.
36. Galiński, M., Lewandowski, A., and Stępniak, I., Ionic liquids as electrolytes, *Electrochim. Acta*, 51, 5567, 2006.
37. Kubo, W. et al., Quasi-solid-state dye-sensitized solar cells using room temperature molten salts and a low molecular weight gelator, *Chem. Commun.*, 374, 2002.
38. Sudhir, N.V.K.A., Brennecke, J.F., and Samanta, A., How polar are room-temperature ionic liquids?, *Chem. Commun.*, 413, 2001.
39. Dyson, P.J. et al., Determination of hydrogen concentration in ionic liquids and the effect (or lack of) on rates of hydrogenation, *Chem. Commun.*, 2418, 2003.
40. Endres, F. and El Abedin, S.Z., Air and water stable ionic liquids in physical chemistry, *Phys. Chem. Chem. Phys.*, 8, 2101, 2006.
41. Pernak, J. et al., Synthesis and properties of chiral imidazolium ionic liquids with a ($1R,2S,5R$)-(-)-menthoxymethyl substituent, *New J. Chem.*, 31, 879, 2007.
42. Pernak, J. and Feder-Kubis, J., Synthesis and properties of chiral ammonium-based ionic liquids, *Chem. Eur. J.*, 11, 4441, 2005.
43. Pernak, J. and Feder-Kubis, J., Chiral pyridinium-based ionic liquids, containing the ($1R,2S,5R$)-(-)-menthyl group, *Tetrahedron: Asymmetry*, 17, 1728, 2006.
44. Okoturo, O.O. and Van der Noot, T.J., Temperature dependence of viscosity for room temperature ionic liquids, *J. Electroanal. Chem.*, 568, 167, 2004.
45. Heintz, A., Recent developments in thermodynamics and thermophysics of non-aqueous mixtures containing ionic liquids. A review, *J. Chem. Thermodyn.*, 37, 525, 2005.
46. Shobukawa, H. et al., Ion transport properties of lithium ionic liquids and their ion gels, *Electrochim. Acta*, 50, 3872, 2005.
47. Barton, A.F.M., *CRM Handbook of Solubility Parameter*, CRC Press, Boca Raton, FL, 1985.
48. Letcher, T.M. and Reddy, P., Ternary (liquid + liquid) equilibria for mixtures of 1-hexyl-3-methylimidazolium (tetrafluoroborate or hexafluorophosphate) + benzene + an alkane at $T = 298.12$ K and $p = 0.1$ MPa, *J. Chem. Thermodyn.*, 37, 415, 2005.
49. Lee, S.H. and Lee, S.B., The Hildebrand solubility parameters, cohesive energy densities and internal energies of 1-alkyl-3-methylimidazolium-based room temperature ionic liquids, *Chem. Commun.*, 3469, 2005.
50. Domańska, U., Pobudkowska, A., and Wiśniewska, A., Solubility and excess molar properties of 1,3-dimethylimidazolium methylsulfate, or 1-butyl-3-methylimidazolium methylsulfate, or 1-butyl-3-methylimidazolium octylsulfate ionic liquids with *n*-alkanes and alcohols: analysis in terms of the PFP and FBT models, *J. Solution Chem.*, 35, 311, 2006.
51. Domańska, U., Bogel-Łukasik, E., and Bogel-Łukasik, R., 1-Octanol/water partition coefficients of 1-alkyl-3-methylimidazolium chloride, *Chem. Eur. J.*, 9, 3033, 2003.
52. Domańska, U. and Bogel-Łukasik, R., Solubility of ethyl-(2-hydroxyethyl)-dimethylammonium bromide in alcohols (C_2-C_{12}), *Fluid Phase Equilib.*, 233, 220, 2005.
53. Domańska, U., Thermophysical properties and thermodynamic phase behavior of ionic liquids, *Thermochim. Acta*, 448, 19, 2006.
54. Domańska, U. and Casás, L.M., Solubility of phosphonium ionic liquid in alcohols, benzene and alkylbenzenes, *J. Phys. Chem. B*, 111, 4109, 2007.
55. Blanchard, L.A., Gu, Z., and Brennecke, J.F., High-pressure phase behavior of ionic liquid/CO_2 systems, *J. Phys. Chem. B*, 105, 2437, 2001.

56. Seddon, K.R., Stark, A., and Torres, M.J., Influence of chloride, water, and organic solvents on the physical properties of ionic liquids, *Pure Appl. Chem.*, 72, 2275, 2000.
57. Zhang, S. et al., Determination of physical properties for the binary systems of 1-ethyl-3-methylimidazolium tetrafluoroborate + H_2O, *J. Chem. Eng. Data*, 49, 760, 2004.
58. Zafarani-Moattar, M.T. and Shekaari, H. Volumetric and speed of sound of ionic liquid, 1-butyl-3-methylimidazolium hexafluorophosphate with acetonitrile and methanol at $T = $ (298.15 to 318.15) K, *J. Chem., Eng. Data*, 50, 1694, 2005.
59. Wang, J. et al., Excess molar volumes and excess logarithm viscosities for binary mixtures of the ionic liquid 1-butyl-3-methylimidazolium hexafluorophosphate with some organic solvents, *J. Solution Chem.*, 34, 585, 2005.
60. Pereiro, A.B. and Rodriguez, A., Thermodynamic properties of ionic liquids in organic solvents from (298.15 to 303.15) K, *J. Chem. Eng. Data*, 52, 600, 2007.
61. Treszczanowicz, A.J. and Benson, G.C., Excess volumes of alkanols + alkane systems in terms of an association model with a Flory contribution term, *Fluid Phase Equilib.*, 23, 117, 1985.
62. Heintz, A. et al., Excess molar volumes and liquid-liquid equilibria of the ionic liquid 1-methyl-3-octyl-imidazolium tetrafluoroborate mixed with butan-1-ol and pentan-1-ol, *J. Solution Chem.*, 34, 1135, 2005.
63. Gołdon, A., Dąbrowska, K., Hofman, T., Densities and excess volumes of the 1,3-dimethylimidazolium methylsulfate + methanol system at temperatures from (313.15 to 333.15) K and pressures from (0.1 to 25) MPa. *J. Chem. Eng. Data*, 52, 1830, 2007.
64. Kato, R., Krummen, M., and Gmehling, J., Measurement and correlation of vapor-liquid equilibria and excess enthalpies of binary systems containing ionic liquids and hydrocarbons, *Fluid Phase Equilib.*, 224, 47, 2004.
65. Kato, R. and Gmehling, J., Systems with ionic liquids: measurements of VLE and γ^{∞} data and prediction of their thermodynamic behavior using original UNIFAC, mod. UNIFAC(Do) and COSMO-RS(01), *J. Chem. Thermodyn.*, 37, 603, 2005.
66. Najdanovic-Visak, V. et al., Pressure, isotope, and water co-solvent effects in liquid-liquid equilibria of (ionic liquid + alcohol) systems, *J. Phys. Chem. B*, 107, 12797, 2003.
67. Letcher, T.M., Łachwa, J., and Domańska, U., The excess molar volumes and enthalpies of (*N*-methyl-2-pyrrolidinone + an alcohol) at $T = $ 298.15 K and the application of the ERAS theory, *J. Chem. Thermodyn.*, 33, 1169, 2001.
68. Gomes de Azevedo, R. et al., Thermophysical and thermodynamic properties of ionic liquids over an extended pressure range: [bmim][NTf$_2$] and [hmim][NTf$_2$], *J. Chem. Thermodyn.*, 37, 888, 2005.
69. Hofman, T. et al., Densities, excess volumes, isobaric expansivity, and isothermal compressibility of the (1-ethyl-3-methylimidazolium ethylsulfate + methanol) system at temperatures (298.15 to 333.15) K and pressures from (0.1 to 35) MPa, *J. Chem. Thermodyn.*, 40, 580, 2008.
70. Dzyuba, S.V. and Bartsch, R.A., Influence of structural variation in 1-alkyl(aralkyl)-3-methylimidazolium hexafluorophosphate and bis(trifluoro methylsulfonyl)imides on physical properties of the ionic liquids, *Chem. Phys. Chem.*, 3, 161, 2002.
71. Gordon, J.E. and SubbaRao, G.N., Fuse organic salts. 8. Properties of molten straight-chain isomers of tetra-*n*-pentylammonium salts, *J. Am. Chem. Soc.*, 100, 7445, 1978.

72. Cipriano, B.H., Raghavan, S.R., and McGuiggan, P.M., Surface tension and contact angle measurements of a hexadecyl imidazolium surfactant adsorbed on a clay surface, *Colloids and Surfaces A: Physicochem. Eng. Aspects*, 262, 8, 2005.

73. Anderson, J.L. et al., Surfactant solvation effects and micelle formation in ionic liquids, *Chem. Comm.*, 2444, 2003.

74. Sirieix-Plénet, J., Gaillon, L., and Letellier, P., Behaviour of a binary solvent mixture constituted by an amphiphilic ionic liquid, 1-decyl-3-methylimidazolium bromide and water. Potentiometric and conductimetric studies, *Talanta*, 62, 979, 2004.

75. Beyaz, A., Oh, W.S., and Reddy, V.P., Ionic liquids as modulators of the critical micelle concentration of sodium dodecyl sulfate, *Colloids Surf. B*, 35, 119, 2004.

76. Modaressi, A. et al., CTAB aggregation in aqueous solutions of ammonium based ionic liquids; conductimetric studies, *Colloids and Surfaces A: Physicochem. Eng. Aspects*, 296, 104, 2007.

77. Pernak, J., Sobaszkiewicz, K., and Foksowicz-Flaczyk, J., Ionic liquids with symmetrical dialkoxymethyl-substituted imidazolium cations, *Chem. Eur. J.*, 10, 3479, 2004.

78. Domańska, U. and Marciniak, A., Liquid phase behaviour of 1-hexyloxy-3-methyl-imidazolium-based ionic liquids with hydrocarbons: the influence of anion, *J. Chem. Thermodyn.*, 31, 577, 2005.

79. Domańska, U. and Bogel-Łukasik, R., Physicochemical properties and solubility of alkyl-(2-hydroxyethyl)-dimethylammonium bromide, *J. Phys. Chem. B*, 109, 12124, 2005.

80. Pernak, J. et al., Long alkyl chain quaternary ammonium-based ionic liquids and potential applications, *Green Chem.*, 8, 798, 2006.

81. Fredlake, C.P. et al., Thermophysical properties of imidazolium-based ionic liquids, *J. Chem. Eng. Data*, 49, 954, 2004.

82. Bonhôte, P. et al., Hydrophobic, highly conductive ambient-temperature molten salts, *Inorg. Chem.*, 35, 1168, 1996.

83. Domańska, U., Kozłowska, M.K., and Rogalski, M., Solubilities, partition coefficients, density, and surface tension for imidazoles + octan-1-ol + water or + *n*-decane, *J. Chem. Eng. Data*, 47, 456, 2002.

84. Domańska, U. and Kozłowska, M.K., Solubility of imidazoles in ethers, *J. Chem. Eng. Data*, 48, 557, 2003.

85. Domańska, U. and Kozłowska, M.K. Solubility of imidazoles in ketones, *Fluid Phase Equilib.*, 206, 253, 2003.

86. Domańska, U. and Pobudkowska, A., Solubility of 2-methylbenzimidazole in ethers and ketones, *Fluid Phase Equilib.*, 206, 341, 2003.

87. Domańska, U. and Bogel-Łukasik, E., Solubility of benzimidazoles in alcohols, *J. Chem. Eng. Data*, 48, 951, 2003.

88. Domańska, U., Pobudkowska, A., and Rogalski, M. Solubility of imidazoles, benzimidazoles and phenylimidazoles in different organic solvent, *J. Chem. Eng. Data*, 49, 1082, 2004.

89. Domańska, U. and Marciniak, A. Experimental liquid-liquid equilibria of 1-methylimidazole with hydrocarbons and ethers, *Fluid Phase Equilib.*, 238, 37, 2005.

90. MacDonald, J.C., Dorrestein, P.C., and Pilley, M.M., Design of supramolecular layers via self-assembly of imidazole and carboxylic acids, *Cryst. Growth Des.*, 1, 29, 2001.

91. Domańska, U. and Bogel-Łukasik, E., Measurements and correlation of the (solid + liquid) equilibria of [1-decyl-3-methylimidazolium chloride + alcohols (C$_2$-C$_{12}$)], *Ind. Eng. Chem. Res.*, 42, 6986, 2003.

92. Domańska, U., Bogel-Łukasik, E., and Bogel-Łukasik, R., Solubility of 1-dodecyl-3-methylimidazolium chloride in alcohols (C$_2$-C$_{12}$), *J. Phys. Chem. B*, 107, 1858, 2003.

93. Domańska, U. and Bogel-Łukasik, E., Solid-liquid equilibria for systems containing 1-butyl-3-methylimidazolium chloride, *Fluid Phase Equilib.*, 218, 123, 2004.

94. Rogalski, M. et al., Surface and conductivity properties of imidazoles solutions, *Chem. Phys.*, 285, 355, 2002.

95. Domańska, U. and Mazurowska, L., Solubility of 1,3-dialkylimidazolium chloride or hexafluorophosphate or methylsulfonate in organic solvents. Effect of the anions on solubility, *Fluid Phase Equilib.*, 221, 73, 2004.

96. Domańska, U. and Marciniak A., Solubility of 1-alkyl-3-methylimidazolium hexafluorophosphate in hydrocarbons, *J. Chem. Eng. Data*, 48, 451, 2003.

97. Domańska, U. and Marciniak A., Solubility of ionic liquid [emim][PF$_6$] in alcohols, *J. Phys. Chem. B*, 108, 2376, 2004.

98. Domańska, U. and Marciniak A., Phase behaviour of 1-hexyloxymethyl-3-methyl-imidazolium and 1,3-dihexyloxymethyl-imidazolium based ionic liquids with alcohols, water, ketones and hydrocarbons: The effect of cation and anion on solubility, *Fluid Phase Equilib.*, 260, 9, 2007.

99. Domańska, U., Bąkała, I., and Pernak, J., Phase equilibria of an ammonium ionic liquid with water and organic solvents, *J. Chem. Eng. Data*, 52, 309, 2007.

100. Domańska, U., Ługowska, K., and Pernak, J., Phase equilibria of didecyldimethylammonium nitrate ionic liquid with water and organic solvents, *J. Chem. Thermodyn.*, 39, 729, 2007.

101. Domańska, U. and Morawski, P., Influence of high pressure on solubility of ionic liquids: experimental data and correlation, *Green Chem.*, 9, 361, 2007.

102. Domańska, U., Żołek-Tryznowska. Z., and Królikowski, M., Thermodynamic phase behavior of ionic liquids, *J. Chem. Eng. Data*, 52, 1872, 2007.

103. Wilson, G.M. Vapour-liquid equilibrium. XI. A new expression for excess free energy of mixing, *J. Am. Chem. Soc.*, 86, 127, 1964.

104. Nagata, I., On the thermodynamics of alcohol solutions. Phase equilibria of binary and ternary mixtures containing any number of alcohols, *Fluid Phase Equilib.*, 19, 153, 1985.

105. Nagata, I. et al., Ternary liquid-liquid equilibria and their representation by modified NRTL equations, *J. Thermochim. Acta*, 45, 153, 1981.

106. Kato, R. and Gmehling, J., Measurements and correlation of vapor-liquid equilibria of binary systems containing the ionic liquids [EMIM][(CF$_3$SO$_2$)$_2$N], [BMIM][(CF$_3$SO$_2$)$_2$N], [MMIM][(CH$_3$)$_2$PO$_4$] and oxygenated organic compounds respectively water, *Fluid Phase Equilib.*, 231, 38, 2005.

107. Wong, D.S.H. et al., Phase equilibria of water and ionic liquids [emim][PF$_6$] and [bmim][PF$_6$], *Fluid Phase Equilib.*, 194–197, 1089, 2002.

108. Anthony, J.L., Maginn, E.J., and Brennecke, J.F., Solution thermodynamics of imidazolium-based ionic liquids and water, *J. Phys. Chem. B*, 105, 10942, 2001.

109. Seddon, K.R., Stark, A., and Torres, M-J., Influence of chloride, water, and organic solvents on the physical properties of ionic liquids, *Pure Appl. Chem.*, 72, 2275, 2000.

110. Lin, H. et al., Solubility of selected dibasic carboxylic acid in water, in ionic liquid of [Bmim][BF₄], and in aqueous [Bmim][BF₄] solutions, *Fluid Phase Equilib.*, 253, 130, 2007.
111. Fortunato, R. et al., Supported liquid membranes using ionic liquids: study of stability and transport mechanism, *J. Membr. Sci.*, 242, 197, 2004.
112. Domańska, U. and Marciniak, A., Liquid phase behaviour of 1-butyl-3-methylimidazolium 2-(2-methoxyethoxy)-ethylsulfate with organic solvents and water, *Green Chem.*, 9, 262, 2007.
113. Cammarata, L. et al., Molecular states of water in room temperature ionic liquids, *Phys. Chem. Chem. Phys.*, 3, 5192, 2001.
114. Sandler, S.I. and Orbey, H., The thermodynamics of long-lived organic pollutants, *Fluid Phase Equilib.*, 82, 63, 1993.
115. Ropel, L. et al., Octanol-water partition coefficients of imidazolium-based ionic liquids, *Green. Chem.*, 7, 83, 2005.
116. Choua, Ch.-H. et al., 1-Octanol/water partition coefficient of ionic liquids, 15th Symposium of Thermophysical Properties, Boulder, CO, USA, June, 22–27, 2003.
117. Stepnowski, P., Solid-phase extraction of room temperature imidazolium ionic liquids from aqueous environmental samples, *Anal. Bioanal. Chem.*, 381, 189, 2005.
118. Stepnowski, P., Application of chromatographic and electrophoretic methods for the analysis of imidazolium and pyridinium cations as used in ionic liquids, *Int. J. Mol. Sci.*, 7, 497, 2006.
119. Jastorff, B. et al., Progress in evaluation of risk potential of ionic liquids-basis for an eco-design of sustainable products, *Green Chem.*, 7, 362, 2005.
120. Stepnowski, P., Mrozik, W., and Nichthauser, J., Adsorption of alkylimidazolium and alkylpyridinium ionic liquids onto natural soils, *Environ. Sci. Technol.*, 41, 511, 2007.
121. Yang, M. et al., Solid-liquid phase equilibria in binary 1-octanol plus *n*-alkane mixtures under high pressure. Part 2 (1-Octanol + *n*-octane, *n*-dodecane) systems, *Fluid Phase Equilib.*, 204, 55, 2003.
122. Huddleston, J.G. et al., Room-temperature ionic liquids as novel media for clean liquid-liquid extraction, *Chem. Commun.*, 1765, 1998.
123. Meindersma, G.W., Podt, A.J.G., and de Haan, A.B., Selection of ionic liquids for the extraction of aromatic mixtures, *Fuel Proc. Technol.*, 87, 59, 2005.
124. Diedrich, A. and Gmehling, J., Measurement of heat capacities of ionic liquids by differential scanning calorimetry, *Fluid Phase Equilib.*, 244, 68, 2006.
125. Lei, Z., Arlt, W., and Wasserscheid, P., Separation of 1-hexene and *n*-hexane with ionic liquids, *Fluid Phase Equilib.*, 241, 290, 2006.
126. Arlt, W. et al., DE Pat. No 10160518.8
127. Domańska, U., Pobudkowska, A., and Eckert, F., (Liquid + liquid) phase equilibria of 1-alkyl-3-methylimidazolium methylsulfate with alcohols, ethers, ketones, *J. Chem. Thermodyn.*, 38, 685, 2006.
128. Domańska, U., Pobudkowska, A., and Eckert, F., Liquid-liquid equilibria in the binary systems (1,3-dimethylimidazolium or 1-butyl-3-methylimidazolium methylsulfate + hydrocarbons), *Green Chem.*, 8, 268, 2006.
129. Marsh, K.N. et al., Room temperature ionic liquids as replacements for conventional solvents—A review, *Kor. J. Chem. Eng.*, 19, 357, 2002.
130. Wu, C-T. et al., Liquid-liquid equilibria of room temperature ionic liquids and butan-1-ol, *J. Chem. Eng. Data*, 48, 486, 2003.

131. Sahandzhieva, K. et al., Liquid-liquid equilibrium in mixtures of the ionic liquid 1-*n*-butyl-3-methylimidazolium hexafluorophosphate and an alcohol, *J. Chem. Eng. Data*, 51, 1516, 2006.

132. Heintz, A., Lehman, J.K., and Wertz, Ch., Thermodynamic properties of mixtures containing ionic liquids. 3. Liquid-liquid equilibria of binary mixtures of 1-ethyl-3-methylimidazolium bis(trifluoromethylsulfonyl)imide with propan-1-ol, butan-1-ol, and pentan-1-ol, *J. Chem. Eng. Data*, 48, 472, 2003.

133. Crosthwaite, J.M. et al., Liquid phase behavior of imidazolium-based ionic liquids with alcohols: effect of hydrogen bonding and non-polar interactions, *Fluid Phase Equilib.*, 228–229, 303, 2005.

134. Crosthwaite, J.M. et al., Liquid phase behavior of ionic liquids with alcohols: experimental studies and modeling, *J. Phys Chem. B*, 110, 9354, 2006.

135. Domańska, U., Marciniak, A., and Bogel-Łukasik, R., Phase equilibria (SLE, LLE) of *N, N*-dialkylimidazolium hexafluorophosphate or chloride, in *Ionic Liquids IIIA: Fundamentals, Progress, Challenges, and Opportunities Properties and Structure*, R.D. Rogers and K.R. Seddon (Eds), Washington, D.C., 2005; ACS, NY, 2003.

136. Domańska, U., Solubilities and thermophysical properties of ionic liquids, *Pure Appl. Chem.*, 77, 543, 2005.

137. Łachwa, J. et al., Condensed phase behaviour of ionic liquid-benzene mixtures; congruent melting of a [emim][NTf$_2$] C$_6$H$_6$ inclusion crystal, *Chem. Commun.*, 2445, 2006.

138. Łachwa, J. et al., Changing from an unusual high-temperature demixing to a UCST-type in mixtures of 1-alkyl-3-methylimidazolium bis{(trifluoromethyl)sulfonyl}amide and arenes, *Green Chem.*, 8, 262, 2006.

139. Łachwa, J. et al., Evidence for lower critical solution behavior in ionic liquid solutions, *J. Am. Chem. Soc.*, 127, 6542, 2005.

140. Renon, H. and Prausnitz, J.M., Local composition in thermodynamic excess functions for liquid mixtures, *AIChE J.*, 14, 135, 1968.

141. Klamt, A. et al., Refinement and parametrization of COSMO-RS, *J. Phys. Chem. A*, 102, 5074, 1998.

142. Klamt, A. and Eckert, F., COSMO-RS: a novel and efficient method for a priori prediction of thermophysical data of liquids, *Fluid Phase Equilib.*, 172, 43, 2000.

143. Arlt, W., Spuhl, O., and Klamt, A., Challenges in thermodynamics, *Chem. Eng. Proc.*, 43, 221, 2004.

144. Jork, C. et al., Tailor-made ionic liquids, *J. Chem. Thermodyn.*, 37, 537, 2005.

145. Trindade, J.R. et al., Salting-out effect in ionic liquid solutions, *J. Phys. Chem. B*, 11, 4737, 2007.

146. Arlt, W., Seiler, M., and Jork, C., New classes of compounds for chemical engineering: ionic liquids and hyperbranched polymers, The Sixth Italian Conference on Chemical and Process Engineering, IChea P-6, Pisa, Italy, 8–11 June, 2003, AIDIC Conference Series, 3, 1257, 2003.

147. Arce, A., Rodriguez, O., and Soto A., *tert*-Amyl ethyl ether separation from its mixtures with ethanol using the 1-butyl-3-methylimidazolium trifluoromethanesulfonate ionic liquid: liquid-liquid equilibrium, *Ind. Eng. Chem. Res.*, 43, 8323, 2004.

148. Arce, A., Rodriguez, O., and Soto A., Experimental determination of liquid-liquid equilibrium using ionic liquids: *tert*-amyl ethyl ether + ethanol + 1-octyl-3-methylimidazolium chloride system at 298.15 K, *J. Chem. Eng. Data*, 49, 514, 2004.

149. Arce, A., Rodriguez, O., and Soto A., A comparative study on solvents for separation of *tert*-amyl ethyl ether and ethanol mixtures. New experimental; data for 1-ethyl-3-methylimidazolium ethyl sulfate ionic liquid, *Chem. Eng. Sci.*, 61, 6929, 2006.
150. Arce, A., Rodriguez, O., and Soto A., Effect of anion fluorination in 1-ethyl-3-methylimidazolium as solvent for the liquid extraction of ethanol from ethyl *tert*-butyl ether, *Fluid Phase Equilib.*, 242, 164, 2006.
151. Arce, A., Rodriguez, O., and Soto A., Use of a green and cheap ionic liquid to purify gasoline octane booster, *Green Chem.*, 9, 247, 2007.
152. Arce, A., Rodriguez, O., and Soto A., Purification of ethyl *tert*-butyl ether from its mixtures with ethanol by using an ionic liquid, *Chem. Eng. J.*, 115, 219, 2006.
153. Pereiro, A.B. et al., HMImPF$_6$ ionic liquid that separates the azeotropic mixture ethanol + heptane, *Green Chem.*, 8, 307, 2006.
154. Selvan, M.S. et al., Liquid-liquid equilibria for toluene + heptane + 1-ethyl-3-methylimidazolium triodide and toluene + heptane + 1-butyl-3-methylimidazolium triodide, *J. Chem. Eng. Data*, 45, 842, 2000.
155. Arce, A. et al., Separation of aromatic hydrocarbons from alkanes using the ionic liquid 1-ethyl-3-methylimidazolium bis{(trifluoromethyl)sulfonyl}amide, *Green Chem.*, 9, 70, 2007.
156. Letcher, T.M. and Deenadayalu, N., Ternary liquid-liquid equilibria for mixtures of 1-methyl-3-octylimidazolium chloride + benzene + an alkane at $T = 298.2$ K and 1 atm, *J. Chem. Thermodyn.*, 35, 67, 2003.
157. Letcher, T.M. and Reddy, P., Ternary liquid-liquid equilibria for mixtures of 1-hexyl-3-methylimidazolium (tetrafluoroborate or hexafluoroborate) + benzene + an alkane at $T = 298.2$ K and 1 atm, *J. Chem. Thermodyn.*, 37, 415, 2005.
158. Meindersma, G.W., Podt, A.J.G., and deHaan, A., Ternary liquid-liquid equilibria for mixtures of toluene + *n*-heptane + an ionic liquid, *Fluid Phase Equilib.*, 247, 158, 2006.
159. Arce, A. et al., Separation of benzene and hexane by solvent extraction with 1-alkyl-3-methylimidazolium bis{(trifluoromethyl)sulfonyl}amide ionic liquids: effect of the alkyl-substituent length, *J. Phys Chem. B*, 111, 4732, 2007.
160. Domańska, U., Pobudkowska, A., and Królikowski, M., Separation of aromatic hydrocarbons from alkanes using ammonium ionic liquid C$_2$NTf$_2$ at $T = 298.15$ K, *Fluid Phase Equilib.*, 259, 173, 2007.
161. Domańska, U., Pobudkowska, A., and Żołek-Tryznowska, Z., Effect of ionic liquid's cation on the ternary system (IL + *p*-xylene + hexane) at $T = 298.15$ K, *J. Chem. Eng. Data*, 52, 2345, 2007.
162. Visser, A.E. et al., Liquid/liquid extraction of metal ions in room temperature ionic liquids, *Sep. Sci. Technol.*, 36, 785, 2001.
163. Visser, A.E. et al., Task-specific ionic liquids for the extraction of metal ions aqueous solutions, *Chem. Commum.*, 135, 2001.
164. Visser, A.E. et al., Metal ion separations in aqueous biphasic systems and room temperature ionic liquids, PhD thesis, University of Alabama, Tuscaloosa, AL, USA, 2002.
165. Wei, G.-T., Yang, Z., and Chen, C.-J., Room temperature ionic liquid as a novel medium for liquid/liquid extraction of metal ions, *Anal. Chim. Acta*, 488, 183, 2003.
166. Dietz, M.L., Ionic liquids as extraction solvents: Where do we stand? *Sep. Sci. Technol.*, 41, 2047, 2006.

167. Dai, S., Ju, Y.H., and Barnes, C.E., Solvent extraction of strontium nitrate by a crown ether using room-temperature ionic liquids, *J. Chem. Soc., Dalton Trans.*, 1201, 1999.
168. Verevkin, S.P. et al., Thermodynamic properties of mixtures containing ionic liquids. Vapor pressures and activity coefficients of *n*-alcohols and benzene in binary mixtures with 1-methyl-3-butyl-imidazolium bis(trifluoromethyl-sulfonyl)imide, *Fluid Phase Equilib.*, 236, 222, 2005.
169. Vasiltsova, T.V. et al., Thermodynamic properties of mixtures containing ionic liquids. Activity coefficients of ethers and alcohols in 1-methyl-3-ethyl-imidazolium bis(trifluoromethyl-sulfonyl)imide using the transpiration method, *J. Chem. Eng. Data*, 50, 142, 2005.
170. Vasiltsova, T.V. et al., Thermodynamic properties of mixtures containing ionic liquids. Activity coefficients of aliphatic and aromatic esters and benzylamine in 1-methyl-3-ethyl-imidazolium bis(trifluoromethyl-sulfonyl)imide using the transpiration method, *J. Chem. Eng. Data*, 51, 213, 2006.
171. Anthony, J.L., Maginn, E.J., and Brennecke, J.F., Solution thermodynamics of imidazolium-based ionic liquids and water, *J. Phys. Chem. B*, 105, 10942, 2001.
172. Rebelo, L.P.N. et al., On critical temperature, normal boiling point, and vapor pressures of ionic liquids, *J. Phys. Chem. B*, 109, 6040, 2005.
173. Earle, M.J. et al., The distillation and volatility of ionic liquids, *Nature*, 439, 831, 2006.
174. Heintz, A., Kulikov, D.V., and Verevkin, S.P., Thermodynamic properties of mixtures containing ionic liquids. 1. Activity coefficients at infinite dilution of alkanes, alkenes, and alkylbenzenes in 4-methyl-*n*-butylpyridinium tetrafluoroborate using gas-liquid chromatography, *J. Chem. Eng. Data*, 46, 1526, 2001.
175. Heintz, A., Kulikov, D.V., and Verevkin, S.P., Thermodynamic properties of mixtures containing ionic liquids. Activity coefficients at infinite dilution of polar solutes in 4-methyl-N-butyl-pyridinium tetrafluoroborate using gas-liquid chromatography, *J. Chem. Thermodyn.*, 34, 1341, 2002.
176. Heintz, A., Kulikov, D.V., and Verevkin, S.P., Thermodynamic properties of mixtures containing ionic liquids. 2. Activity coefficients at infinite dilution of hydrocarbon and polar solutes in 1-methyl-3-ethyl-imidazolium bis(trifluoromethyl-sulfonyl)amide and in 1,2-dimethyl-3-ethyl-imidazolium bis(trifluoromethyl-sulfonyl)amide using gas-liquid chromatography, *J. Chem. Eng. Data*, 47, 894, 2002.
177. Krummen, M., Wasserscheid, P., and Gmehling, J., Measurements of activity coefficients at infinite dilution in ionic liquids using the dilutor technique, *J. Chem. Eng. Data*, 47, 1411, 2002.
178. David, W. et al., Activity coefficients of hydrocarbon solutes at infinite dilution in the ionic liquid, 1-methyl-3-octyl-imidazolium chloride from gas-liquid chromatography, *J. Chem. Thermodyn.*, 35, 1335, 2003.
179. Letcher, T.M. et al., Determination of activity coefficients at infinite dilution of solutes in the ionic liquid 1-hexyl-3-methylimidazolium tetrafluoroborate using gas-liquid chromatography at the temperatures 298.15 K and 323.15 K, *J. Chem. Eng. Data*, 48, 1587, 2003.
180. Kato, R. and Gmehling, J., Activity coefficients at infinite dilution of various solutes in the ionic liquids [MMIM][CH_3SO_4], [MMIM][$CH_3OC_2H_4SO_4$], [MMIM][$(CH_3)_2PO_4$], [$C_5H_5NC_2H_5$][$(CF_3SO_2)_2N$] and [C_5H_5NH][$C_2H_5OC_2H_4OSO_3$], *Fluid Phase Equilib.*, 37–44, 226, 2004.

181. Heintz, A. et al., Thermodynamic properties of mixtures containing ionic liquids. 5. Activity coefficients at infinite dilution of hydrocarbons, alcohols, ester and aldehydes in 1-methyl-3-butyl-imidazolium bis(trifluoromethylsulfonyl) imide using gas-liquid chromatography, *J. Chem. Eng. Data*, 50, 1510, 2005.

182. Letcher, T.M. et al., Activity coefficients at infinite dilution measurements for organic solutes in the ionic liquid 1-butyl-3-methylimidazolium 2-(2-methoxyethoxy)ethyl sulfate using glc at T = (298.15, 303.15 and 308.15) K, *J. Chem. Thermodyn.*, 37, 587, 2005.

183. Letcher, T.M. et al., Activity coefficients at infinite dilution measurements for organic solutes in the ionic liquid 1-hexyl-3-methylimidazolium bis(trifluoro methylsulfonyl)-imide using glc at T = (298.15, 303.15 and 308.15) K, *J. Chem. Thermodyn.*, 37, 1327, 2005.

184. Deenadayalu, N., Letcher, T.M., and Reddy, P., Determination of activity coefficients at infinite dilution of polar and nonpolar solutes in the ionic liquid 1-ethyl-3-methylimidazolium bis(trifluoromethylsulfonyl)-imide using gas-liquid chromatography at the temperature 303.15 K or 318.15 K, *J. Chem. Eng. Data*, 50, 105, 2005.

185. Letcher, T.M. et al., Determination of activity coefficients at infinite dilution of solutes in the ionic liquid 1-butyl-3-methylimidazolium octyl sulfate using gas-liquid chromatography at the temperature 298.15 K, 313.15 K or 328.15 K, *J. Chem. Eng. Data*, 50, 1294, 2005.

186. Letcher, T.M. and Reddy, P., Determination of activity coefficients at infinite dilution of organic solutes in the ionic liquid, trihexyl(tetradecyl)-phosphonium tris(pentafluoroethyl)trifluorophosphate by gas-liquid chromatography, *Fluid Phase Equilib.*, 235, 11, 2005.

187. Mutelet, F. and Jaubert, J-N., Measurements of activity coefficients at infinite dilution in 1-hexadecyl-3-methylimidazolium tetrafluoroborate ionic liquid, *J. Chem Thermodyn.*, 39, 1144, 2007.

188. Balyes, J.W., Letcher, T.M., and Moolan, W.C., The determination of activity coefficients at infinite dilution using GLC with moderately volatile solvents, *J. Chem Thermodyn.*, 25, 781, 1993.

189. Eike, D.M., Brennecke, J.F., and Maginn, E.J., Predicting infinite-dilution activity coefficients of organic solutes in ionic liquids, *Ind. Eng. Chem. Res.*, 43, 1039, 2004.

190. Belvèze, L.S., Brennecke, J.F., and Stadtherr, M.A., Modeling of activity coefficients of aqueous solutions of quaternary ammonium salts with the electrolyte-NRTL equation, *Ind. Eng. Chem. Res.*, 43, 815, 2004.

191. Magee, J., Heat capacity and enthalpy of fusion for 1-butyl-3-methyl-imidazolium hexafluorophosphate. 17th IUPAC Conference on Chemical Thermodynamics (ICCT 2002), Rostock, Germany, 2002.

192. Holbrey, J.D. et al., Heat capacities of ionic liquids and their applications as thermal fluids, in *Ionic Liquids as Green Solvents*, K.R. Seddon and R.D. Regers (Eds), ACS Symposium Series 856, American Chemical Society, Washington, D.C., 2003, 121.

193. Troncoso, J. et al., Thermodynamic properties of imidazolium-based ionic liquids: densities, heat capacities, and enthalpies of fusion of [bmim][PF$_6$] and [bmim][NTf$_2$], *J. Chem. Eng. Data*, 51, 1856, 2006.

194. Waliszewski, D. et al., Heat capacities of ionic liquids and their heats of solution in molecular liquids, *Thermochim. Acta*, 433, 149, 2005.

195. Van Valkenburg, M.E. et al., Ionic liquid heat transfer fluids, 15th Symposium on Thermophysical Properties, Boulder, CO, USA, 2003.

196. Marczak, W., Verevkin, S.P., and Heintz, A., Enthalpies of solution of organic solutes in the ionic liquid 1-methyl-3-ethyl-imidazolium bis-(trifluoromethylsulfonyl)amide, *J. Solution Chem.*, 32, 519, 2003.
197. Ohno, H., *Electrochemical Aspects of Ionic Liquids*, Wiley-Interscience, Hoboken, NJ, 2005.
198. Watanabe, M., Yamada, S-I., and Ogata, N., Ionic conductivity of polymer electrolytes containing room temperature molten salts based on pyridinium halide and aluminium chloride, *Electrochim. Acta*, 40, 2285, 1995.
199. Watanabe, M. and Mizumura, T., Conductivity study on ionic liquid/polymer complexes, *Solid State Ionics*, 86–88, 353, 1996.
200. Nakai, Y., Ito, K., and Ohno, H., Ion conduction in molten salts prepared by terminal-charged PEO derivatives, *Solid State Ionics*, 113–115, 199, 1998.
201. Ohno, H., Nakai, Y., and Ito, K., Ionic conductivity of molten salts formed by polyether/salt hybrids, *Chem. Lett.*, 15, 1998.
202. Ohno, H. and Ito, K., Room-temperature molten salt polymers as matrix for fast ion conduction, *Chem. Lett.*, 751, 1998.
203. Yoshizawa, M. and Ohno, H., Molecular brush having molten salt domain for fast ion conduction, *Chem. Lett.*, 889, 1999.
204. Ito, K., Nishina, N., and Ohno, H., Enhanced ion conduction in imidazolium-type molten salts, *Electrochim. Acta*, 45, 1295, 2000.
205. McFarlane, D.R. et al., High conductivity molten salts based on the imide ion, *Electrochim. Acta*, 45, 1271, 2000.
206. Yoshizawa, M. et al., Ion conduction in zwitterionic-type molten salts and their polymers, *J. Mater. Chem.*, 11, 1057, 2001.
207. Ogihara, W. et al., Ionic conductivity of polymer gels deriving from alkali metal ionic liquids and negatively charged polyelectrolytes, *Electrochim. Acta*, 49, 1797, 2004.
208. Hayamizu, K. et al., Ionic conduction and ion diffusion in binary room-temperature ionic liquids composed of [emim][BF$_4$] and LiBF$_4$, *J. Phys. Chem. B*, 108. 19527, 2004.
209. Xu, H. et al., Conductivity and viscosity of 1-allyl-3-methyl-imidazolium chloride + water and + ethanol from 293.15 K to 333.15 K., *J. Chem. Eng. Data*, 50, 133, 2005.
210. Widegren, J.A. et al., Electrolytic conductivity of four imidazolium-based room-temperature ionic liquids and the effect of a water impurity, *J. Chem. Thermodyn.*, 37, 569, 2005.
211. Jarosik, A. et al., Conductivity of ionic liquids in mixtures, *J. Mol. Liq.*, 123, 43, 2006.
212. Lewandowski, A. and Galiński, M., *General Properties Of Ionic Liquids As Electrolytes For Carbon-based Double Layer Capacitors, New Carbon Based Materials for Electrochemical Energy Storage Systems*, Barsukov et al. (Eds), Springer, The Netherlands, 2006, 73–83.
213. Hanke, C.G. and Lynden-Bell, R.M., A simulation study of water-dialkylimidazolium ionic liquid mixtures, *J. Phys. Chem. B*, 107, 10878, 2003.
214. Santos, L.M.N.B.F. et al., Ionic liquids: first direct determination of their cohesive energy, *J. Am. Chem. Soc.*, 129, 284, 2007.
215. Valderrama, J.O. and Robles, P.A., Critical properties, normal boiling temperatures, and acentric factors of fifty ionic liquids, *Ind. Eng. Chem. Res.*, 46, 1338, 2007.

chapter two

Experimental and theoretical structure of ionic liquids

Tristan Gerard Alfred Youngs, Christopher Hardacre, and Claire Lisa Mullan

Contents

2.1 Introduction

An understanding of the microscopic structure of liquids is paramount in understanding the macroscopic properties of the system. Given the sheer number of potential ionic liquids, quantifying the interactions in the liquid state is of enormous benefit when selecting candidate ionic liquids, or at least narrowing any potential search to perhaps less than a hundred candidates rather than many tens of thousands. However, the complex interplay between species in the liquid state presents a significant challenge even for simple molecular solvents. In the case of ionic liquids, the introduction of charged species provides even more variety and directionality to the interactions and deepens the problem. What are the local environments of the individual species? Which are the important interactions between them? How, in the case of ionic liquids, can the individual ions be tailored to adjust the properties of choice? Experimentally, several techniques have begun to answer these questions but remain restricted to relatively simple systems. Nevertheless, combined with theoretical insights, a detailed and useful picture of the microscopic structure in these systems may be built.

2.2   Structure of simple molten salts

The structure of simple molten salts based on atomic ions, for example the alkali halide salts, has been studied extensively for the last 30 years, since it became feasible to apply experimental methods such as neutron diffraction and extended x-ray absorption fine structure (EXAFS) at the high temperatures associated with the melt phase. The pioneering work of Edwards et al. on molten NaCl provides perhaps the best-known example [1], and illustrates the standard template for the liquid structure of a purely ionic binary melts. Figure 2.1 shows the radial distribution functions (RDFs) obtained from the study for all pairs of ions, and clearly shows the charge ordering present in the system. The profiles are dominated by the unlike-ion (g_{NaCl}) curve located at ~2.7 Å. A minimum occurs thereafter in g_{NaCl} at ~4.3 Å, which corresponds to the maximum in the like-ion curves (g_{NaNa} and g_{ClCl}). It is important to note that there is no overlap above $g = 1$ between g_{NaCl} and either of the like-ion traces, giving a clear indication that concentric shells of anions and cations exist with no penetration of one into the other. Furthermore it is evident that this structuring persists out to at least the third Na···Cl shell. One important quantity that can be obtained from experiments such as these is the coordination number (CN) of the ions. From an integration of the first peak in the g_{NaCl} and g_{NaNa}/g_{ClCl} RDFs $CN_{like} = 5.8 \pm 0.1$ and $CN_{unlike} = 13.0 \pm 0.5$ are found. For CN_{like} this suggests an octahedral 6-coordinate arrangement of ions about a central site, and for CN_{unlike} a less-ordered, but still quantifiable, shell of 13. In addition, this bears considerable resemblance to the solid-state structure of the salt.

Since the work of Edwards et al. a vast array of molten salts has been studied, including mixtures and eutectics [2–5]. Beyond the binary ionic melt there are two important factors affecting the local order of the system, namely, the covalency of the interaction between unlike ions and the stoichiometry

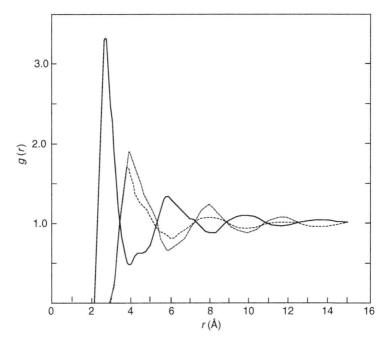

Figure 2.1 Radial distribution functions for an NaCl melt at 1148 K, as found by neutron diffraction. Concentric, mutually exclusive shells of anions and cations can clearly be seen extending to large distances relative to the sizes of the ions. (From Edwards, F.G., Enderby, J.E., Howe, R.A., and Page, D.I., *J. Phys. C: Solid State Phys.*, 8, 3483–3490, 1975. With permission.)

of the ions in the system. We will not discuss the effect of stoichiometry since all the ionic liquids considered later are simple 1:1 salts, but the nature of the interaction between ions is of particular importance. For example, consider the series of silver halides from AgCl to AgI, where both neutron diffraction [6,7] and EXAFS [8–10] techniques have been employed to investigate the changes in liquid structure when progressing from the smaller, less-polarizable chloride anion to the larger, more-polarizable iodide anion. A roughly tetrahedral arrangement of ions is found for all salts near their respective melting points, with CNs calculated to be 3.8, 3.9, and 4.6 for AgCl, AgBr, and AgI, respectively [6]. On further heating the local order is seen to decrease for Cl$^-$ and Br$^-$ salts resulting in first shell CNs whereas for AgI little significant change in the local structure occurs. This has been attributed to the effect of the increased covalent character associated with the Ag$^+$/I$^-$ ion pair, compared with essentially ionic interactions found for the Ag$^+$/Cl$^-$ and Ag$^+$/Br$^-$. Kawakita et al. also noted that the molten structures of AgCl and AgBr are different from the mineral structure with respect to the cation–anion distance [7]. In contrast, for AgI the cation–anion distances between the solid and the melt phases are almost identical, again indicating that the cause is increased covalency between the Ag^+/I^- ion pair.

2.3   Structure of pure ionic liquids

Even the simplest ionic liquids are a significant departure from the simple binary molten salt, with both the cation and the anion often being poly-atomic, asymmetric organic or inorganic ions. Although it is possible to make the approximation that both species are just (slightly larger) ionic spheres with differing polarizability this is an oversimplification. The introduction of atomically fine charge distributions means this approach is unlikely to adequately describe the structure and properties of ionic liquids, even those containing ions that are roughly pseudo-spherical such as tetrafluoroborate or hexafluorophosphate. The distribution of the charge over the ions provides the variety of interactions observed within the liquid state, and should not be oversimplified if an accurate picture is desired.

Owing partly to the vast number of potential ionic liquids, experimentally derived structural data for the liquid state is still scarce. 1,3-Dialkylimidazo-lium ionic liquids are among the most common family of solvents currently under study, since they were the first to be synthesized and put to use as room-temperature molten salt solvents, and have received the most atten-tion both experimentally and theoretically. Molecular dynamics and Monte Carlo techniques have been particularly important in predicting properties of the liquid state of these materials, while neutron diffraction has been able to elucidate structural descriptions from experiments. In combination, these last two techniques make up a powerful, complementary theory/experiment combination which is able to provide a detailed examination (and justifica-tion) of the microscopic structure of the liquid state. This was realized in the original work on NaCl with references made to theoretical studies of the same system [11], and, since then, the two combined methodologies have been used extensively for many different systems [12–14]. Since the prac-tice of simulating ionic liquids is also still relatively young (the first such study being that of Hanke et al. in 2001) [15] the evolution of forcefields and experiments have been complementary, with experimental results allowing the proper evaluation of forcefields, and by return the creation of more accurate forcefields with which to model and compare with the experi-mental data.

2.3.1   1,3-Dimethylimidazolium chloride

The simplest alkylimidazolium ionic liquid* 1,3-dimethylimidazolium chlo-ride ([C_1C_1Im]Cl, numbering of the cation atoms is shown in Figure 2.2) and

---

* Note that [C_1C_1Im]Cl is not a room temperature molten salt since its melting point is just above 450 K.

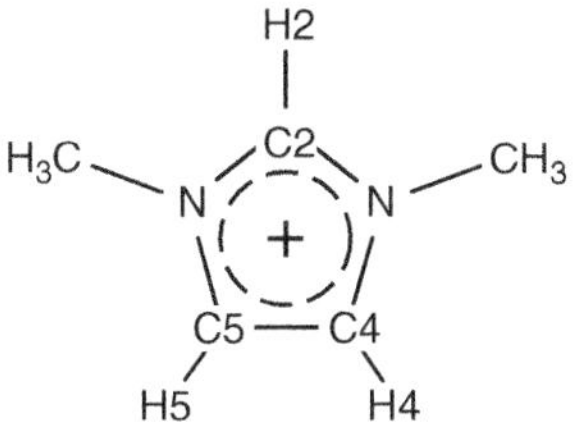

Figure 2.2   Labeling of atoms in the 1,3-dimethylimidazolium cation ($[C_1C_1Im]^+$).

has served as a model ionic liquid for many theoretical studies owing to its simplicity and comparatively rich experimental characterization. For much the same reason it was the first ionic liquid studied by neutron diffraction by Hardacre et al. in 2003 [16]. Figure 2.3 shows g_{+-} and g_{++} as determined from the empirical potential structure refinement (EPSR) process [17–19]. From Figure 2.3 it is clear that the RDFs possess many features in common with those for NaCl given earlier. Notably, the cation–anion and cation–cation curves overlap very little above $g = 1$, and there is clear structure beyond the first coordination shell. Integration of the RDFs up to the position of the first minimum in g_{+-} gives a CN of ~6, again showing much similarity with NaCl. The first peak in the g_{++} curve occurs at ~5.5 Å indicating the separation between pairs of cations and reflecting the width of the primary anion shell. However, considering only the spherically-averaged RDFs misses out much of the importance of the cation–anion interaction. For example, if taken in isolation it might be tempting to conclude that the ions are arranged around each other in a regular manner similar to the packing of charged spheres as found in the case of NaCl. Chemical intuition forces a consideration that strong directional character of the ion–ion interactions will exist. The ions do not, in general, have spherical symmetry, and the electrostatic surface of the cation will contain chemically distinct sites with varying Lewis character. To understand the structure in detail, this directionality must be fully considered.

From neutron diffraction experiments and molecular dynamics/Monte Carlo simulations three-dimensional, spatial distributions of the liquid can be determined by considering the positions of species relative to a given site in the liquid, usually located at the center-of-mass of one of the ions and defined by a set of local axes (which may only be uniquely defined for a nonlinear arrangement of three atoms or more). Taken as an ensemble average over time, the resulting distributions represent the probability of finding a given site at a given position around a central species. Depicted graphically, contour surfaces represent regions of a given probability, and are either measured relative to the probability of finding a species in the bulk (as with normalization of RDFs) or as a percentage of the number of molecules of a given type occurring within some distance.

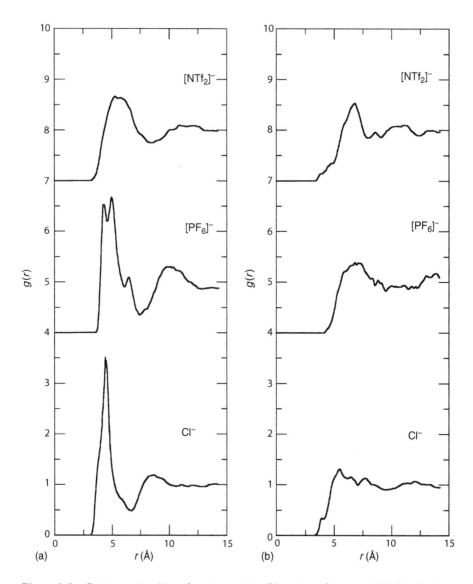

Figure 2.3 Cation–anion (a) and cation–cation (b) center-of-mass radial distribution functions determined by empirical potential structure refinement from neutron diffraction studies of [C_1C_1Im]Cl, [C_1C_1Im][PF_6], and [C_1C_1Im][Tf_2N]. (From *Acc. Chem. Res.*, 40, 1146–1155, 2007. With permission.)

Figure 2.4 shows the distributions of anions about cations for [C_1C_1Im]Cl as determined from EPSR at two different probabilities. The direction-ality of the interactions is clear, with the highest probability regions of anions located along the vectors of the aromatic C–H bonds of the ring and

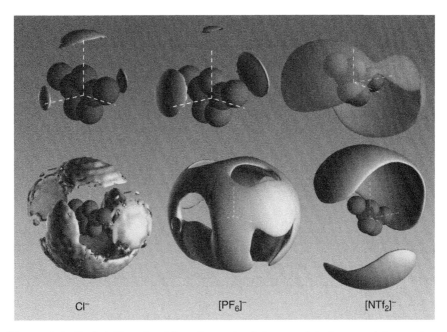

Figure 2.4 Probability densities determined by empirical potential structure refinement from neutron diffraction studies of [C₁C₁Im]Cl, [C₁C₁Im][PF₆], and [C₁C₁Im][Tf₂N]. Surfaces represent anions (top row) and cations (bottom row) about a central cation, drawn to encompass the top 10% of ions within 8 Å (From *Acc. Chem. Res.*, 40, 1146–1155, 2007. With permission.)

secondary interactions in the regions above and below the ring. Interestingly, in the liquid state, no significant association of the anions with the methyl groups of the cation is seen. The crystal structure of [C₁C₁Im]Cl is dominated by H–Cl contacts, involving both hydrogens of the imidazolium ring and those of the methyl groups [20]. In the liquid phase only the aromatic hydrogens of the ring are able to interact strongly enough with the anions to promote any significant structuring, owing to their increased acidity compared with the aliphatic methyl hydrogens, nevertheless the general similarity with the crystal structure should be noted.

No less than four classical forcefields have been used in simulation studies of [C₁C₁Im]Cl [21–24], and show general agreement with the neutron diffraction data. Simulated RDFs follow the experimental results closely, but tend to display a *multiplet* rather than a single, well-defined first peak [22], Figure 2.5. It is possible that the more complex structure arises from too-strong interactions between ions, resulting in an RDF more akin to the solid than the liquid. The individual features of the simulation peaks may be uniquely assigned to specific H–Cl interactions within the liquid, as

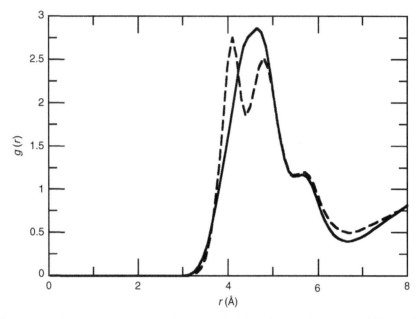

Figure 2.5 Cation–anion center-of-mass radial distribution functions of [C_1C_1Im]Cl calculated using two different classical forcefields: (a) that of Canongia-Lopes [21] (solid line) and (b) that of Liu et al. [22] (dashed line). Both have a small shoulder at ~5.6 Å, and the model of Liu et al. displays a splitting of the primary peak.

detailed in Figure 2.5. One discrepancy between simulation and experiment has been the probability of finding anions above or below the imidazolium ring, which is clearly evident from the experiment but which is not agreed upon by simulations. Figure 2.6 shows examples of simulated probability distributions from two different forcefields where one predicts the association of anions with the aromatic ring and one does not. Reproduction of the experimental features by the second model has been obtained by modification of the atomic charges in the system, with the end result that the aromatic ring has an overall slight positive charge, compared with a slight negative charge in the alternative model. It is enough to note that, while simulations are able to reproduce the general features of the experiment fairly easily, the finer aspects require more thought regarding the potentials employed and are particularly sensitive to the representation of charges on the ions and the distribution of total charge [25–27]. [C_1C_1Im]Cl has also been the subject of several *ab initio* molecular dynamics studies [28–30] that are consistent with the neutron data, but again do not show the presence of anions above and below the imidazolium ring. Typically, in both forcefield and *ab initio* simulations the regions above and below the

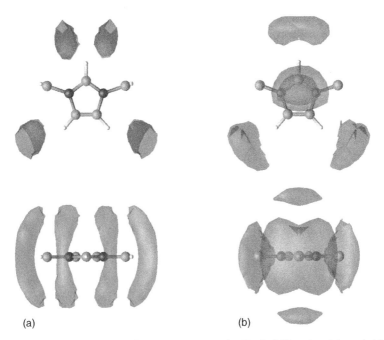

(a)   (b)

Figure 2.6 **(See color insert following page 224.)**   Probability densities of chloride anions about a central $[C_1C_1Im]^+$ calculated from MD simulations using two different classical forcefields: (a) that of Canongia-Lopes [21], which has an overall negatively charged imidazolium ring (C and N atoms only) and (b) a model that has an overall positively charged imidazolium ring. Both surfaces are drawn at five times the bulk density of anions.

ring are populated by cations; however, these cations are positioned far enough away from the plane of the imidazolium ring such that they do not penetrate the first solvation shell.

2.3.2   *1,3-Dimethylimidazolium hexafluorophosphate*

1,3-Dimethylimidazolium hexafluorophosphate ($[C_1C_1Im][PF_6]$) presents an interesting contrast to $[C_1C_1Im]Cl$ owing to the marked differences in the anion. The chloride anion is small and nonpolarizable and will readily form hydrogen bonds while $[PF_6]^-$ anion is large, a poor nucleophile, and weakly coordinating. These properties serve to reduce the association between cation and anion, decreasing the melting point of $[C_1C_1Im][PF_6]$ to 362 K. As found for the analogous chloride salt, neutron diffraction studies on the liquid [31] again show the alternating shells of cations and anions in the RDFs, Figure 2.3, with the distance between cations increased to 6.3 Å to accommodate the larger anion. Integration of the first g_{+-} peak gives a CN of 6.8, higher than for the chloride-containing liquid. In comparison with the first g_{+-} peak in [C$_1$C$_1$Im]Cl, here there are three associated maxima instead of one, indicating that the cation–anion interaction in [C$_1$C$_1$Im][PF$_6$] is more complex despite the reduced association between the anions and the cations. Probability distributions of [PF$_6$]$^-$ around the cation, Figure 2.4 show that, as with [C$_1$C$_{21}$I$_2$m]Cl, there are three regions for which the highest probability of finding anions exists. Anion density above and below the plane of the imidazolium ring clearly exists, but since no bonding interaction is possible between anions in this position and the cation the distribution is a result of electrostatic interactions only, and are favored over C–H$\cdots$F hydrogen-bonding interactions which are weak owing to the chemical nature of the anion. The most significant portion of the top 2.5% are positioned along the C2–H2 vector at the top of the imidazolium ring, while increasing the scope of the plot to encompass the top 20% of molecules within 9 Å reveals a region below the H4 and H5 hydrogens. On first inspection these regions might be considered to be the result of hydrogen-bonding interactions between ring hydrogens and the fluorine atoms of the anion, but it may be more instructive to understand these as a consequence of maximizing the favorable electrostatic interactions. Figure 2.7 shows probability distributions of cations about [PF$_6$]$^-$ where it is interesting that cations are not located along P–F bond vectors, but rather occupy positions at the corners of a cube where the P–F bonds point toward the faces of the cube. The [PF$_6$]$^-$ anions themselves are positioned directly along the P–F vectors, which seems to be a

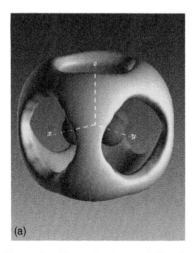

(a)   (b)

Figure 2.7 (**See color insert following page 224.**)   Probability densities determined by empirical potential structure refinement from neutron diffraction studies of [C$_1$C$_1$Im][PF$_6$]. Densities are for (a) [PF$_6$]$^-$ and (b) [C$_1$C$_1$Im]$^+$ around a central anion.

remarkably high-energy configuration, but is just an arrangement dictated by the intermediate cations and lead to the anions and cations occupying mutually-exclusive positions.

Classical simulations by Hanke et al. [15] employed a model without methyl hydrogens and which showed excellent agreement with the experimental RDFs. In particular, the main feature of g_{+-} at ~4.8 Å has two peaks which are also evident in the EPSR-derived curve, although the secondary peak at ~6.3 Å is missing in the simulated data. Other studies have been performed on the related ionic liquid 1-butyl-3-methylimidazolium hexafluorophosphate ([C_4C_1Im][PF_6]), and some relevant results can be extracted. For example, Morrow and Maginn used an explicit-hydrogen model [32] and showed from atom–atom RDFs that the [PF_6]$^-$ anion prefers to coordinate with the hydrogen at the C2 position over the hydrogens at the C4/C5 positions, in line with the probability densities in Figure 2.4 above. Interestingly, the g_{+-} resulting from the use of the explicit-hydrogen model displays a *singlet* primary peak, whereas another model containing no hydrogens employed by Shah et al. [33] again displays the fine structure observed by experiment and the model of Hanke et al. This perhaps reaffirms the fact that the distribution of anions about cations is more influenced by Coulombic forces and packing rather than directional interactions provided by the presence of explicit hydrogens. Probability densities calculated for [C_1C_1Im][PF_6] by Hanke et al. [15] via classical molecular dynamics and for [C_4C_1Im][PF_6] by Bhargava and Balasubramanian [34] via *ab initio* molecular dynamics are very similar and reproduce the essential features of the experiment with high-probability regions above and below the imidazolium ring and along the H4 and H5 bond vectors, although neither shows any significant region along the C2–H2 bond vector as was found in the experiment and also observed in the crystal structure [35].

2.3.3 *1,3-Dimethylimidazolium bis{(trifluoromethyl)sulfonyl}imide*

Recently, ionic liquids based on the bis{(trifluoromethyl)sulfonyl}imide anion ([Tf_2N]$^-$) have grown in popularity owing to their hydrophobicity and high thermal stability. The [Tf_2N]$^-$ anion marks a significant departure from those already studied, not least because of the inherent flexibility of its structure. It is also much more weakly coordinating than either the chloride of hexafluorophosphate anions studied owing to its increased size over which the negative charge is spread. This weakens the Coulombic interactions between the ions and results in reduced viscosity compared with chloride- and hexafluorophosphate-based ionic liquids, for example, see Ref. 36.

Neutron diffraction studies performed by Deetlefs et al. in conjunction with theoretical investigations were conducted as part of the data analysis process [37]. Figure 2.3 shows the RDFs extracted from EPSR of the neutron data and once more shows the (by now expected) oscillatory structure for the individual curves revealing the concentric shell structure of anions and cations. However, the oscillations are far weaker than observed for the other anions and dampen more quickly indicating less structuring of the liquid relative to a given ionic center. The primary g_{++} peak is located at around 7.0 Å reflecting a further expansion of the liquid effected by the increased size of the anion. Strikingly, though, there is significant overlap between the g_{+-} and g_{++} RDFs indicating that the primary solvation shell about a cation is not an anion-exclusive zone, and instead contains both types of ion. For the second shell the effect is enhanced, the g_{+-} and g_{++} curves being almost coincident. As a result of the softer ionic binding between the anions and the cations in the case of $[Tf_2N]^-$, the cations are able (or are forced) to approach each other more closely and produce intermolecular contacts that are at least as favorable as those between cations and anions, despite the obvious repulsive Coulombic forces between the cation species.

Probability densities obtained from EPSR are shown in Figure 2.4. Here there is a clear preference for the anions to sit above and below the ring, and also to associate with the methyl groups forming a band of anions around the cation. As the hydrogen-bonding ability of the anion has decreased, the probability of finding anions above and below the ring has steadily increased inferring that in the present salt it is more a case of packing forces rather than specific interactions dominating the liquid structure. Indeed, the cations may now be found in positions that in the two other ionic liquids were dominated by anions—that is, apparently associated with the aromatic hydrogens of the imidazolium ring. Again, however, this is a result of packing rather than of a geometry enforced by intermolecular contacts.

The crystal structure of $[C_1C_1Im][Tf_2N]$ bears little resemblance to the liquid [38] in contrast to those for $[C_1C_1Im]Cl$ and $[C_1C_1Im][PF_6]$. Stacks of cations resulting from π–π interactions are clearly seen in the solid state, yet in the liquid the relevant positions above and below the imidazolium ring are occupied solely by anions. Also, the conformation of the $[Tf_2N]^-$ in the crystal is almost exclusively cis, but in the liquid a mixture of trans and cis conformers is observed in the ratio of 4:1. Even aspects as subtle as this are picked up by the complementary forcefield simulations that predict a similar ratio.

2.3.4 Summary

The structure of ionic liquids in the liquid state is determined by the interplay of two interactions. Firstly, a general Coulombic interaction that, in the absence of any others, would result in a concentric shell structure of ions similar to those observed for simple molten salts. Secondly, directional interactions between ions arising from charge distribution over the molecular surface. If strong enough, the latter will determine the major features of the ion distributions about a given center, as is the case for $[C_1C_1Im]Cl$ and, to a large extent, $[C_1C_1Im][PF_6]$. In these cases the presence of hydrogen bonds formed between the weakly acidic hydrogens on the imidazolium ring and either the chloride or the fluoride atoms of the anion serve to direct the positions of the unlike-charged species, with the like-charged species tending to order as a result of this preliminary structure. When the hydrogen-bonding ability of the anion is reduced to such an extent that no hydrogen bonds between cation and anion are formed (as is the case for the $[Tf_2N]^-$ anion) then the influence of charge ordering and packing is much greater on the liquid structure. Pronounced sharing of the coordination shells by cations and anions occurs for the $[Tf_2N]^-$-based liquid as a result of the reduction in association between the cation and the anion through reduced hydrogen-bonding ability, with packing determining the relative positions of ions.

CNs for the primary shell increase as the size of the anion increases, since the shell must expand to account for the larger anions and create a larger volume. For the Cl^-, $[PF_6]^-$ and $[Tf_2N]^-$-based salts the CNs are 6, 6.8, and 8.8, respectively, but it is not always possible to define exclusive contacts or localities which translate to the presence of individual ions in the primary coordination shell (as is possible, for instance, for liquid water which shows a CN of 4 and four well-defined lobes in the distribution, or for a simple binary salt such as NaCl which has a CN of 6 and may be legitimately conferred to be an octahedral arrangement). For example, the probability density of anions about cations in $[C_1C_1Im][PF_6]$ visually shows that there are five well-defined regions of high likelihood of finding an anion (above and below the plane of the imidazolium ring, and along each of the aromatic C–H bond vectors). However, the CN is 6.8, and so the remaining 1.8 ions in the primary coordination shell exist as disordered, weakly interacting anions. While likely interaction sites may be identified, care should be taken not to overinterpret probability densities in this kind of situation.

2.4 Solvation environments of small molecules in ionic liquids

Typical ionic liquids consist of organic cations paired with inorganic anions, complicating the rule of thumb that *like dissolves like*. This combination of properties potentially offers a panacea with respect to dissolution and results in ionic liquids which can dissolve and solvate both organic and inorganic species. It is interesting to understand (given the strong charge-ordered structure seen in the ionic liquids already studied) how solutes physically *fit in* to the bulk liquid and how much the overall structure is affected.

2.4.1   Benzene: 1,3-dimethylimidazolium hexafluorophosphate

From the empirical observation that the simple 1,3-dialkylimidazolium-based ionic liquids dissolve aromatic compounds more readily than aliphatic compounds, this presents an interesting system which may be used to probe the solvation interactions by both theory and experiment. Deetlefs et al. performed neutron diffraction experiments on $[C_1C_1Im][PF_6]$ containing 33 and 67 mol% concentrations of benzene and found that as with pure ionic liquids, significant charge ordering of the system resulted [39]. To accommodate the additional solute, the structure of the liquid changes as expected with the primary coordination shell around the cation showing reduced anion coordination. For example, in the pure ionic liquid, the first shell contains 6.8 anions whereas 5.5 and 4.1 are found for the 33 and 67 mol% systems, respectively. The benzene molecules are predominantly associated with $[C_1C_1Im]^+$, with 3.3 and 9.1 (33 and 67 mol%, respectively) found in the primary solvation shell with anions being replaced in favor of the aromatic solute. This region, however, is still anion-exclusive with little-to-no presence of cations, as found in the pure liquid. Furthermore, the relatively high number of aromatic molecules within the primary coordination shell indicates that the benzene is readily accommodated by the ionic liquid, whilst still retaining a large proportion of the expected charge-shell structure. Figure 2.8 shows the probability densities of molecules around a central cation from EPSR analysis of the neutron data. For $[PF_6]^-$ anions, Figure 2.8a, the shapes of the distributions are remarkably similar to those of the pure liquid (Figure 2.4), suggesting that those anions replaced by benzene in the primary coordination shell were not directly interacting with the cation, and instead were disordered about the remaining volume of the shell. However, this is clearly not always the case as in Figure 2.8b that shows regions of high probability for finding benzene molecules about the cation that coincide with some of those for the anion. Therefore, while the averaged anion distribution about the cation remains largely unchanged from the pure liquid, benzene molecules may be found to replace anions in several positions. The distribution of cations about the cation, (Figure 2.8c) shows the largest difference with respect to the pure liquid structure (Figure 2.4) with the density associated with the methyl groups absent for the benzene mixture (and replaced by benzene molecules—compare regions in Figures 2.4 and 2.8c). Overall, we may summarize that:

1. The benzene molecules are largely associated with $[C_1C_1Im]^+$.
2. The cation–anion distribution is relatively undisturbed by the presence of the solute, even at high concentrations.

Harper and Lynden-Bell [40] studied 1:1 mixtures of benzene, 1,3,5-trifluorobenzene, and hexafluorobenzene in $[C_1C_1Im][PF_6]$. For benzene it was found that cations are concentrated above and below the plane of

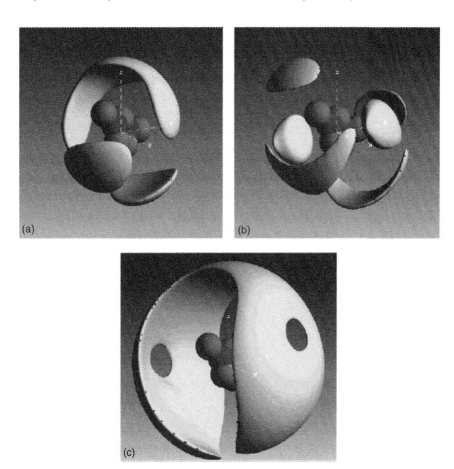

Figure 2.8 **(See color insert following page 224.)**   Probability densities determined by empirical potential structure refinement from neutron diffraction studies of $[C_1C_1Im][PF_6]$ containing 33 mol% benzene. Densities are for (a) $[PF_6]^-$, (b) benzene, and (c) $[C_1C_1Im]^+$ around a central cation. Surfaces are drawn to encompass the top 25% of ions within 8 Å for the anion and benzene and 10 Å for the cation. (From Deetlefs, M., Hardacre, C., Nieuwenhuyzen, M., Sheppard, O., and Soper, A. K., *J. Phys. Chem. B*, 109, 1593–1598, 2005. With permission.)

the aromatic solute while anions prefer the equatorial region of the ring, and observed charge ordering extending to several shells of ions which is supported by the experimentally determined structure. For hexafluorobenzene the association is also strong and well-defined, but effectively reversed. The influence on the ionic liquid is due to the quadrupole moment present on the benzene and its fluorinated derivative. In the former the quadrupole provides a negative electrostatic field above and below the plane of the molecule, hence these are cation-rich regions. After fluorination the quadrupole is reversed, hence the cation-rich regions become anion-rich, and vice versa. However, subsequent analysis of energetics of various interactions reveals that the benzene–cation and hexafluorobenzene–anion interactions (i.e., those between the aromatic solute and the ions in those regions above and below the plane of the solute) are repulsive in nature. Instead, it is the benzene–anion and hexafluorobenzene–cation interactions (i.e., those between the aromatic solute and the ions in the equatorial region) that are energetically-favorable—thus, the presence of ions above and below the plane of the solutes is purely a result of the strong cation–anion interactions in the ionic liquid. In other words, ionic liquid ions about the equator of the solute are directly associated, while those above and below the aromatic ring are the result of secondary structuring by the equatorial ions. Thus, while a probability density might appear to suggest strong interactions between a solute and both the anion and the cation of a given liquid, it is more likely that only one of the ions strongly associates with the solute and the presence of the other is a side effect of this association. In this instance it is found experimentally that benzene will readily dissolve in the ionic liquid, whereas hexafluorobenzene does not, suggesting that the primary interaction found between solvent and solute in the classical simulations is not strong enough to facilitate dissolution.

2.4.2   *Glucose: 1,3-dimethylimidazolium chloride*

The dissolution of cellulose in ionic liquids is currently being investigated as a replacement for volatile, environmentally unfriendly solvents [41–43]. The nature of interactions between glucose and cellobiose (the monomer and dimer subunits of cellulose) in $[C_1C_1Im]Cl$ has been probed by $^{35/37}Cl$ and ^{13}C nuclear magnetic resonance (NMR) experiments [44]. Therein, it was found that the $^{35/37}Cl$ signal is strongly dependent on the concentration of sugar dissolved in the ionic liquid suggesting that, as might be expected, the predominant solvent/solute interaction is between the anion and the sugar, a result of hydrogen bonding between hydroxyl groups and the chloride. Qualitative methods were used to determine the stoichiometry of this interaction and a value of 1:5 glucose:chloride was proposed, that is, each hydroxyl group is associated with a single chloride anion. It has been suggested [45] since this work that the actual interaction ratio is closer to 1:4, and that the initial study measured the number of OH–Cl interactions, but not the number of chlorides involved. Recent molecular dynamics simulations by Youngs et al. have enabled characterization of the nature of the glucose/solvent interaction more fully, and are shown to be consistent with measured experimental neutron data [46]. On average, the glucose is most often coordinated to four chloride anions via hydrogen bonding through the hydroxyl groups, with one chloride being shared by two hydroxyl groups.

It should be noted that the 1:5 case previously proposed by Moyna et al. is observed, but accounts for <12% of the total number of considered glucose environments. At the other extreme, it is also possible that the glucose can be fully solvated (i.e., its hydrogen-bonding requirements satisfied) by as few as three chlorides, with two pairs of hydroxyl groups bridged by common anions.

Figure 2.9 shows the glucose–cation and glucose–anion RDFs for an isolated sugar molecule in the ionic liquid. A sharp peak at ~5.0 Å exists for the glucose–anion curve reflecting the strong association of chlorides with hydroxyl groups, while the glucose–cation curve shows a broad and shallow peak at ~6.5 Å. The data strongly suggests that the cations do not occupy much of the primary coordination shell since the overlap between the g_{++} and g_{+-} curves is only slight. Again, oscillations indicating a shell-like ordering of ionic liquid ions about the glucose are clearly visible. For a 16.7 mol% solution of glucose in [C₁C₁Im]Cl it is interesting to observe that, despite the size of the solute, the cation–anion RDF is relatively unperturbed compared with the pure ionic liquid or the ionic liquid containing a low concentration of glucose. Within the first solvation shell little change is visible, while for the second peak it is possible to infer a slight decrease indicating

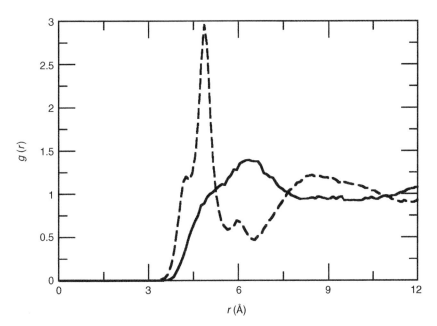

Figure 2.9 Glucose–cation (solid line) and glucose–anion (dashed line) center-of-mass radial distribution functions calculated from molecular dynamics simulations of a single sugar molecule in [C₁C₁Im]Cl.

a reduced number of anions in the second solvation shell. The most notable change occurs for the anion–anion RDFs since at the higher concentration the number of chlorides associated with glucose molecules becomes more significant. There also exist chlorides that have no association with cations, that is, those forming bridges with two glucose molecules simultaneously. Despite the particularly strong association with Cl⁻ it is possible for the glucose molecule to undergo conformation change within the ionic liquid, and several different conformers are identified as existing for significant periods of time in the work of Youngs et al.

2.4.3   *PCl₃/POCl₃: 1,3-dimethylimidazolium bis{(trifluoromethyl)sulfonyl}imide*

Phosphorous trichloride (PCl₃) and phosphorous oxychloride (POCl₃) are moisture-sensitive reagents used in a variety of important synthetic routes; however, because of their sensitivity to water their handling and storage is an issue. Recently, hydrophobic ionic liquids with bis[(trifluoromethyl)sulfonyl] imide anions have been reported to both readily solubilize PCl₃ and POCl₃ and stabilize them with respect to hydrolysis if exposed to moisture [47]. This suggests a valuable use of ionic liquids as a combined reaction medium and storage solution for these reactive reagents.

Neutron diffraction experiments have been performed [48] on mixtures of the two compounds with $[C_1C_1Im][Tf_2N]$ but the raw $S(Q)$ data showed very little difference between the pure ionic liquid and that containing up to 14 mol% of solute. Although it may seem surprising that, despite the presence of the solute, the structure is not notably affected, some precedent exists with both benzene and glucose solutes also providing similar scenarios. It is tempting to consider the idea that, since the experiment does not offer a *positive* result, the data is at fault in some way. Simulations in this instance represent a useful way to validate the experimental findings one way or the other (provided we have confidence in the forcefield). Figure 2.10 plots the simulated $S(Q)$ data for a pure system of $[C_1C_1Im][Tf_2N]$ against the differences between it and the simulated $S(Q)$ data for 1:6 mixtures of both $PCl_3:[C_1C_1Im][Tf_2N]$ and $POCl_3:[C_1C_1Im][Tf_2N]$. With the exception of some small deviations associated with the first broad peak in the low-Q region for each of the substitutions, little significant difference may be found from the various simulations. It should be noted that the small changes observed in the simulations between the solute + IL systems and the pure ionic liquid are within the uncertainty of experimental measurements and, therefore, would not be observed.

Molecular simulations of 14 mol% solutions reveal an interesting contrast between the two solutes. On the one hand, PCl₃ tends to form microdomains where high concentrations of solute exist, whereas POCl₃ is more

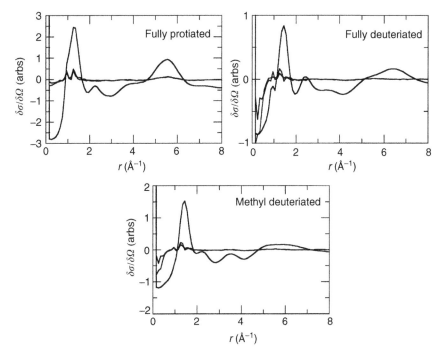

Figure 2.10 **(See color insert following page 224.)** Simulated $S(Q)$ data for a pure system of [C$_1$C$_1$Im][Tf$_2$N] (black line) and differences between this and data for 14% mole fractions of PCl$_3$ (red line) and POCl$_3$ (blue line) for fully-protiated, fully-deuteriated, and methyl-deuteriated systems.

molecularly diffuse throughout the solvent. This is illustrated through probability densities calculated over nanosecond slices of a 15 ns simulation (Figures 2.11 and 2.12) where at $t = 0$ the solutes are randomly distributed throughout their respective systems. After a few nanoseconds it is clear that PCl$_3$ molecules reduce their association with the ionic liquid ions, with few high-probability regions existing. Conversely, the degree of association with other PCl$_3$ molecules increases until the only high-probability region surrounds the molecule (neglecting areas opposite faces of the PCl$_3$ pyramid). With POCl$_3$ there is relatively little change in the probability distributions as time progresses, the solute forming contacts with ionic liquid anions and other POCl$_3$ molecules at all times. The strongest association is between [C$_1$C$_1$Im]$^+$ and oxygen, as might be expected, while association with anions is much weaker. In the latter the [Tf$_2$N]$^-$ anions are found in the primary coordination shell and are located in opposite faces of the POCl$_3$ tetrahedron. The remainder of the primary solvation shell is largely occupied by other solute molecules, replacing anions on to the tetrahedral face regions.

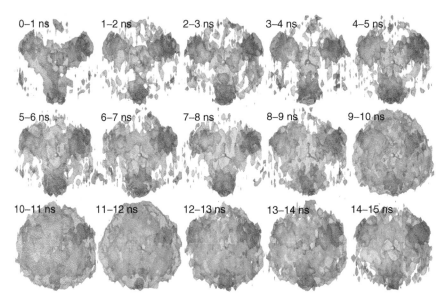

Figure 2.11 **(See color insert following page 224.)** Probability densities determined by MD studies of 14 mol% solutions of PCl_3 in $[C_1C_1Im][Tf_2N]$, calculated over consecutive nanosecond intervals in a continuous 15 ns simulation. Densities are for $[Tf_2N]^-$ (red), $[C_1C_1Im]^+$ (blue), and PCl_3 (green) around a central PCl_3. Surfaces are drawn at twice ($[C_1C_1Im]^+$ and $[Tf_2N]^-$) and three times (PCl_3) the bulk densities of the individual species.

2.4.4 Summary

Ionic liquids can clearly dissolve a range of different solutes, and may be driven by specific interaction with either the cation or the anion, or neither. For glucose in $[C_1C_1Im]Cl$ the association is clearly with the anion owing to hydrogen bonding from the hydroxyl groups, but nevertheless glucose–cation interactions, although relatively weak, may still exert some influence on the dissolution process. In contrast, energetic analysis has shown that benzene (hexafluorobenzene) is clearly and directly associated with the anions (cations) around its equator, with the density of the other ions above and below the plane of the ring purely a result of ordering arising from the ionic liquid cation–anion interaction. Comparison of simulations between $[C_1C_1Im]Cl$ and $[C_1C_1Im][PF_6]$ have shown that this is not necessarily the case for glucose. Hexafluorobenzene and $POCl_3$, on the other hand, interact primarily with the cation rather than the anion since their predominant interaction sites or regions are electronegative rather than electropositive. PCl_3, with little polarization along the P–Cl bonds, presents only a weak external electrostatic field to the solvent. This, coupled with the strong association between ionic liquid anions, forces the solute to segregate or self-associate

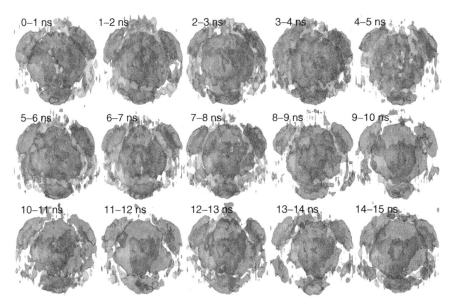

Figure 2.12 **(See color insert following page 224.)** Probability densities determined by MD studies of 14 mol% solutions of $POCl_3$ in $[C_1C_1Im][Tf_2N]$, calculated over consecutive nanosecond intervals in a continuous 15 ns simulation. Densities are for $[Tf_2N]^-$ (red), $[C_1C_1Im]^+$ (blue), and $POCl_3$ (green) around a central $POCl_3$. Surfaces are drawn at twice ($[C_1C_1Im]^+$ and $[Tf_2N]^-$) and three times ($POCl_3$) the bulk densities of the individual species.

into microdomains. Nevertheless, such structuring by the ionic liquid is still sufficient to considerably slow reaction of the compound to air- or solvent-based moisture.

The Lewis donor ability of the solute is an important defining aspect of the dissolution process. If the solute is capable of forming reasonably strong interactions with either (or both) of the ionic liquid ions then dissolution of solutes may proceed up to reasonably high solute/solvent ratios. Neutron diffraction and molecular dynamics simulation can be used to probe the exact interactions occurring and influencing the salvation process, but the indication of energetically favorable interactions with either ion is sometimes not enough to guarantee a comparable experimental result—for example, for hexafluorobenzene the strength of interaction is not strong enough to permit dissolution.

2.5 Conclusions

Pure ionic liquids, when composed of cations coupled with strong- or moderately-coordinating anions (e.g., Cl^- or $[PF_6]^-$) tend to exhibit a liquid structure that (on average) bears a great deal in common with the solid-phase structure.

A common observation with all the solutes studied is that, no matter how well-defined are the solute–cation and solute–anion interactions, the ionic liquid cation–anion structure is relatively unaffected—cations are still surrounded by a bulk-like arrangement of anions, and vice versa.

Discussions here have focused on the more well-characterized salts based on the $[C_1C_1Im]^+$ cation, but this is a somewhat model system, and may accentuate the interactions observed in many of these examples. Where longer alkyl chains are present on the cation liquid and solid-state packing are disrupted and can lead to the formation of ionic/nonionic microdomains in the liquid and which may have a considerable influence on the dissolution of solutes [49].

References

1. Edwards, F.G., Enderby, J.E., Howe, R.A., and Page, D.I., The structure of molten sodium chloride, *J. Phys. C: Solid State Phys.*, 8, 3483–3490, 1975.
2. Blander, M., Bierwagen, E., Calkins, K.G., Curtiss, L.A., Price, D.L., and Saboungi, M.-L. Structure of acidic haloaluminate melts: Neutron diffraction and quantum chemical calculations, *J. Chem. Phys.*, 97, 2733–2741, 1992.
3. Lee, Y.-C., Price, D.L., Curtiss, L.A., Ratner, M.A., and Shriver, D.F., Structure of the ambient temperature alkali metal molten salt AlCl3/LiSCN, *J. Chem. Phys.*, 114, 4591–4594, 2001.
4. Takahashi, S., Suzuya, K., Kohara, S., Koura, N., Curtiss, L.A., and Saboungi, M.-L., Structure of 1-ethyl-3-methylimidazolium chloroaluminates: Neutron diffraction measurements and ab initio calculations, *Z. Phys. Chem.*, 209, 209–221, 1999.
5. Trouw, F.R., and Price, D.L., Chemical applications of neutron scattering, *Ann. Rev. Phys. Chem.*, 50, 571–601, 1999.
6. Inui, M., Takeda, S., Shirakawa, Y., Tamaki, S., Waseda, Y., and Yamaguchi, Y., Structural study of molten silver halides by neutron diffraction, *J. Phys. Soc. Jap.*, 60, 3025–3031, 1991.
7. Kawakita, Y., Enosaki, T., Takeda, S., and Maruyama, K., Structural study of molten Ag halides and molten AgCl–AgI mixture, *J. Non-Cryst. Solids*, 353, 3035–3039, 2007.
8. Di Cicco, A., and Minicucci, M., Solid and liquid short-range structure determined by EXAFS multiple-scattering data analysis, *J. Synchrotron Rad.*, 6, 255–257, 1999.
9. Di Cicco, A., Taglienti, M., Minicucci, M., and Filipponi, A., Short-range structure of solid and liquid AgBr determined by multiple-edge X-ray absorption spectroscopy, *Phys. Rev. B*, 62, 12001–12013, 2000.
10. Inui, M., Takeda, S., Maruyama, K., Shirakawa, Y., and Tamaki, S., XAFS measurements on molten silver-halides, *J. Non-Cryst. Solids*, 192–193, 351–354, 1995.
11. Lantelme, F., Turq, P., Quentrec, B., and Lewis, J.W.E., Application of the molecular dynamics method to a liquid system with long range forces (Molten NaCl), *Mol. Phys.*, 28, 1537–1549, 1974.

12. Tosi, M.P., Pastore, G., Saboungi, M.L., and Price, D.L., Liquid structure and melting of trivalent metal, *Physica Scripta*, T39, 367–371, 1991.
13. Lai, S.K., Li, W., and Tosi, M.P., Evaluation of liquid structure for potassium, zinc, and cadmium, *Phys. Rev. A*, 42, 7289–7303, 1990.
14. Mason, P.E., Neilson, G.W., Enderby, J.E., Saboungi, M.-L., and Brady, J.W., Structure of aqueous glucose solutions as determined by neutron diffraction with isotopic substitution experiments and molecular dynamics calculations, *J. Phys. Chem. B*, 109, 13104–13111, 2005.
15. Hanke, C.G., Price, S.L., and Lynden-Bell, R.M., Intermolecular potentials for simulations of liquid imidazolium salts, *Mol. Phys.*, 99, 801–809, 2001.
16. Hardacre, C., Holbrey, J.D., McMath, S.E.J., Bowron, D.T., and Soper, A.K., Structure of molten 1,3-dimethylimidazolium chloride using neutron diffraction, *J. Chem. Phys.*, 118, 273–278, 2003.
17. Soper, A.K., Empirical potential Monte Carlo simulation of fluid structure, *Chem. Phys.*, 202, 295–306, 1996.
18. Soper, A.K., The radial distribution functions of water and ice from 220 to 673 K and at pressures up to 400 MPa, *Chem. Phys.*, 258, 121–137, 2000.
19. Soper, A.K., Tests of the empirical potential structure refinement method and a new method of application to neutron diffraction data on water, *Mol. Phys.*, 99, 1503–1516, 2001.
20. Arduengo, I.M., Dias, H.V.R., Harlow, R.L., and Kline, M., Electronic stabilization of nucleophilic carbenes, *J. Am. Chem. Soc.*, 114, 5530–5534, 1992.
21. Canongia Lopes, J.N., Deschamps, J., and Padua, A.A.H., Modeling ionic liquids using a systematic all-atom force field, *J. Phys. Chem. B*, 108, 2038–2047, 2004.
22. Liu, Z.P., Huang, S.P., and Wang, W.C., A refined force field for molecular simulation of imidazolium-based ionic liquids, *J. Phys. Chem. B*, 108, 12978–12989, 2004.
23. Urahata, S.M., and Ribeiro, M.C.C., Structure of ionic liquids of 1-alkyl-3-methylimidazolium cations: A systematic computer simulation study, *J. Chem. Phys.*, 120, 1855–1863, 2004.
24. Youngs, T.G.A., Del Pópolo, M.G., and Kohanoff, J., Development of complex classical forcefields through force matching to ab initio data: Application to a room-temperature ionic liquid, *J. Phys. Chem. B*, 110, 5697–5707, 2006.
25. Lynden-Bell, R.M., and Youngs, T.G.A., Using DL_POLY to study the sensitivity of liquid structure to potential parameters, *Mol. Sim.*, 32, 1025–1033, 2006.
26. Youngs, T.G.A., and Hardacre, C., Application of static charge transfer within an ionic liquid forcefield and its effect on structure and dynamics, submitted to *Chem. Phys. Chem.*, in press.
27. Bhargava, B.L., and Balasubramanian, S., Refined potential model for atomistic simulations of ionic liquid, *J. Chem. Phys.*, 127, 114510, 2007.
28. Del Pópolo, M.G., Lynden-Bell, R.M., and Kohanoff, J., Ab initio molecular dynamics simulation of a room temperature ionic liquid, *J. Phys. Chem. B*, 109, 5895–5902, 2005.
29. Bühl, M., Chaumont, A., Schurhammer, R., and Wipff, G., Ab Initio Molecular dynamics of liquid 1,3-dimethylimidazolium chloride, *J. Phys. Chem. B*, 109, 18591–18599, 2005.
30. Bhargava, B.L., and Balasubramanian, S., Intermolecular structure and dynamics in an ionic liquid: A Car-Parrinello molecular dynamics simulation study of 1,3-dimethylimidazolium chloride, *Chem. Phys. Lett.*, 417, 486–491, 2006.

31. Hardacre, C., McMath, S.E.J., Nieuwenhuyzen, M., Bowron, D.T., and Soper, A.K., Liquid structure of 1,3-dimethylimidazolium salts, *J. Phys.-Cond. Mat.*, 15, S159–S166, 2003.

32. Morrow, T.I., and Maginn, E.J., Molecular dynamics study of the ionic liquid 1-*n*-butyl-3-methylimidazolium hexafluorophosphate, *J. Phys. Chem. B*, 106, 12807–12813, 2002.

33. Shah, J.K., Brennecke, J.F., and Maginn, E.J., Thermodynamic properties of the ionic liquid 1-*n*-butyl-3-methylimidazolium hexafluorophosphate from Monte Carlo simulations, *Green Chem.*, 4, 112–118, 2002.

34. Bhargava, B.L., and Balasubramanian, S., Insights into the structure and dynamics of a room-temperature ionic liquid: Ab initio molecular dynamics simulation studies of 1-*n*-butyl-3-methylimidazolium hexafluorophosphate ([bmim][PF$_6$]) and the [bmim][PF$_6$]-CO$_2$ mixture, *J. Phys. Chem. B*, 111, 4477–4487, 2007.

35. Holbrey, J.D., Reichert, W.M., Nieuwenhuyzen, M., Sheppard, O., Hardacre, C., and Rogers, R.D., Liquid clathrate formation in ionic liquid-aromatic mixtures, *Chem. Commun.*, 476–477, 2003.

36. Matsumoto, H., Matsuda, T., and Miyazaki, Y., Room temperature molten salts based on trialkylsulfonium cations and bis(trifluoromethylsulfonyl)imide, *Chem. Lett.*, 12, 1430–1431, 2000.

37. Deetlefs, M., Hardacre, C., Nieuwenhuyzen, M., Padua, A.A.H., Sheppard, O., and Soper, A.K., Liquid structure of the ionic liquid 1,3-dimethylimidazolium bis(trifluoromethyl)sulfonylamide, *J. Phys. Chem. B*, 110, 12055–12061, 2006.

38. Holbrey, J.D., Reichert, W.M., and Rogers, R.D., Crystal structures of imidazolium bis(trifluoromethanesulfonyl)imide ionic liquid salts: The first organic salt with a *cis*-TFSI anion conformation, *Dalton Trans.*, 2267–2271, 2004.

39. Deetlefs, M., Hardacre, C., Nieuwenhuyzen, M., Sheppard, O., and Soper, A.K., Structure of ionic liquid-benzene mixtures, *J. Phys. Chem. B*, 109, 1593–1598, 2005.

40. Harper, J.B., and Lynden-Bell, R.M., Macroscopic and microscopic properties of solutions of aromatic compounds in an ionic liquid, *Mol. Phys.*, 102, 85–94, 2004.

41. Swatloski, R.P., Spear, S.K., Holbrey, J.D., and Rogers, R.D., Dissolution of cellulose with ionic liquids, *J. Am. Chem. Soc.*, 124, 4974–4975, 2002.

42. Zhang, H., Wu, J., Zhang, J., and He, J., 1-Allyl-3-methylimidazolium chloride room temperature ionic liquid: A new and powerful nonderivatizing solvent for cellulose, *Macromolecules*, 38, 8272–8277, 2005.

43. Schlufter, K., Schmauder, H.P., Dorn, S., and Heinze, T., Bacterial cellulose in the ionic liquid 1-n-butyl-3-methylimidazolium chloride, *Macromol. Rapid Commun.*, 27, 1670–1676, 2006.

44. Remsing, R.C., Swatloski, R.P., Rogers, R.D., and Moyna, G., Mechanism of cellulose dissolution in the ionic liquid 1-*n*-butyl-3-methylimidazolium chloride: a ^{13}C and [35/37]Cl NMR relaxation study on model systems, *Chem. Commun.*, 1271–1273, 2006.

45. Youngs, T.G.A., Holbrey, J.D., Deetlefs, M., Nieuwenhuyzen, M., Gomes, M.F.C., and Hardacre, C., A molecular dynamics study of glucose solvation in the ionic liquid 1,3-dimethylimidazolium chloride, *Chem. Phys. Chem.*, 7, 2279–2281, 2006.

46. Youngs, T.G.A, Holbrey, J.D., and Hardacre, C., A molecular dynamics study of glucose solvation in the ionic liquid 1,3-dimethylimidazolium chloride, *J. Phys. Chem. B*, 111, 13765–13774, 2007.
47. Amigues, E., Hardacre, C., Keane, G., Migaud, M., and O'Neill, M., Ionic liquids – media for unique phosphorus chemistry, *Chem. Commun.*, 72–74, 2006.
48. Holbrey, J.D., Hughes, K., Youngs, T.G.A., and Hardacre, C., Unpublished results.
49. Blesic, M., Marques, M.H., Plechkova, N.V., Seddon, K.R., Rebelo, L.P.N., and Lopes, A., Self-aggregation of ionic liquids: Micelle formation in aqueous solution, *Green Chem.*, 9, 481–490, 2007.

chapter three

Ionic liquid advances in optical, electrochemical, and biochemical sensor technology

Sheila N. Baker, Taylor A. McCarty, Frank V. Bright,
William T. Heller, and Gary A. Baker

Contents

3.1   Introduction

The salt, sodium chloride, familiar to the layperson as table salt, exhibits a very high-melting point near 800°C. Given this fact alone, envisioning this or comparable molten inorganic salt environments serving as viable platforms for chemical and especially biosensor development is inconceivable. However, the recent emergence of ionic liquids (ILs), generally defined as organic salts with unusually low melting points below 100°C (many are in fact liquid at/below ambient temperature), makes this an intriguing possibility. Generally based on inorganic or organic anions paired with large, usually asymmetric, organic cations, ILs have been touted as potentially *green* replacements for traditional molecular solvents as they are nonvolatile, nonflammable, thermally stable, and recyclable [1–11]. Although ILs per se have been known for over a century—one of the earliest examples known is the protic IL ethylammonium nitrate $[C_2H_5NH_3][NO_3]$ which melts near 14°C—they have come under intense scrutiny only more recently due to implications for their use in a diverse range of solvent applications. Several hundreds of different ILs are now known to exist in the liquid (or supercooled) state at room temperature or below, making handling of these fluids convenient for a host of applications.

In addition to the features just mentioned, ILs exhibit a wide electrochemical window, high ionic conductivity, a broad temperature range of the liquid state, and frequently possess excellent chemical inertness as well. Moreover, the physical properties of ILs—including density, melting point, conductivity, polarity, Lewis acidity, viscosity, and enthalpy of vaporization—can all be tuned by changing the cation and anion pairing. Thus one can, in principle, design an IL for a specific task (e.g., extraction, separation, reaction) simply by manipulating its key physicochemical properties as a result of appropriate cation/anion pairings. The dual nature (discrete ions) of ILs allows for the compartmentalized molecular-level design of a wide range of versatile molten systems. All of these features, particularly their tunable property sets, have proved to be important drivers in the areas of electrochemistry, separation sciences, chemical synthesis, catalysis, energetic materials, pharmaceutics, biotechnology, lubricants, heat transfer fluids, nanochemistry, and analytical chemistry, among others. To date, ILs have been used in numerous chemical applications but they have, of course, been most often discussed as possible substitutes for volatile organic solvents in synthesis and catalysis, industrial processing, electrochemistry, and separation technologies. Most recently, interest surrounding their distinctive potential in analytical chemistry, particularly in sensor and device technology, is gaining extraordinary momentum [10–14]. This chapter focuses on such contemporary developments into the application of ILs in optical, and electrochemical- biosensors and, to a lesser extent, in actuator and microfluidic devices.

Full names, molecular structures, and shorthand nomenclature for all ILs discussed within this chapter are provided in Table 3.1.

3.2   Optical sensors

This section focuses on the implementation of IL technology to the optical sensing field. This IL application is relatively unexplored and this section provides a summary of the very few reports that have been published on the topic to date.

Pandey and coworkers explored the use of 1-butyl-3-methylimidazolium hexafluorophosphate, $[C_4C_1Im][PF_6]$, as a solvent for polycyclic aromatic hydrocarbon (PAH) analysis [15]. The researchers investigated the steady-state emission behavior of the two different PAH types (alternate versus nonalternate) in the presence of the fluorescence quencher nitromethane dissolved in either acetonitrile, $[C_4C_1Im][PF_6]$, or 90 wt% glycerol in water. The authors discovered that, in the above systems, nitromethane selectively quenches the emission of alternant PAHs and thus one can distinguish between alternate and nonalternate PAHs using this method. The nitromethane quenching obeys simple Stern–Volmer-type behavior for the five alternate PAH compounds studied in acetonitrile, $[C_4C_1Im][PF_6]$, and 90 wt% glycerol in water. The nitromethane quenching efficiency of the alternate PAHs in $[C_4C_1Im][PF_6]$ and 90 wt% glycerol is, in fact, very similar and collectively lower in comparison to that in acetonitrile. This behavior is attributed to the much higher viscosity of $[C_4C_1Im][PF_6]$ and 90 wt% glycerol as compared to acetonitrile, a conclusion underscoring the need for developing ILs with superior fluidity.

Baker and coworkers [16] reported on a self-referencing luminescent thermometer designed around the temperature-dependent intramolecular excimer formation/dissociation of the molecular probe 1,3-*bis*(1-pyrenyl)propane (BPP) dissolved in 1-butyl-1-methylpyrrolidinium *bis*(trifluoromethyl-sulfonyl)imide, $[C_4C_1pyr][Tf_2N]$. Upon an increase in temperature, and hence a decrease in the IL's bulk viscosity, the excimer-to-monomer fluorescence

Table 3.1 Additional Ionic Liquids Discussed within Chapter 3 and their Corresponding Chemical Structures and Abbreviations

Full name	Structure	Abbreviation
1-(2-Hydroxyethyl)-3-methylimidazolium tetrafluoroborate		$[OHC_2C_1Im][BF_4]$
1-Butyl-3-ethylimidazolium camphorsulfonate		$[C_4C_2Im][CamSO_3]$
1-Butyl-3-methylimidazolium camphorsulfonate		$[C_4C_1Im][CamSO_3]$
1-Butyl-3-methylpyrrolidinium bis(trifluoromethylsulfonyl)imide		$[C_4C_1pyr][Tf_2N]$
1-Butyl-3-trimethylsilylylimidazolium hexafluorophosphate		$[C_4Me_3SiIm][PF_6]$
1-Hexyl-3-methylimidazolium hydroxide		$[C_6C_1Im][OH]$
1-Pentyl-3-methylimidazolium tris(pentafluoroethyl) trifluorophosphate		$[C_5C_1Im][3CF_3PF_3]$

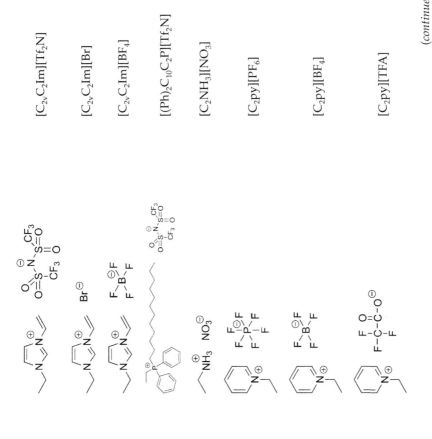

1-Vinyl-3-ethylimidazolium *bis*(trifluoromethylsulfonyl)imide — [C₂ᵥC₂Im][Tf₂N]

1-Vinyl-3-ethylimidazolium bromide — [C₂ᵥC₂Im][Br]

1-Vinyl-3-ethylimidazolium tetrafluoroborate — [C₂ᵥC₂Im][BF₄]

Dodecylethyldiphenyl-phosphonium *bis*(trifluoromethylsulfonyl)imide — [(Ph)₂C₁₀C₂P][Tf₂N]

Ethylammonium nitrate — [C₂NH₃][NO₃]

N-ethylpyridinium hexafluorophosphate — [C₂py][PF₆]

N-ethylpyridinium tetrafluoroborate — [C₂py][BF₄]

N-ethylpyridinium trifluoroacetate — [C₂py][TFA]

(continued)

Table 3.1 (Continued)

Full name	Structure	Abbreviation
N-octylpyridinium hexafluorophosphate		[C₈py][PF₆]
Tetraheptylammonium dodecylbenzenesulfonate		[(C₇)₄N][C₁₂BeSO₃]
Tetrahexylphosphonium camphorsulfonate		[(C₆)₄P][CamSO₃]
Tetraoctylphosphonium dodecylbenzenesulfonate		[(C₈)₄P][C₁₂BeSO₃]

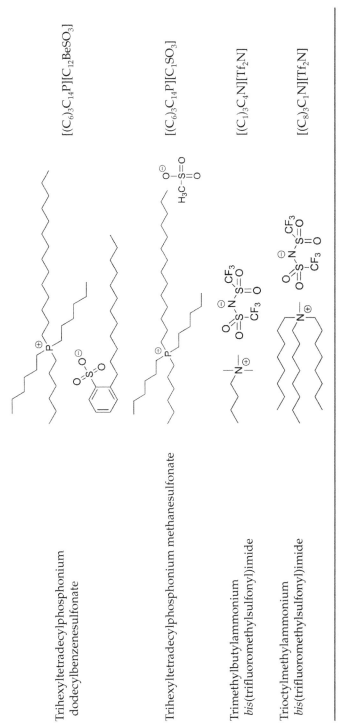

Trihexyltetradecylphosphonium
dodecylbenzenesulfonate

[(C₆)₃C₁₄P][C₁₂BeSO₃]

Trihexyltetradecylphosphonium methanesulfonate

[(C₆)₃C₁₄P][C₁SO₃]

Trimethylbutylammonium
bis(trifluoromethylsulfonyl)imide

[(C₁)₃C₄N][Tf₂N]

Trioctylmethylammonium
bis(trifluoromethylsulfonyl)imide

[(C₈)₃C₁N][Tf₂N]

Note: Although the abbreviations used may differ slightly from popular convention, an editorial decision was implemented to improve fluency and uniformity amongst the various chapters.

intensity ratio I_E/I_M for the probe increases, allowing temperature to be tracked optically (see Figure 3.1). The process is completely reversible over extended heating and cooling cycles, and displays no hysteresis in an operational range spanning 25–140°C. The luminescent thermometer is also among the most precise available, with an average uncertainty in temperature estimation below 0.35°C in the 25–100°C range.

ILs, specifically $[C_4C_1Im][BF_4]$, and $[C_4C_1Im]Br$, have also been employed as optical sensor matrices for the detection of gaseous and dissolved CO_2. Recently, Ertekin and coworkers developed a new optical CO_2 sensor that is based on the spectrophotometric signal changes of the ion pair, bromothymol blue/tetraoctylammonium $(BTB^-/[(C_8)_4N]^+)$ [17]. The authors report pK_a values

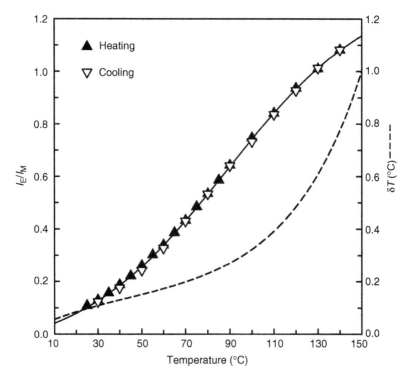

Figure 3.1 Analytical working curve for a self-indexed luminescent thermometer based on the ratio between the measured excimer (E, 475 nm) and monomer (M, 375 nm) emission bands of 1,3-*bis*(1-pyrenyl)propane in $[C_4C_1pyr][Tf_2N]$. The optical thermometer is perfectly reversible in the temperature range shown and highly precise, with the measured uncertainties in the ratio (I_E/I_M) falling well within the symbol dimensions. The dashed curve represents the temperature uncertainty predicted from explicit differentiation of a sigmoidal fit to the calibration profile: $\delta T \cong |\partial T/\partial R| \cdot \delta R$ where $R = I_E/I_M$. (Reprinted from Baker, G.A., Baker, S.N., and McCleskey, T.M., *Chem. Commun.*, 2932–2933, 2003. Copyright 2003 Royal Society of Chemistry. With permission.)

for BTB in $[C_4C_1Im][BF_4]$ and $[C_4C_1Im]Br$ of 9.68 and 9.74, respectively, demonstrating that the $BTB^-/[(C_8)_4N]^+$ ion pair can be used as an absorption-based CO_2 indicator when dissolved in ILs. Upon addition of HCO_3^-, there is a significant decrease in the peak intensity around 420 nm with a corresponding increase in the peak intensity near 625 nm. The reported limits of detection are 1.4% for CO_2 (g) and 10^{-6} M HCO_3^- for dissolved CO_2.

Ertekin and coworkers developed an additional optical CO_2 sensor based on the fluorescence signal intensity changes of the pH-sensitive fluorescent dye 8-hydroxypyrene-1,3,6-trisulfonic acid trisodium salt (HPTS) dissolved in ILs [18]. When HCO_3^- was added to HPTS solution, the fluorescence intensity of the peak centered around 520 nm decreased by 90% in $[C_4C_1Im]$ $[BF_4]$ and by 75% in $[C_4C_1Im]Br$. The reported detection limit for CO_2 (g) was 1.4% while the detection limit for dissolved CO_2 was 10^{-8} M HCO_3^-. The sensor exhibited excellent stability and repeatability over a time period >7 months.

Li and coworkers synthesized the novel IL 1-butyl-3-trimethylsilylimidazolium hexafluorophosphate and demonstrated its utility for liquid/liquid extraction of inorganic mercury. Using o-carboxyphenyl diazoamino p-azobenzene as a chelator to form a stable neutral complex with the metal ion, the authors demonstrated selective extraction into the hydrophobic IL phase [19]. When sodium sulfide was added to the IL phase, the mercury ion was back-extracted into the aqueous layer, providing an avenue for recycling the IL. The authors report extraction and back-extraction efficiencies of 99.9 and 100.1%, respectively, for a 5.0 μg/L aqueous mercury standard. The mercury detection limit was 0.01 ng/mL in water and the method was successfully applied to detecting trace mercury in natural water samples.

ILs have been utilized as capillary coatings for capillary zone electrophoresis with optical detection. Li and coworkers developed a method combining capillary zone electrophoresis and potential gradient detection to separate 11 different metal ions including alkali and alkaline-earth metals, nickel, lead, and ammonium ions [20]. In this work, 1-hexyl-3-methylimidazolium hydroxide IL was covalently coated onto a capillary surface. The system contained the IL as a background electrolyte, lactic acid as chelating reagent, 18-crown-6-ether as an inclusion reagent for analytes, and α-cyclodextrin for modulating the mobility of the IL. When compared to the separation efficiency of a bare silica capillary, the IL-based capillary afforded higher separation efficiencies.

3.3   Electrochemical sensors incorporating ionic liquids

A variety of electrochemical sensors have been developed that use ILs as functional media, where the unique properties of the solvents serve vital roles that an aqueous or organic media could not fulfill. This emerging class of electrochemical sensors is dealt with in this section, reserving discussion of IL-based bioelectrochemical sensors for Section 3.4.

3.3.1   Quartz crystal microbalance-based sensors

Liang and coworkers developed a quartz crystal microbalance (QCM) device that incorporates ILs as the sensing medium [21]. In this case, a membrane of ILs (1-R_1-2-R_2-3-methylimidazoliums paired with Tf_2N^- or BF_4^-, where R_1 is ethyl, propyl, or methyl and R_2 is H or methyl) was deposited on the surface of a 12.5 mm diameter, 6MHz AT-cut crystal using a conventional spin coater. Exposure of the IL membranes to organic vapors, which are generally quite soluble in ILs, altered the resonance frequency of the QCM. Importantly, the responses of the various ILs studied were functions of the gaseous organic species to which the membranes were exposed. While the responses under these idealized conditions demonstrated a clear IL and analyte vapor dependence arising from changes in the adsorbed mass and viscosity of the organic-saturated membrane, the use of a multi-QCM array under realistic conditions might allow for the detection and differentiation of multiple species by comparing the frequency response of the array against a library of known vapors. The rapid response of the membranes to the presence of organic vapor (a few seconds), makes such characterization possible in real time.

Jin and coworkers developed a high-temperature sensor array using a QCM-based sensor [22]. Thin films of seven ILs were employed to provide sensitivity to concentrations of various flammable organic vapors (i.e., ethanol, dichloromethane, benzene, heptane). The sensor array operated at a high temperature of 120°C. Much like the design followed by Liang and coworkers earlier [21], the change in the resonance frequency of various IL-coated QCM crystals was a function of the analyte, its concentration, and the IL as summarized in Figure 3.2. By subjecting the response of the array to linear discrimination analysis (LDA), it is possible to infer the species of organic vapor present with 96% accuracy. The team further characterized the system to understand the detection limitations of the design by measuring the adsorption enthalpy and entropy in the ILs used for the sensor, which were found to be of the same order of magnitude as the enthalpy and entropy of vaporization (30–40 kJ/mol and 85–112 J/(mol K), respectively).

Schäfer and coworkers [23] developed a QCM-IL sensor for use as an artificial nose using the ubiquitous [C_4C_1Im][PF_6]. The IL was spin coated onto the surface of a 10 MHz AT-cut quartz crystal with gold electrodes. The work specifically studied the response of the sensor to ethyl acetate. The deposition of the IL on the surface of the electrode decreased the resonance frequency of the QCM by 2017 Hz. Exposure to increasing amounts of ethyl acetate vapor produced a linear increase in frequency, which was attributed to a progressive decrease in viscosity of the IL upon adsorption of the analyte. The response time, given as the time to full saturation of the

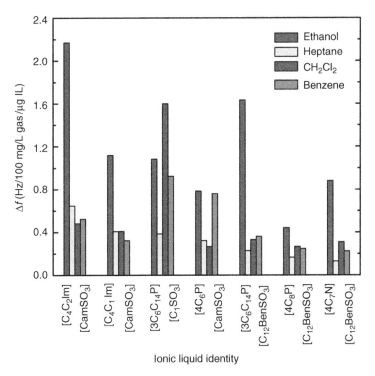

Figure 3.2 Signal patterns from 80% organic vapor samples at 120°C. Responses are normalized to the surface loading of ionic liquids and the concentrations of vapors in the carrier gas. (Reprinted from Jin, X., Yu, L., Garcia, D., Ren, R.X., and Zeng, X., *Anal. Chem.*, 78, 6980–6989, 2006. Copyright 2006 American Chemical Society. With permission.)

response of the sensor, was found to be 360 s, which is a great deal higher than expected simply based on diffusion of the ethyl acetate into the IL. The disparity was attributed to the design of the entire sensor assembly and may warrant further study.

Seyama and coworkers incorporated ILs into plasma-polymerized films (PPFs) deposited on QCMs for constructing sensory arrays to detect alcohol vapors from methanol, ethanol, *n*-propanol, and *n*-butanol [24]. The PPFs were generated from D-phenylalanine, which was chosen because such films are coarse with submicron cracks into which ILs can penetrate. The particular IL selected for their investigation was 1-ethyl-3-methylimidazolium tetrafluoroborate, $[C_2C_1Im][BF_4]$, because it shares stability toward water and air with many common ILs while also having a relatively low viscosity. The sensory array, consisting of 8 mm diameter, 9 MHz AT-cut crystals, incorporated $[C_2C_1Im][BF_4]$ into the PPFs at concentrations of 0–18 mM. The detection level of the array was 16–84 ppm with frequency responses being a function of alcohol species and film thickness, making chemical identification possible using principle components analysis (PCA). Response times were not particularly short, and frequency shifts reported in the paper were collected after 180 min, except for one test using only 40 min of gas adsorption. This is an interesting alternative to gas discrimination by means of arrays containing different ILs.

Demonstrating the popularity of QCMs for sensor applications, Goubaid-oulline and coworkers opted to entrap IL within nanoporous alumina deposited on the surface of the QCM to eliminate dewetting of the QCM and solving the problem of the soft IL surface [25]. The sensors that employed the ILs $[C_4C_1Im][Tf_2N]$, and $[3C_8C_1N][Tf_2N]$ were designed to detect organic vapors including acetonitrile, cyclohexane, isooctane (2,2,4-trimethylpentane), methanol, tetrahydrofuran, and toluene. The nanoporous alumina, with pore diameters of 120 nm, resulted in a more robust sensor. The sensor maintained a better response when ILs were added at less than the full capacity of the alumina pores. The detection limits for the organic vapors ranged from 321 to 7634 mg/m^3. Sensor response times were on the order of minutes.

3.3.2   Ionic liquid alternatives for membrane-based electrodes

Buzzeo and coworkers employed ILs as the electrolyte medium in a simplified, microelectrode electrochemical sensor that uses the robust nature of ILs to produce an improved device [26]. By leveraging the fact that ILs have a negligible vapor pressure, they eliminated the gas–porous membrane traditionally used to maintain the reservoir of electrolyte above the electrode. The ILs $[C_4C_1Im][Tf_2N]$, $[C_4C_1C_1Im][Tf_2N]$), $[C_8C_1Im][Tf_2N]$, $[C_{10}C_1Im][Tf_2N]$, and 1-pentyl-3-methylimidazolium tris(pentafluoroethyl)trifluorophosphate were tested. The design was determined to have comparable performance to more traditional sensor designs that employ a 1 μm gas-permeable polytetrafluoroethylene (PTFE) membrane using either water or dimethylsulfoxide (DMSO) as the medium, with the time to 95% maximum response ranging from 13 to 63 ms. This response is an order of magnitude slower than membrane-independent sensor designs in which the diffusion zone to the electrode does not overlap with the PTFE membrane.

A solid-state sensor for molecular oxygen was developed by Wang and coworkers [27] based on the IL $[C_2C_1Im][BF_4]$ as an alternative to both amperometric sensors using a metallized gas–permeable membrane to encapsulate the electrolyte solution and gas sensors based on a solid polymer electrolyte film, such as Nafion®. Unlike the solid membrane or an encapsulated electrolyte, the IL retains excellent electrochemical properties over a wide range of environmental conditions. The IL-based sensor employs three electrodes: glassy carbon (GC), silver and platinum as working, pseudoreference, and auxiliary electrodes, respectively, in an arrangement similar to that of a solid-state electrochemical cell.

In addition to responding to the oxidation and reduction of molecular O$_2$, the sensor can be used to determine the concentration of the gas by measuring the transient and steady-state currents over a range from 0 to 100% O_2. The response time of the sensor was estimated at 2.5 min, most of which is due to requirements for changing the O_2 environment. Improved sensitivity is obtained using chronoamperometry instead of performing a steady-state measurement. It is also expected to extend the lifetime of the sensor by oxidizing the reactive species $O_2^{\bullet-}$.

Wang and coworkers [28] reported a similar gaseous O_2 sensor using $[C_4C_1Im][PF_6]$ which is more hydrophobic than $[C_4C_1Im][BF_4]$. The sensor was tested with both dry and water-saturated O_2, which produces a two-electron reduction instead of a one-electron reduction, and gave similar performance to $[C_4C_1Im][BF_4]$.

In a slightly different approach, Shvedene and coworkers utilized three ILs ($[C_4C_1Im][PF_6]$, $[C_4C_1C_1Im][Tf_2N]$, and dodecylethyldiphenylphosphonium *bis*(trifluoroethanesulfon)imide as plasticizers to construct ion-selective membrane electrodes using poly(methyl methacrylate) and poly(vinyl chloride) as ion responsive media [29]. By simple mixing of powdered polymer and IL within a glass or Teflon dish, homogeneous and flexible membranes were produced in most cases. The membranes were preconditioned with the desired analyte for up to several hours before use. In addition to simple salts, the researchers studied common surfactant salts and amino acid-based salts. The $[C_4C_1Im][PF_6]$ plasticized membranes did not perform particularly well, having no sensitivity to anions and losing sensitivity to ions over a period of a few days. In total, 18 different combinations of ILs and polymers were tested. The sensors were found to have a roughly pH-independent response with good reproducibility and response times <20 s.

3.3.3   Ionic liquid-modified electrodes

Maleki and coworkers developed an IL-based carbon composite electrode with good electrochemical properties that would be well-suited for use in sensor and biosensor applications [30]. (Note: Electrochemical biosensors based on modified electrodes (MEs) are further fleshed out in the subsequent section.) The IL used was *N*-octylpyridinium hexafluorophosphate which served as a binder in carbon paste electrodes (CPEs) using graphite powder. The IL binding agent resulted in more uniform electrodes than traditional CPEs resulting, in part, to a lower resistance of $8 \pm 2\ \Omega$ and wetting angle of $45 \pm 3°$. The IL CPE has a wider potential window (anodic limit, $V_{an} \sim 1.3$ V; cathodic limit, $V_{cat} = -1.0$ V) with an improved current density relative to a bare GC electrode.

The electrode has several other advantages as well. For instance, it combines the advantages of edge plane characteristics of both carbon nanotubes and pyrolytic graphite electrodes together with the low cost of CPEs and the robustness of metallic electrodes. It further provides a remarkable increase in the rate of electron transfer of different organic and inorganic electroactive compounds, offers a marked decrease in the overvoltage for biomolecules (e.g., nicotinamide adenine dinucleotide (NADH), dopamine, ascorbic acid), and resists NADH surface fouling effects. Depending upon the choice of electrolyte, the electrode can feature ion-exchange properties and adsorptive characteristics similar to clay-MEs. Due to its mechanical strength, the IL CPE could be applied as an effective flow-through detector in flowing streams, and since a mixture of IL and graphite is easily moldable, the fabrication of different electrode geometries is completely feasible. The favorable electrochemical response, high reversibility, sensitivity, and selectivity observed for these electrodes toward biomolecules together with its resistance to electrode fouling make it an excellent candidate for the construction of a new generation of biosensors. In the next section, we discuss such IL MEs and their utility in assorted biosensing formats.

3.4   Biosensors based on ionic liquids

3.4.1   Optical-based assays

One of the first reported biosensors using ILs involved the detection of the organophosphate paraoxon [31]. Because of the acutely toxic effects of organophosphates, systems for detection and degradation of such compounds are in high demand. Enzymatic analysis is often used for organophosphate detection due to its high selectivity and simplicity. However, organic solvent modifiers, often required due to the low water solubility of most organophosphates, can decrease or fully wipe out the enzyme activity at the core of such biosensors. Inspired by earlier successful biocatalysis within ILs, Malhotra and Zhang investigated the use of ILs to replace organic modifiers in an acetylcholinesterase (AChE)-based bioassay for paraoxon detection [31]. Investigation of the ILs N-ethylpyridinium tetrafluoroborate ([C$_2$py][BF$_4$]), N-ethylpyridinium hexafluorophosphate ([C$_2$py][PF$_6$]), N-ethylpyridinium trifluoroacetate ([C$_2$py][TFA]), and [C$_4$C$_1$Im][BF$_4$] in pH 7.2 phosphate buffer (PB) established that AChE activity was retained for low concentrations of ILs, with the greatest activity (~87%) seen at 2.5% [C$_4$C$_1$Im][BF$_4$]. Therefore, the effects of these ILs and the organic modifiers acetone and cyclohexane were compared at 2.5% modifier concentration in pH 7.2 PB (with the exception of [C$_2$py][PF$_6$] being studied at 0.125 M due to its limited solubility). After a set incubation period, the absorbance of the AChE-catalyzed acetylcholine hydrolysis products at 405 nm, in the presence and absence of paraoxon, was measured to determine the relative inactivation of AChE for each modifier system relative to neat buffer. In these studies, the use of 0.125 M [C$_2$py][PF$_6$] as buffer modifier afforded the largest percentage drop-in AChE activity (i.e., the highest sensitivity) when challenged with paraoxon. Inhibition kinetics followed a first-order model and it was revealed that AChE inactivation by paraoxon was 22-fold more efficient in 0.125 M [C$_2$py][PF$_6$] compared with inactivation in neat buffer. Such studies suggest the potential of ILs as practical modifiers in colorimetric-based enzyme assays, particularly toward analytes having restricted water solubilities.

The first demonstration that antibody-based immunoassays are feasible in IL media was provided by Baker and coworkers [32]. Using a fluorescence quenching assay based on the binding of BODIPY® FL with polyclonal anti-BODIPY FL antibodies, these authors showed that free antibody binding affinity could be almost fully retained for a substantial volume fraction (75 vol%) of the IL [C_4C_1Im][BF_4] in PB (100 mM, pH 8.0). Measured equilibrium binding affinities (K_f), despite being somewhat lower relative to the benchmark value in PB, remained higher at 50 vol% [C_4C_1Im][BF_4] ($K_f = 5.30 \times 10^6$ M^{-1}) than those determined for water-miscible organic solvents at the same volume fraction, including acetonitrile, ethanol, and DMSO. In fact, the K_f value for 50% [C_4C_1Im][BF_4] was over three decades higher than the value in 50% acetonitrile in which the anti-BODIPY was significantly inactivated. Moreover, for an increase from 50 to 75 vol% [C_4C_1Im][BF_4], K_f increased unexpectedly to 1.52×10^7 M^{-1}. However, IL concentrations exceeding 75 vol% [C_4C_1Im][BF_4] could not be studied in a homogeneous format due to solubility limits. Thus, to study antibody–hapten binding in a neat IL system, the authors instead developed a heterogeneous assay utilizing an array of anti-BODIPY *capture* antibodies immobilized onto a (3-aminopropyl)dieth oxymethyl silane self-assembled monolayer pattern, as summarized in Figure 3.3. The anti-BODIPY bioarrays were incubated in [C_4C_1Im][X] (X = BF_4^-, Tf_2N^-, TfO^-, or PF_6^-) containing the target BODIPY FL, followed by rinsing with the given IL. In each case, a high-contrast fluorescence image resulted, demonstrating that immobilized antibodies are able to maintain their binding integrity at analytically relevant levels in IL. Overall, these results hold compelling possibilities for advancing biosensors targeting a range of analyte species including traditionally difficult lipophilic targets of key importance in pharmacology, petrochemistry, food science, biomedicine, environmental monitoring, and homeland security applications.

3.4.2 Modified electrode biosensors

By far the most highly investigated aspect of IL-based biosensors is in the realm of electrochemistry, with most based on IL MEs [33–47,48–57] wherein the IL typically serves as both binder and conductor. Common attributes observed when incorporating ILs into electrodes include higher conductivity, good biocatalytic ability, long-term stability (including stability at elevated temperature), superior sensitivity, improved linearity, better selectivity, and the ability to fabricate third-generation biosensors with direct (mediator-less) electron transfer between protein and electrode. The remainder of Section 3.4.2 is further broken into specific subsections treating different types of IL MEs and these subsections appear in roughly chronological order, based on their appearance in the open literature. Most investigations to date have centered on the amperometrically determined biocatalytic activity of

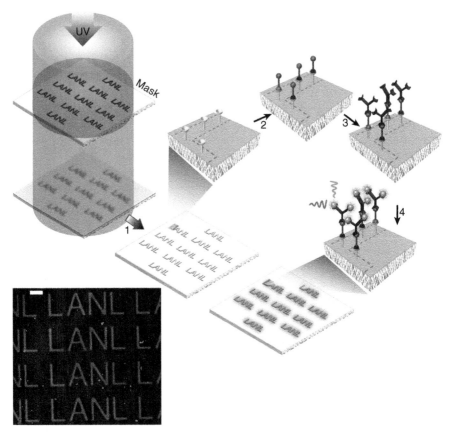

Figure 3.3 **(See color insert following page 224.)**   Schematic illustration of self-assembled monolayer (SAM) photolithographic patterning via masked exposure to Hg lamp irradiation and subsequent antibody conjugation to fabricate immunosurfaces for use in an ionic liquid-based immunoassay. Shown in the inset (lower left) is a representative fluorescence image showing spatially selective BODIPY binding within neat $[C_4C_1Im][Tf_2N]$. (Reprinted from Baker, S.N., Brauns, E.B., McCleskey, T.M., Burrell, A.K., and Baker G.A., *Chem. Commun.*, 2851–2853, 2006. Copyright 2006 Royal Society of Chemistry. With permission.)

common enzymes such as glucose oxidase (GOx), horseradish peroxidase (HRP), or various other heme proteins (e.g., Hb, Mb, cyt *c*) incorporated into electrodes. Table 3.2 provides a summary of the performance characteristics of various MEs discussed within the chapter.

3.4.2.1   Carbon nanotube modified electrodes

The electrochemical detection of the neurotransmitter dopamine is complicated by the high concentration of biologically coexisting ascorbic acid, which has an oxidation potential lying very close to dopamine's at solid

Table 3.2 Performance Characteristics of Ionic Liquid-Based Modified Electrodes

Ionic liquid	Additional component(s)	Coated electrode	Enzyme(s)	Analyte(s) detected	Detection range (M)	LOD (M)	Method	Ref.
$[C_8C_1Im][PF_6]$	Multiwalled carbon nanotube	GCE	N/A	Dopamine	10^{-4} to 10^{-6}	10^{-7}	Amperometry	33
$[C_4C_1Im][BF_4]$	Acid-treated multiwalled carbon nanotube	GCE	Hemin, Hb horseradish peroxidase	H_2O_2	NR	NR	Cyclic voltammograms	34
$[C_4C_1Im][PF_6]$	Multiwalled carbon nanotube	GCE	GOx	Glucose	5×10^3 to 20×10^{-3}	NR	Chrono-amperometry	35
$[C_4C_1Im][PF_6]$	Multiwalled carbon nanotube	Carbon microfiber	N/A	Dopamine	18×10^{-6} to $180 \times 10{-6}$	NR	Osteryoung square wave voltammetry	36
$[C_4C_1Im][BF_4]$	Ferrocene/sol–gel	GCE	Horseradish peroxidase	H_2O_2	0.02×10^{-3} to 0.26×10^{-3}	1.1×10^{-6}	Amperometry	37
$[C_4C_1Im][BF_4]$	Nafion/sol–gel	Basal plane graphite	Hb	O_2	0.14×10^{-6} to 1.82×10^{-6}	3.2×10^{-9}	Amperometry	38
Poly[veim][Br]	NA	Platinum	GOx	Glucose	2.5×10^{-5} to 2.0×10^{-3} (aqueous); 9.0×10^{-5} to 5.0×10^{-3} (nonaqueous)	3.8×10^{-6} (aqueous); 2.0×10^{-5} (nonaqueous)	Amperometry	40
Poly[veim] $[BF_4]$	NA	Platinum	GOx	Glucose	2.5×10^{-4} to 2.5×10^{-3} (aqueous); 2.5×10^{-5} to 4.0×10^{-4} (nonaqueous)	2.5×10^{-4} (aqueous); 6.0×10^{-6} (nonaqueous)	Amperometry	40
Poly[veim] $[Tf_2N]$	NA	Platinum	GOx	Glucose	2.5×10^{-4} to 8.0×10^{-4} (aqueous); 6.0×10^{-4} to $1.1 \times 10{-3}$ (nonaqueous)	3.8×10^{-6} (aqueous); 1.4×10^{-4} (nonaqueous)	Amperometry	40
$[C_4C_1Im][BF_4]$	Chitosan	GCE	Hb	O_2, Trichloroacetic acid	NR; 0.4×10^{-3} to 56×10^{-3}	NR	Amperometry	41
$[C_4C_1Im][BF_4]$	Chitosan	GCE	Horseradish peroxidase	H_2O_2	0.75×10^{-6} to 135×10^{-6}	NR	Amperometry	42

(continued)

Table 3.2 (Continued)

Ionic liquid	Additional component(s)	Coated electrode	Enzyme(s)	Analyte(s) detected	Detection range (M)	LOD (M)	Method	Ref.
$[C_4C_1Im][PF_6]$	Bentonite clay	Basal plane pyrolytic graphite electrode	Hb	H_2O_2	8.0×10^{-6} to 2.0×10^{-4}	NR	Amperometry	43
				Trichloroacetic acid	NR			
				Nitrite	2×10^{-3} to 21×10^{-3}			
$[OHC_2mim][BF_4]$	NA	Basal plane graphite	Mb	O_2	$0-60 \times 10^{-6}$	2.3×10^{-8}	Amperometry	44
$[2\text{-}C_1C_4py][BF_4]$	NA	Basal plane graphite	Cytc	H_2O_2	1×10^{-6} to 16×10^{-6}	5.16×10^{-8}	Differential pulse voltammetry	45
$[C_4C_1Im][PF_6]$	Nafion	GCE	Horseradish peroxidase	H_2O_2, O_2	NR	NR	Cyclic voltammograms	46
$[C_4C_1Im][PF_6]$	Paraffin and carbon powder	Teflon electrode holder	Mb Hb Horseradish peroxidase	H_2O_2	9.04×10^{-5} to 1.53×10^{-3} 1.29×10^{-4} to 1.90×10^{-3} 7.57×10^{-5} to 1.04×10^{-3}	2.7×10^{-5} 3.9×10^{-5} 2.3×10^{-5}	Amperometry	47
$[C_4C_1Im][PF_6]$	Carbon powder	Electrode tube	Hb	H_2O_2 Nitrite	1.0×10^{-6} to 100×10^{-6} NR	1.0×10^{-6} 2.0 NR	Amperometry	48
$[C_4C_1Im][PF_6]$	Gold nanoparticles/ N-dimethylformamide	GCE	GOx	Glucose	$0.1^{-1} \times 10^{-6}$ and $2-20 \times 10^{-6}$	NR	Amperometry	49
$[C_4C_1Im][PF_6]$	Polyaniline synthesized in $[C_4C_1Im]$ $[PF_6]$	Platinum sheet	Uricase	Uric acid	1×10^{-6} to 1.0×10^{-3}	NR	Amperometry	50

Note: NR — not reported; GCE — glassy carbon electrode.

electrodes. Various approaches have been investigated to circumvent this intrinsic interference. One promising approach involves the use of multi-walled carbon nanotube–IL(MWCN-IL)-modified (GC) electrodes [33]. In this work, Zhao et al. mixed MWCNs with the $[C_8C_1Im][PF_6]$, by grinding them together in a mortar to create a gel-like paste which was then applied to the surface of a cleaned GC electrode. Using a platinum wire and a saturated calomel electrode as auxiliary and reference electrodes, respectively, cyclic voltammograms (CVs) were measured for dopamine in PB for both the MWCN-IL-modified GC electrode and a bare GC electrode. In both cases, two pairs of redox peaks characteristic to dopamine were observed. An immediate advantage of the MWCN-IL ME was a larger peak current with smaller peak separations, an indication of faster electron transport to the electrode surface. Similar measurements for ascorbic acid and uric acid revealed that the anodic peak potentials were respectively shifted more negative (by ~0.31 V) and more positive (~0.02 V), when employing the MWCN-IL ME compared with GC. This feature helps to eliminate overlap between dopamine's anodic peak and the anodic peaks of ascorbic acid and uric acid as occurs at a GC electrode. Further resolution was achieved by using differential pulse voltammetry. At pH 7.08, the ascorbic acid and uric acid peaks are separated from dopamine by ca. 0.20 and 0.15 V, respectively. Hence dopamine could be determined in the presence of uric acid and ascorbic acid in 100-fold excess, as revealed in Figure 3.4. The detection limit of dopamine was determined to be 1.0×10^{-7} M with a linear dynamic range up to 1.0×10^{-4} M.

Tao et al. [34] compared the interaction of $[C_4C_1Im][BF_4]$ with three different types of carbonaceous material for use in MEs: acid-treated MWCNs (AMWCNs), pristine MWCNs (PMWCNs), and pyrolytic graphite powder (PGP). To prepare the MEs, each carbon material was ground with the water-miscible IL $[C_4C_1Im][BF_4]$, resulting in AMWCN-$[C_4C_1Im][BF_4]$, PMWCN-$[C_4C_1Im][BF_4]$, and PGP-$[C_4C_1Im][BF_4]$ composites which were then applied to a polished GC electrode surface. The composite films presumably attached through electrostatic adsorption between the negatively charged surface and the positive component of the IL. The conductivities and optical properties of each composite were compared using ac impedance and Raman spectroscopy. $[C_4C_1Im][BF_4]$ formed a gel when ground with MWCNs, whereas a viscous liquid resulted from the admixture of $[C_4C_1Im][BF_4]$ with PGP. Raman spectroscopy demonstrated that both AMWCNs and PMWCNs electro-statically interact with $[C_4C_1Im][BF_4]$ while no such interaction occurs between PGP and $[C_4C_1Im][BF_4]$, accounting for the fact that PGP-$[C_4C_1Im][BF_4]$ blend fails to gel. Data from ac impedance and CV measurements (using potassium ferricyanide, $K_3[Fe(CN)_6]$) demonstrated that the ME conductivity increases in the following order: PGP-$[C_4C_1Im][BF_4]$ <PMWCN-$[C_4C_1Im][BF_4]$ <AMWCN-$[C_4C_1Im][BF_4]$. Thus, the AMWCN-$[C_4C_1Im][BF_4]$ ME system was selected for additional bioelectrochemical studies. For these experiments, the redox biocatalyst (i.e., hemin, hemoglobin (Hb), or HRP) was initially

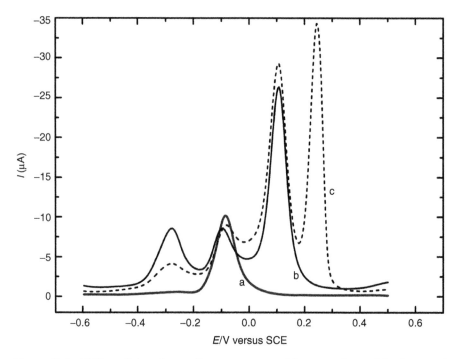

Figure 3.4 Differential pulse voltammograms with correction of background current for (a) 0.4 mM ascorbic acid, (b) 0.4 mM ascorbic acid + 0.05 mM dopamine, and (c) 0.4 mM ascorbic acid + 0.05 mM dopamine + 0.05 mM uric acid at a multiwalled carbon nanotube-ionic liquid/GC electrode. The total weight of the gel on the multiwalled carbon nanotube-ionic liquid modified electrode is 0.1 mg. Scan rate = 20 mVs. (Reprinted from Zhao, Y., Gao, Y., Zhan, D., Liu, H., Zhao, Q., Kou, Y., Shao, Y., Li, M., Zhuang, Q., and Zhu, Z., *Talanta*, 51–57, 2005. Copyright 2005 Elsevier. With permission.)

mixed with the IL prior to grinding with AMWCNs, and the ensuing gel fixed to the GC electrode by rubbing. CV measurements showed a pair of reversible peaks attributed to the heme Fe^{III}/Fe^{II} redox couple for all three MEs. The hemin-, Hb-, and HRP-AMWCN-$[C_4C_1Im][BF_4]$ MEs all demonstrated electrocatalytic behavior toward H_2O_2, with the fastest electron transfer rate observed for the hemin-based ME. While the authors did not elucidate a dynamic range or detection limit for their ME biosensors, the MEs were reproducibly fabricated and reportedly exhibited good stability, retaining ≥60% of the peak current after 20 cycles.

In related work, Dong and coworkers reported using the hydrophobic IL $[C_4C_1Im][PF_6]$ mixed with MWCNs to form two types of MEs [35,36]. In one report, MWCNs were mixed with $[C_4C_1Im][PF_6]$ and applied to a GC electrode [35]. Environmental scanning electron microscopy (ESEM) images of the films showed homogeneously dispersed island-like particles that open up into a number of rope-like 3-D networks. They incorporated GOx into

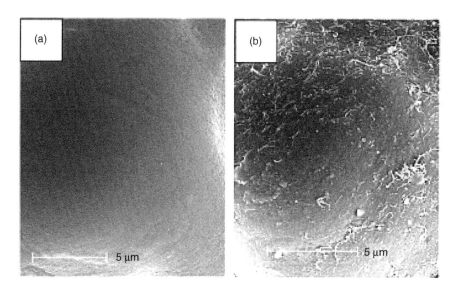

Figure 3.5 Scanning electron microscope images of the surfaces of (a) a bare carbon fiber microelectrode and (b) a multiwalled carbon nanotube -[C₄C₁Im][PF₆]-modified carbon fiber microelectrode. (Reprinted from Liu, Y., Zou, X., and Dong, S., *Electrochem. Commun.*, 8, 1429–1434, 2006. Copyright 2006 Elsevier. With permission.)

the ME by electrode immersion in a PB solution containing the enzyme for 12 h. Using chronoamperometry at an electrode potential of 0.7 V, they were able to detect glucose with a linear dynamic range of 5–20 mM using this MWCN-[C₄C₁Im][PF₆] ME.

In an earlier study, the same group used a similar approach to modify a carbon fiber microelectrode with an MWCN-[C₄C₁Im][PF₆] layer [36]. This electrode modification markedly increased the electron transfer rate, as evidenced by higher current responses for CV measurements of $K_3[Fe(CN)_6]$ compared with an unmodified carbon fiber microelectrode. These results can be understood by comparing ESEM images for a naked carbon fiber microelectrode with an MWCN-[C₄C₁Im][PF₆]-modified one. As shown in Figure 3.5, the surface of the unmodified electrode is relatively smooth, however, after MWCN-[C₄C₁Im][PF₆] modification and immersion in water, a network of randomly tangled MWCN bundles can be plainly observed. Formation of these networks not only increases the electrode surface area, but also greatly improves the electrochemical properties of the microelectrode. Because carbon nanotubes are known to be catalytic in the electrooxidation of dopamine, ascorbic acid, and NADH, the authors tested the modified microelectrode for detection of these three analytes. Using Osteryoung square wave voltammetry, the detection of 18–180 μM dopamine in the presence of 600 μM ascorbic acid was possible, due to fortuitous separation of the oxidation peak potentials, a result reminiscent of that observed earlier in Zhu et al. [33].

3.4.2.2   Sol–gel composite ionic liquid modified electrodes

Liu and coworkers investigated a hybrid material combining an IL with a silica-based sol–gel to generate novel MEs for amperometric biosensing [37]. In this work, a mixture of tetraethylorthosilicate (TEOS, 2 mL), $[C_4C_1Im][BF_4]$ (1 mL), water (1 mL), and 1.0 M HCl (0.05 mL) was stirred for 3 h, followed by storage under ambient conditions for 1 h to generate a pre-hydrolyzed sol. To fabricate MEs, an ethanolic solution of ferrocene was first applied to a GC electrode and allowed to dry. Next, HRP in PB was gently mixed with the sol, and 10 µL of this final solution was cast onto the ferrocene-coated GC electrode, prior to drying at 4°C for 24 h. The redox peak separation for ferrocene in the CV was found to be small (64 mV), indicating a fast one-electron reversible process. Subsequently, the ME was tested for catalytic behavior toward H_2O_2 using amperometric methods. Recording current–time response curves for increasing H_2O_2 concentrations, it was found that the electrode could achieve steady-state conditions within 10 s, presumably a result of $[C_4C_1Im][BF_4]$ facilitating electron transfer. The linear response range was 0.02–0.26 mM H_2O_2 and the detection limit was ~1 µM. The calculated Michaelis–Menten constant of 2 mM was smaller than that reported for HRP in a conventional sol–gel, indicating the higher affinity of the hybrid enzyme electrode toward H_2O_2. Of the interferents tested (e.g., ethanol, glucose, sucrose, oxalic acid, uric acid, and ascorbic acid), only the latter interfered, but only to a small extent. For 10 successive assays, the standard deviation was just ±3.1%, and electrode-to-electrode reproducibility ($n = 6$) was excellent (±5.1%). When stored at 4°C over 20 days, 95% of the initial current response was retained with no loss in sensitivity. The high stability was ascribed to favorable sol–gel/IL interactions, with IL loss curtailed by sufficiently small sol–gel pores in the material. It is important to point out that ferrocene is used as the electron mediator in this configuration, making it a *second-generation* biosensor. Because the bulk of the IL MEs discussed in this chapter require no mediator, the feasibility of this approach in the absence of ferrocene also seems high.

Wei et al. reported on the immobilization of the IL $[C_4C_1Im][BF_4]$ onto the surface of a basal plane graphite (BPG) electrode by sandwiching the IL between layers of Nafion and silica sol–gel [38]. For fabrication, a Nafion in ethanol solution was first applied to a cleaned BPG electrode surface and allowed to dry. Next, an IL solution in ethanol was applied and, after drying, an aliquot of a TEOS-based sol was added and allowed to gel. Finally, the ME was soaked in a buffer solution containing Hb to imbibe the protein. Initial CV measurements revealed the small peak separation of the characteristic redox peaks from the heme, consistent with a fast electron transfer reaction. When $[C_4C_1Im][BF_4]$ was omitted from the electrode assembly, the redox peaks were absent, revealing the IL's importance in providing an environment suitable for electron transfer between Hb and the BPG surface. Moreover, when either the sol–gel or Nafion layer was omitted, although the redox peaks were weaker they were still observable. The ME was tested for catalytic activity in the reduction of O_2. Amperometric current responses reached steady-state within 5 s upon changes in O_2 concentration and, compared with a bare BPG electrode, the O_2 overpotential for the ME was decreased by about 0.35 V. The reported analytical operating range was 0.14–1.82 μM with a limit of detection of 3.2 nM at pH 7.0. The electrode was remarkably stable during the course of 500 continuous scans, and when stored in pH 7.0 PB, retained its activity for 2 weeks.

3.4.2.3 Ionic liquid-based particles for modified electrodes

The López-Ruiz and Mecerreyes groups have worked together to make IL-based microparticles for incorporation into MEs [39,40]. The microparticles were synthesized using a novel two-step route [39]. First, 1-vinyl-3-ethylimidazolium bromide ($[C_{2v}C_2Im][Br]$) microparticles with different amounts of bisacrylamide crosslinker were prepared by concentrated emulsion polymerization. In the second step, the bromide ion was exchanged in water by metathesis with salts of various anions including BF_4^-, $CF_3SO_3^-$ (TfO^-), Tf_2N^-, and dodecylbenzenesulfonate (DBS). Microparticles of poly$[C_{2v}C_2Im][Br]$, poly$[C_{2v}C_2Im][Tf_2N]$, and poly$[C_{2v}C_2Im][BF_4]$ containing GOx were investigated for making modified biosensor electrodes [40]. The incorporation of GOx was carried out by adding the enzyme in the aqueous phase of the emulsion polymerization step. The authors also optimized the pH of the synthetic medium, monomer concentration, and crosslinker concentration. Typically, the resulting particles had a diameter between 2 and 12 μm with a mean particle diameter close to 5 μm.

ME formation was accomplished by placing the microparticles on a platinum electrode and covered with a dialysis membrane. Using amperometric detection, glucose was successfully detected with the exact detection range depending on the type and amount of particles used; that is, in the 2.5×10^{-5} to 8×10^{-3} M range for aqueous solution and 9.0×10^{-5} to 1.1×10^{-3} M in 80:20 (v/v) acetonitrile–buffer. The limits of detection spanned 10^{-4} to 10^{-6} M. The sensor response was both pH and temperature dependent with the maximum sensor response achievable at a pH range of 6–7 and with increasing current response for increasing temperature up to 40°C. In addition, a Nafion membrane could be added to reduce the interference caused by negatively-charged species. When stored frozen in PB, these MEs could completely retain their initial activity even after 150 days.

3.4.2.4 Chitosan/ionic liquid modified electrodes

A chitosan/IL ME incorporating Hb for the electrocatalytic detection of oxygen and trichloroacetic acid (TCA) was investigated by Li et al. [41]. The Hb/chitosan/$[C_4C_1Im][BF_4]$ MEs were prepared by casting a 6 μL droplet of a buffer solution containing 1.2 mg/mL chitosan, 2.4 mg/mL Hb, and 5 vol% $[C_4C_1Im][BF_4]$ onto a polished GC electrode and allowing the film to air dry. CV measurements in PB (50 mM, pH 7.0) showed a pair of well-defined redox peaks that were ascribed to direct electron transfer between Hb and the underlying electrode. When the $[C_4C_1Im][BF_4]$ component was excluded from the ME preparation no such redox peaks were observed. Once again, this indicates that the IL is responsible for facilitating electron transfer between the enzyme and the electrode. Further CV measurements at increasing scan rates between 0.04 and 1.0 V/s resulted in progressively increasing peak currents, revealing the electron transfer between the Hb and GC electrode surface to be a confined electrochemical process. The authors demonstrated stability for their MEs over 50 consecutive cycles and 4 h with only a 1.6% reduction in peak current. The electrocatalytic properties were further tested for reduction of O_2 and TCA. The Hb/chitosan/$[C_4C_1Im][BF_4]$ ME showed higher electrocatalytic activity toward O_2 than Hb/chitosan, Hb, or bare GC electrodes, supporting the role of the IL in facilitating electron transfer. The Hb/chitosan/$[C_4C_1Im][BF_4]$ ME was also biocatalytically active toward TCA. The reduction peak currents increased linearly for a TCA concentration range from 0.4 to 56 mM.

The Li group also studied a chitosan/$[C_4C_1Im][BF_4]$ ME using HRP as the incorporated enzyme [42]. UV–Vis measurement of a drop cast film of HRP/chitosan/$[C_4C_1Im][BF_4]$ on quartz showed a Soret band (403 nm) consistent with that of the heme group in native HRP. CV measurements showed a pair of stable, well-defined redox peaks for the Fe^{III}/Fe^{II} transformation of the heme moiety for direct electron transfer between the HRP and the electrode. As before, removal of $[C_4C_1Im][BF_4]$ during the ME preparation gave no redox peaks. Increasing the scan rate from 0.04 to 1.0 V/s also resulted in concurrent increase in the peak current, as was the case for Hb. The HRP/chitosan/$[C_4C_1Im][BF_4]$ ME was stable over 50 consecutive cycles and for 6 h, with only a 2.2% reduction of peak current. The cathodic peak current for the HRP/chitosan/$[C_4C_1Im][BF_4]$ ME was linear in H_2O_2 concentration from 0.75–135 μM. The ME was able to detect 60 μM H_2O_2 in the presence of ethanol, glucose, sucrose, and uric acid, but sulfide and ascorbic acid both interfered with H_2O_2 detection. To test long-term stability, the ME was stored in the dry state at 4°C and measured twice a week for 30 days and, over that period, only a 5% decrease in current response to 60 μM H_2O_2 was seen.

3.4.2.5 Clay/ionic liquid modified electrodes

Sun investigated the use of Hb/clay/IL composite MEs for the electrocatalytic detection of H_2O_2, TCA, and nitrite [43]. The IL $[C_4C_1Im][PF_6]$ (2 mL) was stirred with bentonite clay (1 mg) for 1 h after which 1 mg of Hb was dispersed into the mixture. A volume of this dispersion was then cast onto a freshly polished basal plane pyrolytic graphite (BPPG) electrode. UV–Vis studies revealed the position of the Hb Soret band to be 410 and 412 nm for dry and buffer-immersed films, respectively. This is consistent with a near-native environment surrounding the heme within Hb in the Hb/clay/

$[C_4C_1Im][PF_6]$ composite. CV measurements of the ME showed that the Fe^{III}/Fe^{II} redox couple peaks exhibited a quite small separation of 40 mV at a sweep rate of 0.2 V/s, revealing a very fast electron transfer. The electrochemical response remained unchanged after 30 consecutive cycles. After storage at 4°C for 2 weeks, the peak currents lost only 2% of the initial response. The Hb/clay/$[C_4C_1Im][PF_6]$ ME showed good electrocatalytic activities toward H_2O_2, TCA, and nitrite. The amplitudes of the reduction peak for Hb/clay/$[C_4C_1Im][PF_6]$ films in CV scans were linear with H_2O_2 concentration in the range of 8.0×10^{-6} to 2.0×10^{-4} M and for nitrite spanning 2–21 mM.

3.4.2.6   Ionic liquid-coated modified electrodes

Examples of MEs wherein the IL itself is coated unmodified onto the working electrode, with no additional components, save for enzyme, also exist [44,45]. For instance, by using the IL 1-(2-hydroxyethyl)-3-methylimidazolium tetrafluoroborate, $[HOC_2C_1Im][BF_4]$, as the supporting electrolyte (0.17 M in H_2O), well-defined and quasi-reversible redox peaks for myoglobin (Mb) were obtained at a BPG electrode with fast electron transfer, as evidenced by the narrow separation between anodic and cathodic peak potentials (66 mV) [44]. Such behavior was absent when PB was used in place of the IL solution. The peak current was found to slowly increase over time following immersion of the BPG electrode in $[HOC_2C_1Im][BF_4]$/Mb/H_2O solution, reaching a maximum after 2.5 h. A plausible explanation for this behavior is the slow chemisorption of Mb onto the BPG. The resulting ME could be stored for up to 2 weeks in an IL solution during which the Mb retained it activity. This Mb/$[HOC_2C_1Im][BF_4]$ ME showed excellent catalytic activity for the reduction of O_2 and could be used to directly detect up to 60 μM O_2 in aqueous solution with a detection limit of 2.3×10^{-8} M.

Ding et al. immobilized cytochrome c (cyt *c*) and 2-methyl-*n*-butyl-pyridinium BF_4, $[2-C_1C_4py][BF_4]$, on the surface of a BPG electrode [45]. Studies of BPG electrode immersion into $[2-C_1C_4py][BF_4]$ solution containing cyt *c* illustrated that 5 h was required to reach optimal enzyme loading. Differential pulse voltammetry was applied to detect H_2O_2 in the range of 1–16 μM with a 5.16×10^{-8} M detection limit using this ME. In addition, the ME was stable for over 400 cycles and, when stored in buffer at 4°C, boasts a shelf life of 2 weeks.

3.4.2.7   Nafion/ionic liquid modified electrodes

Doping Nafion with IL yields a bilateral advantage in the construction of MEs. That is, ILs boost the conductivity of Nafion and, in turn, the Nafion membrane allows for better structural integrity and adherence to the electrode surface. Chen et al. prepared such MEs by sonicating equal volumes of Nafion solution and $[C_4C_1Im][PF_6]$, and coating the homogenous mixture onto a GC electrode [46]. They verified the attachment of the material to the electrode through X-ray photoelectron spectroscopy (XPS) measurements. Further characterization of the electrode through ac impedance and CV measurements, using $[Fe(CN)_6]^{3-}$ as a probe, demonstrated that the IL promotes electron transfer, counterbalancing the blocking of electron and mass transfer by Nafion. These electrode materials proved to be very stable, with only slight changes in the CV spectra observed after 30 cycles; HRP-incorporated MEs made using this approach were electrocatalytic in the reduction of O_2 and H_2O_2.

3.4.2.8   Carbon paste/ionic liquid modified electrodes

CPEs display shortcomings associated with poor fabrication reproducibility and mechanical fragility compared to bare metal electrodes, limiting their overall utility. Typically, paraffin is used as a liquid binder, however, it is nonconductive and thus weakens the electrochemical response. To circumvent this hurdle, Wang et al. used a $[C_4C_1Im][PF_6]$/paraffin oil mixture as binder [47]. Unmodified CPEs were made using an 85:15 (w/w) mix of carbon paste to paraffin oil as a base, with varying volumes of $[C_4C_1Im][PF_6]$ included for IL/CPE construction. The optimal ratio was found to be 70:12:18 carbon paste/paraffin oil/$[C_4C_1Im][PF_6]$. This mixture was packed into a Teflon electrode holder, using a copper wire for electrical contact. Enzymes studied for electrocatalytic acitivity in this work included Mb, Hb, and HRP. Incorporation of $[C_4C_1Im][PF_6]$ resulted in improved sensitivity, dynamic range, and response time. SEM images reveal that, in paraffin oil, the carbon powders form a granular surface where the individual granules can be discerned. By contrast, $[C_4C_1Im][PF_6]$ addition produces a more uniform topography in which the carbon granules are presumably bridged by IL ions, vastly improving conductivity in the paste. The electrocatalytic responses of the incorporated enzymes were investigated and all three showed activity for H_2O_2 reduction in amperometric studies. The linear ranges were 9.04×10^{-5} to 1.53×10^{-3} M (Mb), 1.29×10^{-4} to 1.90×10^{-3} M (Hb), and 7.57×10^{-5} to 1.04×10^{-3} M (HRP). The associated limits of detection were 2.7×10^{-5}, 3.9×10^{-5}, and 2.3×10^{-5} M, respectively. The authors also determined Michaelis–Menten constants of 3.43, 0.21, and 0.10 mM for Mb, Hb, and HRP, respectively. These values are at least an order of magnitude lower than typical literature values, providing evidence that ILs can provide a microenvironment favorable for electron transfer and protein-substrate affinity. Not only did the IL/CPE sensor exhibit high catalytic efficiency, but it took only 5 s to achieve steady-state current upon additions of H_2O_2. Additionally, the researchers found that varying the pH from 3.5 to 9 had relatively little effect on enzyme performance, suggesting a remarkable shielding of the enzyme by the hydrophobic IL. They also showed that the electrocatalytic response increased as the temperature was increased to 50°C, contrary to conventional bioelectrodes which are frequently characterized by optimal operating temperatures closer to ambient temperature. Finally, the IL/CPEs remained stable for 3 months when kept dry at 4°C while stability decreased during buffer storage, suggesting that dry storage methods should be considered when further developing IL-based bioelectrodes.

In a similar study, Wei et al. [58] substituted $[C_4C_1Im][PF_6]$ for paraffin as the binder for a CPE. In this work, the enzyme Hb was not directly incorporated in the carbon paste/$[C_4C_1Im][PF_6]$ mixture but was instead entrapped within a sodium alginate (SA) hydrogel film on the surface of the CPE. Once again, the presence of the IL increased the electron transfer rate, provided a biocompatible surface, and allowed for direct electron transfer without the requirement for a traditional mediator. SEM images of the ME suggest that $[C_4C_1Im][PF_6]$ ions bridge the carbon flakes, giving rise to higher conductivity. The electrocatalytic behavior of Hb was investigated in the reduction of H_2O_2. Catalytic peak currents were linear for 1–100 μM H_2O_2 and a limit of detection of 1 μM was determined. The electrocatalytic reduction of nitrite was also reported, however, the useful range and detection limit were not given.

3.4.2.9   Gold nanoparticle/ionic liquid-based modified electrodes

Zeng and coworkers combined the advantages of gold nanoparticles (GNPs), such as high surface-to-volume ratio and biocompatibility, with those of $[C_4C_1Im][PF_6]$ to create a ME for glucose detection in serum and in beer [48]. These researchers combined citrate-capped GNPs, N, N-dimethylformamide (DMF), $[C_4C_1Im][PF_6]$, and GOx and transferred a volume of the mixture to the surface of a polished GC electrode. In the absence of DMF (or replacing DMF with acetonitrile, ethyl acetate, or acetone), the redox peaks of GOx were weak, unstable, or absent altogether. This suggests a synergistic interaction between GNPs, $[C_4C_1Im][PF_6]$, and DMF, allowing electron transport to the electrode surface. The ME was electrocatalytic toward glucose and peak current was linear in glucose concentration in the 0.1–1 μM and 2–20 μM regimes. The authors successfully detected glucose within serum and beer samples at micromolar concentrations, demonstrating the real world capability of GNP/IL MEs.

3.4.2.10   Polyaniline modified electrodes

Kan and coworkers electropolymerized aniline to form polyaniline (PANI) on the surface of a platinum sheet using $[C_2C_1Im][C_2SO_4]$, as the electrolyte [49]. Uricase was electrically doped into the PANI film by immersing the ME in a buffer solution containing uricase and sweeping the potential to 0.6 V, thus pulling in the negatively charged uricase. The resulting uricase/ PANI ME was electrocatalytically responsive toward uric acid, with the measured current depending upon uric acid concentration, pH (2–12), and temperature (up to 40°C). The linear dynamic range for uric acid detection was 10^{-6} to 10^{-3} M. Potential interferents such as acetaminophen, glutathione, l-cysteine, and ascorbic acid had no adverse effects on uric acid detection. The PANI sensor polymerized in $[C_2C_1Im][C_2SO_4]$ also exhibited far superior stability during storage compared to PANI prepared in HCl solution. In fact, when stored at 4°C, the current response decreased by only 18 and 50% after 157 and 260 days, respectively, whereas the HCl-prepared uricase/PANI electrode lost 39% of its activity within just 4 h.

3.4.3 Bioelectrochemical sensors using ionic liquids as electrolyte medium

As mentioned previously, ILs can act as excellent electrolyte media. When considering their tunable physicochemical properties and their reported ability to stabilize biomolecules [50], it comes as no surprising that researchers have begun to investigate their use as electrolytes in bioelectrochemical sensors [51–53]. Pang et al. first immobilized HRP onto a GC electrode using an aragose hydrogel [51]. When using dry $[C_4C_1Im][BF_4]$ as the electrolyte, no obvious redox peaks for the HRP/agarose electrode could be observed. However, with the addition of 4.5% water, the HRP became electrocatalytically active toward H_2O_2. The addition of DMF to the agarose hydrogel containing HRP further increased its activity, suggesting the benefit of a mediator in this instance. The activity of the HRP/agarose ME in water-containing $[C_4C_1Im][BF_4]$ increased with increasing temperature up to 65°C, and the reduction peak current was linear with H_2O_2 concentration from the limit of detection 6.10×10^{-7} M to 1.32×10^{-4} M. The same group subsequently extended this investigation to include the proteins Hb, Mb, and catalase (cat), with similar results [52].

The direct electrochemistry of heme protein-based MEs employing *dry* $[C_4C_1Im][BF_4]$ as the supporting electrolyte was realized by Xiong and coworkers using DMF/chitosan MEs containing the entrapped redox proteins [53]. In this research, they entrapped the heme-containing proteins Hb, Mb, cat, and cyt *c* in DMF/chitosan organohydrogel films on a GC electrode and, when immersed in water-free $[C_4C_1Im][BF_4]$, redox peaks corresponding to the heme center were observable for each protein studied. Electrocatalytic reactions toward H_2O_2 were followed for each protein and, in general, responses were linear in the micro- to millimolar H_2O_2 range. Glucose, uric acid, ascorbic acid, l-cysteine, and L-tyrosine caused no interference due to their low solubility. When stored in $[C_4C_1Im][BF_4]$ at 4°C, the electrodes retained 98% of their initial response for 36 h.

3.5 Ionic liquids in actuators, microfluidics, and microreactor devices

3.5.1 Electroactive actuators

Ionic polymer transducers, such as those based on Nafion, offer several advantages over other electrochemical transducers, including the ability to generate large strains under small applied voltages, compatibility with conformal structures, and high sensitivity to motion when used in charge sensing mode. However, one primary disadvantage of this type of transducer is dehydration and the corresponding loss in performance of these materials when operated in open air. It has been shown previously that this dehydration problem can be overcome by using ILs as the electrolyte.

In addition, ILs posses a wider electrochemical stability window in comparison to water and as a result, larger actuation voltages can be applied, in theory. The method for plating metal electrodes is known to have a significant impact on actuator performance. Combining these two ideas, Akle et al. have developed a new method for fabricating high-strain ionomeric/IL actuators based on the use of ILs in concert with metal powder–painted electrodes [54]. Their direct assembly plating method involved the uptake of $[C_2C_1Im][TfO]$, by a Nafion membrane, followed by painting the membrane with a polymer/metal mixture containing a solution of 5% (w/v) Nafion (47 wt%), glycerol (47 wt%), and metal powder (6 wt%). To assess the effects of particle conductivity and relative surface area on performance, electrodes were fabricated using a mixture of Au (low surface area-to-volume, high conductivity) and RuO_2 (high surface area, low conductivity). By varying the ratio of these two metal powders, the electrode composition was correlated to the transducer properties, including strain and electrical impedance. By studying various electrode compositions, the authors found that the RuO_2 electrode is superior to Au in producing strain at an equivalent applied voltage. Results for electrical impedance indicate that there is a tradeoff between capacitance and resistance. That is, the RuO_2-only electrode produced a higher capacitance because of its larger surface area, and resulted in lower impedance at low frequencies. However, it also produced higher impedance values at frequencies above 10 Hz because of the higher resistivity of RuO_2 compared with Au. Thus, adding small amounts of Au to the electrode decreased the high-frequency resistance but it also decreased the low-frequency capacitance due to its smaller specific area. In a second set of experiments, the electrode composition was varied whilst maintaining a constant metallic volume fraction (40%) in the electrode. The authors found that the strain response followed the same trend as that obtained by varying the metal weight percent. That is, the 100% RuO_2 device exhibited the highest low-frequency capacitance and Au addition maintained the low-frequency capacitance while also decreasing the high-frequency impedance to ~30 Ω. Further increase in the Au content reduced the capacitance, causing a decrease in strain generation. The authors also tested the repeatability of the plating process by testing three identical transducers created by painting four layers of 100% RuO_2 with a metal-to-polymer weight ratio of 2.5:1, as this composition was determined to give the most optimal performance. The authors claimed little variation in the strain/volt frequency response but no specific standard deviation was provided. The researchers also determined the long-term stability of the three identical transducers. When the samples were actuated with 1 and 2 V peak potentials, they did not exhibit any notable degradation in response after more than 250,000 cycles. In comparison, water-swollen ionic polymer transducers operated for 3600 cycles resulted in a 96% decrement in motion due to solvent loss via evaporation. These results vividly demonstrate the long-term stability of ionopolymer transducers feasible simply by eliminating water and using IL in its

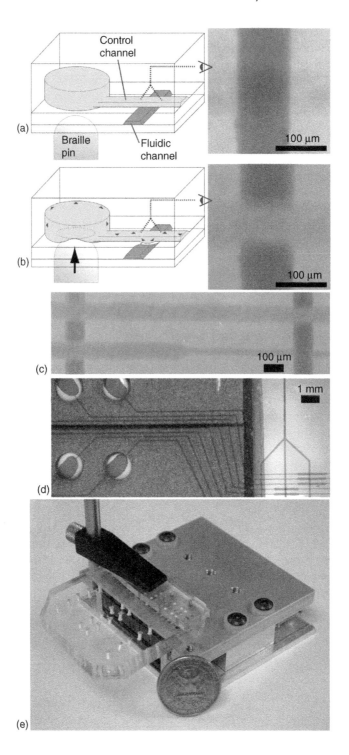

place. We note that, when samples were actuated with 3 V, there was a decrease in performance which is likely tied to degradation of $[C_2C_1Im][TfO]$.

3.5.2 Microfluidic devices

ILs are attractive liquids for use in microfluidic devices because they possess negligible vapor pressure, doing away with unwelcome evaporation effects. Takayama and coworkers recently designed a novel hydraulic microfluidic valve system employing an IL as the hydraulic fluid (Figure 3.6) [55]. The device is similar to a multilayer soft lithography (MSL) pneumatic valve system but, in this case, the valves are pressurized mechanically by movable Braille pins as opposed to externally delivered, high-pressure gas. The pins are aligned below the flow channels and are separated in-between by a flexible barrier. Each pin movement compresses an on-chip piston that pressurizes the connected control channel. Because of the reversible pressurization of the control channel, it can close and reopen areas of the fluidic channels located directly below. In this device, $[C_4C_1Im][BF_4]$ served as the piston fluid and was shown to play a vital role in the reliable operation of the hydraulic system. Because the hydraulic channel volume is relatively small, volume losses due to evaporation and gas permeation effects can significantly alter the system's operation. The authors overcame these issues by utilizing an IL as the hydraulic fluid. As anticipated, no IL volume change within the microfluidic channels (or when open to the atmosphere) occurred over the course of more than a week. One drawback presented by this system is its slower opening and closing response times (~0.3 to 2 s) compared with pneumatic valves, however, the authors showed that minimizing the control channel length could improve response times.

Figure 3.6 **(See color insert following page 224.)**   (a) Schematic of the hydraulic valve and a top-down view with an open valve. Control channels are in pink and fluidic channels are in blue. The clear bulk material is polydimethylsiloxane (PDMS), including the flexible membrane between the control and fluidic channels at their intersection (schematics are not drawn to scale). The piston has an average diameter of ~910 μm and a height of 152 μm, whereas valve intersections are typically 100×100 μm² with 9 μm high fluidic channels and ~16 μm high control channels. The valve and piston can be centimeters apart. (b) The same schematic with a vertical translation of a piezoelectrically driven Braille pin and a top-down view with a close valve and a pressurized control channel. (c) A top-down view of four intersections of pressurized control (red) and fluidic (blue) channels. All channels are 9 μm high and 100 μm wide except for the lower right control channel that is 40 μm wide. (d) A top-down view of the Braille pins aligned underneath pistons (left) and microfluidic valves (right). (e) A PDMS device with multiple hydraulic valves mounted onto a palm-sized USB-powered and controlled Braille display module with 64-pin actuators. (Reprinted from Gu W., Chen H., Tung, Y.C., Meiners, J.C., and Takayama, S., *Appl. Phys. Lett.*, 90, 033505–033508, 2007. Copyright 2007 American Institute of Physics. With permission.)

De Mello et al. implemented ILs as a temperature control system for a microfluidic device based on Joule heating [56]. This device consisted of a working channel structure surrounded by a separate, but parallel, heating channel. Initially, the researchers filled the heating channel with a concentrated KCl solution and applied a dc current. It was found that heating caused point boiling and arcing and the resulting chlorine plasma rapidly etched the glass microstructures. The authors then turned to ILs as a potential heating medium because of their high conductivity and excellent thermal and electrochemical stability. Using $[C_4C_1Im][Tf_2N]$ and $[C_4C_1Im][PF_6]$, they first applied a dc current to the IL-based heating system and found that the applied voltage had a significant effect on the heating rate. Applying voltages under 3 kV produced the predicted hyperbolic relationship between rates of heat generation and loss, whereas higher voltages (>3.5 kV) resulted in S-shaped temperature responses with time, a result attributed to a significant decrease in IL viscosity with increasing temperature. Constant current use was found to cause fouling and degradation of the electrode, however, the use of ac current (50 Hz) produced smooth and controllable heating with minimal degradation of the heating medium near the electrodes. It was found that variable heating rates were achievable with a precision between 0 and 5°C min^{-1}. In addition, through simple feedback control, the authors showed that the temperature could be controlled to within 0.2°C between room temperature and 140°C. Furthermore, the researchers demonstrated that the microchannel elements could be maintained at a fixed temperature for extended periods (10 min). Overall, the authors reported that microfluidic devices incorporating the IL-based heater are simple to construct, require minimal volume (~750 pL per channel), and are optically transparent.

Kim and coworkers implemented ILs in a microfluidic device designed for the enantioselective separation of (S)-ibuprofen [57]. In this study, a three-phase flow microfluidic device was used for the separation of (S)-ibuprofen from the racemic drug using a lipase-facilitated IL flow (ILF). Schematic views of the enantioselective transport and the microfluidic device are shown in Figure 3.7. The aqueous feeding phase consisted of racemic ibuprofen and *Candida rugosa* lipase (CRL), an enzyme that selectively catalyzes the esterification of (S)-ibuprofen. The middle phase is made up of the $[C_6C_1Im][PF_6]$ in which the resulting ester from the CRL-catalyzed esterification selectively dissolves. The aqueous receiving phase consists of porcine pancreas lipase (PPL) which catalyzes the back-hydrolysis of the ester, producing native ibuprofen and ethanol, both of which are water soluble. The authors showed that ILF played an essential role in dissolving the resulting ester and separating the aqueous feeding and receiving phases. The thickness of the ILF could be precisely controlled by the applied flow rate. The ILF flow rate also affected the transport of (R,S)-ibuprofen.

Specifically, as the IRF flow rate was increased, the transport of (R,S)-ibuprofen from the ILF to the receiving phase decreased. In both the feeding and receiving phases, the transport ratio of (S)-ibuprofen was calculated to

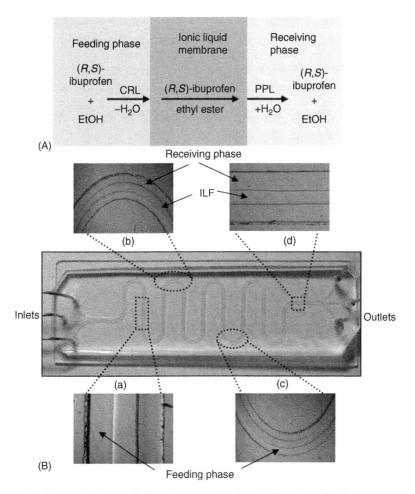

Figure 3.7 **(See color insert following page 224.)**   (A) Enantioselective transport of (S)-ibuprofen by means of ionic liquid flow within a microfluidic device. (B, Bottom) Photographs of the three-phase flow in the microchannel: (a) center near the inlets of the microchannel, (b and c) arc of the microchannel, and (d) center near the outlets of the microchannel. Flow rates of the aqueous phase and the ionic liquid flow phase in (a–d) were 1.5 and 0.3 mL/h, respectively. (Reprinted from Huh, Y.S., Jun, Y.S., Hong, Y.K., Hong, W.H., and Kim, D.H., *J. Mol. Catal.* B, 43, 96–101, 2006. Copyright 2006 Elsevier. With permission.)

be twice that of (R)-ibuprofen, demonstrating that the lipase-catalyzed reactions drove the selective transport of (S)-ibuprofen through the IL. Additionally, the authors showed that the optical resolution ratio in the feeding phase was higher as compared to that of the receiving phase, indicating that CRL selectively catalyzed the esterification of (S)-ibuprofen and then preferentially transported the (S)-ibuprofen ethyl ester from the feeding phase into the ILF. The enantiomeric excess in the receiving phase was reported to be about 77% at different ILF rates. The authors also compared the microfluidic separation to that of a supported liquid membrane (SLM)-based system. It was found that the SLM system allowed for a greater amount of (R,S)-ibuprofen to be transported from the feed phase to the receiving phase because of turbulence flow and sufficient reaction time (20–40 h) in both the feeding and receiving phases. However, the microfluidics-based system separated (S)-ibuprofen efficiently for a shorter working time (30–60 s).

Garrell and coworkers reported the ability to manipulate IL droplets, along with organic solvents and aqueous surfactant solutions, in air on a digital microfluidic platform [59]. The device was composed of two plates: the bottom plate consisted of an array of conductive electrodes, a dielectric layer, and a hydrophobic Teflon-AF coating; the top plate had a single indium tin oxide (ITO) electrode and Teflon-AF coating. The liquid droplet was sandwiched between the two plates at a particular spacing. The researchers found that, for organic solvents, the feasibility of droplet actuation is loosely correlated with the liquid's dipole moment, dielectric constant, and conductivity. In general, liquids were found to be movable if their dipole moments were >0.9 D, dielectric constants >3, and/or conductivities >10^{-9} S m^{-1}. The authors also demonstrated that $[C_4C_1Im][PF_6]$ and $[C_4C_1Im][BF_4]$ could be translated in their digital microfluidic device. They also found that the movability of the droplet depends on the ac frequency used for actuation. In particular, they determined that the movability depends on a liquid's complex permittivity (>8×10^{11}), as well as the interplate distance. The actuation of IL droplets in microfluidic devices may lead to novel microfluidic reactions, sensors, separations, and extractions, particularly given the recent introduction of magnetic ILs [60,61].

3.5.3   Ionic liquid microreactor

Vaultier and coworkers have developed an *open* digital microfluidic system in which electrowetting on dielectric (EWOD) actuation is used to displace IL droplets [62]. ILs are implemented instead of volatile organic solvents to overcome the problem of solvent evaporation which makes necessary a blanket or oil covering. Chemical synthesis using minute amounts of reagents (<1 µL) was accomplished using this approach. The final products were analyzed either by external detection using mass spectrometry (MS) or high performance liquid chromatography (HPLC) (off-line) or directly on-chip by electrochemical measurement. The chip was comprised of a network of Au electrodes structured on silicon. A dielectric layer was incorporated to avoid electrochemical reactions during the actuation of electrodes along with a hydrophobic layer, which provided a surface compatible with electrowetting. An Au microcatenary enabled the polarization of the droplet and its guidance during displacement. Through the use of the Lippmann–Young equation, the authors confirmed that the electrowetting of ILs is less efficient in comparison to aqueous salt solutions. However, the efficiency of electrowetting of the ILs is on the order of deionized water, and the anion and cation choice markedly influences the electrowetting efficiency. A hysteresis study was conducted from the advancing and receding angles on the electrowetting curves. The amount of hysteresis was found to depend on the nature of the anion and the values were determined to be 2–3° for $[C_4 3C_1 N][Tf_2 N]$; ~7° for $[C_4 C_1 Im][BF_4]$; and 8° for $[C_4 C_1 Im][PF_6]$. ILs have smaller contact angles (θ_o = 70–94° for ILs; 110° for water) and lower surface tensions (34–46 mN^{-1}m for ILs; 72 mN^{-1}m for water) in comparison to water [63]. On this basis, the authors expected different behaviors for ILs compared to water. In fact, the ILs did display a wider working interval and the IL droplet motion differed from water droplet motion in that the entire water droplet translated together, whereas for the IL droplet the face of the droplet began to move prior to mobilization of the back of the droplet when the next electrode was switched on. The authors attribute this disparity to the high viscosity of the ILs along with their relatively high wettability and lower surface tension. The speed of IL droplet movement could be controlled with the applied potential but was limited to 1–10 mm^{-1}s, while water droplets were more mobile (up to 120 mm^{-1}s). The authors applied their digital microfluidic setup to perform Greico's reaction, comparing results to that from a macrovolume setup. The final products were analyzed using HPLC and MS. The conversion rate was reported to be 98–100% and the results correlated well to reactions performed in macrovolume. On-line chip analysis was also performed and a cathodic peak at –1.46 V was obtained versus the Au reference electrode, corresponding to the reduction of the final product.

3.6   Concluding remarks

A rapidly emerging field in analytical research involves the development of sensors, electrodes, and diagnostic devices centered around ILs as alternatives to molecular solvents and conventional materials. In this chapter, we have reviewed and discussed some recent research results for various sensor formats that derive significant advantage from the incorporation of an IL component. Examples from optrodes to actuators, including modified enzyme electrodes and lab-on-chip efforts are discussed. To date, work has focused mainly on ILs as drop-in substitutes—active medium, membrane, electrolyte, or coating—in an existing analytical method or sensor construct. Additionally, ILs are just beginning to open the door to novel analytical methods that are not possible using classical solvent systems. Although the molecular basis for the design of ILs with desired properties lags somewhat behind these efforts, ultimately, the term *task-specific IL* will find meaningful relevance in the analytical sciences. Most excitingly, novel applications may emerge in areas that were not even considered in the original concept. The foundation is being laid but perhaps the best is yet to come! We certainly believe this to be so.

References

1. Wasserscheid, P., and Welton, T., Eds., *Ionic Liquids In Synthesis*, 2nd Ed., Wiley-VCH, Weinheim, 2007.
2. Seddon, K.R., Ionic liquids for clean technology, *J. Chem. Technol. Biotechnol.*, 68, 351–356, 1997.
3. Brown, R.A., Pollet, P., McKoon, E., Eckert, C.A., Liotta, C.L., and Jessop, P.G., Asymmetric hydrogenation and catalyst recycling using ionic liquid and supercritical carbon dioxide, *J. Am. Chem. Soc.*, 123, 1254–1255, 2001.
4. Smietana, M., and Mioskowski, C., Preparation of silyl enol ethers using (bistrimethylsilyl)acetamide in ionic liquids, *Org. Lett.*, 7, 1037–1039, 2001.
5. Liu, F., Abrams, M.B., Baker, R.T., and Tumas, W., Phase-separable catalysis using room temperature ionic liquids and supercritical carbon dioxide, *Chem. Commun.*, 433–434, 2001.
6. Brennecke, J.F., and Maginn, E.J., Ionic liquids: Innovative fluids for chemical processing, *AIChE J.*, 47, 2384–2389, 2001.
7. Marsh, K.N., Deev, A., Wu, A., Tran, E., and Klamt, A., Room temperature ionic liquids as replacements for conventional solvents: A review, *Korean J. Chem. Eng.*, 19, 357–362, 2002.
8. Welton, T., Ionic liquids in catalysis, *Coord. Chem. Rev.*, 248, 2459–2477, 2004.
9. Earle, M., Forestier, A., Olivier-Bourbigou, H., and Wasserscheid, P., *Ionic Liquids in Synthesis*, Wiley-VCH, Verlag, 2003.
10. Baker, G.A., Baker, S.N., Pandey, S., and Bright, F.V., An analytical view of ionic liquids, *Analyst*, 130, 800–807, 2005.
11. Pandey, S., Analytical applications of room-temperature ionic liquids: A review of recent efforts, *Anal. Chim. Acta*, 556, 38–45, 2006.
12. Koel, M., Ionic liquids in chemical analysis, *Crit. Rev. Anal. Chem*, 35, 177–192, 2005.
13. Dietz, M.L., Ionic liquids as extraction solvents: Where do we stand? *Sep. Sci. Technol.*, 41, 2047–2063, 2006.
14. Anderson, J.L., Armstrong, D.W., and Wei, G.T., Ionic liquids in analytical chemistry, *Anal. Chem.*, 78, 2893–2902, 2006.
15. Fletcher, K.A., Pandey, S., Storey, I.K., Hendricks, A.E., and Pandey, S., Selective fluorescence quenching of polycyclic aromatic hydrocarbons by nitromethane within room temperature ionic liquid 1-butyl-3-methylimidazolium hexafluorophosphate, *Anal. Chim. Acta*, 453, 89–96, 2002.
16. Baker, G.A., Baker, S.N., and McCleskey, T.M., Noncontact two-color luminescence thermometry based on intramolecular luminophore cyclization within an ionic liquid, *Chem. Commun.*, 2932–2933, 2003.
17. Oter, O., Ertekin, K., Topkaya, D., and Alp, S., Room temperature ionic liquids as optical sensor matrix materials for gaseous and dissolved CO_2, *Sens. Actuators B*, 117, 295–301, 2006.
18. Oter, O., Ertekin, K., Topkaya, D., and Alp, S., Emission-based optical carbon dioxide sensing with HPTS in green chemistry reagents: Room-temperature ionic liquids, *Anal. Bioanal. Chem.*, 386, 1225–1234, 2006.
19. Li, Z., Wei, Q., Yuan, R., Zhou, X., Liu, H., Shan, H., and Song, Q., A new room temperature ionic liquid 1-butyl-3-trimethylsilylimidazolium hexafluorophosphate as a solvent for extraction and preconcentration of mercury with determination by cold vapor atomic absorption spectrometry, *Talanta*, 71, 68–72, 2007.

20. Qin, W., and Li, S.F.Y., Determination of ammonium and metal ions by capillary electrophoresis–potential gradient detection using ionic liquid as background electrolyte and covalent coating agent, *J. Chromatogr. A.*, 1048, 253–256, 2004.

21. Liang, C., Yuan, C.Y., Warmack, R.J., Barnes, C.E., and Dai, S., Ionic liquids: A new class of sensing materials for detection of organic vapors based on the use of a quartz crystal microbalance, *Anal. Chem.*, 74, 2172–2176, 2002.

22. Jin, X., Yu, L., Garcia, D., Ren, R.X., and Zeng, X., Ionic liquid high-temperature gas sensor array, *Anal. Chem.*, 78, 6980–6989, 2006.

23. Shäfer, T., Di Francesco, F., and Fuoco, R., Ionic liquids as selective depositions on quartz crystal microbalances for artificial olfactory systems: A feasibility study, *Microchem. J.*, 85, 52–56, 2007.

24. Seyama, M., Iwasaki, Y., Tate, A., and Sugimoto, I., Room-temperature ionic-liquid-incorporated plasma-deposited thin films for discriminative alcohol-vapor sensing, *Chem. Mater.*, 18, 2656–2662, 2006.

25. Goubaidoulline, I., Vidrich, G., and Johannsmann, D., Organic vapor sensing with ionic liquids entrapped in alumina nanopores on quartz crystal resonators, *Anal. Chem.*, 77, 615–619, 2005.

26. Buzzeo, M.C., Hardacre, C., and Compton, R.G., Use of room temperature ionic liquids in gas sensor design, *Anal. Chem.*, 76, 4583–4588, 2004.

27. Wang, R., Okajima, T., Kitamura, F., and Ohsaka, T., A novel amperometric O_2 gas sensor based on supported room-temperature ionic liquid porous polyethylene membrane-coated electrodes, *Electroanalysis*, 16, 66–72, 2004.

28. Wang, R., Hoyano, S., and Ohsaka, T., O_2 gas sensor using supported hydrophobic room-temperature ionic liquid membrane-coated electrode, *Chem. Lett.*, 33, 6–7, 2004.

29. Shvedene, N.V., Chernyshov, D.V., Khrenova, M.G., Formanovsky, A.A., Baulin, V.E., and Pletnev, I.V., Ionic liquids plasticize and bring ion-sensing ability to polymer membranes of selective electrodes, *Electroanal.*, 18, 1416–1421, 2006.

30. Maleki, N., Safavi, A., and Tajabadi, F., High-performance carbon composite electrode based on an ionic liquid as a binder, *Anal. Chem.*, 78, 3820–3826, 2006.

31. Zhang C., and Malhotra, S.V., Increased paraoxon detection by acetylcholinesterase inactivation with ionic liquid additives, *Talanta*, 67, 560–565, 2005.

32. Baker, S.N., Brauns, E.B., McCleskey, T.M., Burrell, A.K., and Baker G.A., Fluorescence quenching immunoassay in an ionic liquid, *Chem. Commun.*, 2851–2853, 2006.

33. Zhao, Y., Gao, Y., Zhan, D., Liu, H., Zhao, Q., Kou, Y., Shao, Y., Li, M., Zhuang, Q., and Zhu, Z., Selective detection of dopamine in the presence of ascorbic acid and uric acid by a carbon nanotubes-ionic liquid gel modified electrode, *Talanta*, 66, 51–57, 2005.

34. Tao, W., Pan, D., Liu, Q., Yao, S., Nie, Z., and Han, B., Optical and bioelectrochemical characterization of water-miscible ionic liquids based composites of multiwalled carbon nanotubes, *Electroanalysis*, 18, 1681–1688, 2006.

35. Liu, Y., Liu, L., and Dong, S., Electrochemical characteristics of glucose oxidase adsorbed at carbon nanotubes modified electrode with ionic liquid as binder, *Electroanalysis*, 19, 55–59, 2007.

36. Liu, Y., Zou, X., and Dong, S., Electrochemical characteristics of facile prepared carbon nanotubes-ionic liquid gel modified microelectrode and application in biochemistry, *Electrochem. Commun.*, 8, 1429–1434, 2006.

37. Liu, Y., Shi, L., Wang, M., Li, Z., Liu, H., and Li, J., A novel room temperature ionic liquid sol–gel matrix for amperometric biosensor application, *Green Chem.*, 7, 655–658, 2005.

38. Zhao, G.C., Xu, M.Q., Ma, J., and Wei, X.W., Direct electrochemistry of hemoglobin on a room temperature ionic liquid modified electrode and its electrocatalytic activity for the reduction of oxygen, *Electrochem. Commun.*, 9, 920–924, 2007.

39. Marcilla, R., Sanchez-Paniagua, M., López-Ruiz, B., López-Cabarcos, E. Ochoteco, E., Grande, H., and Mecerreyes, D., Synthesis and characterization of new polymeric ionic liquid microgels, *J. Polymer Sci. A.*, 44, 3958–3965, 2006.

40. López, M.S.P., Mecerreyes, D., López-Carbarcos, E., and López-Ruiz B., Amperometric glucose biosensor based on polymerized ionic liquid microparticles, *Biosensors and Bioelectronics*, 21, 2320–2328, 2006.

41. Lu, X., Hu, J., Yao, X., Wang, Z., and Li, J., Composite system based on chitosan and room temperature ionic liquid: Direct electrochemistry and electrocatalysis of hemoglobin, *Biomacromol*, 7, 975–980, 2006.

42. Lu, X., Zhang, Q., Zhang, L., and Li, J., Direct electron transfer of horseradish peroxidase and its biosensor based on chitosan and room temperature ionic liquid, *Electrochem. Commun.*, 8, 874–878, 2006.

43. Sun, H., Direct electrochemical and electrocatalytic properties of heme protein immobilized on ionic liquid-clay-nanoparticle-composite films, *J. Porous Mater.*, 13, 303–397, 2006.

44. Ding, S.F., Xu, M.Q., Zhao, G.C., and Wei, X.W., Direct electrochemical response of Myoglobin using a room temperature ionic liquid, 1-(2-hydroxyethyl)-3-methylimidazolium tetrafluoroborate, as supporting electrolyte, *Electrochem. Commun.*, 9, 216–220, 2007.

45. Ding, S.F., Wei, W., and Zhao, G.C., Direct electrochemical response of cytochrome c on a room temperature ionic liquid, N-butylpyridinium tetrafluoroborate, modified electrode, *Electrochem. Commun.*, 9, 2203–2207, 2007.

46. Chen, H., Wang, Y., Liu, Y., Wang, Y., Qi, L., and Dong, S., Direct Electrochemistry and electrocatalysis of horseradish peroxidase immobilized in Nafion-RTIL composite film, *Electrochem. Commun.*, 9, 469–474, 2007.

47. Wang, S.F., Xiong, H.Y., and Zeng, Q.X., Design of carbon paste biosensors based on the mixture of ionic liquid and paraffin oil as a binder for high performance and stabilization, *Electrochem. Commun.*, 2, 807–812, 2007.

48. Li, J. Yu, J., Zhao, F., and Zheng, B., Direct electrochemistry of glucose oxidase entrapped in nanogold particles-ionic liquid N,N-dimethylformamide composite film on glassy carbon electrode and glucose sensing, *Anal. Chim. Acta*, 587, 33–40, 2007.

49. Jiang, Y., Wang, A., and Kan, J., Selective uricase biosensor based on polyaniline synthesized in ionic liquid, *Sensors Actuators B*, 124, 529–534, 2007.

50. Baker, S.N., McCleskey, T.M., Pandey, S., and Baker, G.A., Fluorescence studies of portein thermostability in ionic liquids, *Chem. Commun.*, 940–941, 2004.

51. Wang, S.F., Chen, T., Zhang, Z.L., and Pang, D.W., Activity and stability of Horseradish Peroxidase in hydrophilic room temperature ionic liquid and its application in non-aqueous biosensing, *Electrochem. Commun.*, 9, 1337–1342, 2007.

52. Wang, S.F., Chen, T., Zhang, Z.L., Pang, D.W., and Wong, K.Y., Effects of hydrophobic room-temperature ionic liquid 1-butyl-3-methylimidazolium tetrafluoroborate on direct electrochemistry and biocatalysis of heme proteins entrapped in agarose hydrogel films, *Electrochem. Commun.*, 9, 1709–1714, 2007.

53. Xiong, H.Y., Chen, T., Zhang, X.H., and Wang, S.F., Electrochemical property and analysis application of biosensors in miscible nonaqueous media–room temperature ionic liquid, *Electrochem. Commun.*, 9, 1648–1654, 2007.

54. Akle, B.J., Bennett, M.D., and Leo D.J., High-strain ionomeric-ionic liquid electroactive actuators, *Sensors Actuators A*, 126, 173–181, 2006.

55. Gu W., Chen H., Tung, Y.C., Meiners, J.C., and Takayama, S., Multiplexed hydraulic valve actuation using ionic liquid filled soft channels and Braille displays, *Appl. Phys. Lett.*, 90, 033505–033508, 2007.

56. de Mello, A.J., Habgood, M., Lancaster, N.L., Welton, T., and Wooton, R.C.R., Precise temperature control in microfluidic devices using Joule heating of ionic liquids, *Lab on a Chip*, 4, 417–419, 2004.

57. Huh, Y.S., Jun, Y.S., Hong, Y.K., Hong, W.H., and Kim, D.H., Microfluidic separation of (S)-ibuprofen using enzymatic reaction, *J. Mol. Catal. B*, 43, 96–101, 2006.

58. Wei, S., Dandan, W., Ruifang, G., and Kui, J., Direct electrochemistry and electrocatalysis of hemoglobin in sodium alginate film on a bmim PF_6 modified carbon paste electrode, *Electrochem. Commun.*, 9, 1159–1164, 2007.

59. Chatterjee, D., Hetayothin, B., Wheeler, A.R., King, D.J., and Garrell, R.L., Droplet-based microfluidics with nonaqueous solvents and solutions, *Lab Chip*, 6, 199–206, 2006.

60. Hayashi, S., Saha, S., and Hamaguchi, H.-O., A new class of magnetic fluids: bmim[FeCl$_4$] and nbmim[FeCl$_4$] ionic liquids, *IEEE Trans. Magn.*, 42, 12–14, 2006.

61. Del Sesto, R.E., McCleskey, T.M., Burrell, A.K., Baker, G.A., Thompson, J.D., Scott, B.L., Wilkes, J.S., and Williams, P., Structure and magnetic behavior of transition metal based ionic liquids, *Chem. Commun.*, 447–449, 2008.

62. Dubois, P., Marchand, G., Fouillet, Y., Berthier, J., Douki, T., Hassine, F., Gmouh, S., and Vaultier, M., Ionic liquid droplet as e-microreactor, *Anal. Chem.*, 78, 4909–4917, 2006.

63. Page, P.M., McCarty, T.A., Baker, G.A., Baker, S.N., and Bright, F.V., Comparison of dansylated aminopropyl controlled pore glass solvated by molecular and ionic liquids, *Langmuir*, 23, 843–849, 2007.

chapter four

Ionic liquids as stationary phases in gas chromatography

Jared L. Anderson

Contents

4.1   Introduction

Ionic liquids (ILs) are nonmolecular solvents that have captured the interest of many in academics and are currently being introduced into a number of industrial processes worldwide. The widespread interest in ILs has drawn scientists of an interdisciplinary nature resulting in an enhanced understanding of these unique solvents. ILs possess a plethora of unique physicochemical and solvation properties that can be varied and tuned for specific applications. The unique solvation properties have produced exciting results when employed as replacements of traditional molecular solvents [1,2]. In addition, most ILs possess negligible vapor pressure and high thermal stability, which designates them as important green solvents that produce few volatile organic compounds. Furthermore, IL-based solvent systems typically exhibit enhanced reaction kinetics resulting in the efficient use of time and energy [1].

The exciting possibilities of producing unique task-specific ILs (TSILs) exist by simple modification of the cation and anion structure [3,4]. The resulting IL can be used to carry out designated applications or to serve a functional purpose in a method or device. In the field of gas chromatography (GC), new stationary phases that exhibit unique separation selectivity, high efficiency, and high thermal stability are in high demand and are particularly sought after. ILs possess many properties that allow them to function as multipurpose stationary phases that exhibit an unique dual-nature retention selectivity.

This chapter is broken into two sections that synergistically have been responsible for the development of IL-based GC stationary phases. The first section describes how the IL stationary phase can be probed to characterize ILs in terms of their multiple solvation interactions and thermodynamic properties to better understand solute–solvent interactions and mixed organic solvent–IL interactions, respectively. The second section of this chapter addresses the systematic approach to develop ILs as a new and viable class of achiral and chiral GC stationary phases. Much of this development has been focused on coated and immobilized IL stationary phases that produce highly efficient separations while exhibiting unique separation selectivity and low bleed at high temperatures. Sections detailing how the combination of cations and anions can be tuned to add further selectivity for more complex separations will also be presented. Properties including viscosity, thermal stability, and surface tension largely dictate the quality and integrity of the stationary phase coating and are additional characteristics that will be discussed. A flow diagram guides the structural development of ILs for the separation of analytes based on desired selectivities and temperature requirements.

4.2 Examination of ionic liquid solvation interactions and thermodynamic properties using gas–liquid chromatography

4.2.1 Ionic liquids as stationary phases in gas chromatography

Three general approaches have been used to characterize ILs in terms of their solvation interactions with solute molecules as well as the determination of important thermodynamic properties of IL/organic solvent mixtures. In almost all cases, the characterization method utilizes the IL as the stationary phase by either coating it on the wall of a capillary column (e.g., wall-coated open tubular column) or by coating the IL on a support to produce a packed column. Probe molecules capable of undergoing known interactions with the stationary phase are then used to probe the interaction capabilities of the stationary phase. The probe molecules' retention times (t_R), retention volumes (V_N), or retention factors (k) can be chromatographically determined to understand the type(s) and magnitude(s) of solute–solvent interactions. For example, a stationary phase containing phenyl groups can undergo π–π interactions with aromatic analytes, thereby increasing their retention. In another example, retention times of acidic analytes are higher on basic stationary phases and vice versa. Each method provides distinct advantages compared to other spectroscopic methods of evaluating solute–solvent interactions, particularly in cases where the IL's cost and purity are important considerations. The manufacturing of a GC column for evaluation typically requires small quantities of the IL (~10 to 100 mg). In addition, the interaction capabilities and thermodynamic parameters can be easily determined at various temperatures by simply examining the retention behavior of probe molecules at the desired temperature(s). Furthermore, the IL-based stationary phase can be conditioned at high temperatures to remove water and any trace impurities thereby allowing for evaluation of the neat IL.

4.2.2 Rohrschneider–McReynolds classification system

To fully characterize and categorize the solute selectivities of GC stationary phases, Rohrschneider and McReynolds pioneered one of the earliest characterization methods [5,6]. The Rohrschneider–McReynolds system is the oldest and widely accepted stationary phase classification systems that is based on the retention of five probe molecules; namely, benzene, butanol, 2-pentanone, nitropropane, and pyridine. Each probe molecule is used to represent a distinct or a combination of interactions with the stationary phase. Benzene measures dispersive interactions with weak proton acceptor properties; butanol measures dipolar interactions with both proton donor and proton acceptor capabilities; 2-pentanone measures dipolar interactions with proton acceptor but not proton donor capabilities; nitropropane measures weak dipolar interactions; and pyridine measures weak dipolar interactions with strong proton acceptor but not proton donor capabilities.

Equation 4.1 describes the Rohrschneider–McReynolds system in terms of the five probes and their corresponding phase constants; namely, benzene (X′), butanol (Y′), 2-pentanone (Z′), nitropropane (U′), and pyridine (S′) with the overall difference in the Kovats retention index (ΔI).

$$\Delta I = aX' + bY' + cZ' + dU' + eS' \qquad (4.1)$$

The value of each phase constant (i.e., X′, Y′, Z′, U′, and S′) is determined by subtracting the retention index of the probe on a squalane stationary phase (I_{SQ}) from the retention index of the probe on the stationary phase being characterized (I_{TP}). For example, the phase constant of benzene (X′) would be calculated as shown in Equation 4.2.

$$X' = \Delta I \text{ (benzene)} = I_{TP}\text{(benzene)} - I_{SQ}\text{(benzene)} \qquad (4.2)$$

To complete the determination of all constants, a value of 1 is assigned to each of the test solutes. In the case of benzene, X′, its coefficient from Equation 4.1 would have the value of a = 1, whereas the remaining coefficients would be set equal to zero (i.e., b = 0, c = 0, d = 0, and e = 0). This process is repeated for the remaining solutes. The magnitude of each phase constant indicates the importance of the interaction in solute retention. Additionally, the overall polarity of the stationary phase can be determined by taking the average of all five phase constants.

Armstrong and coworkers determined the Rohrschneider–McReynolds constants for two imidazolium-based ILs: [C_4C_1Im]Cl and [C_4C_1Im][PF_6] [7]. The structures and physicochemical properties of these two ILs are shown in Table 4.1. A comparison of the phase constants for [C_4C_1Im][PF_6] and [C_4C_1Im]Cl to two common commercial GC stationary phases, namely, DB-5 (phenylmethylpolysiloxane; 5% phenyl) and OV-22 (phenylmethyldiphenyl-polysiloxane; 65% phenyl) indicates that the average polarity of the two ILs is very similar to that of the OV-22 stationary phase. Both ILs exhibited significant proton accepting and dipolar interactions with solute molecules. It was also observed that different anions influenced the magnitude of the phase constants, but did not affect the overall polarity of the stationary phase. Despite the fact that the overall polarities of the studied ILs were similar to each other and other polysiloxane stationary phases, their separation selectivities were very different. Interestingly, both ILs separated nonpolar analytes like a nonpolar stationary phase and polar analytes like a polar stationary phase. This *dual-nature* selectivity makes them unique and powerful stationary phases for separating a wide variety of different analytes. In the case of the [C_4C_1Im]Cl, analytes such as alcohols and carboxylic acids were tenaciously retained, owing to the hydrogen bond basic properties of the anion.

The Rohrschneider–McReynolds approach is helpful in illustrating differences between ILs in terms of the types of interactions they exhibit with

Table 4.1 Names, Structures, and Physicochemical Properties of Selected Ionic Liquids Evaluated as Gas–Liquid Chromatographic Stationary Phases

Number	Ionic Liquid	Melting Point (°C)	System Constants From Solvation Parameter Model	Viscosity (cP)	Surface Tension (dyne/cm)	Thermal Stability	Comments
1	[C$_4$C$_1$Im]Cl (structure) Cl$^-$ 1-butyl-3-methylimidazolium chloride	Solid 65 [45]	40°C [8]: $r = 0.24$, $s = 2.25$, $a = 7.03$, $b = 0$, $l = 0.63$ 70°C [8]: $r = 0.29$, $s = 2.00$, $a = 5.23$, $b = 0$, $l = 0.45$	Solid	54 (28°C)	~145°C [21]	(1) Exhibits poor coating characteristics on bare wall glass capillary columns (2) Tenaciously retains alcohols and carboxylic acids (3) PAF* = 1.45
2	[C$_4$C$_1$Im][Tf$_2$N] (structure) (CF$_3$SO$_2$)$_2$N$^-$ [NTf$_2^-$] 1-butyl-3-methylimidazolium bis[(trifluoromethyl)sulfonyl]imide	−4 [45]	40°C [8]: $r = 0$, $s = 1.89$, $a = 2.02$, $b = 0.36$, $l = 0.63$ 70°C [8]: $r = 0$, $s = 1.67$, $a = 1.75$, $b = 0.38$, $l = 0.56$	52 (20°C) [45]		~185°C [21]	(1) Exhibits tailing of alcohols and carboxylic acids (2) PAF* = 2.50
3	[C$_4$C$_1$Im][TfO] (structure) CF$_3$SO$_3^-$ 1-butyl-3-methylimidazolium trifluoromethanesulfonate (triflate)	16 [45]	40°C [8]: $r = 0$, $s = 1.86$, $a = 3.02$, $b = 0$, $l = 0.61$ 70°C [8]: $r = 0$, $s = 1.73$, $a = 2.71$, $b = 0$, $l = 0.52$	90 (20°C) [45]		~175°C [21]	PAF* = 1.08
4	[C$_4$C$_1$Im][PF$_6$] (structure) PF$_6^-$ 1-butyl-3-methylimidazolium hexafluorophosphate	−8 [45]	40°C[8]: $r = 0$, $s = 1.91$, $a = 1.89$, $b = 0$, $l = 0.62$ 70°C[8]: $r = 0$, $s = 1.70$, $a = 1.58$, $b = 0$, $l = 0.52$	312 (20°C) [45]	42.9 (63°C) [46]	170°C [21]	(1) Ionic liquid produces hydrogen fluoride (HF) due to hydrolytic decomposition of PF$_6^-$ [24] (2) PAF* = 1.31

(continued)

Table 4.1 (Continued)

Number	Ionic Liquid	Melting Point (°C)	System Constants From Solvation Parameter Model		Viscosity (cP)	Surface Tension (dyne/cm)	Thermal Stability	Comments
5	(1S, 2R)-(+)-N,N-dimethylephedrinium bis[(trifluoromethyl)sulfonyl]imide	54 [37]	NA		Solid	NA	NA	(1) First chiral ionic liquid to be used as a gas chromatographic chiral stationary phase (2) IL begins to racemize at temperatures ≥140°C and loses enantioselectivity for some alcohols
6	[BeC$_1$Im][TfO] 1-benzyl-3-methylimidazolium triflate	27 [21]	70°C [21] $r = 0.21$ $s = 1.80$ $a = 2.62$ $b = 0.18$ $l = 0.48$	100°C [21] $r = 0$ $s = 1.70$ $a = 2.41$ $b = 0$ $l = 0.47$	NA	NA	~220°C [21]	Exhibits high selectivity for separation of polycyclic aromatic hydrocarbons, polychlorinated biphenyls, and aromatic sulfoxides
7	[MPC$_1$Im][TfO] 1-(4-methoxyphenyl)imidazolium triflate	45 [21]	70°C [21] $r = 0.54$ $s = 2.06$ $a = 2.83$ $b = 0.59$ $l = 0.40$	100°C [21] $r = 0.28$ $s = 2.05$ $a = 2.63$ $b = 0.16$ $l = 0.38$	NA	NA	~250°C [21]	Exhibits high selectivity for separation of polycyclic aromatic hydrocarbons, polychlorinated biphenyls, and aromatic sulfoxides

8		>−8, <0 [22]	40°C[22] $r = 0.27$ $s = 1.71$ $a = 1.98$ $b = 0.32$ $l = 0.62$	70°C [22] $r = 0.34$ $s = 1.52$ $a = 1.65$ $b = 0.35$ $l = 0.48$	502 (30°C) [22]	42.2 (23°C) [22]	>400°C [22]	Dicationic IL exhibiting high thermal stability
	NTf$_2^-$							
	1,9-di(N-methylpyrrolidinium)nonane *bis*[(trifluoromethyl)sulfonyl]imide							
9	[C$_6$vIm][Tf$_2$N] (CF$_3$SO$_2$)$_2$N$^-$	NA	NA	NA	NA	NA	NA	(1) Used as a monomer for producing immobilized ionomer films for high-temperature separations (2) Monomer readily undergoes polymerization in presence of light and heat
	1-vinyl-3-hexylimidazolium *bis*[(trifluoromethyl)sulfonyl]imide							
10	[C$_9$(vim)$_2$]2[Tf$_2$N] NTf$_2^-$ NTf$_2^-$	NA	Partially crosslinked stationary phase [42] 0.10% C$_9$(vim)–NTf$_2$; 0.10% C$_9$(vim)$_2$–NTf$_2$; 40°C $r = 0$ $s = 1.60$ $a = 1.84$ $b = 0.45$ $l = 0.71$ 70°C $r = 0$ $s = 1.57$ $a = 1.53$ $b = 0.37$ $l = 0.60$	NA	NA	NA	NA	(1) Used as a crosslinker for producing immobilized ionomer films for high-temperature separations (2) Crosslinker readily undergoes polymerization in presence of light and heat
	1,9-di(3-vinylimidazolium)nonane *bis*[(trifluoromethyl)sulfonyl]imidate							

probe molecules. However, the model has deficiencies that preclude it from fully characterizing individual solvation interactions. The method utilizes probe molecules that are too volatile (e.g., benzene, 2-pentanone, and nitropropane) and often elute with the dead volume of the column or exhibit short retention times. In addition, the retention of probe molecules is not dictated by a single solvation interaction but is most often due to several *simultaneous* interactions. Therefore, using this method, it is not possible to deconvolute the phase constants into individual solvation interactions.

4.2.3   Solvation parameter model

To further understand the solvation interactions imparted by the cation and anion, a method must be capable of describing more than a single polarity, solvent strength, or phase constant. A single parameter polarity scale has the disadvantage of describing a weighted average of all solute–solvent interactions [8]. The solvation parameter model (linear solvation free energy relationship [LSFER] model), developed by Abraham [9,10], has been used to characterize liquid- or gas-phase interactions between solute molecules and liquid phases. The solvation parameter model, as described by Equation 4.3, is a linear free energy relationship that involves the following three-step solvation process of a solute: (1) a cavity of suitable size is created in the solvent (IL); (2) the solvent molecules reorganize around the formed cavity; and (3) the solute molecule is introduced into the cavity followed by solute–solvent interactions. The model possesses several inherent characteristics that make it a valuable tool for characterizing complex solvation interactions such as those that take place between a solute molecule and an IL. The model utilizes a large number of probe molecules that are capable of undergoing a multitude of solvation interactions with the stationary phase solvent. Each probe molecule is broken down into individual solute descriptors which define the magnitude of each possible interaction. Solute descriptors are currently available for hundreds of probe molecules [9]. These solute descriptors are defined as: E is an excess molar refraction calculated from the solute's refractive index; S is the solute dipolarity/polarizability; A and B are the solute hydrogen bond acidity and hydrogen bond basicity, respectively; and L is the solute gas–hexadecane partition coefficient at 298 K. In gas–liquid chromatography (GLC), the dependent variable is $\log k$, the adjusted relative retention time. A judicious selection of probe molecules with overlapping interactions allows for the examination of all possible solute–solvent interactions.

$$\log k = c + eE + sS + aA + bB + lL \tag{4.3}$$

The solute descriptors for each analyte and the corresponding retention factor (measured at a specific temperature) are subjected to multiple linear regression analysis (MLRA) to obtain the system constant coefficients of each solute descriptor in Equation 4.3 (i.e., e, s, a, b, and l). The system constants from Equation 4.3 are defined as: e is the ability of the IL to interact via solute π- and n-electrons; s is a measure of the dipolarity/polarizability of the IL; a defines the IL hydrogen bond acidity; b is a measure of the hydrogen bond basicity of the IL; and l describes the dispersion forces. The magnitude of the dispersion forces term indicates how well the IL will separate homologs in a homologous series (e.g., n-alkanes).

In initial work, a total of 17 different ILs were evaluated by the solvation parameter model [8]. Ten of these ILs were comprised of imidazolium or pyrolidinium cations paired with different anions. Many of these compounds represent the *traditional class* of IL solvents that have been used extensively in organic synthesis reactions or in other analytical uses. The remaining seven ILs consisted of substituted ammonium cations that have proven to be successful analyte matrices in matrix-assisted laser desorption ionization (MALDI) mass spectrometry [11].

To delineate the effect of the anion on IL–solute interactions, ILs with different anions and the same cations were examined. Table 4.1 lists the system constants obtained for ILs composed of the $[C_4C_1Im]$ cation paired with various anions (ILs 1–4). The most dominant solvation interactions observed in IL solvents include strong dipolarity (s), hydrogen bond basicity (a), and moderate-to-high cohesive forces (l) which were relatively constant for all ILs examined, but were slightly larger for ILs containing longer alkyl substituents. The hydrogen bond basicity (a) was extremely high for the $[C_4C_1Im]Cl$. This observation is fully supported by the fact that $[C_4C_1Im]Cl$ tenaciously retains analytes that are proton donors (i.e., alcohols and carboxylic acids) [7].

By changing the nature of the cation while maintaining the same anion, the hydrogen bond basicity was observed to remain constant, indicating that this solvation interaction is predominantly anion-controlled. None of the common ILs studied exhibited appreciable hydrogen bond acidity (b-term). The extent of solute interactions via π–π and n–π electrons (r-term) with the IL is primarily dependent on two factors: (1) the nature of the cation and anion combination (e.g., $[C_4C_1Im]Cl$) and (2) the length of the alkyl chain incorporated on the cation (e.g., 1-octyl-3-methylimidazolium and 1-hexyl-3-methylimidazolium cations). The latter ILs exhibited sufficient abilities at retaining analytes that contained electron-rich aromatic systems.

4.2.4   Measurement of ionic liquid thermodynamic parameters

GC employing IL stationary phases has proven to be a powerful tool for the determination of important thermodynamic properties involving mixtures of ILs and organic solvents. Heinz and Verevkin have published an extensive number of thermodynamic parameters for pyridinium- and imidazolium-based ILs using a variety of organic solutes [12–14]. Paramount of all thermodynamic parameters is the activity coefficient at infinite dilution (γ^∞), which provides information about the intermolecular energy between a solute and a solvent. Activity coefficients at infinite dilution are important for characterizing the behavior of liquid mixtures, predicting the existence of azeotropes, estimation of mutual solubilities, and the selection of solvents for extraction and extractive distillation.

To determine the activity coefficient at infinite dilution, packed columns were prepared containing the IL-coated chromosorb support [12–14]. The examined mixture of probe molecules was then injected under infinite dilution conditions. The value of γ^∞ can be calculated from Equation 4.4, as defined by Cruickshank and coworkers [15]:

$$\ln \gamma_{i,3}^\infty = \ln \left(\frac{n_3 RT}{V_N p_1^0} \right) - \frac{B_{11} - V_1^0}{RT} p_1^0 + \frac{2B_{12} - V_1^\infty}{RT} J p_0 \qquad (4.4)$$

where $\gamma_{i,3}^\infty$ is the activity coefficient of solute i at infinite dilution in the stationary phase (index 3); p_1^0 is the vapor pressure of the pure liquid solute; p_0 is the outlet pressure; n_3 is the number of moles of the stationary phase on the column; V_N is the standardized retention volume; B_{11} and B_{12} are the second viral coefficient of the solute and the mixed viral coefficient of the solute (1) with the carrier gas (2), respectively; V_1^0 and V_1^∞ are the liquid molar volume of pure solute and the partial molar volume of solute in the IL at infinite dilution; and J is the compressibility factor.

The partial molar excess enthalpy at infinite dilution, $H_i^{E,\infty}$, can be determined by Equation 4.5.

$$\left(\frac{\partial \ln \gamma_i^\infty}{\partial (1/T)} \right) = \frac{H_i^{E,\infty}}{R} \qquad (4.5)$$

The magnitude of γ^∞ is highly dependent on the structural composition of the IL. Notable trends of these two thermodynamic parameters have been observed for various solute mixtures in IL solvents.

- *Alkanes, alkenes, and alkylbenzenes in [4-C_1C_4py][BF_4].* Activity coefficients of linear *n*-alkanes increase with increasing chain length. The branching of the alkane skeleton reduces γ^∞ in comparison to equivalent normal hydrocarbons. Values of γ^∞ for benzene and alkylbenzene are lower in comparison to those of alkanes and alkenes. Intermolecular interactions between the IL and the solute become stronger with an increasing number of polarizable electrons present in double bonds and aromatic rings [12].

- *Polar solutes in [C₂C₁Im][Tf₂N] and [C₂C₁C₁Im][Tf₂N].* γ^{∞} for *n*-alkanols increase with increasing chain length. Branching of the alkane structure reduces the value of γ^{∞} compared to the linear alcohol. The more polar the solute is, the higher the solubility in the IL. Values of $H_i^{E,\infty}$ increase with increasing alkyl chain length. The introduction of double bonds decreases the value of $H_i^{E,\infty}$. Molecules containing an aromatic ring possess negative values of $H_i^{E,\infty}$, but values increase when larger alkyl groups are substituted on the aromatic ring. Values for $H_i^{E,\infty}$ are negative for acetone, ethyl acetate, and trichloromethane [13].

4.3   Chromatographic characteristics of ionic liquid stationary phases: a comprehensive guide into tuning the structure of ILs for gas chromatographic separations

Poole and coworkers were the first to evaluate liquid organic salts as stationary phases in GLC in the early 1980s [16–19]. Ethylpyridinium bromide, possessing a melting point of 110°C, was used above this temperature to selectively separate various organic compounds containing large dipoles or functional groups capable of hydrogen bonding. Below the melting point, this stationary phase acted as an adsorbent and exhibited poor efficiency and peak asymmetry [16]. The temperature range of the molten salts was expanded by evaluating new classes of stationary phases including tetra-*n*-hexylammonium benzoate, 1-methyl-3-ethylimidazolium chloride, tri-*n*-butylbenzylphosphonium chloride, and tetra-*n*-butylammonium tetrafluoroboroate [18]. Whisker-walled and sodium chloride modified surfaces were used in wall-coated open tubular (WCOT) columns to improve the wetting ability of the molten salts on the glass surface [19].

In addition to providing highly selective separations, there are a multitude of other desired characteristics that a gas chromatographic stationary phase should possess. These properties include high viscosity, low surface tension allowing for wetting of the fused silica capillary wall, high thermal stability, and low vapor pressure at elevated temperatures. The stationary phase solvent should also not exhibit unusual mass transfer behavior.

The remaining sections of this chapter will address the recent development of imidazolium- and pyrrolidinium-based ILs as new classes of high-stability and high-selectivity stationary phases. The structural makeup of the cation/anion and their unique combination will be discussed in detail to provide insight into how IL stationary phases can be developed for specific separations and applications. While this discussion will focus primarily on imidazolium- and pyrrolidinium-based IL stationary phases that have been presented in the literature, a working knowledge of how to *design* or *tune* an

IL to possess desired physical properties in addition to providing optimal chromatographic results will be gained.

4.3.1   Thermal stability characteristics of ionic liquids

One of the most chromatographically useful properties inherent to many ILs lies with their high thermal stabilities. In GC, the thermal stability of the stationary phase is one of the most important considerations as it governs the onset of column bleed and ultimately dictates the lifetime of the stationary phase. Column bleed refers to the volatilization/decomposition of the stationary phase at high temperatures. The characteristic rising baseline and plateau effect observed in gas chromatographic separations at high temperature is attributed to the bleeding of the stationary phase. While column bleed is observed on all stationary phases at high temperatures, the degree of bleeding ultimately determines the lifetime of the column and, when excessive, has a detrimental effect on the efficiency of the column.

It is well known that halide anions lower the thermal stability of ILs due to their nucleophilic nature and their ability to decompose by S_N1 or S_N2 nucleophilic decomposition [20]. ILs that are not thoroughly purified and examined using ion chromatography for the presence of trace levels of halide anions will produce significant column bleed at relatively low-column temperatures. Excessive decomposition/volatilization of $[C_4C_1Im]Cl$ has been observed starting at 145°C [21].

For ILs based on the 1,3-dialkyl imidazolium cation, it was found that the thermal stability can be increased to 220–250°C by incorporating large, bulky cations paired with triflate anions [21]. The structure of these two ILs is shown in Table 4.1 as structures 6 and 7. In addition to exhibiting higher thermal stabilities compared to imidazolium-based ILs employing short alkyl chains, improved control over the separation selectivity was demonstrated for solutes containing nonbonding and π-electrons as well as hydrogen bond basic character. This is likely due to the ability of the electron-rich phenyl ring in IL 7 being able to interact strongly via $\pi-\pi$ interactions with aromatic solutes, whereas the phenyl ring in IL 6 is insulated from the imidazolium ring by the methylene group, slightly decreasing its propensity for $\pi-\pi$ interactions. This example nicely demonstrates the ability of tuning two ILs for enhanced thermal stability while also producing two stationary phases that exhibit quite unique selectivities.

Most recently, it was found that the thermal stability of geminal dicationic ILs is considerably higher than their monocationic analogs [22]. Figure 4.1 illustrates the increase in thermal stability of dicationic ILs containing the $[Tf_2N]$ anion (traces D–G) compared to the monocationic 1-butyl-3-methylimidazolium ILs containing Cl^-, PF_6^-, and Tf_2N^- anions.

Shreeve and coworkers have recently synthesized a class of polyethylene glycol functionalized dicationic ILs that exhibit high thermal stabilities and unique tribological behavior [23].

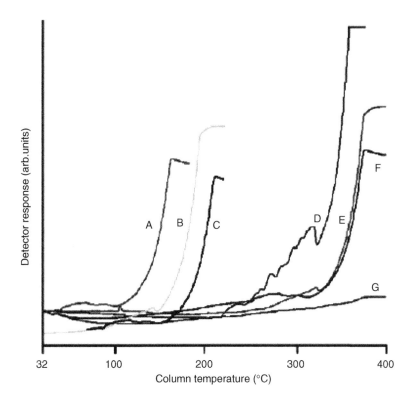

Figure 4.1 Thermal stability diagram obtained by coating seven ILs onto the capillary wall followed by heating and detection of volatilization/decomposition products using a flame ionization detector. The geminal dicationic ILs (D–G) have higher thermal stabilities than those of the traditional monocationic ILs (A–C): (A) $[C_4C_1Im]Cl$; (B) $[C_4C_1Im][PF_6]$; (C) $[C_4C_1Im][Tf_2N]$; (D) $[C_9(C_4pyr)_2]2[Tf_2N]$; (E) $[C_9(C_1Im)_2]2[Tf_2N]$; (F) $[C_{12}(BeIm)_2]2[Tf_2N]$; (G) $[C_9(C_1pyr)_2]2[Tf_2N]$. (From Anderson, J.L., Ding, R., Ellern, A., and Armstrong, D.W., *J. Am. Chem. Soc.*, 127, 593–604, 2005. With permission.)

4.3.2 Development of highly viscous stationary phases

Ideal stationary phases in GC possess high viscosities as well as exhibit small changes in viscosity with large ranges in temperature. To attain high-viscosity ILs that exhibit high thermal stability, several trade-offs must be considered. As observed in Table 4.1, ILs possessing the highest viscosities typically contain halide anions that exhibit superior hydrogen bond coordinating behavior. However, these ILs tend to exhibit lower thermal stabilities (see Section 4.3.1) and tenaciously retain analytes capable of hydrogen bonding to the basic stationary phase. ILs possessing $[PF_6]$ anions tend to be viscous, but these salts tend to undergo partial hydrolytic decomposition to produce hydrogen fluoride (HF) and other complex oxo- and hydroxo-phosphates [24]. Therefore, ILs containing the triflate and *bis*[(trifluoro-methyl)sulfonyl]imide anions tend to be the most popular due to the fact that they typically exhibit high thermal stabilities. Despite the lower viscosities and melting points of ILs comprised of these two anions, the viscosities can typically be increased by slight modifications to the cation structure. An example of such a modification is illustrated by ILs 2 and 8 in Table 4.1 where an order of magnitude increase in viscosity is observed for the dicationic IL compared to the monocationic IL.

4.3.3   Surface tension modifications for producing wall-coated open tubular columns

To carry out gas chromatographic separations in WCOT columns, it is vital that the stationary phase be evenly coated as a thin film on the wall of the capillary column. In addition, the stationary phase should not form droplets during the coating process or chromatographic separation when the oven temperature is rapidly varied. Obtaining highly efficient columns using the static coating method is truly an art, but matching the polarity of the IL to that of the capillary wall is the most important step to producing success-fully coated capillary columns. Fortunately, the tunability of ILs allows for structural modifications that make them highly soluble in the coating solvent in addition to the ability to wet the surface of glass. The surface tension of the IL is crucial in forming a thin and homogeneous coated layer. To obtain optimum wetting of the capillary wall for untreated fused silica, ILs should possess surface tension values in the range of 30–50 dyne/cm [25]. According to Table 4.1, ILs containing halide anions (e.g., [C_4C_1Im]Cl) exhibit the highest values of surface tension whereas ILs containing the [Tf_2N] anion gener-ally exhibit the lowest surface tension values. Indeed, it has been observed experimentally that two separate WCOT columns incorporating [C_4C_1Im]Cl and [C_4C_1Im][Tf_2N] stationary phases can differ in efficiencies by up to 1500 plates/m [7,21]. ILs that have high surface tensions and do not sufficiently wet the capillary wall may be coated onto capillaries that have been subjected to various salt pretreatment methods. Such pretreatment methods have shown promise in producing higher-efficiency columns [7,19].

   Another approach to reducing the surface tension as well as incorporating additional solvation properties into the IL can be carried out by dissolving surfactants in the IL [26,27]. Above the critical micelle concentration (CMC), it has been shown that surfactants form micellar aggregates which can be used to provide unique chromatographic selectivity [26,28]. This approach is described in more detail in Section 4.5 of this chapter. The surfactant can be added at concentrations of up to three times the CMC of the particular IL/surfactant mixture resulting in a lowering of the surface tension to nearly 35–38 dyne/cm, constituting an approximate 30% decrease in the surface tension for most ILs. This reduction in surface tension plays a key role in producing coated capillary columns with homogeneous films that exhibit column efficiencies typically over 3250 plates/m [26,28].

4.3.4 Anion and cation effects on chromatographic selectivity

Separation selectivity is one of the most important characteristics of any chromatographic stationary phase. The functionality of the cation and anion and their unique combinations result in ILs with not only tunable physicochemical properties (i.e., viscosity, thermal stability, and surface tension), but also unique separation selectivities. Although the selectivity for different analytes is dominated by the solvation interactions imparted by the cation and anion, all ILs exhibit an apparent and unique *dual-nature* selectivity that is uncharacteristic of other popular nonionic stationary phases. Dual-nature selectivity provides the stationary phases the ability to separate nonpolar molecules like a nonpolar stationary phase but yet separate polar molecules like a polar stationary phase [7,8]. Typically, GC stationary phases are classified in terms of their polarity (see Section 4.2.2) and the polarity of the employed stationary phase should closely match that of the analytes being separated. ILs possess a multitude of different but simultaneous solvation interactions that give rise to unique interactions with solute molecules. This is illustrated by Figure 4.2 in which a mixture of polar and nonpolar analytes are subjected to separation on a 1-benzyl-3-methylimidazolium triflate ([BeC$_1$Im][TfO]; IL 6 in Table 4.1) column [21].

From this chromatogram, the nonpolar alkanes (shown as peaks 4, 6, 9, and 10), which presumably are only able to interact via dispersion interactions, are well separated with high efficiencies. However, solutes such as aromatics, alcohols, and carboxylic acids are also separated with high selectivity but exhibit longer retention times with lower efficiencies.

A thorough and systematic evaluation of imidazolium-based ILs with different combinations of cations and anions revealed that the anion contributes the most to the unique selectivity exhibited by the IL stationary phase [21]. In addition, the anion is observed to play a critical role in the separation efficiency of various solutes. This behavior is accentuated for solutes capable of hydrogen bonding. ILs containing anions that are moderate to highly hydrogen bond basic (i.e., Cl$^-$ and [Tf$_2$N]$^-$) produce considerable peak asymmetry when separating analytes such as alcohols and carboxylic acids [21]. A measure of peak asymmetry is given by the peak asymmetry factor (PAF). A PAF value of 1.0 is assigned to a peak exhibiting no asymmetry. The PAF of *p*-cresol at 60°C is shown in Table 4.1 for four ILs: [C$_4$C$_1$Im]Cl, [C$_4$C$_1$Im][Tf$_2$N], [C$_4$C$_1$Im][TfO], and [C$_4$C$_1$Im][PF$_6$]. Although ILs based on the [Tf$_2$N] anion typically exhibit high thermal stabilities, large PAF values are common for analytes such as alcohols and acids when they are subjected to stationary phases employing this anion. ILs possessing the triflate or PF$_6^-$ anion typically exhibit acceptable PAFs for similar solutes.

Recent work by Davis and coworkers [29] has shown that sulfone and sulfoxide functionalized imidazolium cations are able to enhance particular

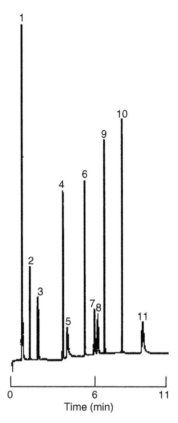

Figure 4.2 Chromatogram illustrating the dual-nature separation selectivity exhibited by IL stationary phases in the separation of a mixture of polar and nonpolar molecules. 1, CH_2Cl_2; 2, methyl caproate; 3, octyl aldehyde; 4, dodecane; 5, octanol; 6, tridecane; 7, naphthalene; 8, nitrobenzene; 9, tetradecane; 10, pentadecane; and 11, octanoic acid. Conditions: 80°C for 3 min, 10°C/min to 130°C on 10 m [BeC$_1$Im][TfO] column. (From Anderson, J.L. and Armstrong, D.W., *Anal. Chem.*, 75, 4851–4858, 2003. With permission.)

solvation interactions with solute molecules, most notably the hydrogen bond acidity and basicity. Therefore, it may be possible to use the same approach to design ILs in which particular solvation interactions are not dominated independently by the cation or anion but are rather imparted in a cooperative fashion by both components.

4.3.5   *Ionic liquid purity and chromatographic performance*

It is well established that small levels of halide impurities in ILs can limit their applications in various catalytic processes [1,30,31]. The minimization of halide and water content in the IL constitute two vital purification steps that must be considered prior to coating or immobilizing the stationary phase. The presence of contaminating halide anions will result in the tenacious retention of alcohols and carboxylic acids as well as to lower the thermal stability of the IL. ILs that are subjected to coating or immobilization should possess water content below 30–50 ppm, as typically measured by Karl Fischer titration. The presence of water during the coating or immobilization step may produce pooling of the stationary phase as well as alter the wetting ability of the IL to the capillary wall. Moreover, it has been observed that when ILs with high water content are coated onto WCOT columns by the static coating method, the evaporation interface will cease movement, resulting in a failure in the coating process. Trace levels of any molecular solvents (e.g., isopropanol, ethyl acetate, and ethanenitrile) commonly used in the preparation or cleanup of ILs will result in poorly coated capillary columns exhibiting inferior separation efficiency.

Synthetic routes employed to obtain the IL product generally take several forms: (1) quaternization of a tertiary amine (e.g., 1-methylimidazole and tributylamine) by an alkylhalide followed by metathesis exchange of the anion; (2) treatment of precursor halide from method 1 above with a silver salt to form the water-immiscible silver halide; (3) acid–base neutralization; (4) direct alkylation of a phosphine and amine by an alkyl triflate or dialkyl sulfate. Methods 3 and 4 typically produce halide-free ILs, but the range of ILs that can be made by these steps are limited. To eliminate the presence of halides, especially in methods 1 and 2 above, it is imperative that the IL be extensively purified [32] and characterized prior to its application as a stationary phase. Halide impurities can be quantitatively determined by the Volhard method using chloride selective electrodes [32], spectrophotometric methods using fluorescent probes [33], ion chromatography [8,34,35], and electrochemistry [36]. Hardacre and coworkers have demonstrated the limit of quantitation for chloride to be <8 ppm using ion chromatography with suppressed conductivity detection [35].

4.4   Ionic liquids as chiral stationary phase solvents

The ability to design chiral ILs in which the cation and anion is of fixed chirality represents additional tuning features of ILs. Two approaches have incorporated ILs as new stationary phases for chiral GC. One method involves the use of chiral ILs as stationary phases in WCOT GC [37]. In the second approach, chiral selectors (e.g., cyclodextrins) were dissolved in an achiral IL and the mixture coated onto the wall of the capillary column [38]. Both approaches can separate a variety of different analytes, but the observed enantioselectivities and efficiencies do not rival those observed with commercially available chiral stationary phases (CSPs).

Armstrong, Welton, and coworkers [37] demonstrated the first application of a chiral IL (see IL 5 in Table 4.1) as a CSP. The chiral ILs evaluated were based on the *N,N*-dimethylephedrinium cation in which the configuration

of the two stereogenic centers were varied to give rise to the (1S,2R)-(+)-
N,N-dimethylephedrinium, (1R,2S)-(−)-N,N-dimethylephedrinium, and (1S,2S)-
(+)-N,N-dimethylephedrinium analogs paired with the *bis*[(trifluromethyl)
sulfonyl]imide anion. The CSPs were found to exhibit enantioselectivity for
four classes of analytes: chiral alcohols (including diols), chiral sulfoxides,
chiral epoxides, and acetylated amines. As expected, the elution order of the
chiral analytes could be reversed by employing the opposite enantiomer of
the chiral IL. It was observed that the configuration of the chiral center played
a significant role in the observed enatioselectivity. For example, the (1S,2S) IL
exhibited no enantioselective separation of chiral alcohols but exhibited similar enantioselectivity to the (1S,2R) IL in the separation of chiral sulfoxides.
It was found that prolonged exposure of the IL to oven temperatures equal to
and more than 140°C resulted in enantioselectivity losses for some classes of
molecules whereas other analytes exhibited no changes. These enantioselectivity losses were attributed to the formation of a dehydration product. The
stereochemically fixed hydroxyl group was found to be imperative in separating all chiral analytes but is not sufficient alone for the enantioseparation
of chiral alcohols, epoxides, and acetylated amines.

A chiral selector can also be dissolved in the IL solvent and be subsequently
coated on the capillary wall [38]. In this approach, the achiral $[C_4C_1Im]Cl$
was used to dissolve permethylated β-cyclodextrin (β-PM) and dimethylated
β-cyclodextrin (β-DM). The chromatographic separations obtained from
these two columns were compared to two commercially available CSPs
based on β-PM and β-DM dissolved in polydimethylsiloxane. From a set of
64 chiral molecules separated by the commercial β-PM column, only 21 of
the molecules were enantioresolved by the IL-based β-PM column. Likewise,
from a collection of 80 analytes separated by the β-DM column, only 16
analytes could be separated on the IL-based β-DM column. The authors
also noted a considerable enhancement in the separation efficiency of the
IL-based CSPs. This result, coupled to the loss of enantioselectivity for most
separations, suggests that the imidazolium cation may occupy the cavity of
the cyclodextrin preventing the analyte–cyclodextrin inclusion complexation that is crucial for chiral recognition. The ability for ILs to form inclusion complexes with cyclodextrin molecules has been recently studied by
Tran and coworkers using near-infrared spectrometry [39].

4.5   Ionic liquid-based micellar gas chromatography

In the early work by Davis and co-workers, it was shown that fluorous
surfactants based on imidazolium cations could be developed to promote
the formation and stabilization of perfluorocarbons in conventional ILs [40].
It was later demonstrated that ILs exhibit solvatophobic interactions with
nonionic, anionic, and zwitterionic surfactant molecules prompting their
aggregation within the solvent [26–28]. Interestingly, the CMC values for the

studied surfactants were all higher than the CMC values for the same surfactants in aqueous solutions. The solvation properties of the IL–micellar solvents were determined by evaluating these phases using the solvation parameter model [26,28]. It was found that solute molecules partition to the micellar IL stationary phase in three ways: (1) partitioning to the monomer surfactants on the surface of the IL; (2) partitioning with the micelle in the bulk IL; (3) partitioning with the IL itself. To further evaluate the partitioning of analytes to the micelles in the IL, a three-phase model was developed treating the carrier gas (G), IL, and micelle (M) as the three phases in which an analyte can partition [28]. This allows for the determination of three partition coefficients (K_{GIL}, K_{GM}, and K_{ILM}) that can be determined based on the chromatographic data. Sodium dodecylsulfate (SDS) and dioctylsulfosuccinate (docSS), examples of anionic and zwitterionic surfactants, respectively, were evaluated in the [C_4C_1Im]Cl. In addition, the solvation effects of two nonionic surfactants, polyoxyethylene-100-stearylether (Brij 700) and polyoxyethylene-23-laurylether (Brij 35) were examined in [C_4C_1Im][PF_6]. To fully characterize and delineate the role of the micellar aggregates in altering the selectivity of the stationary phase, a large number of probe molecules capable of undergoing a variety of solvation interactions were used. It was found that most probe molecules examined exhibited highest partitioning to the micellar phase. Therefore, these analytes exhibit a proportional increase in retention as the number of micelles in the stationary phase is increased. It was also observed that the SDS and docSS surfactants increased the hydrogen bond basicity and dispersion-type interactions of the [C_4C_1Im]Cl. In an analogous manner, π–π, n–π, dipolar, and hydrogen bond basicity interactions were enhanced following the dissolution of Brij 35 and Brij 700 in [C_4C_1Im][PF_6]. The ability to modulate the type and magnitude of solvation interactions through the addition of surfactants further extends the utility of ILs as highly selective, dual-nature stationary phases.

4.6 Binary mixtures of ionic liquids as high-selectivity stationary phases

Mixed stationary phases have been widely used in GC due to the unique selectivity that is mostly not achieved with neat stationary phases. Although neat IL-based stationary phases exhibit unique selectivity compared to many nonionic stationary phases, it is not always possible to completely resolve all analytes during the separation, particularly in complex mixtures.

To examine the utility of stationary phases composed of IL mixtures, a complex mixture of alcohols (both cyclic and aliphatic) and analytes with aromatic functionality were subjected to separation [41] on a stationary phase consisting of the [C_4C_1Im][Tf_2N]. Under optimized conditions, the stationary phase was selective for most molecules, but exhibited poor resolution. Owing to the fact that most ILs containing the Tf_2N^- anion are weak hydrogen bond bases (see Table 4.1), columns were prepared by forming IL mixtures in which the stationary phase was enriched with different weight percentages of $[C_4C_1Im][Cl]$ and $[C_4C_1Im][Tf_2N]$. It was found that the hydrogen bond basicity increased linearly as the concentration of chloride anion (in the form of $[C_4C_1Im]Cl$) was increased. In addition, the retention factors of short-chained alcohols increased by as much as 1100% when the separation was carried out on the column containing the highest percentage of chloride anion compared to that of the neat $[C_4C_1Im][Tf_2N]$. Through tuning the composition of the stationary phase mixture, the separation selectivity and resolution of most analytes were varied. In addition, most alcohols exhibited a reversal of elution order as the stationary phase hydrogen bond basicity increased. The ability to tune the separation selectivity, resolution, and elution order of analytes in a systematic and predictable manner by utilizing stationary phases based on IL mixtures further extends the utility of ILs in gas chromatographic separations.

4.7   Immobilized ionic liquid-based stationary phases for high-temperature separations

The discussion of IL-based stationary phases up to this point has centered around ILs that are either coated as a thin film on a capillary wall or on a solid support. Although ILs exhibit a variety of properties that allow them to be unique stationary phases, their most significant drawback lies with their drop in viscosity with increasing temperature. This results in an increased propensity for flowing of the IL within the capillary, which often produces pooling of the stationary phase and nonuniform film thickness throughout the column. These factors often contribute to diminished analyte retention time reproducibility as well as detrimental effects on separation efficiency.

In an attempt to preserve the unique dual-nature selectivity of ILs while producing a stationary phase that is resilient to flowing at high temperatures, a method of immobilizing ILs as thin films in WCOT columns has been developed [42]. Figure 4.3 illustrates the steps used to form the immobilized IL stationary phase.

The desired IL monomers are mixed with a free-radical initiator (azobisisobutyronitrile, AIBN) in dichloromethane and coated onto the wall of the capillary column using the static coating method. The capillary is then sealed at both ends and heated to initiate polymerization. Finally, the capillary is unsealed and placed in a gas chromatograph and subsequently conditioned to remove any excess AIBN that did not completely decompose or react.

Three types of stationary phases based on the free-radical reaction of monocationic and dicationic vinyl-substituted imidazolium cations were studied [42]. Two examples of such monomers are given by ILs 9 and 10 in Table 4.1. The formation of a linear IL polymer stationary phase was performed by the free radical polymerization of monocationic monomers.

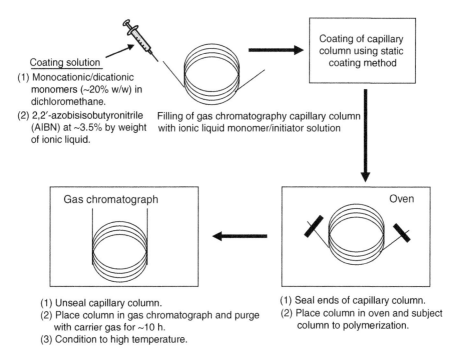

Figure 4.3 Scheme illustrating the development of immobilized ionic liquids by thermally induced free radical polymerization of vinyl-substituted imidazolium-based monocationic and dicationic monomers.

These stationary phases exhibited the lowest thermal stability, producing significant column bleed and dramatic decreases of efficiency above 300°C. Partially crosslinked stationary phases were synthesized by polymerizing a blend of monocationic and dicationic crosslinking monomers. Using this approach, the stationary phase consisting of an equal ratio (w/w) of ILs 9 and 10 (see Table 4.1) exhibited the best thermal stability up to ~280°C. The third matrix consisted of a completely crosslinked IL polymer formed by the polymerization of geminal dicationic IL monomers. This matrix exhibited high stability and enhanced efficiencies at separation temperatures over 350°C. By carefully choosing the appropriate monomers and AIBN-initiator concentration, stationary phases can be easily designed for separations encompassing various temperature ranges. For example, monocationic IL matrixes are best suited for low-temperature separations, whereas partially crosslinked and fully crosslinked stationary phases are ideal for moderate (100–280°C) and high (200–380°C) temperature gas chromatographic separations, respectively. An evaluation of the polymerized IL stationary phases by the solvation parameter model revealed that the solvation interaction parameters were largely unchanged compared to their monomeric analogs [42].

4.8   Choosing the appropriate ionic liquid stationary phase

The extensive range of available cations and anions could provide up to 10^{18} different ILs [43]. Indeed, this is a daunting number of ILs. This chapter has highlighted the important physicochemical and solvation properties that can be tuned by the employed cation/anion combination or structural changes to the cation and anion. This approach can be beneficial to those who are developing IL-based stationary phases for unique applications such as multidimensional GC, in which the IL column provides the unique separation selectivity [44].

Of the imidazolium-based IL stationary phases (both coated and immobilized forms) that have been presented in the literature and discussed in this chapter, their utility is dependent on either the application or the desired analytes to be separated. Figure 4.4 displays a flow diagram that can be followed to choose the appropriate IL from Table 4.1.

The fact that the ILs presented in this chapter have been largely developed for high-temperature separations in GC makes them also viable solvent systems in other high-temperature applications.

4.9   Conclusions and future directions

The various properties exhibited by ILs make them ideal stationary phases in GLC. ILs exhibit a unique dual-nature selectivity that allows them to separate polar molecules like a polar stationary phase and nonpolar molecules like a nonpolar stationary phase. In addition, the combination of cations and anions can be tuned to add further selectivity for more complex separations. Viscosity, thermal stability, and surface tension are vital properties that dictate the quality and integrity of the stationary phase coating and are additional characteristics that can be controlled when custom designing and synthesizing ILs. Furthermore, thermal stability and the integrity of stationary phase film can be improved by immobilizing the IL by free radical polymerization to form stationary phases suitable for low-, moderate-, and high-temperature separations. Chiral ILs have been shown to enantioresolve chiral analytes with reasonable efficiency.

The study of ILs in GLC has yielded important information regarding solute–solvent interactions providing valuable insights into their complex solvation interactions and thermodynamic properties for mixed solvent systems. Moreover, ILs have proven to be an important new class of stationary phases for the separation of a wide variety of different analytes. IL stationary phases will soon be commercially available which will inevitably promote further improvements in separation selectivity, thermal stability, immobilization bonding chemistry/stationary phase stability, and will broaden the range of separated compounds. IL-based stationary phases also hold great promise in GC mass spectrometry where the dual-nature selectivity of the stationary phase eliminates the need for frequent changing of columns.

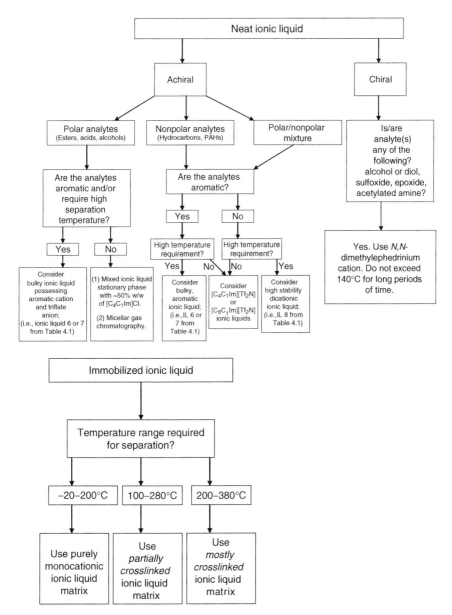

Figure 4.4 Flow diagram for choosing the appropriate neat ionic liquid or immobilized ionic liquid composition for a particular analyte separation. Note that the most important characteristics for choosing the appropriate stationary phase are separation selectivity and thermal stability. Both of these properties can be effectively tuned and optimized by controlling the cation and anion combination.

In addition, stationary phases that exhibit low bleed are extremely important for mass spectrometric applications as well as high-temperature separations.

Further research on mixed IL stationary phases will allow for the chromatographer to tune the stationary phase composition to provide enhanced control over the separation selectivity and analyte elution order, particularly for complicated analyte mixtures. The development of models that correlate analyte retention with the IL composition will prove useful for multidimensional GC. Micellar GC utilizing IL solvents presents an exciting class of highly selective stationary phases. The development of CSPs will likely mature as more chiral ILs are synthesized and evaluated from the chiral pool.

References

1. Welton, T., Room temperature ionic liquids. Solvents for synthesis and catalysis, *Chem. Rev.*, 99, 2071–2083, 1999.
2. Chiappe, C., and Pieraccini, D., Kinetic study of the addition of trihalides to unsaturated compounds in ionic liquids. Evidence of a remarkable solvent effect in the reaction of ICl_2, *J. Org. Chem.*, 69, 6059–6064, 2004.
3. Visser, A.E., Swatloski, R.P., Reichert, W.M., Mayton, R., Sheff, S., Wierzbicki, A., Davis, J.H., and Rogers, R.D., Task-specific ionic liquids incorporating novel cations for the coordination and extraction of Hg2+ and Cd2+: Synthesis, characterization, and extraction studies, *Environ. Sci. Technol.*, 36, 2523–2529, 2002.
4. Soutullo, M.D., Odom, C.I., Wicker, B.F., Henderson, C.N., Stenson, A.C., and Davis, J.H., Reversible CO_2 capture by unexpected plastic-, resin-, and gel-like ionic soft materials discovered during the combi-click generation of a TSIL library, *Chem. Mater.*, 19, 3581–3583, 2007.
5. Rohrschneider, L., Method of characterization for gas chromatographic separation of liquids, *J. Chromatogr.*, 22, 6–22, 1966.
6. McReynolds, W.O., Characterization of some liquid phases, *J. Chromatogr. Sci.*, 8, 685–691, 1970.
7. Armstrong, D.W., He, L., and Liu, Y.-S., Examination of ionic liquids and their interaction with molecules, when used as stationary phases in gas chromatography, *Anal. Chem.*, 71, 3873–3876, 1999.
8. Anderson, J.L., Ding, J., Welton, T., and Armstrong, D.W., Characterizing ionic liquids on the basis of multiple solvation interactions, *J. Am. Chem. Soc.*, 124, 14247–14254, 2002.
9. Abraham, M.H., Scales of solute hydrogen-bonding: Their construction and application to physicochemical and biochemical processes, *Chem. Soc. Rev.*, 22, 73–83, 1993.
10. Kollie, T.O., Poole, C.F., Abraham, M.H., and Whiting, G.S., Comparison of two free energy of solvation models for characterizing selectivity of stationary phases used in gas-liquid chromatography, *Anal. Chim. Acta*, 259, 1–13, 1992.
11. Armstrong, D.W., Zhang, L.K., He, L., and Gross, M.L., Ionic liquids as matrixes for matrix-assisted laser desorption/ionization mass spectrometry, *Anal. Chem.*, 73, 3679–3686, 2001.
12. Heintz, A., Kulikov, D.V., and Verevkin, S.P., Thermodynamic properties of mixtures containing ionic liquids. 1. activity coefficients at infinite dilution of alkanes, alkenes, and alkylbenzenes in 4-methyl-*n*-butylpyridinium tetrafluoroborate using gas-liquid chromatography, *J. Chem. Eng. Data*, 46, 1526–1529, 2001.

13. Heintz, A., Kulikov, D.V., and Verevkin, S.P., Thermodynamic properties of mixtures containing ionic liquids. 2. Activity coefficients at infinite dilution of hydrocarbons and polar solutes in 1-methyl-3-ethyl-imidazolium bis(trifluoromethyl-sulfonyl)amide and in 1,2-dimethyl-3-ethyl-imidazolium bis(trifluoromethyl-sulfonyl)amide using gas-liquid chromatography, *J. Chem. Eng. Data*, 47, 894–899, 2002.

14. Heintz, A., Kulikov, D.V., and Verevkin, S.P., Thermodynamic properties of mixtures containing ionic liquids. Activity coefficients at infinite dilution of polar solutes in 4-methyl-*N*-butyl-pyridinium tetrafluoroborate using gas-liquid chromatography, *J. Chem. Thermodyn.*, 34, 1341–1347, 2002.

15. Cruickshank, A.J.B., Windsor, M.L., and Young, C.L., The use of gas-liquid chromatography to determine activity coefficients and second virial coefficients of mixtures, *Proc. R. Soc. A*, 295, 259–270, 1966.

16. Pacholec, F., and Poole, C.F., Stationary phase properties of the organic molten salt ethylpyridinium bromide in gas chromatography, *Chromatographia*, 17, 370–374, 1983.

17. Poole, C.F., Butler, H.T., Coddens, M.E., Dhanesar, S.C., and Pacholec, F., Survey of organic molten salt phases for gas chromatography, *J. Chromatogr.*, 289, 299–320, 1984.

18. Dhanesar, S.C., Coddens, M.E., and Poole, C.F., Evaluation of tetraalkylammonium tetrafluoroborate salts as high-temperature stationary phases for packed and open-tubular column gas chromatography, *J. Chromatogr.*, 349, 249–265, 1985.

19. Dhanesar, S.C., Coddens, M.E., and Poole, C.F., Surface roughening by sodium chloride deposition for the preparation of organic molten salt open tubular columns, *J. Chromatogr. Sci.*, 23, 320–324, 1985.

20. Awad, W.H., Gilman, J.W., Nyden, M., Harris, R.H., Sutto, T.E., Callahan, J.H., Trulove, P.C., De Long, H.C., and Fox, D.M., Thermal degradation studies of alkyl-imidazolium salts and their application in nanocomposites, *Thermochim. Acta*, 409, 3–11, 2004.

21. Anderson, J.L., and Armstrong, D.W., High stability ionic liquids. A new class of stationary phases for gas chromatography, *Anal. Chem.*, 75, 4851–4858, 2003.

22. Anderson, J.L., Ding, R., Ellern, A., and Armstrong, D.W., Structure and properties of high stability germinal dicationic ionic liquids, *J. Am. Chem. Soc.*, 127, 593–604, 2005.

23. Jin, C.-M., Ye, C., Phillips, B.S., Zabinski, J.S., Liu, X., Liu, W., and Shreeve, J.M., Polyethylene glycol functionalized dicationic ionic liquids with alkyl or polyfluoroalkyl substituents as high temperature lubricants, *J. Mater. Chem.*, 16, 1529–1535, 2006.

24. Swatloski, R.P., Holbrey, J.D., and Rogers, R.D., Ionic liquids are not always green: hydrolysis of 1-butyl-3-methylimidazolium hexafluorophosphate, *Green Chem.*, 5, 361–363, 2003.

25. Alexander, G., and Rutten, G.A.F.M., Surface characteristics of treated glasses for the preparation of glass capillary columns in gas-liquid chromatography, *J. Chromatogr.*, 99, 81–101, 1974.

26. Anderson, J.L., Pino, V., Hagberg, E., Sheares, V.V., and Armstrong, D.W., Surfactant solvation effects and micelle formation in ionic liquids, *Chem. Commun.*, 2444–2445, 2003.

27. Fletcher, K.A., and Pandey, S., Surfactant aggregation within room-temperature ionic liquid 1-ethyl-3-methylimidazolium bis(trifluoromethylsulfonyl)imide, *Langmuir*, 20, 33–36, 2004.

28. Lantz, A.W., Pino, V., Anderson, J.L., and Armstrong, D.W., Determination of solute partition behavior with room-temperature ionic liquid based micellar gas-liquid chromatography stationary phases using the pseudophase model, *J. Chromatogr. A*, 1115, 217–224, 2006.

29. Sharma, N.K., Tickell, M.D., Anderson, J.L., Kaar, J., Pino, V., Wicker, B.F., Armstrong, D.W., Davis, J.H., and Russell, A.J., Do ion tethered functional groups affect IL solvent properties? The case of sulfoxides and sulfones, *Chem. Commun.*, 646–648, 2006.

30. Chauvin, Y., Mussmann, L., and Olivier, H., A novel class of versatile solvents for two-phase catalysis: hydrogenation, isomerization, and hydroformylation of alkenes catalyzed by rhodium complexes in liquid 1,3-dialkylimidazolium salts, *Angew. Chem. Int. Ed.*, 34, 2698–2700, 1996.

31. Wasserscheid, P., and Keim, W., Ionic liquids—new "solutions" for transition metal catalysis, *Angew. Chem. Int. Ed.*, 39, 3772–3790, 2000.

32. Seddon, K.R., Stark, A., and Torres, M.J., Influence of chloride, water, and organic solvents on the physical properties of ionic liquids, *Pure Appl. Chem.*, 72, 2275–2287, 2000.

33. Anthony, J.L., Maginn, E.J., and Brennecke, J.F., Solution thermodynamics of imidazolium-based ionic liquids and water, *J. Phys. Chem. B*, 105, 10942–10949, 2001.

34. Billard, I., Moutiers, G., Labet, A., El Azzi, A., Gaillard, C., Mariet, C., and Lutzenkirchen, K., Stability of divalent europium in an ionic liquid: spectroscopic investigations in 1-methyl-3-butylimidazolium hexafluorophosphate, *Inorg. Chem.*, 42, 1726–1733, 2003.

35. Villagran, C., Banks, C.E., Hardacre, C., and Compton, R.G., Electroanalytical determination of trace chloride in room-temperature ionic liquids, *Anal. Chem.*, 76, 1998–2003, 2004.

36. Xiao, L., and Johnson K.E., Electrochemistry of 1-butyl-3-methyl-1*H*-imidazolium tetrafluoroborate ionic liquid, *J. Electrochem. Soc.*, 150, E307–E311, 2003.

37. Ding, J., Welton, T., and Armstrong, D.W., Chiral ionic liquids as stationary phases in gas chromatography, *Anal. Chem.*, 76, 6819–6822, 2004.

38. Berthod, A., He, L., and Armstrong, D.W., Ionic liquids as stationary phase solvents for methylated cyclodextrins in gas chromatography, *Chromatographia*, 53, 63–68, 2001.

39. Tran, C.D., and De Paoli Lacerda, S.H., Determination of binding constants of cyclodextrins in room-temperature ionic liquids by near-infrared spectrometry, *Anal. Chem.*, 74, 5337–5341, 2002.

40. Merrigan, T.L., Bates, E.D., Dorman, S.C., and Davis, J.H., New fluorous ionic liquids function as surfactants in conventional room-temperature ionic liquids, *Chem. Commun.*, 2051–2052, 2000.

41. Baltazar, Q.Q., Leininger, S.K., and Anderson, J.L., Binary ionic liquid mixtures as gas chromatography stationary phases for improving the separation selectivity of alcohols and aromatics compounds, *J. Chromatogr. A.*, 1182, 119–127, 2008.

42. Anderson, J.L., and Armstrong, D.W., Immobilized ionic liquids as high-selectivity/high-temperature/high stability gas chromatography stationary phases, *Anal. Chem.*, 77, 6453–6462, 2005.

43. Carmichael, A.J., and Seddon, K.R., Polarity study of some 1-alkyl-3-methylimidazolium ambient-temperature ionic liquids with the solvatochromic dye, Nile Red, *J. Phys. Org. Chem.*, 13, 591–595, 2000.

44. Lambertus, G.R., Crank, J.A., McGuigan, M.E., Kendler, S., Armstrong, D.W., and Sacks, R.D., Rapid determination of complex mixtures by dual-column gas chromatography with a novel stationary phase combination and spectrometric detection, *J. Chromatogr. A*, 1135, 230–240, 2006.
45. Poole, C.F., Chromatographic and spectroscopic methods for the determination of solvent properties of room temperature ionic liquids, *J. Chromatogr. A*, 1037, 49–82, 2004.
46. Law, G., and Watson, P.R., Surface tension measurements of *n*-alkylimidazolium ionic liquids, *Langmuir*, 17, 6138–6141, 2001.

chapter five

Ionic liquids in liquid chromatography

Apryll M. Stalcup

Contents

5.1   Introduction

Ionic liquids (ILs) are increasingly being used in analytical chemistry [1–5]. Although the foundations for the application of ILs in gas chromatography has been fairly well-established through the pioneering work of Poole [6–10] and Armstrong [11–13], the applications of ILs in liquid chromatography (LC) are thus far fairly limited. This review will focus on liquid chromatographic applications related to ILs with an emphasis on imidazolium- or pyridinium-based ILs.

The analysis of ILs may afford considerable insight into the physico-chemical properties underlying the rich potential interaction chemistries of ILs [14] and suggest possibilities for future applications. Simultaneously, the unique features of ILs provide some intriguing new possibilities in the area of separations that have yet to be realized. Hence, topics to be covered in this chapter include analysis of ILs by LC, applications of ILs in liquid-phase microextraction (LPME), in high-performance LC (HPLC) as mobile-phase additives, and in capillary electrophoresis (CE) as buffer additives as well as applications of surface-confined ILs (SCIL) as novel stationary phases for LC.

5.2   Structural features of ionic liquids relevant to liquid chromatographic applications

As noted by Boehm and coworkers [15], with the exception of size exclusion chromatography, most chromatographic processes are ultimately the culmination of averaging and amplification of all the potential intermolecular interactions that a solute can undergo with a stationary phase. Hence, intelligent application of novel chromatographic media requires an inventory of all the potential structural features of both analytes and bonded ligands likely to play a role in these interactions between the analytes and the ligands. In most of the liquid chromatographic applications of SCILs, the emphasis has been on ILs containing dialkylimidazolium moieties. In this case, the IL cation can be thought of as a *fuzzy* ion with a delocalized charge. Thus, the presence of the cation/anion pair introduces the potential for electrostatic interactions. However, as will be shown, the imidazolium aromatic ring also promotes interactions with neutral aromatic species through hydrophobic and π–π interactions under aqueous conditions. Further, the imidazolium ring can also serve as a scaffold for introducing additional functionality (e.g., alkyl versus sulfonate), thereby providing additional interactions such as hydrophobic, steric, or electrostatic contributions. Thus, these materials offer a broad range of analyte–selector interaction chemistries that may afford more universal separation schemes to be realized than are presently available.

Thus far, emphasis in characterizing the role of the IL in separations has been in assessing the role of cation. Examination of the role of the anion on various physicochemical properties (e.g., viscosity and water solubility) [16] of ILs with identical cations suggest that the anions may be important as well but thus far, their role in chromatographic properties is less well understood.

5.3 Analysis of ionic liquids

As discussed in Chapter 1, the substitution of anions or cations can dramatically impact the chemical and physical properties of ILs. Further, the presence of small amounts of contaminants (e.g., water and halides) [17] can also dramatically alter the properties of ILs. Thus, assays for the analysis of ILs are critically important for their characterization. Given their nonvolatile nature, it is not too surprising that HPLC and CE play prominent roles in this area [18,19]. Although this topic is covered in considerably more detail elsewhere in this book, the emphasis here will be on what the analysis of ILs reveals about their potential intermolecular interactions which may be exploited in their liquid chromatographic applications.

5.3.1 Liquid chromatographic analysis of ionic liquids

5.3.1.1 Reversed-phase liquid chromatographic analysis of ionic liquids

Not surprisingly, ILs have been analyzed using alkyl-based reversed phase [20] as well as cation-exchange chromatography [21,22]. However, the rich potential interaction chemistries of ILs may also be inferred from the retention behavior of imidazolium- and pyridinium-based ILs on phenyl-based stationary phases which can supply aromatic π–π interaction capability [23]. The multiple modes of interaction between the imidazolium and the phenyl phase are illustrated in Figure 5.1. In this study, it was reported that the role of aromatic π–π interactions in the separations of IL cations could be mediated by the addition of acetonitrile.

Generally, IL assays focus on determining either the cation or the anion, independent of the counter-ion. However, it has been reported that in reversed-phase chromatography, different ILs sharing the same cation but different anions can be separated in salt-free mobile phases [20]. The differential interactions responsible for these separations could arise from a variety of mechanisms including differences in either the ion pair formation constants or ion pair lipophilicities. Differences in anion sorption into the stationary phase were also suggested to play a role as evidenced by the fact that retention of different ILs sharing the same cation but different anions correlated with the lipophilicity of the anion (e.g., hexafluorophosphate salt retained longer than the tetrafluoroborate). The reported elution order seemed to parallel the aqueous solubilities of the corresponding salts [24]. Nevertheless, the authors of this study cautioned against using this method for quantitation of ILs.

In reversed-phase chromatography, the addition of buffers or salts to the mobile phase reduced the separation of a mixture of ILs with different cations and anions to a separation based on the hydrophobicity of the cation [25]. Plots

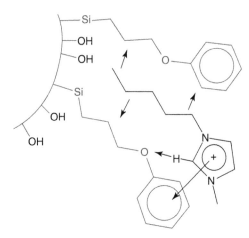

Figure 5.1 Scheme illustrating potential interactions between methylimidazolium cation and phenyl-based reversed-phase stationary phase. Arrows indicate hydrogen bonding, hydrophobic, and π–π interactions. Anion is not shown for clarity. (Adapted from Stepnowski, P., Nichthauser, J., Mrozik, W., and Buszewski, B., *Anal. Bioanal. Chem.*, 385, 1483–1491, 2006.)

of log k' versus carbon number in a homologous series of *n*-alkyl substituted imidazoliums were linear [20]. The impact of increasing the acetonitrile concentration in the mobile phase on retention of imidazolium ILs also followed classical reversed-phase behavior (e.g., reduced retention), consistent with another report on a reversed-phase chromatographic assay in which the addition of trifluoroacetic acid to the mobile phase was required to promote ion pair formation. Not surprisingly, pH had little impact on retention of these IL cations because their ionization is independent of pH.

5.3.1.2   *Ion-exchange liquid chromatographic analysis of ionic liquids*

Paralleling reports by Anderson and coworkers [26] in gas chromatographic applications of IL stationary phases, analysis of IL cations using a propyl sulfonate bonded phase on a silica-based support exhibited multimodal retention [21]. For instance, the influence of acetonitrile on retention for a series of related imidazolium and pyridinium cations was consistent with a reversed phase–type mechanism at low acetonitrile concentrations (e.g., decreasing retention with increasing acetonitrile concentration) but the influence of buffer concentration was consistent with an ion-exchange mechanism (e.g., reduced retention at high buffer concentration), particularly at high organic compositions [21].

5.4   *Ionic liquids in liquid-phase microextraction*

A variety of microextraction techniques have evolved in recent years to address the challenges inherent in the analysis of samples derived from

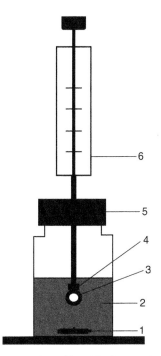

Figure 5.2 Schematic representation of liquid-phase microdroplet extraction setup. (1) Stir bar; (2) sample solution; (3) ionic liquid microdroplet; (4) polytetrafluoro-ethylene (PTFE) tube; (5) septum; (6) microsyringe. (Adapted from Liu, J.-F., Chi, Y.-G., Jiang, G.-B., Tai, C., Peng, J.-F., and Hu, J.-T., *J. Chromatogr. A*, 1026, 143–147, 2004.)

environmental [27] or biological sources [28]. Solid-phase microextraction (SPME) [29], one of the most commonly used and versatile of the micro-extraction techniques, employs a polymeric liquid-like phase coated onto a fused silica rod. SPME sample preparation techniques offer a number of advantages including the ability to sample both headspace and liquid samples as well as the ability to directly inject sample into gas or liquid chromatographic systems. Typical polymeric coatings used in SPME include polydimethylsiloxane and Carbowax/divinylbenzene.

Thus far, SPME methods have had only limited success in isolating polar organics (e.g., chlorophenols [30] and formaldehyde [31]) or ions [32,33] from aqueous mixtures. However, the tunable hydrophobicity and multimodal potential interaction chemistries of ILs suggest potential applications in LPME. Further, their high viscosity coupled with their minimal vapor pressure promotes stable droplet formation. Figure 5.2 illustrates the experimental setup for LPME. In addition, analyte recovery can be performed simply by injecting the droplet onto a liquid chromatographic column.

While work in the application of ILs to LPME is thus far limited, it was reported that although the viscosity of ILs generally allows for larger droplet size, partitioning of analytes into the droplet was inhibited because of their lowered diffusion as analytes enter the viscous IL droplet [34]. In addition, while increasing exposure of the IL microdroplet to aqueous solutions led to increased concentration of the analytes in the droplet, eventual dissolution of the droplet in the aqueous solution ultimately limited extraction times to ~30 min [31,34]. The finite solubility of the IL also limited the aqueous sample volume size.

5.5   Ionic liquids as mobile-phase additives in liquid chromatography

The intrinsic viscosity of ILs and their spectral properties largely preclude their application as neat mobile phases in LC. Thus, the trend has been to apply these materials as mobile-phase additives at low concentrations (e.g., 1–10 mM) [35] in the absence or presence of other organic modifiers (e.g., acetonitrile). Arguably, once the *IL* is dissolved in another solvent, it no longer constitutes a true *IL*. However, the IL paradigm offers a fresh approach for looking at intermolecular interactions in retention processes. Further, solution phase microheterogeneity is known to play a role in reversed-phase retention processes for nonpolar analytes in acetonitrile/water systems [36]. Solution-phase microheterogeneity manifested as ion pair formation certainly seems to be playing a role in the reversed-phase separation of ILs using salt-free mobile phases noted earlier [20]. However, ILs also offer the potential of solution-phase microheterogeneities through clathrate formation. Liquid clathrates containing ILs were recently reported to have formed between aromatic hydrocarbons and imidazolium salts [37]. Such host–guest interactions, creating an extended structure, can occur in the liquid state and could also play a role in separation processes using ILs as mobile-phase additives.

The IL mobile-phase additives are thought to play a variety of conventional roles in facilitating chromatographic separations, as well. In the case of reversed-phase chromatography, it is thought that ILs dynamically coat the stationary support, thereby masking residual silanols [38,39]. A related reversed-phase thin layer chromatographic study revealed that imidazolium tetrafluoroborate IL additives to the mobile phase were more effective at masking surface silanol effects on the chromatographic behavior of basic drugs than triethylamine, dimethyloctylamine, or ammonium hydroxide [40]. The greater efficacy of IL additives, relative to these amines, was ascribed to more effective sorption onto the stationary phase. The enhanced uptake was attributed to a combination of electrostatic interactions between the cation and the surface silanols augmented by hydrophobic interactions between the alkyl substituents on the IL cation with the alkyl ligands of the stationary phase. Indeed, increased lipophilicity of the IL cation provided better silanol-suppressing activity which was manifested in better peak shape, reduced tailing, and enhanced resolution (Figure 5.3). This approach

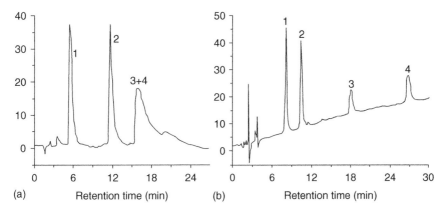

Figure 5.3 Chromatograms of amines with mobile phases containing (a) water adjusted to pH 3 with HCl and (b) 30 mmol/L 1-ethyl-3-methylimidazolium tetrafluoroborate. Chromatographic conditions: C18 column (5 μm, 150 × 4.6 mm ID); rate flow: 1.0 mL/min; detection at 254 nm. Peaks: (1) benzylamine; (2) benzidine; (3) *N,N*-dimethylaniline; and (4) *N*-ethylaniline. (Adapted from Xiaohua, X., Liang, Z., Xia, L., and Shengxiang, J., *Anal. Chim. Acta*, 519, 207–211, 2004.)

was found to be particularly advantageous for polar organics such as organic amines [39,41,42] and phenols [43]. However, this effect was mediated by the nature of the counter-ion [39].

Sorption of the IL cation:anion partners also modifies the stationary phase which can introduce an ion-exchange type of retention. Further, either of the IL partners in the bulk mobile phase can serve as an ion-pairing agent for ionized analytes [44]. The extent to which any of these roles contribute to overall retention likely depends on the structure of the analytes as well as the lipophilicity of the cation, charge diffusivity of the anion, and concentration of the IL in the mobile phase.

5.6   Ionic liquids in capillary electrophoresis

Many of the properties noted for surface silanols on silica-based chromatographic supports (e.g., protein adsorption, tailing of amines, and hydrolytic instability) are relevant for the fused silica surfaces of the capillaries used in CE. Hence, it is not too surprising that some of the strategies reported earlier have been adapted to CE. For instance, polyvinylimidazolium coatings were shown to improve separation robustness for basic proteins by inhibiting sorption onto the capillary walls [45]. However, ILs can also be found in numerous applications of CE, particularly for polar organic analytes, which are well described in the next chapter. In most of these studies, the rich potential interaction chemistries of ILs permit several likely roles for the IL additives including background electrolyte, bulk solution pseudostationary phase [46,47], and capillary wall modifier. Some of these potential roles are illustrated in Figure 5.4 [48,49].

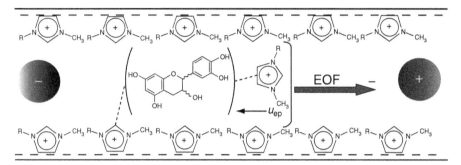

Figure 5.4 Scheme illustrating the mechanism for separation of polyphenols using 1-alkyl-3-methylimidazolium-based ionic liquids. (Adapted from Yanes, E. G., Gratz, S. R., Baldwin, M. J., Robison, S. E., and Stalcup, A. M., *Anal. Chem.*, 73, 3838–3844, 2001.)

5.7 Surface-confined ionic liquids as stationary phases in liquid chromatography

Analogous to the use of ILs as mobile-phase additives, it can be argued that once the IL is immobilized on a surface, it no longer constitutes a true IL. However, the IL paradigm offers a fresh approach for looking at inter-molecular interactions in retention processes while simultaneously offering potential for solving some of the most challenging problems confronting analytical separations.

5.7.1 Determination of void volumes for surface-confined ionic liquids as stationary phases in high-performance liquid chromatography

One unique challenge in using SCILs as stationary phases in HPLC is the determination of the column-void or mobile-phase volume. Accurate determination of retention factors, k', requires measurement of t_0, the void volume.

$$k' = \frac{t_r - t_0}{t_0}$$

In conventional chromatography, t_0 is usually determined by measuring the time required to elute an analyte known not to interact with the stationary phase. A variety of strategies have evolved [50,51], each with their own strengths and weaknesses. For instance, in reversed-phase chromatography, polar compounds such as uracil have been used. However, when using SCILs, identification of a solute which does not interact with the multimodal stationary phase may prove elusive. An alternative strategy that has been used is to fill and weigh the column with two different solvents having very different densities (e.g., dichloromethane: 1.32 g/mL at 25°C; hexane: 0.66 g/mL at 25°C) [52]. The difference in column weight, Δw, can be related by the differences in solvent densities, Δd, to the void volume, V_m, of the column through

$$V_m = \frac{\Delta w}{\Delta d}$$

It should be noted that this approach provides only a static estimate of the void volume of the column. It does not take into account any differences in stationary-phase solvation that is likely to occur in the presence of different solvent systems [53].

5.7.2   Surface-confined ionic liquids as reversed-phase stationary phases in liquid chromatography

One of the earliest reports of SCILs in LC concerned the analysis of ephedrines [54] and tropane alkaloids [55]. In these studies, the N,N-dialkylimidazolium motif was covalently attached through a 3-mercaptopropylsilane linker. Retention of the analytes seemed to be governed predominantly through reversed-phase interactions. In comparison with conventional reversed-phase media, the IL-based stationary phase provided better peak shape and resolution while simultaneously allowing for reduction in the organic content of the mobile phase.

Colón and coworkers [56] also used a propyl linker to covalently attach methyl- and butylimidazolium moieties to a silica support. Overall, retention for organic acids was slightly higher on the butylimidazolium phase which was ascribed to differences in the hydrophobicity contributed by the imidazolium substituent. In contrast to retention increases for acidic compounds with increasing pH on ion-exchange columns, the authors report a decrease in retention which was ascribed to masking of surface silanols by reorganization of the imidazolium motif in compensation for increased silanols deprotonation (Figure 5.5) [56]. However, retention of the organic acids also seemed to rely on ion-exchange interactions.

In an investigation of the retention properties of a butylimidazolium phase covalently attached through an alkyl spacer to a silica support, the chromatographic retention factors for a training set of 28 aromatic analytes were determined using a range of methanol/water [57] or acetonitrile/water [58] mobile phases. In this work, the authors assumed that retention could be modeled as a linear combination of molecular interactions (e.g., hydrophobic, hydrogen bond acidity/basicity, dipole–dipole, and dispersive). The chromatographic data were subjected to a multiple linear regression analysis to extract the linear solvation free energy relationship (LSFER) coefficients [59]. Plots of experimental retention data versus the retention predicted (based on the LSFER coefficients) demonstrated that the LSFER model, with the selected molecular descriptors, did an excellent job of accounting for the

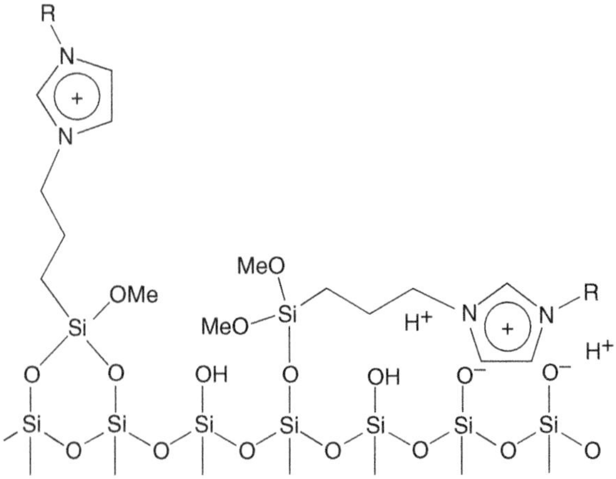

Figure 5.5 Scheme illustrating potential reorientation of bonded imidazolium ligands in response to deprotonation of residual silanols. Anion is not shown for clarity. (Adapted from Wang, Q., Baker, G. A., Baker, S. N., and Colón, L. A., *Analyst*, 131, 1000–1005, 2006.)

various solute/stationary phase interactions despite the presence of ionic moieties in the stationary phase. Interestingly, the stationary phase retained much of its reversed-phase character despite the incorporation of the imidazolium cation. Comparison of retention for a subset of the probe solutes obtained on the imidazolium column and retention reported in the literature [59] on conventional reversed-phase columns revealed that the butylimidazolium stationary phase behaved somewhat like a phenyl phase, despite the presence of the cation. Thus, while hydrophobic interactions are retained in the imidazolium-based phase, despite the presence of the cationic charge, at least some of the hydrophobic interactions with the stationary phase emanate from the imidazolium aromatic ring. Overall, the butylimidazolium column provided better discrimination between classes of aromatic compounds in acetonitrile/aqueous mixtures than in methanol/water. As noted previously in the case of separation of imidazolium cations on phenyl phases [23], the role of π–π interactions in retention could be mediated by the addition of acetonitrile, thus affording better discrimination based on other molecular features of the selected analytes.

5.7.3   Surface-confined ionic liquids as ion exchange-based stationary phases in liquid chromatography

One of the earliest indications of the potential of IL-based moieties for biomolecule separations may be found by examining the report of a methyl imidazole-derived phase in HPLC [60]. In this study, it was found that at a pH below its pK_a, the methyl imidazole-derived phase demonstrated enhanced affinity for the ionized form of cyclic nucleotides, thus suggesting an electrostatic component to retention. Qui and coworkers also reported ion-exchange behavior that could be mediated by adjusting the mobile-phase pH when using a covalently attached imidazole phase [61]. The elution of organic anions (e.g., sodium benzoate, sodium *p*-toluene sulfonic acid, and potassium hydrogen phthalate) after inorganic anions was interpreted as evidence for simultaneous ion exchange and reversed-phase activity on the imidazole phase. In a related study [62], retention of inorganic anions on a methylimidazolium-based stationary phase showed a dependence on pH when using a phosphate buffer that seemed to be an artifact arising from the changing ionic strength near the pK_a's of the phosphate buffer.

Retention for nucleotides with varying degrees of phosphorylation obtained on a butylimidazolium phase revealed that higher phosphorylation also corresponded to higher retention [63]. In this study, it was found that in the range of 20–60 mM ammonium acetate, increasing buffer concentration correlated with reduced retention, consistent with an ion-exchange type of retention mechanism. Another study found protein peak efficiency on a coated quaternized polyvinylimidazole stationary phase [64] to be dependent on stationary phase thickness and adsorption–desorption kinetics. Further, hydrophobic interactions are known to be important for protein separations even on ion-exchange media [65,66].

Other evidence for the contribution of electrostatic interactions to retention on SCIL phases may be found in a study by Sun and Stalcup [67]. In this work, it was reported that while the LSFER approach successfully accounted for intermolecular interactions responsible for retention of nonpolar solutes, inclusion of ionizable solutes such as pyridine or nitrophenol isomers seriously degraded the correlation between experimental and predicted retention. Successful global application of an LSFER approach for a training set which includes ionizable analytes required incorporation of an additional descriptor to account for the *degree of ionization* [68] of the analytes as well as to account for the impact of electrostatic interactions. The additional descriptor incorporated the mobile-phase pH as well as the acid dissociation constant of the analyte.

5.7.4   *Surface-confined ionic liquid stationary phases in liquid chromatography as model systems for liquid–liquid extraction*

While applications and analysis of ILs may provide some guidance on potential applications of SCIL-based phases in LC, these phases may also provide useful information about ILs. As Poole points out [16], a key requirement for the successful integration of ILs in industrial processes is the ability of being applied to rapid liquid–liquid phase separation systems. Shake-flask methods are commonly used to measure IL/water partition coefficients. However, the high viscosity and cost of these materials coupled with the time and effort required for traditional shake-flask methods render this approach highly undesirable. Retention on reversed-phase HPLC columns has been shown to correlate with octanol/water partition coefficients or log P [69,70]. Supercritical fluid-chromatographic [71] data on IL-stationary phase have been found to correlate with IL/vapor partition data. Further, IL partitioning data have also been shown to correlate with octanol/water partitioning coefficients [72]. Thus, it is not unreasonable to expect retention on SCIL-based phases to also correlate with the corresponding partition data [73]. Indeed, a recent report suggests that this may be another important application of SCILs [63].

5.8   Concluding remarks

SCIL sorbents are a rapidly emerging alternative in LC that are poised to address the critical need for novel and more universal separation strategies. These novel sorbent materials present unique potential separation capabilities applicable to a broad range of compounds (e.g., polar organics and biomolecules/biopolymers). Specific features of these sorbents which allow for this unique separation capability include the presence of a *fuzzy* ion with a delocalized charge which can promote interactions with neutral aromatic species, a scaffold for introducing additional functionality, and ion-exchange capability with potentially tunable hydrophobicity.

Currently, the fastest growing areas of HPLC separations are in the areas of biomolecule separations and LC-mass spectrometry (LC-MS) [74]. Biomolecule separations are particularly challenging [75] because they present a variety of interaction modalities ranging from hydrophobic to polar to electrostatic. As indicated here, ILs can supply all of these interaction modalities and their applications, thus far, in a variety of platforms, suggest that they may be particularly advantageous in addressing the critical challenges of polar organic and biomolecular separations.

Acknowledgments

The author would like to gratefully acknowledge support from the Waters Corporation and the National Institutes of Health for financial support (R01 GM067991-01A2).

References

1. Koel, M., Ionic liquids in chemical analysis, *Crit. Rev. Anal. Chem.*, 35, 177–192, 2005.
2. Baker, G. A., Baker, S. N., Pandey, S., and Bright, F. V., An analytical view of ionic liquids, *Analyst*, 130, 800–808, 2005.
3. Anderson, J. L., Armstrong, D. W., and Wei, G.-T., Ionic liquids in analytical chemistry, *Anal. Chem.*, 78, 2892–2902, 2006.
4. Liu, J.-F., Jönsson, J. Å., and Jiang, G.-B., Application of ionic liquids in analytical chemistry, *TRAC*, 24, 20–27, 2005.

5. Pandey, S., Analytical applications of room-temperature ionic liquids: A review of recent efforts, *Anal. Chim. Acta*, 556, 38–45, 2006.
6. Poole, C. F., Furton, K. G., and Kersten, B. R., Liquid organic salt phases for gas chromatography, *J. Chromatogr. Sci.*, 24, 400–409, 1986.
7. Pacholec, F. and Poole, C. F., Stationary phase properties of the organic molten salt ethylpyridinium bromide in gas chromatography, *Chromatographia*, 17, 370–374, 1983.
8. Pacholec, F., Butler, H. T., and Poole, C. F., Molten organic salt phase for gas-liquid chromatography, *Anal. Chem.*, 54, 1938–1941, 1982.
9. Dhanesar, S. C., Coddens, M. E., and Poole, C. F., Evaluation of tetraalkylammonium tetrafluoroborate salts as high-temperature stationary phases for packed and open-tubular column gas chromatography, *J. Chromatogr.*, 349, 249–265, 1985.
10. Coddens, M. E., Furton, K. G., and Poole, C. F., Synthesis and gas chromatographic stationary phase properties of alkylammonium thiocyanates, *J. Chromatogr.*, 356, 59–77, 1986.
11. Armstrong, D. W., He, L., and Liu, Y. Examination of ionic liquids and their interaction with molecules when used as stationary phases in gas chromatography, *Anal. Chem.*, 71, 3873–3876, 1999.
12. Anderson, J. L. and Armstrong, D. W., High-stability ionic liquids. A new class of stationary phases for gas chromatography, *Anal. Chem.*, 75, 4851–4858, 2003.
13. Armstrong, D. W., Zhang, L.-K., He, L., and Gross, M. L., Ionic liquids as matrixes for matrix-assisted laser desorption/ionization mass spectrometry, *Anal. Chem.*, 73, 3679–3686, 2001.
14. Fletcher, K. A., Baker, S. N., Baker, G. A., and Pandey, S., Probing solute and solvent interactions within binary ionic liquid mixtures, *New J. Chem.*, 27, 1706–1712, 2003.
15. Boehm, R. E., Martire, D. E., and Armstrong, D. W., Theoretical considerations concerning the separation of enantiomeric solutes by liquid chromatography, *Anal. Chem.*, 60, 522–528, 1988.
16. Poole, C. F., Chromatographic and spectroscopic methods for the determination of solvent properties of room temperature ionic liquids, *J. Chromatogr. A*, 1037, 49–82, 2004.
17. Seddon, K. R., Stark, A., and Torres, M.-J., Influence of chloride, water and organic solvents on the physical properties of ionic liquids, *Pure Appl. Chem.*, 72, 2275–2287, 2000.
18. Berthier, D., Varenne, A., Gareil, P., Digne, M., Lienemann, C.-P, Magnac, L., and Olivier-Bourbigouc, H., Capillary electrophoresis monitoring of halide impurities in ionic liquids, *Analyst*, 129, 1257–1261, 2004.
19. Stepnowski, P., Application of chromatographic and electrophoretic methods for the analysis of imidazolium and pyridinium cations as used in ionic liquids, *Int. J. Mol. Sci.*, 7, 497–509, 2006.
20. Ruiz-Angel, M. J. and Berthod, A., Reversed phase liquid chromatography of alkyl-imidazolium ionic liquids, *J. Chromatogr. A*, 1113, 101–108, 2006.
21. Stuff, J. R., Separation of cations in buffered 1-methyl-3-ethylimidazolium chloride + aluminum chloride ionic liquids by ion chromatography, *J. Chromatogr.*, 547, 484–487, 1991.
22. Stepnowski, P. and Mrozik, W., Analysis of selected ionic liquid cations by ion exchange chromatography and reversed-phase high performance liquid chromatography, *J. Sep. Sci.*, 28, 149–154, 2005.

23. Stepnowski, P., Nichthauser, J., Mrozik, W., and Buszewski, B., Usefulness of $\pi \cdots \pi$ aromatic interactions in the selective separation and analysis of imidazolium and pyridinium ionic liquid cations, *Anal. Bioanal. Chem.*, 385, 1483–1491, 2006.

24. Branco, L. C., Rosa, J. N., Ramos, J. J. M., and Afonso, C. A. M., Preparation and characterization of new room temperature ionic liquids, *Chem. Eur. J.*, 8, 3671–3677, 2002.

25. Berthod, A., Ruiz-Angel, M. J., and Huguet S., Nonmolecular solvents in separation methods: Dual nature of room temperature ionic liquids, *Anal. Chem.*, 77, 4071–4080, 2005.

26. Anderson, J. L., Ding, J., Welton, T., and Armstrong, D. W., Characterizing ionic liquids on the basis of multiple solvation interactions, *J. Am. Chem. Soc.*, 124, 14247–14254, 2002.

27. Krutz , L. J., Senseman, S. A., and Sciumbato, A. S., Solid-phase microextraction for herbicide determination in environmental samples, *J. Chromatogr. A*, 999, 103–121, 2003.

28. Musteata, F. M. and Pawliszyn, J., Bioanalytical applications of solid-phase microextraction, *TRAC*, 26, 36–45, 2007.

29. Hinshaw, J. V., Solid phase microextraction, *LC-GC North America*, 21, 1057–1061, 2003.

30. Peng, J.-F., Liu, J.-F., Hu, J.-T., and Jiang, G.-B., Direct determination of chlorophenols in environmental water samples by hollow fiber supported ionic liquid membrane extraction coupled with high-performance liquid chromatography, *J. Chromatogr. A*, 1139, 165–170, 2007.

31. Liu, J.-F., Peng, J.-F., Chi, Y.-G., and Jiang, G.-B., Determination of formaldehyde in shiitake mushroom by ionic liquid-based liquid-phase microextraction coupled with liquid chromatography, *Talanta*, 65, 705–709, 2005.

32. Wu, J., Yu, X., Lord, H., and Pawliszyn, J., Solid phase microextraction of inorganic anions based on polypyrrole film, *Analyst*, 125, 391–394, 2000.

33. Temsamani, K. R., Ceylan, Ö., Yates, B. J., Öztemiz, S., Gbatu, T. P., Stalcup, A. M., Mark, H. B., Jr. and Kutner, W., Electrochemically aided solid phase microextraction: conducting polymer film material applicable for cationic analytes, *J. Solid State Electrochem.*, 6, 494–497, 2002.

34. Liu, J.-F., Chi, Y.-G., Jiang, G.-B., Tai, C., Peng, J.-F., and Hu, J.-T., Ionic liquid-based liquid-phase microextraction, a new sample enrichment procedure for liquid chromatography, *J. Chromatogr. A*, 1026, 143–147, 2004.

35. Polyakova, Y. and Row, K. H., Retention behaviour of N-CBZ-D-phenylalanine and D-tryptophan: Effect of ionic liquid as mobile-phase modifier, *Acta Chromatogr.*, 17, 210–221, 2006.

36. Stalcup, A. M., Martire, D. E., and Wise, S. A., A thermodynamic comparison of monomeric and polymeric C_{18} bonded phases using aqueous methanol and acetonitrile mobile phases. *J. Chromatogr.*, 422, 1–14, 1988.

37. Holbrey, J. D., Reichert, W. M., Nieuwenhuyzen, M., Sheppard, O., Hardacre, C., and Rogers, R. D., Liquid clathrate formation in ionic liquid aromatic mixtures, *Chem. Commun.*, 476–477, 2003.

38. Polyakova, Y., Jin, Y., Zheng, J., and Row, K. H., Effect of concentration of ionic liquid 1-butyl-3-methylimidazolium tetrafuoroborate for retention and separation of some amino and nucleic acids, *J. Liq. Chromatogr.*, 29, 1687–1701, 2006.

39. Marszałł, M. P., Bączek, T., and Kaliszan, R., Evaluation of the silanol-suppressing potency of ionic liquids, *J. Sep. Sci.*, 29, 1138–1145, 2006.

40. Kaliszan, R., Marszałł, M. P., Markuszewski, M. J., Bączek, T., and Pernak, J., Suppression of deleterious effects of free silanols in liquid chromatography by imidazolium tetrafluoroborate ionic liquids, *J. Chromatogr. A*, 1030, 263–271, 2004.

41. Tang, F., Tao, L., Luoa, X., Ding, L., Guoa, M., Nie, L., and Yao, S., Determination of octopamine, synephrine and tyramine in Citrus herbs by ionic liquid improved green chromatography, *J. Chromatogr. A*, 1125, 182–188, 2006.

42. He, L., Zhang, W., Zhao, L., Liu, X., and Jiang, S., Effect of 1-alkyl-3-methyl-imidazolium-based ionic liquids as the eluent on the separation of ephedrines by liquid chromatography, *J. Chromatogr. A*, 1007, 39–45, 2003.

43. Waichigo, M. M., and Danielson, N. D., Comparison of ethylammonium formate to methanol as a mobile-phase modifier for reversed-phase liquid chromatography, *J. Sep. Sci.*, 29, 599–606, 2006.

44. Xiaohua, X., Liang, Z., Xia, L., and Shengxiang, J., Ionic liquids as additives in high performance liquid chromatography. Analysis of amines and the interaction mechanism of ionic liquids, *Anal. Chim. Acta*, 519, 207–211, 2004.

45. Xu, R. J., Vidal-Madjar, C., Sebille, B., and Diez-Masa, J. C., Separation of basic proteins by capillary zone electrophoresis with coatings of a copolymer of vinylpyrrolidone and vinylimidazole, *J. Chromatogr. A*, 730, 289–295, 1996.

46. Miller, J. L., Khaledi, M. G., and Shea, D., Separation of polycyclic aromatic hydrocarbons by nonaqueous capillary electrophoresis using charge-transfer complexation with planar organic cations, *Anal. Chem.*, 69, 1223–1229, 1997.

47. Miller, J. L., Khaledi, M. G., and Shea, D., Separation of hydrophobic solutes by nonaqueous capillary electrophoresis through dipolar and charge-transfer interactions with pyrylium salts, *J. Microcol. Sep.*, 10, 681–685, 1998.

48. Cabovska B., Kreishman, G. P., Wassell, D. F., and Stalcup, A. M., CE and NMR studies of interactions between halophenols and ionic liquid or tetraalkylammonium cations, *J. Chromatogr. A*, 1007, 179–187, 2003.

49. Yanes, E. G., Gratz, S. R., Baldwin, M. J., Robison, S. E., and Stalcup, A. M., Capillary electrophoretic application of 1-alkyl-3-methylimidazolium-based ionic liquids, *Anal. Chem.*, 73, 3838–3844, 2001.

50. Rimmer, C. A., Simmons, C. R., and Dorsey, J. G., The measurement and meaning of void volumes in reversed-phase liquid chromatography, *J. Chromatogr. A*, 965(1–2), 219–232, 2002.

51. Oumada, F. Z., Roses, M., and Bosch, E. Inorganic salts as hold-up time markers in C18 columns, *Talanta*, 53(3), 667–677, 2000.

52. Alhedai, A., Martire, D. E., and Scott, R. P. W., Column dead volume in liquid chromatography, *Analyst*, 114(8), 869–875, 1989.

53. Martire, D. E. and Boehm, R. E., Unified theory of retention and selectivity in liquid chromatography. 2. Reversed-phase liquid chromatography with chemically bonded phases, *J. Phys. Chem.*, 87, 1045–1062, 1983.

54. Liu, S.-J, Zhou, F., Xiao, X.-H., Zhao, L., Liu, X., and Jiang, S.-X., Surface confined ionic liquid—A new stationary phase for the separation of ephedrines in high performance liquid chromatography, *Chin. Chem. Lett.*, 15, 1060–1062, 2004.

55. Liu, S.-J, Zhou, F., Zhao, L., Xiao, X.-H., Liu, X., and Jiang, S.-X., Immobilized 1,3-dialkylimidazolium salts as new interface in HPLC separation, *Chem. Lett.*, 33(5), 496–497, 2004.

56. Wang, Q., Baker, G. A., Baker, S. N., and Colón, L. A., Surface confined ionic liquid as a stationary phase for HPLC, *Analyst*, 131, 1000–1005, 2006.

57. Sun, Y., Cabovska, B., Evans, C. E., Ridgway, T. H., and Stalcup, A. M., Retention characteristics of a new butylimidazolium-based stationary phase, *Anal. Bioanal. Chem.*, 382, 728–734, 2005.
58. Sun, Y. and Stalcup, A. M., Mobile phase effects on retention on a new butyl-imidazolium-based high-performance liquid chromatographic stationary phase, *J. Chromatogr. A*, 1126, 276–282, 2006.
59. Reta, M., Carr, P. W., Sadek, P. C., and Rutan, S. C., Comparative study of hydrocarbon, fluorocarbon and aromatic bonded RP-HPLC stationary phases by linear solvation energy relationships, *Anal. Chem.*, 71, 3484–3496, 1999.
60. Turowski, M., Kaliszan, R., Lullmann, C., Genieser, H. G., and Jastorff, B., New stationary phases for the high performance liquid chromatographic separation of nucleosides and cyclic nucleotides. Synthesis and chemometric analysis of retention data, *J. Chromatogr. A*, 728, 201–211, 1996.
61. Qiu, H., Jiang, S., Liu, X., and Zhao, L., Novel imidazolium stationary phase for high-performance liquid chromatography, *J. Chromatogr. A*, 1116, 46–50, 2006.
62. Qiu, H., Jiang, S., and Liu, X., *N*-Methylimidazolium anion-exchange station-ary phase for high-performance liquid chromatography, *J. Chromatogr. A*, 1103, 265–270, 2006.
63. Van Meter, D., Sun, Y., Parker, K., and Stalcup, A. M., Retention characteristics of a new butylimidazolium-based stationary phase. Part II: Anion-exchange and partitioning, *Anal. Bioanal. Chem.*, 390, 897–905, 2008.
64. Memque, R., Vidal-Madjar, C., Racine, M., Piquion, J., and Sebille, B., Anion-exchange chromatographic properties of a-lactalbumin eluted from a quaternized polyvinylimidazole, *J. Chromatogr.*, 553, 165–177, 1991.
65. Fang, F., Aguilar, M.-I., and Hearn, M. T. W., Temperature induced changes in the bandwidth behavior of proteins separated with cation-exchange adsorbents, *J. Chromatogr. A*, 729, 67–79, 1996.
66. Stalberg, J., Jonsson, B., and Horvath, C., Combined effect of coulombic and van der Waals interactions in the chromatography of proteins, *Anal. Chem.*, 64, 3118–3124, 1992.
67. Sun, Y. and Stalcup, A. M., Application of an extended LSFER model to retention on a butylimidazolium-based column for high performance liquid chromatography, *J. Chromatogr. A*, Unpublished results.
68. Canals, I., Portal, J. A., Roses, M., and Bosch, E., Retention of ionizable compounds on HPLC. Modeling retention for neutral and ionizable compounds by linear solvation energy relationships, *Chromatographia*, 56, 431–437, 2002.
69. Harnisch, M., Mockel, H. J., and Schulze, G., Relationship between log $P_{0/w}$ shake-flask values and capacity factors derived from reversed-phase high performance liquid chromatography for n-alkylbenzenes and some OECD reference standards, *J. Chromatogr.*, 282, 315–332, 1983.
70. Lochmüller, C. H., Reese, C., Aschman, A. J., and Breiner, S. J., Current strategies for prediction of retention in high performance liquid chromatography, *J. Chromatogr.*, 656, 3–18, 1993.
71. Planeta, J. and Roth, M., Partition coefficients of low-volatility solutes in the ionic liquid 1-*n*-butyl-3-methylimidazolium hexafluorophosphate-supercritical CO_2 system from chromatographic retention measurements, *J. Phys. Chem. B*, 108, 11244–11249, 2004.
72. Carda-Broch, S., Berthod, A., and Armstrong, D. W., Solvent properties of the 1-butyl-3-methylimidazolium hexafluorophosphate ionic liquid, *Anal. Bioanal. Chem.*, 375, 191–199, 2003.

73. Shetty, P. H., Poole, S. K., and Poole, C. F., Applications of ethylammonium and propylammonium nitrate solvents in liquid-liquid extraction and chromatography, *Anal. Chim. Acta*, 236, 51–61, 1990.
74. Majors, R. E., New chromatography columns and accessories at the 2002 Pittsburgh Conference. Part I. *LC-GC North America*, 20, 248–266, 2002.
75. Stulic, K., Pacakova, V., Suchankova, J., and Claessens, H. A., Stationary phases for peptide analysis by high performance liquid chromatography, *Anal. Chim. Acta*, 352, 1–19, 1997.

chapter six

Ionic liquids as background electrolyte additives in capillary electrophoresis

Merike Vaher and Mihkel Kaljurand

Contents

6.1 Capillary electrophoresis

Electromigration methods compose a family of analytical separation methods based on differences in the mobilities of charged analytes in the electric field. In this chapter, we discuss mainly such electromigration methods that are performed in thin capillaries with inner diameter (i.d.) <0.1 mm. These methods are commonly known as capillary electrophoretic methods where the most important modes are capillary zone electrophoresis (CZE), micellar electrokinetic capillary chromatography (MEKC), capillary gel electrophoresis (CGE), and capillary electrochromatography (CEC).

The migration of charged species under the influence of an externally applied electric field is known as *electrophoresis*. Differences in the mobility of the analytes due to their average charge, size, shape, and properties of the used electrolyte solution form a basis of a valuable separation method in chemistry.

According to Ref. 1, a Russian physicist Reuss carried out the first separations based on this principle already in 1809. He studied the migration of colloidal clay particles and discovered that the liquid adjacent to the negatively charged surface of the wall migrated toward the negative electrode under the influence of an externally applied electric field. Theoretical aspects of this electrokinetic phenomenon (*electroosmosis* by Reuss) were formulated in 1897 by Kohlrausch [2]. In the late 1800s and early 1900s, electrophoretic separations were carried out by several researchers in so-called *U*-shaped tubes. Starting in 1925 with his PhD thesis on the development of free moving-boundary electrophoresis, Tiselius advanced the analytical aspects of electrophoresis. This resulted in the separation of complex protein mixtures based on differences in electrophoretic mobilities [3]. In 1948, Tiselius was awarded the Nobel Prize for Chemistry for his work on electrophoresis. Possibilities of performing electrophoresis in capillaries were investigated by Hjertén [4], Everaerts [5], and Virtanen [6] but their work did not draw much attention to capillary electrophoresis (CE) until the studies of Jorgenson and Lucas appeared. They separated fluorescent dansylated amino acids in a glass capillary with an i.d. of 75 μm. Applying voltages up to 30 kV, they provided the efficiency of more than 400,000 theoretical plates within 25 min [7]. This efficiency, not seen in separation science before, was mainly due to the fact that at diameters <100 μm the capillary wall dissipated the Joule heating generated in the buffer by electric current. Previously, that is, with capillaries having an i.d. over 100 μm, the analyte peaks were heavily broadened due to the unevenly distributed heat in capillaries. After the decrease of the capillary i.d., broadening was markedly diminished and the determining factor of band broadening has molecular diffusion that, in general, is very low in liquids. After the landmark work by Jorgenson, interest in CE started to grow rapidly [8]. Although CE was initially heralded for its speed and low sample volume, the technique was widely accepted because it is quantitative, can be automated, and will separate compounds that have been traditionally difficult to handle by high performance liquid chromatography (HPLC). CE played a crucial role in determining the human genome sequence and CE is the basis for virtually all microfluidics for *lab-on-a-chip* devices. CE can separate polar substances, which are notoriously difficult to analyze with HPLC. Chiral separations are another area in which the use of CE has expanded. The small sample volumes required for CE can be an advantage with a limited sample amount. An area closely related to CE that has not yet achieved its limits is CEC that combines features of CE with LC by using capillaries packed with chromatographic materials.

Will CE ever overtake HPLC? That is not an appropriate question anymore. CE comes into its own for large molecules and when sample sizes are limited. This appears to give its best applications a biological flavor, and it can powerfully address problems where HPLC has a little chance of success—the Human Genome Project being the obvious example. Now that CE has been officially recognized by several regulatory agencies, the

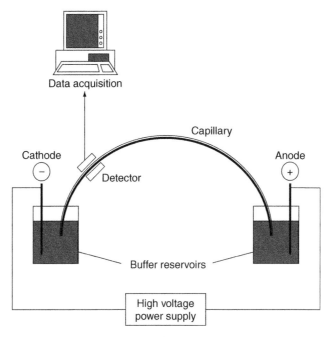

Figure 6.1   A schematic representation of the arrangement of the main components of the capillary electrophoresis instrument.

Food and Drug Administration and the Center for Drug Evaluation and Research among them—it has found a niche in quality assessment and quality control (QA and QC) labs as well. And with CE's hugely successful foray into genomics—it being the undisputed reason the Human Genome Project finished well ahead of schedule—the question is whether it can do the same for proteomics, the next phase of genomic research.

Instrumentation in CE is remarkably simple in design (see Figure 6.1). The heart of the instrument is a thin capillary with the diameter between 10 and 100 μm and length 20–100 cm. The ends of the capillary are placed in a separate buffer reservoir, each containing an electrode. The sample is injected onto the capillary by temporarily replacing one of the buffer reservoirs (normally at the anode) with a sample reservoir and applying either an electric potential or pressure for a few seconds. Electric field is generated by a high-voltage power supply which delivers bipolar voltage up to 30 kV. The applied higher voltages require special precautions for isolation. Also, a detector is needed to monitor the results of the separation processes. An optical or contactless conductivity detector can be placed directly through the capillary wall near the opposite end (normally near the cathode). Usually the absorbance of the analytes is measured by an UV–Vis spectrophotometer but fluorescence, electrochemical and mass spectrometric detectors are needed when detection limit requirements are not met by UV absorbance.

Although conceptually simple, the instrumentation for CE is still rather complex and expensive (about 50 K Euros) due to the high cost of detectors and sampling systems included in contemporary instruments.

6.2   Mobility in capillary electrophoresis

The movement of the charged analyte in the electric field can be described in the simplest way by balancing two forces that influence the ion movement. Ions are accelerated by the force of the electric field F_e equal to $F_e = zE$, where z is the charge of the analyte and E is the electric field strength. This acceleration is balanced by the frictional resistance F_f of the environment, for which a spherical particle equals $F_f = 6\pi\eta rv$, where η is the viscosity of the medium, r is the radius of the analyte and v is the velocity of the analyte. If the two forces are equal, the analyte moves at the constant *electrophoretic velocity* proportional to the strength of the electric field as follows:

$$v = \frac{z}{6\pi\eta r}E \tag{6.1}$$

In Equation 6.1 the proportionality constant

$$\mu = \frac{z}{6\pi\eta r} \tag{6.2}$$

is known as *mobility*.

It must be noted that the separation medium itself moves at a certain velocity. When a voltage is applied to the buffer-filled capillaries, a bulk flow appears, carrying the buffer through the capillary. This phenomena is known as *electroosmotic flow* (EOF) and is the result of the surface charge on the inside of the capillary wall. Most surfaces possess a negative charge which results from the ionization of the surface or the adsorption of ionic species. In fused-silica capillaries both processes occur and result in a negatively charged capillary wall. Due to the negative charge of the capillary a layer of cations builds up near the surface to maintain the charge balance. This creates a double layer of ions near the surface and an electrical potential. This is known as the *zeta* potential. When the voltage is applied across the capillary, the cations forming the double layer are attracted to the cathode. They therefore move through the capillary, and as they are solvated, drag the bulk solution behind them.

Equation 6.1 is valid for a macroscopic particle moving in a continuous medium. In electrophoresis where the analyte ion moves in the media where particle size is comparable with that of the analyte size, this is definitely not the case. Also, analyte ions are not spherical and the term of the ionic radius, the value of which is difficult to estimate, becomes ambiguous. Thus, even in the simplest case in the separation of Li^+, Na^+, and K^+ ions, the elution order does not correspond to that predicted by Equation 6.1. Various improvements of Equation 6.1 have been provided (for discussion see [9]). For a reasonable approximation, the mobility of the analyte seems to be frequently proportional to the following (charge–molar mass) ratio:

$$\mu \propto \frac{z}{\sqrt[3]{M^2}} \tag{6.3}$$

where the proportionality constant has to be determined experimentally and depends on the class of substances.

If the analyte molecules cannot be charged, the electrophoretic separation of them is impossible. However, if some charged ingredient is added into the separation buffer to which the analyte ions have affinity, it is possible to separate them by partitioning the analyte between the free buffer and the ingredient. Assuming a simple mass action law $A + C \leftrightarrow AC$, where A is the analyte and C is the buffer ingredient, the equilibrium constant is $K = [AC]/[A][C]$, where the brackets denote the molar concentrations of A, C, and AC. Now the analyte moves part of the time with the velocity of the separation media v_{EOF} and part of the time with the velocity of the separation media plus the velocity of the charged complex: $v_{EOF} + v_{[AC]}$. The total velocity of the analyte is as follows:

$$v = \frac{[A]}{[A]+[AC]}v_{EOF} + \frac{[AC]}{[A]+[AC]}(v_{EOF} + v_{[AC]}) \tag{6.4}$$

taking into account of the equilibrium, we obtain

$$v = v_{EOF} + \frac{K[C]}{1+K[C]}v_{[AC]} \tag{6.5}$$

The best known CE buffer ingredient is sodium dodecyl sulfate (SDS) proposed by Terabe [10,11]. SDS forms micelles and the separation of neutral analytes is achieved by their partitioning between the buffer and the SDS micelles, that is, by their hydrophobicity. This is the basis of MEKC and the mobility of analytes correlates linearly well with $\log P$ values, where P is the octanol/water partition ratio. Many other buffer ingredients have been proposed. Most of them implement hydrophobic interactions between the analytes and the buffer ingredients but also ciral selectors have been used as well as various affinity probes. Interest in the ILs used as buffer additives in capillary electromigration methods is due to the fact that they could provide an alternative separation mechanism to two currently implemented mechanisms in CE which are based either on the charge to mass ratio or on the hydrophobicity of the analytes.

6.3   Nonaqueous solvents in capillary electrophoresis

The most widely used separation medium in CE is water, that is, an aqueous medium. If an organic solvent has been used instead, the term nonaqueous CE (NACE) is used in order to make the difference. Nonaqueous solvents were first applied to conventional electrophoresis in the early 1950s [12,13]. The first NACE experiments were carried out in 1984 by Walbroehl and Jorgenson [14] and since then they have served as alternative media to the water environment in many electrophoretic applications [15–17].

Nonaqueous background electrolytes (BGEs) broaden the application of CE, displaying altered separation selectivity and interactions between analytes and buffer additives as compared to aqueous BGEs. The interactions that are very weak or nonexistent in amphiprotic solvents may be strong in aprotic solvents. In addition, NACE appears to be ideally suited for online coupling with mass spectrometry (MS) due to high volatility and low surface tension of many organic solvents [18,19].

The major criteria for solvent properties in CE are:

- Wide liquid range
- High relative permittivity, so that the number density of charge carriers is given directly by the nominal concentration of the BGE
- Low viscosity and small molar volume to ensure high mobilities of ions
- Good solvating power for both cations and anions of the BGE and analytes
- Low vapor pressure up to the maximal temperature application
- Chemical stability and nontoxicity

Table 6.1 shows some typical CE solvents with their physical and chemical properties.

Among the solvents, meeting all or most of the above-mentioned criteria are acetonitrile (ACN), formamide, dimethylformamide, methanol, propylene carbonate, dimethylsulfoxide (DMSO), nitromethane, and *N,N*-dimethylacetamide.

Organic solvents can also be classified according to their ability to accept or transfer protons (i.e., their acid–base behavior) [20,21]. Amphiprotic solvents possess donor as well as acceptor capabilities and can undergo autoprotolysis. They can be subdivided into neutral solvents that possess approximately equal donor and acceptor capabilities (water and alcohols), acidic solvents with predominantly proton donor properties (acetic acid, formic acid), and basic solvents with primarily proton acceptor characteristics (formamide, *N*-methylformamide, and *N,N*-dimethylformamide). Aprotic solvents are not capable of autoprotolysis but may be able to accept protons (ACN, DMSO, propylene carbonate). Inert solvents (hexane) neither accept nor donate protons nor are they capable of autoprotolysis.

Table 6.1 Properties of Some Solvents Used in Capillary Electrophoresis

Solvent	Abbreviated symbol	Bp (°C)	Vapor pressure (mmHg)	Density (g cm^{-3})	Viscosity (cP)	Conductivity (S cm^{-1})	Relative permittivity	Dipole moment (D)	pK$_{auto}$	Donor number	Acceptor number	Toxicity[a]
Water	–	100	23.8	0.9970	0.890	6×10^{-8}	78.39	1.85	17.51	–	54.8	–
Formic acid	–	100.6	43.1	1.2141	1.966	6×10^{-5}	58.5	1.82	–	–	83.6	5
Acetic acid	HOAc	117.9	15.6	1.0439	1.130	6×10^{-9}	6.19	1.68	–	–	52.59	10
Methanol	MeOH	64.5	127.0	0.7864	0.551	1.5×10^{-9}	32.7	2.87	16.91	–	41.5	200,T
Ethanol	EtOH	78.3	59.0	0.7849	1.083	1.4×10^{-9}	24.6	1.66	19.10	–	37.1	1000
1-Propanol	1-PrOH	97.2	21.0	0.7996	1.943	9×10^{-9}	20.5	3.09	19.40	–	33.7	200
2-propanol	2-PrOH	82.2	43.3	0.7813	2.044	6×10^{-8}	19.9	1.66	21.08	–	33.5	400
Acetone	Ac	56.1	231	0.7844	0.303	5×10^{-9}	20.6	2.7_{20}	–	17.0	12.5	750
Acetonitrile	ACN	81.6	88.8	0.7765	0.341_{30}	6×10^{-10}	35.9	3.53	32.2	14.1	18.9	40,T
Formamide	FA	210.5	1	1.1292	3.30	2×10^{-7}	111_{20}	3.37_{30}	16.80	–	39.8	10
N-Methyl-formamide	NMF	180–185	0.4_{44}	0.9988	1.65	8×10^{-7}	182.4	3.86	10.74	–	32.1	10
N,N-Dimethyl-formamide	DMF	153	3.7	0.9439	0.802	6×10^{-8}	36.7	3.24	~29	26.6	16.0	10, T
Dimethyl sulfoxide	DMSO	189.0	0.60	1.095	1.99	2×10^{-9}	46.5	4.06	31.8	29.8	19.3	–
Nitro-methane	NM	101.2	36.7	1.1313	0.614	5×10^{-9}	36.7	3.17	–	2.7	20.5	20
Propylene carbonate	PC	241.7	1.2_{55}	1.195	2.53	1×10^{-8}	64.92	4.94	–	15.1	18.3	–
Ethylene carbonate	EC	248.2	3.4_{95}	1.3383	1.9_{40}	$5\times10^{-8}_{40}$	89.8_{40}	4.9	–	16.4	–	–

Note: Unless otherwise stated, the data are at 25°C. The temperatures other than 25°C are shown as subscripts.

[a] The numerical value shows the threshold limit value, which is defined as the maximum permissible vapor concentration that the average person can be exposed to for 8 h/day, 5 days/week without harm, in ppm (cm^3 of solvent per 1 m^3 of air). The mark T shows the solvent has listed in Title III of the Clean Air Act (42 U.S.C. 7401–7626; Public Law 159 from July 14, 1955; 69 Stat. 322) and the Amendments of 1990 as a hazardous air pollutant.

Source: Compton, S. and Brownlee, R., *Biotechniques*, 6, 432–440, 1988. With permission.

6.3.1   Homo- and heteroconjucation

Adequate choice of organic solvents offers a new possibility for CE separations based on interactions that cannot occur or that can hardly be detected in an aqueous medium. For instance, nondissociated Brønsted acids (phenols, carboxylic acids, and alcohols) have been separated in ACN by use of their heteroassociation with small anions like Cl^-, ClO_4^-, and CH_3COO^- (Ac^-). Due to the low electron acceptor (or hydrogen-bond formation) ability of that aprotic solvent, the added anions experience very weak solvation and interact preferably with stronger hydrogen donors, like the Brønsted acids, leading to the formation of heteroconjugated anions [22–24]. The occurrence of this type of complexation would not be favored in a solvent like water, that has a strong solvation ability for hard cations.

In that instance, polyethers are unable to replace the water molecules existing in the solvation shell of the cation. CE separation has been carried out in methanol (MeOH) by Okada [25] based on the complexation of the uncharged analytes with cations such as Na^+, K^+, and NH_4^+, allowing migration of the neutral molecules to the cathode.

In the case of heteroconjugation, the following reaction occurs $X^- + HA \leftrightarrow HAX^-$, where X^- is the BGE anion, HA is an uncharged molecule and AHX^- is a heteroconjugated complex. In order to keep proton transfer negligible, HX must be a much weaker acid than HA. Conjugation constants are much greater in protophobic ACN than in protophilic (MeOH, EtOH, DMSO, H_2O) solvents. This is because HA is greatly stabilized by hydrogen bonding in the protophilic and only poorly so in the protophobic solvents.

Formation constant (K_f) of the heteroconjugated complex (1:1 complexation of analyte-anion) is given by

$$K_f = [AHX^-]/[X^-][HA] \qquad (6.6)$$

where the square brackets indicate the concentrations of the species. The electrophoretic mobility (μ_{ep}) of the complex can be described as

$$\mu_{ep} = (K_f[X^-]/1 + K_f[X^-])\mu_{AHX} \qquad (6.7)$$

where μ_{AHX} is the limiting mobility of the complex. Equation 6.7 is valid only when (HA) is much lower than (X^-) [24].

6.4   Ionic liquids in capillary electrophoresis

The separation medium inside the capillary has to be electrically conductive. It is easy to manipulate the conductivity of aqueous solutions as there are numerous salts soluble in that medium. With organic solvents, the choice of suitable salts/electrolytes is limited. One of the most widely used electrolytes in the NACE is ammonium acetate and to a lesser amount the quaternary ammonium salts due to their favorable solubility in organic solvents. To expand the usability of nonaqueous solvents in CE, a search for new alternative electrolytes compatible with organic solvents is of great interest. One of such compounds might be so-called ILs.

ILs have often been more appropriately called *designer solvents* to indicate the large structural variability of either the cation or the anion of the salt, which accounts for their broad application area. Early application of ILs for analytical separations has been described in several reviews [26–30]. Here we report some recent developments in this field putting them into the context of earlier developments. ILs can be used with aqueous buffers as well as with nonaqueous media buffers. At room temperature these liquids are miscible with ACN (and with many other organic solvents) which makes it easy to use them for adjustment of analyte mobility and separation. The use of nonaqueous solvents in the analysis has become a distinct field of interest in CE and is referred to here as NACE. Organic solvents are of interest in capillary electroseparations because they extend the range of application of CE techniques to more hydrophobic species, addressing one of the main limitations of this method. Their properties are governed by strong, proton donor–acceptor and orientation interactions.

6.4.1 Ionic liquids in nonaqueous capillary electrophoresis

Among BGE additives, dialkylimidazolium-based ILs are more widely used. These liquid organic salts were first proposed by Vaher et al. for the nonaqueous BGE system [31–33]. In these studies ILs were used as an electrolytic ingredient for the BGE. The miscibility of ILs with ACN makes it easy to use them for adjustment of analyte mobility and for separation in NACE. Various ILs [C_nC_1Im] [X] (n = 4,8); X = PF_6^-, CH_3COO^-, CF_3COO^-, $(CF_3SO_2)_2N^-$, $CF_3(CF_2)_2COO^-$) have been used for the separation of insoluble dyes [31], phenols, aromatic acids [32], and polyphenols [34] in ACN. The works of Vaher et al. demonstrated that ILs cannot change the direction of EOF in the capillary (i.e., the bulk flow was toward the cathode) in various organic solvents. The migration direction of the analytes gives an evidence that they are negatively charged because they eluted after the EOF neutral marker (which indicates that the neutral analytes had interaction with IL anions). The studies also revealed that the nature of the anionic part of ILs affects slightly the electrophoretic mobility and the IL concentration influences the general electrophoretic mobility of the separation media (BGE). Separation of the dyes was achieved because they dissociate in the presence of ILs (as well as in that of some phenolic compounds) or associate with the IL (phenolphtalein, thymolphthalein) anion via a process known as *heteroconjugation* [35]. In other words, separation of otherwise neutral analytes is achieved because they become charged in the presence of ILs in the separation media or form a complex with the IL anion, in both ways the mobility of the solute is changed. It was found that the migration order of the analytes is based on the differences in the effective charge to hydrodynamic radius ratio of the analytes and the ability to form heteroconjugates with IL anions. In CE ACN is the main solvent where heteroconjucation based separations are observed and measured [23,35]. As Miller et al. have shown, by the separation of acidic solutes the combination of heteroconjugation and deprotonation can take place [36,16].

It has been shown [37] that choosing MeOH as a solvent, alkylimidazolium salts can be used to separate a metal cation. The separation of Na^+, K^+, Ca^{2+}, and Mg^{2+} ions was achieved by using 20 mM 1-butyl-3-methylimidazolium heptafluorobutanoate ($[C_4C_1Im][C_3F_7COO]$) in MeOH with indirect UV detection in mineral water. Under the described conditions alkali and alkaline earth metal cations are well separated without using the complexing agents.

Borissova et al. [38] demonstrated that contactless conductivity detection may be applied to the monitoring of the separation process in nonaqueous separation media, allowing the use of the UV light absorbing in imidazolium-based electrolyte additives. Attempts to separate cations of ILs in pure ACN have failed. When the sample consists of toluene as a neutral marker, and traces of the IL, then the IL migrates slower than the neutral marker. This suggests that some form of retention against a neutral marker of the IL may occur in pure ACN. This observation may be explained by the adsorption–desorption of the IL on the capillary wall. The IL may migrate in this medium in the form of ion pairs whose cations interact with the capillary surface. The retention of ILs against the neutral marker disappears when their concentrations in the sample are increased. From Figure 6.2a it is seen that the EOF velocity follows a decreasing trend with the increasing additive concentration. As shown in Figure 6.2b, the change in the migration of the analytes with the change of the BGE concentration follows a trend similar to the conductivity of the electrophoretic medium described in Ref. 39. The addition of IL to ACN increases the conductivity of BGE. The conductivity then reaches the maximum and relatively slowly decreases as the IL concentration in the separation medium increases.

As illustrated in Figure 6.3a, the addition of 7 mM 1-ethyl-3-methylimidazolium ethylsulfate ($[C_2C_1Im][C_2SO_4]$) to ACN results in perfect separation of 1-butyl-3-methylimidazolium methylsulfate ($[C_4C_1Im][C_1SO_4]$), 1-octyl-3-methylimidazolium hexafluorophosphate ($[C_8C_1Im][PF_6]$), and 1-decyl-3-methylimidazolium bis(pentafluoroethansulfonyl)imide ($[C_{10}C_1Im][Pf_2N]$) (electropherogram b). In another experiment (Figure 6.3b), the addition of 20 mM of 1-butyl-3-methylimidazolium trifluoroacetate ($[C_4C_1Im][FAcO]$) to ACN led to total resolving of five different cations of ILs. The separation may occur due to the interaction of analytes with IL ions in the separation media. The cations of analytes with different lengths of alkyl side chains interact differently with the BGE additive.

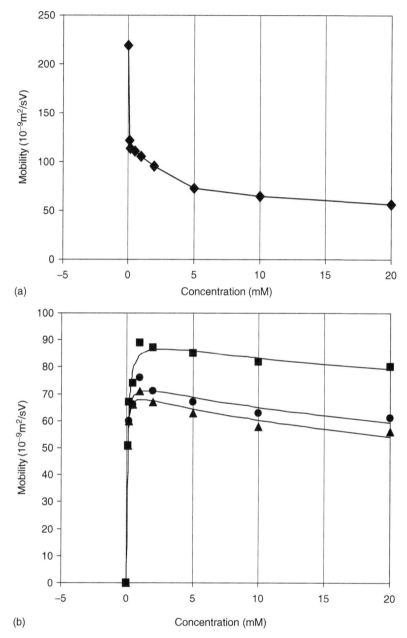

Figure 6.2 The effect of the concentration of added IL ([C$_4$C$_1$Im][FAcO]) to acetonitrile on the mobility of electroosmosis (a) and on the actual electrophoretic mobility of cations of (■)—[C$_2$C$_1$Im] [C$_2$SO$_4$], ●—[C$_8$C$_1$Im] [PF$_6$] and ▲—[C$_{10}$C$_1$Im] [Pf$_2$N] (b) Copyright Wiley-VCH Verlag GmbH & Co. KGaA. (From Borissova, M., Gorbatšova, J., Ebber, A., Kaljurand, M., Koel, M., and Vaher, M., *Electrophoresis*, 28, 3600–3605, 2007. With permission.)

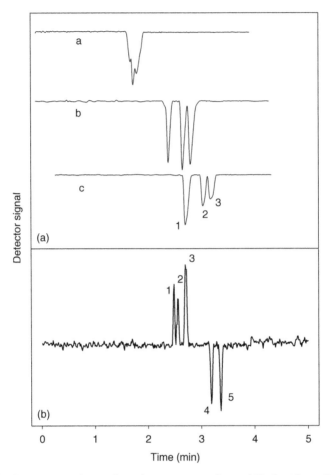

Figure 6.3 Separation of ionic liquids in acetonitrile modified with an IL additive. (Reproduced from Borissova, M., Gorbatšova, J., Ebber, A., Kaljurand, M., Koel, M., and Vaher, M., *Electrophoresis*, 28, 3600–3605, 2007. Copyright Wiley-VCH Verlag GmbH&Co. KGaA. With permission.) (a)—additive is [C_2C_1Im] [C_2SO_4] (a—0.15 mM, b—7 mM, c—11 mM), 1—[C_4C_1Im] [$MeSO_4$], 2—[C_8C_1Im] [PF_6], 3—[$C_{10}C_1Im$] [Pf_2N]; (b)—additive is [C_4C_1Im] [FAcO] (20 mM), 1—[C_1C_1Im] [C_1SO_4], 2—[C_2C_1Im] [C_2SO_4], 3—N-butyl-3-methylpyridinium dicyanamide [C_4C_1py] [dca], 4—[C_8C_1Im] [PF_6], 5—[$C_{10}C_1Im$][Pf_2N]. Applied voltage 18 kV; injection hydrodynamically $h = 15$ cm for 5 s.

Our recent study demonstrate [40] that a thermo-marker makes it possible to observe the change in EOF during the separation process (Figure 6.4) in organic solvents where ILs are used as BGEs. The comparison of the EOF value of a chemical marker with that of a thermo-marker showed that they are almost identical.

The separation mechanism in CE with IL ingredients has recently been studied very thoroughly by help of powerful mathematical tools like

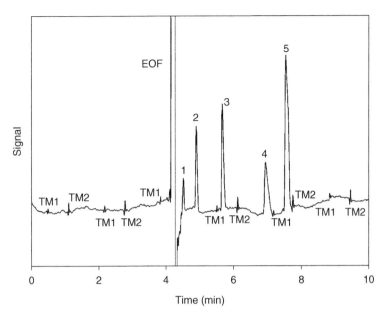

Figure 6.4 Monitoring of electroosmosis stability during separation of derivatives of benzoic acid using a thermo-marker. TM1—first thermo-marker, TM2—second thermo-marker; 1—aminobenzoic acid, 2—benzoic acid, 3—3,5-dihydroxybenzoic acid, 4—salicylic acid, 5—3,5-dinitrobenzoic acid. Background electrolyte: 5 mM $[C_4C_1Im][FAcO]$ in acetonitrile.

experimental design and solvatation linear free energy formalism. The aim of a study by Francois et al. was to elucidate the influence of the IL concentration on the electrophoretic behavior of four arylpropionic acids and to identify the interaction between the analytes and the IL cation [41]. The presence of 1-*n*-butyl-3-methylimidazolium bis(trifluoromethanesulfonyl) imide $([C_4C_1Im][Tf_2N])$ in ACN/alcohol BGEs was investigated in this work. The influence on the mobility of the IL concentration, the nature and the proportion of the organic solvents, and the concentration of the ionic components of the BGE was studied by a four-factor D-optimal experimental design. It provided a deeper insight into analyte interaction with the IL cation present both in the BGE solution and adsorbed onto the capillary wall. It was found that the electrophoretic mobilities of the analyte pass through a maximum for a given IL concentration (~7 mM). Such a behavior was not anticipated from a simple possible interaction between the anionic analytes (profens) and the IL cation present in the BGE since it would have led to a monotonous decrease in the electrophoretic mobilities of profens. Instead, it may rather be explained by competitive interactions between the profen and the IL cation adsorbed onto the capillary wall or into the free IL cation present in the BGE. In effect, for the lowest IL concentrations, a chromatographic like interaction between the profens and the IL cations adsorbed onto the capillary wall should lead to increased migration times which, in the counterelectroosmotic migration mode, may be interpreted as an apparent increase in electrophoretic mobility. Conversely, for IL concentrations higher than ca. 7 mM, the decrease in electrophoretic mobility suggested that an ion-pair interaction between the anionic profen and the free IL cations in the BGE might become prevailing, after the capillary wall had been fully coated with the IL cation.

As it follows from the discussion above, to design the relevant separation protocol for given analytes in CE, it is required to develop a method by choosing suitable ILs or mixtures of ILs and organic solvents, with exactly the right properties for chromatography and electrophoresis, especially in terms of viscosity, conductivity, and absorbance of UV light. Determining these properties in ILs can be rather time-consuming and expensive, requiring individual tests for each property and large sample sizes (several millilitres). To address this issue, a team of chemists from the École Nationale Supérieure de Chemie de Paris, France, led by Pierre Gareil, has developed a single test for measuring these properties, by taking advantage of CE [39]. In their test, a CE tube was filled with an IL and then its UV absorbance was measured using a standard diode array detector (DAD). To measure the viscosity, they introduced a plug of the flow marker (benzyl alcohol) into the tube and then injected the IL until the marker reached the end of the tube and was spotted by the DAD. Using the Hagen–Poiseuille law, the time taken for the flow marker to reach the detector can be used to calculate the viscosity. Finally, the conductivity was calculated by simply applying a voltage across the tube and measuring the electric current after the flow marker had been pushed out. The test, which can be used with both pure ILs and mixtures of ILs and organic solvents, is quick and simple, and requires a sample volume of only 50 μL. The possibility of modeling viscosity of the mixtures of a given IL with various molecular solvents by a unique exponential function was confirmed and extended. For any IL, viscosity variations are practically independent of the organic solvent nature. A tendency was observed according to the solvation ability of the IL anion by the solvent and according to the strength of the anion–cation interaction for the IL cation. The conductivity of IL–molecular solvent mixtures exhibits a nonmonotonous variation whatever the couple tested, due to an overlay of ion-pairing loosening and dilution effects. Additional modeling of conductivity accounting for ion-pairing phenomena is clearly needed. The availability of these data should be a great support in various application areas, including the screening of new separation media for CE.

In the last years, ILs have been applied as matrices for matrix-assisted laser desorption/ionization (MALDI) MS [42], thus expanding the use of MALDI. In Ref. 38 the suitability of alkylammonium- and alkylimidazolium salts of α-cyano-4-hydroxycinnamic acid was investigated as a MALDI matrix and at the same time as the additive of BGE. The alkylammonium salt produced better separation of phenolic compounds than the alkylimidazolium salt. The investigation suggests that it is possible to synthesize ILs suitable for electrophoretic analysis as well as for online MALDI–MS analysis.

6.4.2 Ionic liquids as background electrolyte additives in aqueous media

Yanes and coworkers [43] demonstrated an application of IL for aqueous CE for the separation of phenolic compounds (flavonoids) found in grape seed extracts. By using $[C_nC_1Im]^-$ (n = 2, 4) ILs as additives for the running electrolyte, a simple and reproducible electrophoretic method for the separation of polyphenols was developed. It was speculated that the separation mechanism was based on an association between the imidazolium cations and the polyphenols. The role of the alkyl substituents on the imidazolium cations was investigated and discussed [43]. The anion has little effect on the separation while a related study demonstrated that interaction between phenolic compounds and the IL cations in water occurred through π–π interactions.

The imidazolium ions present in the CE BGE coat the capillary walls, generating an anodic EOF at low pH values (see Figure 6.5). The direction and magnitude of EOF essentially depend on pH of BGE. The change of direction of EOF takes place between pH 7.5 and 8 in the case of 1-ethyl-3-methylimidazolium tetrafluoroborate ([$C_2C_1Im][BF_4$]) [44].

The neutral polyphenols (pK$_a$ = 9.5 − 10.5) can associate with either the imidazolium cations coating the capillary wall or with the free cations in the bulk solution and thus be separated. Experiments show that the association of the polyphenols with the free imidazolium ion seems to dominate [43].

The species of both cation and anion of ILs have also a significant influence on separation. The increase of the carbon chain length of the cationic part of

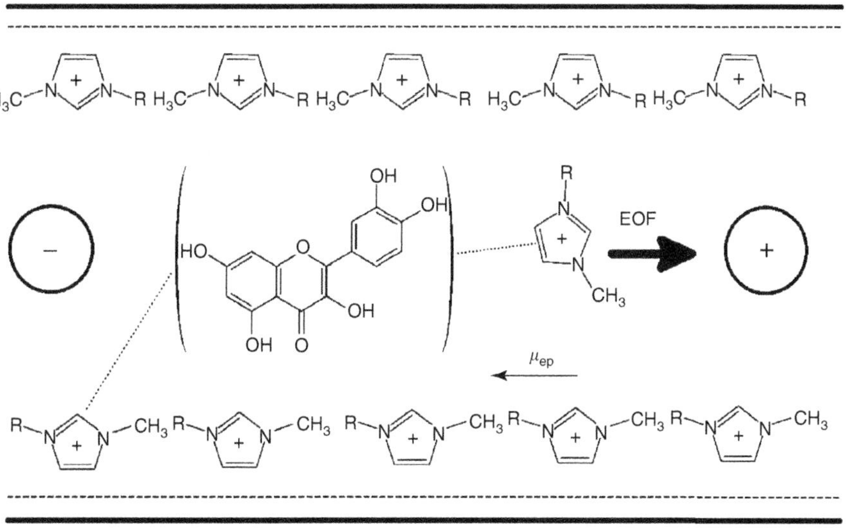

Figure 6.5 Mechanism of polyphenols' separation using 1-alkyl-3-methylimidazolium-based ionic liquids.

the IL [C$_2$C$_1$Im][BF$_4$] versus [C$_4$C$_1$Im][BF$_4$] provided better resolution and a wider separation window, but needed a longer analysis time. The influence of the anionic part for the separation was even more drastic: for the same organic cation, [C$_2$C$_1$Im]$^+$, no separations were obtained with [TfO]$^-$ and NO$_3^-$ as counterions, while the counterion [BF$_4$]$^-$ provided good, reproducible separation.

Further research has supported previous findings by Stalcup et al. [43]. It is well established now that with the presence of ILs as a running buffer ingredient, the imidazolium ions coat the capillary walls. This fact can be advantageously implemented for improving the CE analysis. A common problem in the CE separation of basic proteins is the negatively charged surface, caused by the presence of silanol groups of the fused-silica capillary material, which electrostatically attracts the positively charged sites of proteins. When ILs were added to the running electrolyte, the basic proteins were repelled by the capillary wall because of the surface charge reversal, which improves the separation efficiency and repeatability. Jiang et al. [45] added imidazolium-based ILs into the running electrolyte to dynamically coat the capillary for successful CE separation of basic proteins (lysozyme, cytochrome c, trypsinogen, and α-chymotrypsinogen A). Baseline separation, high efficiency, and symmetrical peaks were obtained during the separation of the four proteins. Again, the separation mechanism that involves association between imidazolium cations and proteins was suggested.

In other applications, Cabovska et al. [46] investigated the CE behavior of monohalogenated phenols in the presence of [C$_2$C$_1$Im][BF$_4$] and compared the results with those obtained with tetraethylammonium tetrafluoroborate ([(C$_2$)$_4$N][BF$_4$]) electrolytes. CE results indicate that the interactions between halophenols and cations of the IL or tetra-alkylammonium salt are similar to those reported previously for polyphenols. The halophenols behave as though carrying a positive charge through association with the cation of the electrolyte. Separation was achieved for some bromo- and iodophenols using both electrolytes, and partial separation was achieved for some chlorophenols. The electrophoretic mobility of the fluorophenols was too close to obtain separation in every case. The affinity toward the cations depends on the size of the halogen as well as the site of the substituents. Iodophenols have the largest negative electrophoretic mobility and the highest affinity for the cations. In all cases, 2-substituted halophenols came out after the 4-substituted halophenols. In both cases, increased halogen size correlated with increased affinity for the electrolyte cation. For isomers, the ortho-substituted isomer exhibited higher affinity than the para isomer.

ILs were tested as additives to the phosphate–acetate buffer for the separation of chlorophenoxy and benzoic herbicide acids [47]. Again, it was found that in the 40 mM phosphate–acetate BGE containing 10% ACN and having pH 4.5, the addition of 10 mM 1-butyl-3-methylimidazoium could reverse EOF. The shoulder-merged peaks of two herbicide acids, 2,4-dichlorobenzoic acid and 3,5-dichlorobenzoic acid were successfully resolved by the addition of the IL cation. Apart from this, the results showed that different IL cations had different influences on the migration behavior of some of the analytes, while IL anions did not introduce any obvious difference to the separation.

In another study, Mwongela and coworkers reported the use of ILs as modifiers in the separation of achiral and chiral analytes in MEKC together with polymeric surfactants [48]. IL appears to assist the separation of hydrophobic mixtures by maintaining adequate background current. ILs ($[C_nC_1Im][X]$ (n = 2,4, X = BF_4^-, PF_6^-, $CF_3SO_3^-$, Cl^-) were applied to separate two achiral mixtures (alkyl aryl ketones and chlorophenols) and one chiral mixture (binaphthyl derivatives). Their investigation showed that the presence of ILs rendered improved resolution and peak efficiency. The polymeric surfactants and ILs were added to a low-conducting buffer solution as a pseudostationary phase and its modifiers, respectively. The separation of the analyte mixture depended on the interaction of the analytes with the polymeric surfactants, whereas the ILs influenced the migration time and peak efficiency.

ILs have been used to separate and determine the purity of anthraquinones. Rapid and sensitive determination of anthraquinones in Chinese herb using 1-butyl-3-methylimidazolium-based IL with β-cyclodextrin (β-CD) as a modifier in CZE was provided by Qi et al. [49]. Successful separation and identification of four anthraquinones of *Paedicalyx attopevensis* Pierre ex Pitard extracts has been achieved. In the running electrolyte the anthraquinones may associate with the imidazolium ions or with the β-CDs. They may be entirely or partly embedded in the cavity of the β-CDs, so the association with the free imidazolium ions in the bulk solution was weak and those analytes, that were not embedded in the cavity of the β-CDs had rather stronger association with the imidazolium ions in the system. The mechanism of separation is illustrated in Figure 6.6.

In another application, the IL was used to determine eight carboxylates as copper complexes [50] when a new IL dimethyldinonylammonium bromide (DMDNAB) was compared to a common flow modifier tetradecyltrimethylammonium bromide (TTAB). As the metal complexes were negatively charged, the measurements were carried out in an anion mode with the flow reversal in the capillary. Better separation was achieved when DMDNAB was used as the flow modifier. In both CE methods all the peaks in the electropherograms were properly separated, the calibration plots gave good correlation coefficients and all eight carboxylates were detected in <7.5 min. The two methods were tested with natural water samples and a paper mill sample, and proved to be feasible.

In the study of Wang et al. [51], an IL ($[C_2C_1Im][BF_4]$) and HP-β-CD as modifiers were added to the buffer to separate hyperoside, luteolin, and chlorogenic acid. Experiments explored the effect of concentration of $[C_2C_1Im][BF_4]$ and HP-β-CD on separation. The results indicated that a simultaneous use of 1.0 mmol/L HP-β-CD and 1%(v/v) $[C_2C_1Im][BF_4]$ in the

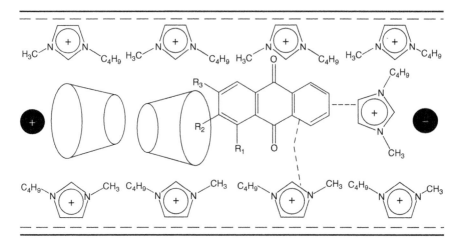

Figure 6.6 Mechanism of separation of anthraquinones using [C₄C₁Im]-based IL and β-CD as BGE additives. (Adapted from Yu, L., Qin, W., and Li, S. F. Y., *Anal. Chim. Acta*, 547, 165–171, 2005. With permission.)

separation buffer enables one to achieve a good compromise in resolution and analysis time.

ILs have been used for MEKC as modifiers for the quantification of the active components of lignans found in the medicinal herbs *Schisandra* species [52]. Preliminary investigations employing SDS alone as a surfactant did not lead to the necessary resolution of the studied compounds but the addition of [C₄C₁Im][BF₄] to the SDS micellar system resulted in the complete separation of all the compounds. The method was successfully applied to determine lignans in extracts of *Schisandra chinensis* (Turcz.) Baill. and *Schisandra henryi* C.B. Clarke in <13 min (5 mM borate–5 mM phosphate buffer in the presence of 20 mM SDS and 10 mM [C₄C₁Im][BF₄]).

Speculating about the effect of ILs on the MEKC separation, the authors suggest that the addition of [C₄C₁Im][BF₄] to typical anionic surfactant systems could dramatically affect MEKC separations. In SDS micelle solutions with only sodium counterions present, these positively charged imidazolium cations could be electrostatically attracted to the negatively charged SDS micelle surface. This would cause a neutralization of the effective head group charge and a reduction of the electrostatic repulsion between the charged hydrophilic headgroups of the surfactant molecules, thus affecting the size and shape of the micelles formed and thereby altering the separation. According to Ref. 53, the lower electrostatic repulsion between the charged hydrophilic headgroups of ionic surfactants has caused a remarkable decrease in the critical micelle concentration (CMC), as compared to that in pure water (8.1 mM). In this study, separation of the compounds could still be achieved even with 5 mM SDS, which is below the CMC of SDS. One possible explanation for the separation was the presence of some micellar aggregate, which indicates the possibility that imidazolium cations have the ability to bind SDS.

Marszałł et al. demonstrated that ILs are suitable modifiers of the BGE for pharmaceutical analysis of the closely related drug analogs [54]. The study demonstrates the use of $[C_2C_1Im][BF_4]$ as modifiers in the separation of nicotinic acid and its structural isomers by CE. The separation mechanism involved the free imidazolium ions which can interact with the inner surface of the capillary wall. At certain IL concentrations (70 mM of IL), it stops the EOF. Increased $[C_2C_1Im][BF_4]$ concentration to 150 mmol/L caused a decrease of migration times of the analytes, an improvement of peak's shape, and an increase in separation performances, however, in contrast to the findings of Yanes and Yu [43,47] this did not reverse the direction of EOF. Neither did the authors consider the possibility of heteroconjugation between the IL cation and the analytes (on improving separation but only on the basis of the charge to mass ratio).

Separation mechanism in CE with IL ingredients was further studied by Yue and Shi in the application of $[C_nC_1Im]$ ILs for the separation of bioactive flavonoids as analytical probes [55]. In this study, a CZE method was established to resolve natural flavonoids, quercetin, kaempferol, and isorhamnetin in the Chinese herbal extract from *Hippophae rhamnoides* and its medicinal preparation (Sindacon Tablet). The effects of the alkyl group, imidazolium counterion (anionic part), along with the concentration of IL were investigated and discussed. When most of the authors have simply stated the association process occurring between the analytes and either IL cation or anion, Yue and Shi moved a bit further in their discussions of the separation mechanism. According to them the separation mechanism seems to be the hydrogen-bonding interaction between the imidazolium cations of the IL and the flavonoids. To test the deduction, $[C_4C_1C_1Im][BF_4]$, in which the hydrogen atom at C_2 of the imidazolium cation was substituted by a methyl group, was added as an additive instead of the above IL. The resolution of flavonoids was destroyed completely. The phenomenon illuminated that the interaction between the H2 and the analytes plays an important role in the separation of all flavonoids. The possibility of formation of a complex between the flavonoids and the borate anions was also studied. The experiments showed that the peaks of kaempferol and quercetin overlapped and migrated together when no IL was added in the buffer. The association of the flavonoids and the imidazolium cation was demonstrated to be dominant. In case the IL concentration had been higher, the formation of micelles could have been the cause for separation. However, in this study, the formation of the pseudostationary phase seemed to be impossible because of the very low concentrations of IL. The resolution was improving with the increasing length of the alkyl group. This suggests that the strength of hydrogen bonding between hydrogen atom at C2 of the imidazolium cation with flavonoid compounds is increasing with the increasing alkyl chain length in the imidazolium cation. The comparison of $[C_nC_1Im][BF_4]$ (n = 2,3,5) and $[C_4C_1Im][BF_4]$ showed that the resolution with the latter additive yielded better separation. The reason could be the steric hindrance that prevented the $[C_5C_1Im][BF_4]$ from forming the hydrogen bonding with analytes. The role of hydrogen atom at C2 position was confirmed further by a recent independent study of Jiang et al. [56]. In this work, $[C_2C_1Im][BF_4]$ was used for the coating of a silica capillary to reduce or invert the EOF in CZE. Excellent separations of amino acids and arylalkanoic acids were obtained. Such separations could not be obtained in a naked (an untreated) capillary in the presence of the cationic surfactants like cetyltrimethylammonium bromide (CTAB) or polycationic polymer hexadimethrine bromide (HDB). The results indicate that $[C_2C_1Im][BF_4]$ does not only modulate the EOF but also acts as a discriminator. Further experiments indicated again that the interaction between hydrogen atom at C2 carbon of imidazolium ring and acid drugs plays an important role in the separation. The following experiments proved the theory assumed above. Instead of $[C_2C_1Im][BF_4]$, $[C_4C_1C_1Im][BF_4]$, in which the hydrogen on the C2 carbon of the imidazolium cation was replaced by methyl, was added to the electrolyte solution. The results showed that the resolution of acid compounds was completely destroyed. The phenomenon revealed that the interaction between the C2 carbon hydrogen and the analytes plays an important role in the separation of acidic drugs. The acceptor of hydrogen interaction between the C2 carbon hydrogen and the analytes was the atom of oxygen in amino acids and in arylalkanoic acids.

Simultaneous determination of bioactive flavone derivatives in Chinese herb extraction by CE by using IL electrolyte systems was proposed by Qi et al. [57]. For comparison, the same study was carried out with borate as an electrolyte. IL was used as the main electrolyte for running buffer and β-CD was used as a modifier. As a result of the study, the difference of ILs and borate in CE was discussed and the advantage of the IL was shown. The separation mechanism was based on the interaction between the negatively charged flavone derivatives and the imidazolium cations as well as the β-CD. Successful identification and separation of four bioactive flavone derivatives of *S. santolinum* (Schrenk) Poljak extraction has been achieved. It was found that $[C_2C_1Im][BF_4]$ and $[C_4C_1Im][BF_4]$, used as the running electrolytes in CE, are better than the borate, especially at high ionic strength.

ILs have been used in the CE-electrochemiluminescence (ECL) method to determine bioactive constituents in Chinese traditional medicine [58]. CE/Tris(2,2-bipyridyl) ruthenium(II) ($Ru(bpy)_3^{2+}$) ECL, CE-ECL, with an IL detection system was established to determine bioactive constituents in Chinese traditional medicine opium poppy, which contains large amounts of coexistent substances. Running buffer containing 25 mM borax–8 mM $[C_2C_1Im][BF_4]$ (pH 9.18) was used, which resulted in significant changes in separation selectivity and obvious enhancement in ECL intensities for those alkaloids with similar structures. Quantitative analysis of four alkaloids was achieved in <7 min. Detection limits of thebaine, codeine, morphine, and narcotine were 0.25, 0.25, 0.001, and 1 μM level, correspondingly. The method was successfully applied to determine the amounts of opium alkaloids in real poppy samples.

The individual solutepseudo-hase interactions of aqueous micellar assemblies of (*N*-alkyl-*N*-methylpyrrolidinium bromide ([C$_n$C$_1$pyr]Br) were evaluated based on the Abraham solvation parameter model correlations approach using MEKC [59]. The IL cation-derived surfactants examined in this study provided highly efficient MEKC separations and, as with conventional surfactants, the magnitudes of the linear solvatation free energy relationship (LSFER) coefficients showed that lipophilicity and hydrogen-bond acidity still play the most important roles in MEKC retention. Using CTAB as a point of reference, however, [C$_n$C$_1$pyr]Br micellar pseudophases provide unique solvent characteristics and are less *hydrophobic* (i.e., have better ability to interact with polar compounds), more cohesive, and less polarizable. Surprisingly, no trends were found with alkyl tail length, showing the primary influence exerted by the nature of the head group on the chemical selectivity. Authors speculate that one can tailor the structure of IL-based surfactants in order to solve different separation problems requiring varied chromatographic selectivity. Further, their utility could be expected to be translated to other fields such as materials engineering and biotechnology. For example, the relatively high cohesivity of [C$_n$C$_1$pyr]Br micelle systems justifies a moderate optimism in regard to their possible application as cationic detergents for the isolation, extraction, and/or solubilization of membrane receptors and proteins or these surfactants and subsequent IL analogs may one day find use as emulsifiers and dispersion agents in a wide range of areas from cosmetics to (possibly) biomedical use.

Last but not least, IL mixtures can also be analyzed by using CE [60]. The method, where citric buffer is used as the running electrolyte, is simple and reproducible. The separation of a standard mixture is in linear accordance with the relative molecular mass (M$_r$) of solutes regardless of the type of substitution (alkyl or aryl). The paper also discussed the applicability of a method for tracking the photo degradation kinetics of an exemplary IL.

6.4.3   Ionic liquids as modifiers of capillary wall

To reverse the EOF direction in the silica capillary, ILs were also covalently bound to the internal capillary surface (by static coating). This approach was applied to separate positively charged drugs [61], DNA [62] and ammonium and metal ions by CE–potential gradient detection using ILs as BGE and covalent coating reagent [63,64]. Experiments showed that the covalently ILs-coated capillary could be used for at least 80 h with relatively stable EOF. Another advantage of covalent coating of capillary inner surface with ILs is the fact that the system becomes compatible with MS; whereas, dynamic coating is not compatible with MS (because the ILs are nonvolatile).

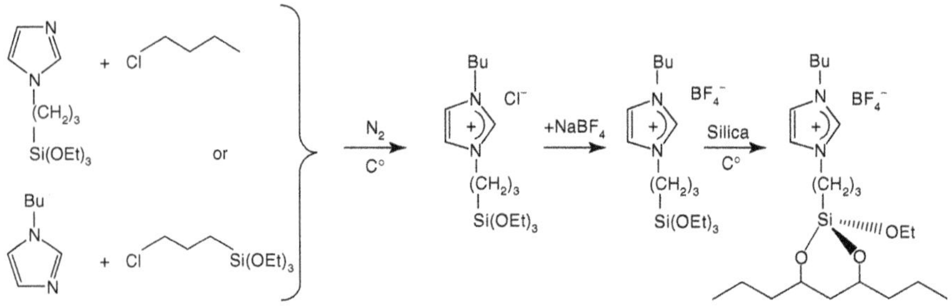

Figure 6.7 The binding of ionic liquid to capillary wall by heterogenous way.

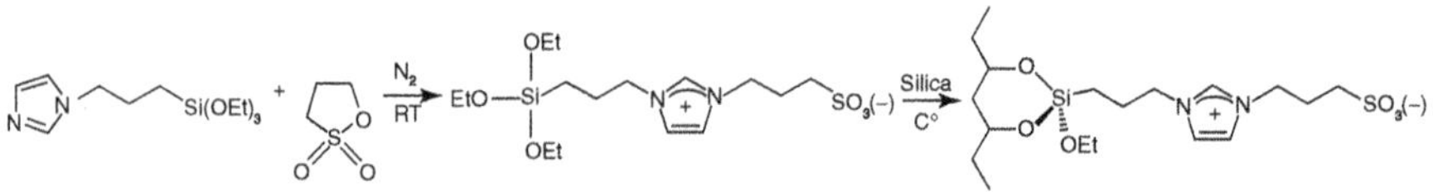

Figure 6.8 Preparation of a capillary coated with zwitterionic salt.

To reduce the interaction between the analytes and the capillary wall, as well as to modify the magnitude and direction of EOF, the imidazolium-based ILs were permanently bounded to the capillary wall in two different ways (Figure 6.7, heterogenous and homogenous) [65].

The effectiveness of the coating has been investigated by separating phenolic compounds in the nonaqueous media. The EOF was found to be anodic and dependent on the pH of the separation buffer. In another study [66], imidazole containing zwitterionic salt (N-3-(-triethoxysilylpropyl)-4,5-dihy-droimidazole) was attached to the silica capillary wall via the formation of a covalent bond (Figure 6.8).

In this case, the EOF was cathodic and its velocity remained almost constant in the pH range of 4–7. The separation performance of IL-coated capillary in an aqueous buffer (sorbic acid) was proved by the separation of seven alkylphosphonic acids and their esters. The separation was accomplished in <6 min.

For more comprehensive coverage of the topic see the corresponding chapter 5 in this book and reviews [26–29].

6.5   Concluding remarks

In conclusion, as follows from the discussion above, the chemistry of ILs as buffer systems in CE is rich and promising. Many questions remain still unanswered and the area of further research is extensive. IL applications have demonstrated its usefulness for separating difficult analytes like positional isomers and for charging neutral hydrophobic analytes in NACE.

Another important feature of ILs is that they can effectively reduce or reverse the EOF in the capillary. The latter is more pronounced if an IL is covalently attached to the capillary wall.

References

1. Compton, S. and Brownlee, R., Capillary electrophoresis, *Biotechniques*, 6, 432–440, 1988.
2. Kohlrausch, F., Über Konzentrations-Verschiebungen durch Elektrolyse im Inneren von Lösungen und Lösungsgemischen, *Ann. Phys. Chem., (Leipzig)*, 62, 209–239, 1897.
3. Tiselius, A. W. K., *The moving-boundary method of studying the electrophoresis of proteins,* Ph.D. thesis, University of Uppsala, Sweden, 1930.
4. Hjertén, S., Free zone electrophoresis, *Chromatogr. Rev.*, 9, 122–219, 1967.
5. Everaerts, F. M. and Hoving-Keulemans, W. M. L., Zone electrophoresis in capillary tubes, *Sci. Tools*, 17, 25–28, 1970.
6. Virtanen, R., Zone electrophoresis in a narrow-bore tube employing potentiometric detection. Theoretical and experimental study, *Acta Polytech. Scand.*, 123, 1–67, 1974.
7. Jorgenson, W. and Lukacs, K. D., High-resolution separations based on electrophoresis and electroosmosis, *J. Chromatogr.*, 218, 209–216, 1981.
8. For example, performing search with SciFinder® Scholar™ program (American Chemical Society) on the keyword: "capillary electrophoresis", results about 30 thousand hits.
9. Jouyban, A. and Kenndler, E., Theoretical and empirical approaches to express the mobility of small ions in capillary electrophoresis, *Electrophoresis*, 27, 992–1005, 2006.
10. Terabe, S., Otsuka, K., Ichikawa, K., Tsuchiya, A., and Ando, T., Electrokinetic separations with micellar solutions and open-tubular capillaries, *Anal. Chem.*, 56, 111–113, 1984.
11. Nishi, H., Tsumagari, N., and Terabe, S., Effect of tetraalkylammonium salts on micellar electrokinetic chromatography of ionic substances, *Anal. Chem.*, 61, 2434–2439, 1989.
12. Hayek, M., Electrophoresis in non-aqueous media, *J. Phys. Colloid. Chem.*, 55, 1527–1533, 1951.
13. Paul, M. H. and Durrum, E. L., Ionophoresis in nonaqueous solvent systems, *J. Am. Chem. Soc.*, 74, 4721–4722, 1952.
14. Walbroehl, Y. and Jorgenson, J. W., On-column UV absorption detector for open tubular capillary zone electrophoresis, *J. Chromatogr.*, 315, 135–143, 1984.
15. Subirats, X., Porras, S. P., Rosés, M., and Kenndler, E., Nitromethane as solvent in capillary electrophoresis, *J. Chromatogr. A*, 1079, 246–253, 2005.
16. Porras, S. P., Capillary zone electrophoresis of same extremely weak bases in acetonitrile, *Anal. Chem.*, 78, 5061–5067, 2006.
17. Qi, S., Ding, L., Tian, K., Chen, X., and Hu, Z., Novel and simple nonaqueous capillary electrophoresis separation and determination bioactive diterpenes in Chinese herb, *J. Pharm. Biomed. Anal.*, 40, 35–41, 2006.
18. Scriba, G. K. E., Nonaqueous capillary electrophoresis-mass spectrometry, *J. Chromatogr. A*, 1159, 28–41, 2007.

19. Sturm, S., Seger, C., and Stuppner, H., Analysis of Central European corydalis species by nonaqeous capillary electrophoresis-electrospray ion trap mass spectrometry, *J. Chromatogr. A*, 1159, 42–50, 2007.
20. Valko, I. E., Siren, H., and Riekkola, M.-L., Capillary electrophoresis in nonaqueous media: An overview, *LC-GC*, 15, 560–564, 1997.
21. Tjornelund, J. and Hansen, S. H., Non-aqueous capillary electrophoresis of drugs: Properties and application of selected solvents, *J. Biochem. Biophys. Methods*, 38, 139–153, 1999.
22. Okada, T., Nonaqueous ion-exchange chromatography and electrophoresis: Approaches to nonaqueous solution chemistry and design of novel separation, *J. Chromatogr. A*, 804, 17–28, 1998.
23. Riekkola, M.-L., Jussila, M., Porras, S. P., and Valk´o, I. E., Nonaqueous capillary electrophoresis, *J. Chromatogr. A*, 892, 155–170, 2000.
24. Porras, S. P., Kuldvee, R., Palonen, S., and Riekkola, M.-L., Capillary electro-phoresis of methyl-substituted phenols in acetonitrile, *J. Chromatogr. A*, 990, 35–44, 2003.
25. Okada, T., Non-aqueous capillary electrophoretic separation of Brønsted acids as heteroconjugated anions, *J. Chromatogr. A*, 771, 275–284, 1997.
26. Liu, J. F., Jonsson, J. A., and Jiang, G. B., Application of ionic liquids in analytical chemistry, *TrAC*, 24, 20–27, 2005.
27. Pandey, S., Analytical applications of room-temperature ionic liquids: A review of recent efforts, *Anal. Chim. Acta*, 556, 38–45, 2006.
28. Schubert, T. J. S., Ionische Flüssigkeiten, *Nachrichten aus der Chemie*, 53, 1222–1226, 2005.
29. Stalcup, A. M. and Cabovska, B., Ionic liquids in chromatography and capillary electrophoresis, *J. Liq. Chromatogr. Relat. Technol.*, 27, 1443–1459, 2004.
30. Chen, X. and Qi, S., The capillary electrophoresis based on ionic liquids, *Current Anal. Chem.*, 2, 411–419, 2006.
31. Vaher, M., Koel, M., and Kaljurand, M., Non-aqueous capillary electrophore-sis in acetonitrile using ionic-liquid buffer electrolytes, *Chromatographia*, 53, S302–S306, 2001.
32. Vaher, M., Koel, M., and Kaljurand, M., Ionic liquids as electrolytes for nonaqueous capillary electrophoresis, *Electrophoresis*, 23, 426–430, 2002.
33. Vaher, M., Koel, M., and Kaljurand, M., Application of 1-alkyl-3-methy-limidazolium-based ionic liquids in non-aqueous capillary electrophoresis, *J. Chromatogr. A*, 979, 27–32, 2002.
34. Vaher, M. and Koel, M., Separation of polyphenolic compounds extracted from plant matrices using capillary electroporesis, *J. Chromatogr. A.*, 990, 225–230, 2003.
35. Kuldvee, R., Vaher, M., Koel, M., and Kaljurand, M., Heteroconjugation-based capillary elctrophoretic separation of phenolic compounds in acetonitrile and propylene carbonate, *Electrophoresis*, 24, 1627–1634, 2003.
36. Miller, J. L., Shea, D., and Khaledi, M. G., Separation of acidic solutes by nonaque-ous capillary electrophoresis in acetonitrile-based media. Combinated effects of deprotonation and heteroconjugation, *J. Chromatogr. A*, 888, 251–266, 2000.
37. Vaher, M. and Koel, M., Specific background electrolytes for nonaqueous capillary electrophoresis, *J. Chromatogr. A*, 1068, 83–88, 2005.
38. Borissova, M., Gorbatšova, J., Ebber, A., Kaljurand, M., Koel, M., and Vaher, M., Non-aqueous capillary electrophoresis using contactless conductivity detection and ionic liquids as background electrolytes in acetonitrile, *Electrophoresis*, 28, 3600–3605, 2007.

39. François, Y., Zhang, K., Varenne, A., and Gareil, P., New integrated measurement protocol using capillary electrophoresis instrumentation for the determination of viscosity, conductivity and absorbance of ionic liquid–molecular solvent mixtures, *Anal. Chim. Acta*, 562, 164–170, 2006.

40. Seiman, A., Vaher, M., and Kaljurand, M., Monitoring of electroosmotic flow of solution of ionic liquid in nonaqueous media using thermal marks. *J. Chromatogr. A*, 1189, 266–273, 2008.

41. Francois, Y., Varenne, A., Juillerat, E., Servais, A.-C., Chiap, P., and Gareil, P., Nonaqueous capillary electrophoretic behavior of 2-aryl propionic acids in the presence of an achiral ionic liquid: A chemometric approach, *J. Chromatogr. A*, 1138, 268–275, 2007.

42. Tholey, A. and Heinzle, E., Ionic (liquids) matrices for matrix-assisted laser desorption/ionization mass spectrometry – applications and perspectives, *Anal Bioanal. Chem.*, 386, 24–37, 2006.

43. Yanes, E. G., Gratz, S. R., Baldwin, M. J., Robison, S. E., and Stalcup, A. M., Capillary electrophoretic application of 1-alkyl-3-methylimidazolium-based ionic liquids, *Anal. Chem.*, 73, 3838–3844, 2001.

44. Zeng, H.-L., Shen, H., Nakagama, T., and Uchiyama, K., Property of ionic liquid in electrophoresis and ots application in chiral separation on microchips, *Electrophoresis*, 28, 4590–4596, 2007.

45. Jiang, T.-F., Gu, Y.-L., Liang, B., Li, J.-B., Shi, Y.-P., and Ou, Q.-Y., Dynamically coating the capillary with 1-alkyl-3-methylimidazolium based ionic liquids for separation of basic proteins by capillary electrophoresis, *Anal. Chim. Acta*, 479, 249–254, 2003.

46. Cabovska, B., Kreishman, G. P., Wassell, D. F., and Stalcup, A. M., Capillary electrophoretic and nuclear magnetic resonance studies of interactions between halophenols and ionic liquid or tetraalkylammonium cations, *J. Chromatogr. A*, 1007, 179–187, 2003.

47. Yu, L., Qin, W., and Li, S. F. Y., Ionic liquids as additives for separation of benzoic acid and chlorophenoxy acid herbicides by capillary electrophoresis, *Anal. Chim. Acta*, 547, 165–171, 2005.

48. Mwongela, S. M., Numan, A., Gill, N. L., Agbaria, R. A., and Warner, I. M., Separation of achiral and chiral analytes using polymeric surfactants with ionic liquids as modifiers in micellar eletrokinetic chromatography, *Anal. Chem.*, 75, 6089–6096, 2003.

49. Qi, S., Cui, S., Chen, X., and Hu, Z., Rapid and sensitive determination of anthraquinones in Chinese herb using 1-butyl-3-methylimidazolium-based ionic liquid with β-cyclodextrin as modifier in capillary zone electrophoresis, *J. Chromatogr. A*, 1059, 191–198, 2004.

50. Laamanen, P.-L., Busi, S., Lahtinen, M., and Matilainen, R., A new ionic liquid dimethyldinonylammonium bromide as a flow modifier for the simultaneous determination of eight carboxylates by capillary electrophoresis, *J. Chromatogr. A*, 1095, 164–171, 2005.

51. Wang, Y. L., Hu, Z. B., and Yuan, Z. B., Ionic liquid and HP-β-CD modified capillary zone electrophoresis to separate hyperoside, luteolin and chlorogenic acid, *Chinese Chem. Lett.*, 17, 231–234, 2006.

52. Tian, K., Qi, S., Cheng, Y., Chen, X., and Hu, Z., Separation and determination of lignans from seeds of Schisandra species by micellar electrokinetic capillary chromatography using ionic liquid as modifier, *J. Chromatogr. A*, 1078, 181–187, 2005.

53. Mata, J., Varade, D., Ghosh, G., and Bahadur, P., Effect of tetrabutylammonium bromide on the micelles of sodium dodecyl sulfate, *Colloids Surf. A: Physicochem. Eng. Aspects,* 245, 69–73, 2004.

54. Marszałł, M. P., Markuszewski, M. J., and Kaliszan, R., Separation of nicotinic acid and its structural isomers using 1-ethyl-3-methylimidazolium ionic liquid as a buffer additive by capillary electrophoresis, *J. Pharm. Biomed. Anal.,* 41, 329–332, 2006.

55. Yue, M.-E. and Shi, Y.-P., Application of 1-alkyl-3-methylimidazolium-based ionic liquids in separation of bioactive flavonoids by capillary zone electrophoresis, *J. Sep. Sci.,* 29, 272–276, 2006.

56. Jiang, T.-F., Wang, Y.-H., and Lv, z.-H., Dynamic coating of a capillary with room-temperature ionic liquids for the separation of amino acids and acid drugs by capillary electrophoresis, *J. Anal. Chem.,* 61, 1108–1112, 2006.

57. Qi, S., Li, Y., Deng, Y., Cheng, Y., Chen, X., and Hu, Z., Simultaneous determination of bioactive flavone derivatives in Chinese herb extraction by capillary electrophoresis used different electrolyte systems—borate and ionic liquids, *J. Chromatogr. A,* 1109, 300–306, 2006.

58. Gao, Y., Xiang, Q., Xu, Y., Tian, Y., and Wang, E., The use of CE-electrochemiluminescence with ionic liquid for the determination of bioactive constituents in Chinese traditional medicine, *Electrophoresis,* 27, 4842–4848, 2006.

59. Schnee, V. P., Baker, G. A., Rauk, E., and Palmer, C. P., Electrokinetic chromatographic characterization of novel pseudo-phases based on N-alkyl-N-methylpyrrolidinium ionic liquid type surfactants, *Electrophoresis,* 27, 4141–4148, 2006.

60. Markuszewski, M. J., Stepnowski, P., and Marszall, P., Capillary electrophoretic separation of cationic constituents of imidazolium ionic liquids, *Electrophoresis,* 25, 3450–3454, 2004.

61. Qin, W. and Li, S. F. Y., An ionic liquid coating for determination of sildenaftyl and UK-103, 320 in human serum by capillary zone electrophoresis – ion trap mass spectrometry, *Electrophoresis,* 23, 4110–4116, 2002.

62. Qin, W. and Li, S. F. Y., Electrophoresis of DNA in ionic liquid coated capillary, *Analyst,* 128, 37–41, 2003.

63. Qin, W., Wei, H., and Li, S. F. Y., 1,3-Dialkylimidazolium-based room-temperature ionic liquids as background electrolyte and coating material in aqueous capillary electrophoresis, *J. Chromatogr. A,* 985, 447–454, 2003.

64. Qin, W. and Li, S. F. E., Determination of ammonium and metal ions by capillary electrophoresis–potential gradient detection using ionic liquid as background electrolyte and covalent coating reagent, *J. Chromatogr. A,* 1048, 253–256, 2004.

65. Borissova, M., Koel, M., and Kaljurand, M., Ionic liquids for silica modification: Assessment by capillary zone electrophoresis, in Rogers, R. D. and Seddon, K. R. (Eds.), *ACS Symposium Series 975, Ionic Liquids: Not Just Solvents Anymore,* ACS, Washington, D.C., 35–46, 2007.

66. Borissova, M., Vaher, M., Koel, M., and Kaljurand, M., Capillary zone electrophoresis on chemically bonded imidazolium based salts, *J. Chromatogr. A,* 1160, 320–325, 2007.

chapter seven

Ionic liquids as stationary phases in countercurrent chromatography

Alain Berthod, Maria-Jose Ruiz-Angel, and Samuel Carda-Broch

Contents

7.1   Introduction

Countercurrent chromatography (CCC) is a separation technique that uses a support-free liquid stationary phase [1]. Since the mobile phase is also liquid, biphasic liquid systems are used. Ionic liquids (ILs), as a new class of solvents, should be evaluated in CCC.

CCC is that uses multipartitioning of solutes between two liquid phases to separate them. The affinity of the solutes for the two liquid phases, called solute partitioning, will be the only physicochemical parameter responsible for solute separation by CCC. Since the CCC technique is not yet well known in the scientific community, it will be briefly described and its capabilities will be exposed: why is it so interesting to work with a liquid stationary phase?

The properties of the ILs are extensively described by experts in other chapters of this book, so, only the specific properties essential when working with CCC will be briefly tackled.

7.2   Countercurrent chromatography

7.2.1   A liquid stationary phase

CCC works with a liquid stationary phase and a biphasic liquid system. Actually this is the main advantage and, at the same time, the main problem of the technique. The liquid stationary phase is used for its loading properties. Since the solutes have access to the volume of the liquid phase, slight overload problems occur. In the preparative liquid chromatography (LC) with the classical solid stationary phase, the solutes exchange between the mobile phase volume and the surface of the solid stationary phase. So this surface is rapidly saturated and overload problems occur. In high performance liquid chromatography (HPLC), the active stationary phase has a very limited volume (Figure 7.1).

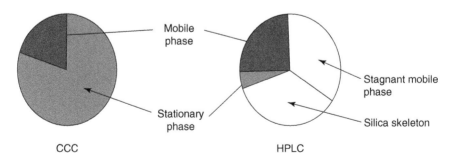

Figure 7.1 Comparison of phase volumes inside CCC and HPLC column. The phase ratio, V_S/V_M is between 0.9 and 9 in a CCC column and 0.01 and 0.1 in a silica-based HPLC column.

In CCC, there are nothing else but mobile and stationary phases inside the column. Since it is possible to work with very high concentrations of solutes in CCC, the technique is mainly used in the preparative conditions. If the liquid nature of the stationary phase is the main advantage of CCC, it is also its main problem. Two frits at both ends of the column are what is needed to maintain a solid stationary phase. A liquid stationary phase, without any support, is difficult to maintain really stationary.

7.2.2   A very simple retention mechanism

Partitioning of the solutes between the two liquid phases is the only chemical mechanism responsible for solute separation. The retention equation is

$$V_R = V_M + K_D V_S \tag{7.1}$$

where V_R, V_M, and V_S denote retention, mobile, and stationary phase volumes, respectively. K_D is the solute distribution ratio or partition coefficient in the biphasic liquid system used for separation. It is expressed as the ratio of the solute concentration in the stationary phase over the solute concentration in the mobile phase. As shown in Figure 7.1, the sum $V_S + V_M$ is actually the column volume V_C. So Equation 7.1 can be equally written as

$$V_R = V_C + (K_D - 1)V_S \tag{7.2}$$

The retention equation allows us to understand the first major difference between the CCC solute retention and the retention obtained with any other chromatographic technique. Usually, in chromatography, the same solute mixture separated on the same column and using the same mobile phase produces the same chromatogram. If it is not the case, it is a sign of column wearing or problems in the hardware (pump, detector, or injector).

Figure 7.2 shows that in CCC, the observed chromatogram significantly depends on the V_S/V_M phase ratio that is expressed using the stationary phase retention parameter Sf as

$$Sf = V_S/V_C \tag{7.3}$$

For any CCC column, Sf is between 0 (no stationary phase is retained) and 1 (no mobile phase inside the column). Often, Sf is expressed in percentage. Figure 7.2 also shows clearly that the resolution between peaks drastically depends on Sf. Naturally, if there is no stationary phase retained in a CCC column, there is no possible separation. All injected solutes will elute together at the column volume (V_C).

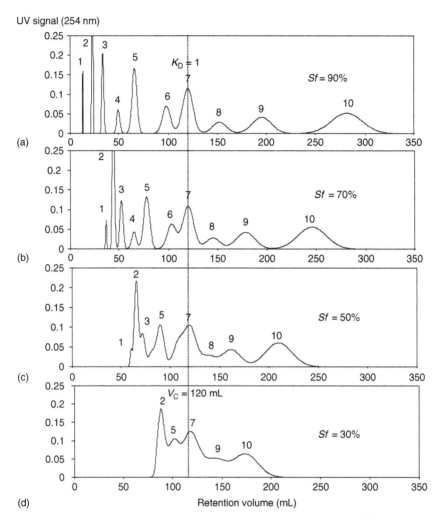

Figure 7.2 Separation of a 10-solute sample with the same liquid system and the same 120 mL CCC column. (a) 108 mL of stationary phase are retained at 1400 rpm and 1 mL/min mobile phase flow rate. (b) V_S = 84 mL at 1100 rpm, 1 mL/min. (c) V_S = 60 mL at 800 rpm, 1 mL/min. (d) V_S = 36 mL at 600 rpm, 1 mL/min. The dotted vertical line corresponds to the column volume and compound 7 with K_D = 1 (Equation 7.2). (Adapted from Berthod, A., *Countercurrent Chromatography: The Support-Free Liquid Stationary Phase*, Elsevier, Amsterdam, 2002.)

7.2.3   *Two types of countercurrent chromatography columns*

As noted previously, it is difficult to maintain a liquid stationary phase without any support, especially when an immiscible liquid mobile phase is passed through it. It is for this reason that centrifugal fields are used in all modern CCC column applications. There are two types of CCC columns: a hydrodynamic

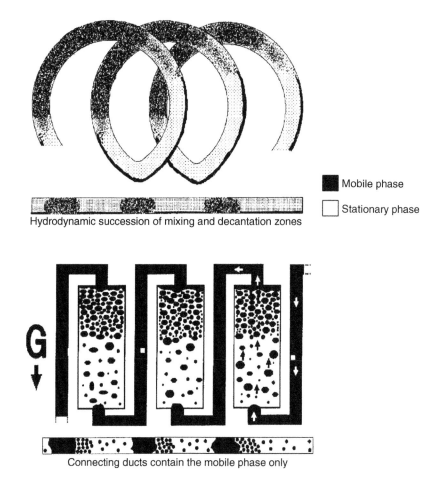

Mobile phase

Stationary phase

Hydrodynamic succession of mixing and decantation zones

Connecting ducts contain the mobile phase only

Figure 7.3   Sketch illustrating the two kinds of CCC columns. *(Top) Hydrodynamic* columns contain a long coiled tube rotating in a planetary way that creates a succession of mixing and decantation zones always having the two liquid phases in contact. *(Bottom)* In *hydrostatic* columns, the stationary phase is contained in channels interconnected by ducts in which there is only mobile phase.

column and a hydrostatic column, because there are two ways to contain a liquid stationary phase. Figure 7.3 is an oversimplified sketch illustrating the two hydrodynamic and hydrostatic principles.

In hydrodynamic columns, there are spools of coiled tube that rotate on themselves and around a central axis. These combined rotations create a planetary motion with a highly variable centrifugal field that produces mixing zones followed by decantation zones (Figure 7.3). The stationary phase is partly retained inside the coils if the mobile phase is flown the right way. The coil rotation produces an Archimedean force that pushes the liquid phases toward one end of the coil called the head (higher pressure).

The other end is called the tail (lower pressure). If the mobile phase is the lighter (upper) liquid, it should be flown entering the CCC column tail side and exiting from the head side. If the mobile phase is the denser and lower liquid phase, it should be flown in the head-to-tail direction [1].

In hydrostatic columns, also called centrifugal partition chromatographs [2], there are disks containing engraved chambers or channels intercon-nected by ducts. The stationary phase is maintained in the channels by a constant centrifugal field produced by the disk rotation. Here also, the mobile phase should be flown in the right direction. Figure 7.3 shows in black the light upper mobile phase that should enter through the channel bottom to move against the constant centrifugal field G in an ascending way. The bottom-to-top ascending way is similar to the tail-to-head way of the hydro-dynamic column. It would be the opposite or descending way with a denser or lower mobile phase [2].

Table 7.1 compares the properties and features of the two types of CCC col-umns and Figure 7.4 shows two examples of such different chromatographs.

The good points of the hydrostatic columns are their silent and smooth running and their ability to retain any biphasic liquid system, including the aqueous two-phase systems (ATPS). However, there is a pressure build-up in hydrostatic columns. Indeed, to force the liquid mobile phase through the liquid stationary phase contained in a particular channel, a small hydro-static pressure is needed. This small pressure depends on the centrifugal field (rotor rotation speed), liquid density difference, and the height of the stationary-phase liquid inside the channel. Since liquids transmit pressure, this small hydrostatic pressure adds up channel after channel and becomes significant inside the hydrostatic column. Eventually, a pumping system

Table 7.1   Main Features of the Two Types of Countercurrent
Chromatography Columns

Countercurrent chromatography column	Hydrostatic	Hydrodynamic
Name and acronym	Centrifugal partition chromatograph (CPC)	High-speed countercurrent chromatograph (HSCCC)
Liquid phase retained in	Channels and ducts	Coiled tubing
Number of axes of rotation	1	2
Centrifugal field	Constant	Variable
Pressure	Moderate (10–60 kg/cm^2)	Low (0.5–5 kg/cm^2)
Noise	Silent centrifuge	Noisy gear assembly
Maintenance	Rotating seals to lubricate and check	Flying leads to change every ~500 h
General comments	Good liquid phase retention, lower plate number or efficiency	Difficult retention of some liquid systems, higher plate number or efficiency

Figure 7.4 **(See color insert following page 224.)** (*Left*) The fast centrifugal partition chromatograph (FCPC) 100 chromatograph with hydrostatic column of 100 mL
with integrated alternating pump and pilot electronics (www.armen-instrument.
com, Vannes, France); the upper inset shows a part of a disk with the engraved double channels and connecting ducts. (*Right*) The Milli chromatograph with hydrodynamic CCC column of 18 mL (www.DynamicExtractions.com, Uxbridge, UK).

able to produce 20/50 kg/cm² is needed to operate the column. Hydrostatic
columns have a limited efficiency (relatively low plate number per column
volume) compared to the hydrodynamic columns that have a better plate
number per column volume and work at a very low pressure (few kilograms
per square centimeters) and no rotary seals. But there are biphasic liquid
systems, such as the ATPS, that hydrodynamic columns have serious problems to retain and these columns generate a significant amount of noise due
to the gear assembly needed to produce the planetary motion. At the moment,
both types of CCC columns have been equally developed and several com-
panies produce different column sizes, from the low analytical volume
(~20 mL) up to the high preparative column volume (25 L).

7.2.4   The biphasic liquid system

A major difference between CCC and HPLC is the choice of chromatographic conditions. In HPLC, the column (=the stationary phase) is selected:
particle size, pore size and volume, bonding chemistry, column length, and
diameter. For example, very often an octadecyl (C_{18})-bonded silica column is
selected when working in the reversed phase LC (RPLC). Next, the mobile
phase is prepared, independently selecting an organic modifier, methanol
or acetonitrile, a pH value, and a working temperature. This is absolutely
not the way to work in CCC. Both the mobile and the stationary phases must
be selected together. Since CCC works with biphasic liquid systems in equilibrium, any change in one phase will induce a change in the other phase.

Biphasic liquid systems used in CCC may have a wide variety of polarities. The most polar systems are the ATPS made by two aqueous-liquid phases, one containing a polymer, for example, polyethylene glycol (PEG), the other one being a salt solution, for example, sodium hydrogen phosphate. The less polar systems do not contain water; there can be two-solvent systems, such as heptane/acetonitrile or dimethylsulfoxide/hexane systems or mixtures of three or more solvents. Intermediate polarity systems are countless since any proportion of three or more solvents can be mixed. Ternary phase diagrams are used when three solvents are mixed together.

7.3   Liquid systems with ionic liquids

7.3.1   A new class of solvents

As abundantly exposed in other chapters of this book, ILs are a new class of solvents. Recently, Seddon stated: *Years ago, I predicted that ionic liquids would change the face of organic chemistry. It is clear now that they have the potential to revolutionize all activities where liquids can be used* [3]. Since CCC uses liquid systems to separate molecules, ILs should be considered.

Pure ILs have a dual nature since they are actually molten salts or a mixture of cations and anions. They were found to have a relatively high solvent polarity, comparable to that of short-chain alcohols [4–5]. Since CCC needs to work with a biphasic liquid system, water-insoluble ILs should be selected if an aqueous phase is desired. 1-butyl-3-methylimidazolium hexafluorophosphate ($[C_4C_1Im][PF_6]$) has limited water solubility (18 g/L or 1.3% or 63 mM [5]) and is easy to synthesize. It was the first IL used in CCC [6].

An important property of ILs is their viscosity often higher than that of common molecular solvents. Liquids having viscosity >5–10 cP (0.005–0.01 Pa s) cannot be handled conveniently by classical chromatographic pumping equipment. Practical flow rates (1–5 mL/min) would generate very high pressure. It is possible to fill the CCC columns with ILs at low flow rates or by gravity and then to work using the ILs as stationary phases pushing the less-viscous liquid mobile phases at reasonable flow rates.

IL viscosity is extremely sensitive to additives [5]. Mixtures of IL and compatible solvents and water may produce biphasic liquid systems usable in CCC. The short-chain alcohol–$[C_4C_1Im][PF_6]$–water and acetonitrile–$[C_4C_1Im][PF_6]$–water ternary phase diagrams have been studied [7]. Alcohols were found to have a tendency to dissolve preferentially in the aqueous upper phase producing an IL lower phase of limited volume and high viscosity [7]. Acetonitrile partitions well between the upper aqueous phase and the lower IL phase, greatly reducing the viscosity of the IL-rich lower phase.

7.3.2   The acetonitrile–$[C_4C_1Im][PF_6]$–water ternary phase diagram

Figure 7.5 shows the acetonitrile–$[C_4C_1Im][PF_6]$–water ternary phase diagram [7]. The three apexes of the right-angle triangle correspond to the

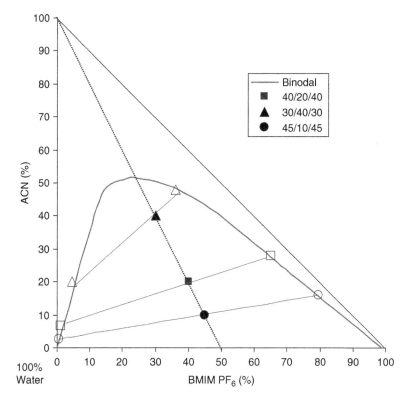

Figure 7.5 The acetonitrile–[C₄C₁Im][PF₆]–water ternary mass phase diagram at room temperature. Closed symbols: compositions belonging to the 50:50 w/w [C₄C₁Im][PF6]–water initial mixtures (dotted line) that separate in two phases whose compositions are listed in Table 7.2 and located by the corresponding open symbols and tie-lines. (Adapted from Berthod, A. and Carda-Broch, S., *J. Liq. Chromatogr. & Rel. Technol.*, 26, 1493–1508, 2003.)

three pure solvents. The thick line, called a binodal line, delineates the compositions producing a single phase and those producing two phases. Three compositions were studied. Each composition separates in two liquid phases.

Table 7.2 shows full compositions and properties. The partition of acetonitrile is about one-third in the upper aqueous phase and two-thirds in the lower IL phase. Then starting with an equal mass of IL and water, the upper phase occupies always a lower volume or has a lower mass than the lower phase. The best composition selected for CCC was water–acetonitrile–[C₄C₁Im][PF₆] (40:20:40 % w/w). This composition produced a good density difference between the two liquid phases (0.13 g/cm³, Table 7.2) as well as an IL lower phase viscosity of 3 cP (0.003 Pa s), low enough to allow for smooth pump operation.

Table 7.2 Compositions of the Acetonitrile/[C₄C₁Im][PF₆]/Water Biphasic Liquid Systems Shown in Figure 7.5

	Point ●			Point ■			Point ▲		
	H₂O	Acetonitrile	[C₄C₁Im][PF₆]	H₂O	Acetonitrile	[C₄C₁Im][PF₆]	H₂O	Acetonitrile	[C₄C₁Im][PF₆]
% w/w	45	10	45	40	20	40	30	40	30
Aqueous upper liquid phase									
Density (g/cm³)		0.975			0.96			0.94	
% w/w	96.8	2.5	0.7	92.2	7.8	1.0	75.5	20.0	4.5
% v/v	96.3	3.2	0.5	90.7	8.6	0.72	72.3	24.6	3.1
Molecular fraction	98.8	1.1	0.04	96.8	3.1	0.07	89.3	10.4	0.33
Ionic liquid-rich lower liquid phase									
Density (g/cm³)		1.19			1.09			0.96	
Viscosity (cP or mPa s)		14			3			n.d	
% w/w	4.5	16.0	79.5	7.0	28.0	65.0	16.0	48.0	36.0
% v/v	5.4	25.0	69.6	7.8	40.1	52.1	15.5	59.5	25.0
Molecular fraction	27.0	42.5	30.5	29.8	52.6	17.6	40.7	53.5	5.8
Upper over lower phase ratio									
In mass		0.785 (44%/56%)			0.639 (39%/61%)			0.333 (25%/75%)	
In volume		0.639 (39%/61%)			0.563 (36%/64%)			0.316 (24%/76%)	
Δd g/cm³		0.215			0.31			0.02	

Note: Viscosity is expressed on centipoises (cP) or millipascal second (mPa s); n.d.: the viscosity was not determined because it was too low for our apparatus (lower than 2 cP or mPa s).

7.3.3 An aqueous two-phase system with an ionic liquid

The two liquid layers of ATPS are aqueous solutions. Since their introduction by Albertson, ATPS have proved extremely useful to separate and purify biological material, especially proteins [8]. Earlier they were used in CCC for this purpose [9–10]. It has been reported recently that [C$_4$C$_1$Im]Cl, a fully water-soluble IL, is able to form two aqueous phases when K$_2$HPO$_4$ is added to its solution [11]. This biphasic system was used to fractionate opium alkaloids, demonstrating its ability to dissolve molecules of intermediate polarity differently [12].

Figure 7.6 shows the [C$_4$C$_1$Im]Cl–K$_2$HPO$_4$–water mass ternary phase diagram at room temperature. In that case there are three different zones: a single phase zone toward the water apex; a biphasic zone containing

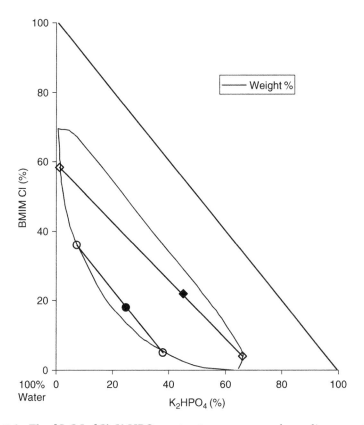

Figure 7.6 The [C$_4$C$_1$Im]Cl–K$_2$HPO$_4$–water ternary mass phase diagram at room temperature. Closed symbols: compositions prepared by weighting proportions that separate in two aqueous phases whose compositions are listed in Table 7.3 and located by the corresponding open symbols and tie-lines. (Adapted from Ruiz-Angel, M.J., Pino, V., Carda-Broch, S., Berthod, A., *J. Chromatogr., A*, 1151, 65–73, 2007. With permission.)

Table 7.3   Chemical Compositions of the Aqueous Two-Phase Systems
[C_4C_1Im]Cl/K_2HPO_4/Water Corresponding to the Two Tie-Lines
Determined and Presented in Figure 7.6

	Point ●			Point ◆		
	[C_4C_1Im]Cl	K_2HPO_4	Water	[C_4C_1Im]Cl	K_2HPO_4	Water
% w/w	18	25	57	22	45	33
Upper phase	○ d = 1.095 (Δd = 0.147 g/mL)			◇ d = 1.072 (Δd = 0.595 g/mL)		
% w/w	36	7.5	56.5	58.5	1.0	40.5
Molecular fraction	6.1	1.3	92.6	12.9	0.2	86.9
Lower phase	○ d = 1.242 g/mL			◇ d = 1.667 g/mL		
% w/w	5	38	57	4	66	30
Molecular fraction	0.8	6.4	92.8	1.1	18.3	80.6
Upper over lower phase ratio						
In mass	0.736 (42.4%/57.6%)			0.50 (33.3%/66.7%)		
In volume	0.834 (45.5%/54.5%)			0.776 (43.7%/56.3%)		

Note: [C_4C_1Im]Cl = 1N-butyl-3N-methylimidazolium chloride.
        Δd is the density difference between the lower and the upper phase.

intermediate amounts of IL and phosphate salt, and a zone with saturated salts and the two liquid phases close to the hypotenuse of the diagram. The bimodal curve delineates the two liquid-phase zone usable in CCC.

Two compositions marked by close symbols were prepared and the two liquid phases obtained were fully analyzed as listed in Table 7.3.

The IL is mainly located in the upper phases of the ATPS. The lower phases contain the phosphate salt. Water is partitioning almost evenly between the two-liquid phases. By definition, all initial compositions belonging to the same tie-line split into two liquid phases of identical composition; only the volume amounts (and the phase ratios) change. Since the tie-lines are, by chance, almost parallel to the water hypotenuse, the phase diagram clearly shows that the chemical composition of the two liquid phases crucially depends on the water content of the system.

7.4   Ionic liquids and countercurrent chromatography

Pure ILs and liquid phases with viscosities higher than 5 cP (0.005 Pa s) cannot be used easily in CCC. The mobile phase has a problem when moving through a viscous stationary phase and vice versa. The two systems presented in Figures 7.5 and 7.6 form the liquid phases of usable viscosities. Their capabilities were studied with CCC columns.

7.4.1   Liquid stationary-phase retention

A necessary condition to separate solutes in CCC is that the column retains some stationary phase. Equation 7.3 defines the stationary phase retention parameter *Sf*. It was demonstrated that *Sf* depends on the CCC column used and the CCC conditions: rotor rotation speed and mobile phase flow rate. The *Sf* parameter depends on the mobile phase flow rate *F* according to

$$Sf = A - BF^{1/2} \tag{7.4}$$

where *A* and *B* are constants that depend on the strength of the centrifugal field (or on the rotor rotation speed). *A* is equal to or very close to unity or 100% [13].

7.4.1.1   *[C₄C₁Im][PF₆]–Acetonitrile–water liquid system*

The denser IL-rich liquid phases were used as the stationary phase and the upper aqueous phase was the mobile phase pumped in the tail-to-head (ascending) direction.

Table 7.4 lists the IL-rich stationary phase retention obtained with a 53 mL hydrodynamic CCC column rotating at difference speeds. As expected, the constant *A* is close to 1, and the   constant *B* decreased from 0.76 to

Table 7.4   Regression Parameters of the $Sf = A - BF^{1/2}$ Lines

[C₄C₁Im][PF₆]–Acetonitrile–water system						
% w/w	Rotor speed[a] (rpm)	*A*	*B*	r^2	*Sf* at 1 mL/ min (%)	$F_{Sf=0}$[b] (mL/min)
40:20:40	519	0.94	0.76	0.998	18	1.5
	626	1.01	0.44	0.989	57	5.3
	727	1.04	0.40	0.989	64	6.8
	829	1.03	0.32	0.981	71	10.4
	928	0.94	0.17	0.975	77	30.6
45:10:45	534	0.98	0.67	0.984	31	2.1
	631	1.04	0.54	0.986	50	3.7
	731	0.96	0.35	0.992	61	7.5
	835	0.91	0.25	0.991	66	13.2
	932	0.86	0.17	0.999	69	25.6
[C₄C₁Im]Cl 15.6–K₂HPO₄ 23.1–water 61.3% w/w aqueous two-phase systems						
Hydrodynamic column[a]	500	0.66	0.36	0.91	30	3.3
	900	0.80	0.41	0.94	39	2.0
Hydrostatic column[c]	500	0.99	0.29	0.988	70	11.6
	900	1.03	0.31	0.991	72	11.0

[a] Hydrodynamic CCC column of 53 mL, IL-rich stationary phase; the flow rate, *F*, was in the 0.5–3 mL/min range in the tail-to-head direction.
[b] $F_{Sf=0} = (A/B)2$ is the *flush away* flow rate.
[c] Hydrostatic CCC column of 101 mL, 1060 channels. The flow rate was in the range of 0.5–3 mL/ min in the head-to-tail (hydrodynamic) or descending (hydrostatic CCC columns).

0.17 mL$^{-1/2}$min$^{1/2}$ as the rotor rotation increased from 520 to 930 rpm. From a practical point of view, it means that the CCC column was able to retain more than 70% of its volume in the stationary phase when it was rotating at more than 800 rpm and when the mobile phase flow rate was ≤1.5 mL/min.

7.4.1.2 *[C$_4$C$_1$Im]Cl–K$_2$HPO$_4$–water aqueous two-phase systems*

With the ATPS, the IL-rich phase is the lighter upper phase (Table 7.3). It implies that the denser mobile phase should be flown in the head-to-tail direction in a hydrodynamic CCC column or in the descending direction in a hydrostatic column. Table 7.4 (bottom) shows that the IL phase retention by our hydrodynamic CCC column was extremely limited (less than 40% at 900 rpm and 1 mL/min). It has always been observed that ATPS with PEG and phosphate salts are difficult to retain in hydrodynamic CCC columns [10,14]. Hydrostatic columns are more efficient in retaining the IL-rich aqueous phase. Fifty percent of the hydrostatic column volume was the IL-rich aqueous phase when the descending phosphate-phase flow rate was ≤1 mL/min and the rotor rotation was at least 500 rpm (Table 7.4).

The study of the IL-rich upper aqueous phase retention of the [C$_4$C$_1$Im]Cl–K$_2$HPO$_4$–water ATPS shows that hydrostatic columns are able to retain liquid systems poorly retained by hydrodynamic CCC columns. A few trials were done using the IL-rich phase as the mobile phase pushed in the ascending way. An acceptable retention of the phosphate-rich aqueous phase (72% at 1 mL/min and 900 rpm) was obtained with the hydrostatic column when no retention (*Sf* = 0%) could be obtained in all conditions with the hydrodynamic column.

7.4.2 *Solute partitioning in ionic liquid-containing liquid systems*

Equation 7.1 shows that the retention volume of a solute can be used to estimate its distribution constant, or the partition coefficient in the liquid system used in the CCC column:

$$K_D = (V_R - V_M)/V_S \qquad (7.5)$$

Table 7.5 lists the partition coefficients $K_{IL/w}$ of a variety of chemicals obtained in the 40/20/40 [C$_4$C$_1$Im][PF$_6$]–acetonitrile–water system [6]. Figure 7.7 is a typical example of the $K_{IL/w}$ measurement for 4-nitrobenzoic acid.

A 53 mL hydrodynamic CCC machine was equilibrated with the 40/20/40 biphasic liquid system. The results at equilibrium and 830 rpm were: V_M = 15 mL and V_S = 53−15 = 38 mL of IL-rich liquid stationary phase retained in the hydrodynamic column. The injection of the 4-nitrobenzoic solution at a mobile phase flow rate of 1 mL/min produced the chromatogram in Figure 7.7. The retention volume of 4-nitrobenzoic acid is 18.6 mL. Equation 7.5 gives the $K_{IL/w}$ coefficient of (18.6−15)/38 = 0.095.

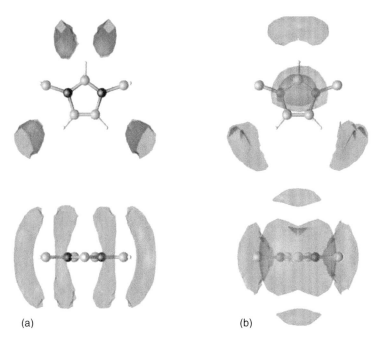

Color Figure 2.6 Probability densities of chloride anions about a central $[C_1C_1Im]^+$ calculated from MD simulations using two different classical forcefields: (a) that of Canongia-Lopes [21], which has an overall negatively charged imidazolium ring (C and N atoms only) and (b) a model that has an overall positively charged imidazolium ring. Both surfaces are drawn at five times the bulk density of anions.

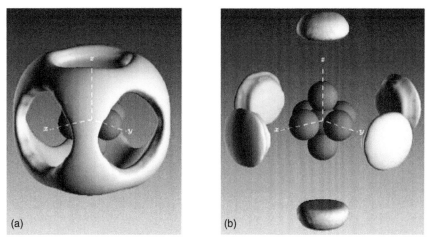

Color Figure 2.7 Probability densities determined by empirical potential structure refinement from neutron diffraction studies of $[C_1C_1Im][PF_6]$. Densities are for (a) $[PF_6]^-$ and (b) $[C_1C_1Im]^+$ around a central anion.

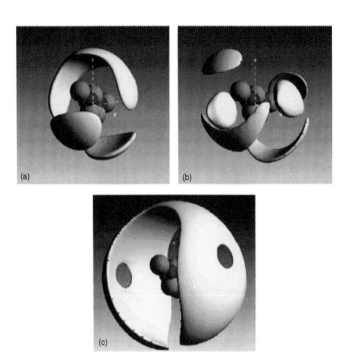

Color Figure 2.8 Probability densities determined by empirical potential structure refinement from neutron diffraction studies of $[C_1C_1Im][PF_6]$ containing 33 mol% benzene. Densities are for (a) $[PF_6]^-$, (b) benzene, and (c) $[C_1C_1Im]^+$ around a central cation. Surfaces are drawn to encompass the top 25% of ions within 8 Å. for the anion and benzene and 10 Å. for the cation. (From Deetlefs, M., Hardacre, C., Nieuwenhuyzen, M., Sheppard, O., and Soper, A. K., *J. Phys. Chem. B*, 109, 1593–1598, 2005. With permission.)

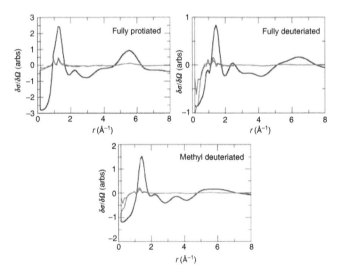

Color Figure 2.10 Simulated $S(Q)$ data for a pure system of $[C_1C_1Im][Tf_2N]$ (black line) and differences between this and data for 14% mole fractions of PCl_3 (red line) and $POCl_3$ (blue line) for fully-protiated, fully-deuteriated, and methyl-deuteriated systems.

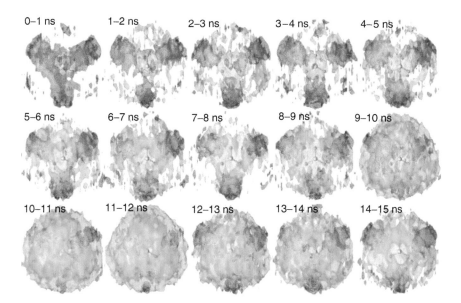

Color Figure 2.11 Probability densities determined by MD studies of 14 mol% solutions of PCl$_3$ in [C$_1$C$_1$Im][Tf$_2$N], calculated over consecutive nanosecond intervals in a continuous 15 ns simulation. Densities are for [Tf$_2$N]$^-$ (red), [C$_1$C$_1$Im]$^+$ (blue), and PCl$_3$ (green) around a central PCl$_3$. Surfaces are drawn at twice ([C$_1$C$_1$Im]$^+$ and [Tf$_2$N]$^-$) and three times (PCl$_3$) the bulk densities of the individual species.

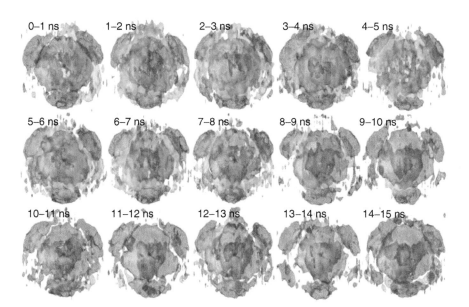

Color Figure 2.12 Probability densities determined by MD studies of 14 mol% solutions of POCl$_3$ in [C$_1$C$_1$Im][Tf$_2$N], calculated over consecutive nanosecond intervals in a continuous 15 ns simulation. Densities are for [Tf$_2$N]$^-$ (red), [C$_1$C$_1$Im]$^+$ (blue), and POCl$_3$ (green) around a central POCl$_3$. Surfaces are drawn at twice ([C$_1$C$_1$Im]$^+$ and [Tf$_2$N]$^-$) and three times (POCl$_3$) the bulk densities of the individual species.

Color Figure 3.3   Schematic illustration of self-assembled monolayer (SAM) photo-lithographic patterning via masked exposure to Hg lamp irradiation and subsequent antibody conjugation to fabricate immunosurfaces for use in an ionic liquid-based immunoassay. Shown in the inset (lower left) is a representative fluorescence image showing spatially selective BODIPY binding within neat $[C_4C_1Im][Tf_2N]$. (Reprinted from Baker, S.N., Brauns, E.B., McCleskey, T.M., Burrell, A.K., and Baker G.A., *Chem. Commun.*, 2851–2853, 2006. Copyright 2006 Royal Society of Chemistry. With permission.)

Color Figure 3.6   (a) Schematic of the hydraulic valve and a top-down view with an open valve. Control channels are in pink and fluidic channels are in blue. The clear bulk material is polydimethylsiloxane (PDMS), including the flexible membrane between the control and fluidic channels at their intersection (schematics are not drawn to scale). The piston has an average diameter of ~910 μm and a height of 152 μm, whereas valve intersections are typically 100×100 μm² with 9 μm high fluidic channels and ~16 μm high control channels. The valve and piston can be centimeters apart. (b) The same schematic with a vertical translation of a piezoelectrically driven Braille pin and a top-down view with a close valve and a pressurized control channel. (c) A top-down view of four intersections of pressurized control (red) and fluidic (blue) channels. All channels are 9 μm high and 100 μm wide except for the lower right control channel that is 40 μm wide. (d) A top-down view of the Braille pins aligned underneath pistons (left) and microfluidic valves (right). (e) A PDMS device with multiple hydraulic valves mounted onto a palm-sized USB-powered and controlled Braille display module with 64-pin actuators. (Reprinted from Gu W., Chen H., Tung, Y.C., Meiners, J.C., and Takayama, S., *Appl. Phys. Lett.*, 90, 033505–033508, 2007. Copyright 2007 American Institute of Physics. With permission.)

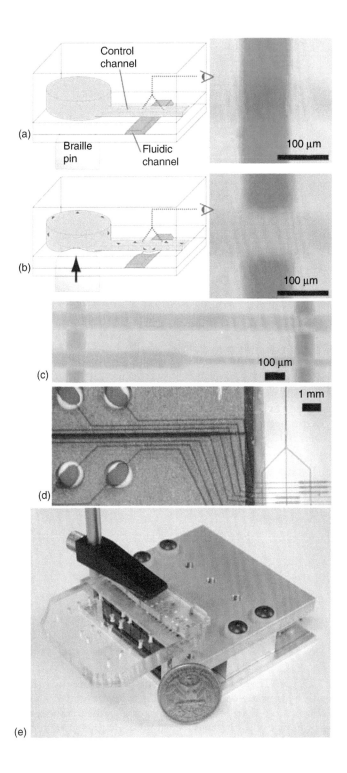

(a) Control channel

Braille pin   Fluidic channel

100 μm

(b)

100 μm

(c)   100 μm

(d)   1 mm

(e)

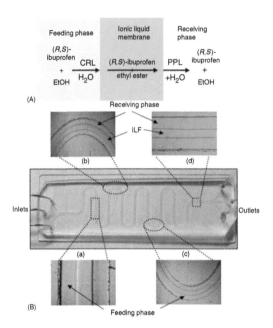

Color Figure 3.7 (A) Enantioselective transport of (S)-ibuprofen by means of ionic liquid flow within a microfluidic device. (B, Bottom) Photographs of the three-phase flow in the microchannel: (a) center near the inlets of the microchannel, (b and c) arc of the microchannel, and (d) center near the outlets of the microchannel. Flow rates of the aqueous phase and the ionic liquid flow phase in (a–d) were 1.5 and 0.3 mL/h, respectively. (Reprinted from Huh, Y.S., Jun, Y.S., Hong, Y.K., Hong, W.H., and Kim, D.H., *J. Mol. Catal.* B, 43, 96–101, 2006. Copyright 2006 Elsevier. With permission.)

Color Figure 7.4 (*Left*) The fast centrifugal partition chromatograph (FCPC) 100 chromatograph with hydrostatic column of 100 mL with integrated alternating pump and pilot electronics (www.armen-instrument.com, Vannes, France); the upper inset shows a part of a disk with the engraved double channels and connecting ducts. (*Right*) The Milli chromatograph with hydrodynamic CCC column of 18 mL (www.DynamicExtractions.com, Uxbridge, UK).

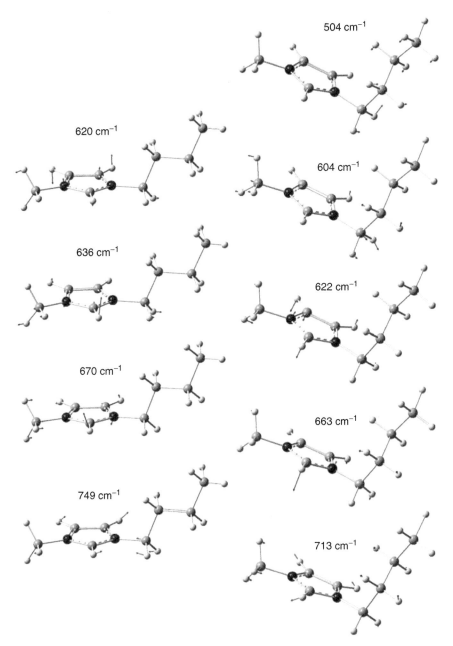

504 cm⁻¹

620 cm⁻¹

604 cm⁻¹

636 cm⁻¹

622 cm⁻¹

670 cm⁻¹

663 cm⁻¹

749 cm⁻¹

713 cm⁻¹

Color Figure 12.11   Some of our calculated normal modes of certain bands of the
AA and GA forms of the $[C_4C_1Im]^+$ cation. The arrows indicate vibrational amplitudes of atoms. As found by Hamaguchi et al. (Hamaguchi, H., and Ozawa, R., *Adv.
Chem. Phys.,* 131, 85–104, 2005) also our C8 methylene CH_2 rocking vibration was coupled to the ring modes only for the gauche-anti conformer (Berg, R. W., Deetlefs, M.,
Seddon, K. R., Shim, I., and Thompson, J. M., *J. Phys. Chem. B,* 109, 19018–19025, 2005)
and (Berg, R. W., Unpublished results, 2006. With permission.)

Table 7.5 Solute Partition Coefficients Measured by Countercurrent Chromatography in Two Ionic Liquid-Containing Biphasic Liquid Systems

Solute	$K_{IL/w}$	Log $K_{IL/w}$	Log K_{oct}
[C₄C₁Im][PF₆]–acetonitrile–water system 40/20/40			
Aniline	12.4	1.09	0.915
Benzonitrile	37.2	1.57	1.575
3-Chlorophenol	28.6	1.46	2.485
Histidine	0.14	−0.85	−3.727
4-Hydroxybenzoic acid	1.3	0.11	1.557
2-Nitroaniline	35.4	1.55	1.798
4-Nitrobenzoic acid	0.095	−1.02	1.838
2-Nitrophenol	40	1.60	1.854
Phenol	11.3	1.05	1.475
Phthalic acid	1.4	0.15	0.732
2-Toluidine	21.6	1.33	1.364
Sulfacetamide	6.9	0.84	−0.976
[C₄C₁Im]Cl 15.6–K₂HPO₄ 23.1–water 61.3 % w/w			
Methanol	4	0.60	−1.00
Ethanol	7	0.85	−0.22
Propanol	13	1.11	0.26
Butanol	30	1.48	0.80
Pentanol	60	1.78	1.30
Hexanol	110	2.04	1.82
Penicillin G	120	2.08	1.83
			Log K_{PEG}
Cytochrome c	90	1.95	−1
Myoglobin	140	2.15	−0.5
Ovalbumin	180	2.26	0.08
Hemoglobin	220	2.34	1.5

Notes: $K_{IL/w}$ is the partition coefficient expressed as the solute concentration in the IL-rich liquid phase over the solute concentration in the other (acetonitrile or phosphate-rich) liquid phase. (Berthod, A. and Carda-Broch, S., *Anal. Bioanal. Chem.*, 380, 168–177, 2004.)

Log K_{oct} is the octanol/water partition coefficient. (Berthod, A. and Carda-Broch, S., *Anal. Bioanal. Chem.*, 380, 168–177, 2004.)

Log K_{PEG} is the protein partition coefficient obtained in the PEG–K₂HPO₄–water ATPS. (Wood, P.L., Hawes, D., Janaway, L., Sutherland, I.A., *J. Liq. Chromatogr. & Rel. Technol.*, 26, 1373–1396, 2003.)

The $K_{IL/w}$ coefficients of nonionizable solutes were all significantly higher than their respective coefficient in the octanol/water reference system. Ionizable compounds have $K_{IL/w}$ coefficients close or lower to their K_{oct} coefficients. However, since the liquid system was not buffered, they may be partly ionized that would decrease the measured coefficient. The ionized forms have a greater affinity for the aqueous phase than for the IL-rich phase. The conclusion of the study done with the [C₄C₁Im][PF₆]–acetonitrile–water system was that the polarity of the IL phase is higher than that of ethanol or propanol [6].

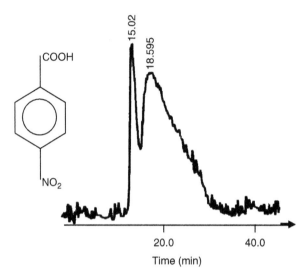

Figure 7.7 CCC chromatogram of the 4-nitrobenzoic elution. Hydrodynamic CCC column 53 mL; rotor speed: 830 rpm; liquid system $[C_4C_1Im][PF_6]$–acetonitrile–water 40/20/40 %w/w; stationary phase IL-rich lower phase, V_S = 38 mL; mobile phase: acetonitrile-rich upper phase, F = 1 mL/min in the tail-to-head direction; injection volume: 500 µL of a mixture of 6 ppm of 4-nitrobenzoic acid (retention time 18.595 min) + 2.5 ppm phenylalanine (retention time 15.02 min) as a hold-up volume tracer; UV detection 254 nm.

Table 7.5 also lists the coefficients obtained in the IL-based ATPS for proteins and short-chain linear alcohols [14]. The protein coefficients are two orders of magnitude higher than those obtained in the classical PEG 1000–K_2HPO_4–water ATPS. The short-chain alcohols, a homologous series, have coefficients about one order of magnitude higher than their K_{oct} coefficient. The partitioning of Penicillin G was recently studied in the same ATPS and similar results were obtained (Table 7.5) [15]. These results indicate that the IL-rich phase of the $[C_4C_1Im]Cl$–K_2HPO_4–acetonitrile–water system has a polarity much lower than the octanol phase of the reference system and also much lower than the PEG aqueous phase obtained with classical ATPS made with PEG 1000 and the same potassium hydrogeno phosphate salt [10,13,16].

7.5 Conclusion

CCC works with biphasic liquid systems. Since ILs are able to form such biphasic systems with a number of solvents, they have some potential uses in CCC. A low viscosity is desirable; this point is a serious problem precluding the use of ILs in CCC: many ILs have a viscosity >5 mPa s. Solvents such as acetonitrile can dramatically reduce ILs viscosity. Other potential drawbacks in the use of ILs in LCs are the problems that could be encountered in continuous detection. Many ILs have a significant UV absorbance (imidazolium or pyridinium rings) that limits the use of the most widely used UV detector. All ILs have a very low volatility that precludes the use of the evaporative light scattering detector and mass spectroscopy since these two detection techniques need to evaporate the mobile phase.

Compositions of the system $[C_4C_1Im][PF_6]$–acetonitrile–water form two liquid phases that were used with hydrostatic and hydrodynamic CCC columns. The polarity of the IL-acetonitrile liquid phase is as high as that of ethanol without being soluble in water. In another example, $[C_4C_1Im]Cl$, a fully water-soluble IL, was able to form an aqueous solution nonmiscible with a K_2HPO_4 aqueous solution. The separation capabilities of these two biphasic liquid systems were not fully exploited in CCC. It was shown that the polarity of the IL-based ATPS was very much different from that of the PEG 1000–K_2HPO_4–water ATPS commonly used in protein separation. Proteins as well as short-chain alcohols greatly favor the IL aqueous phase rather than the phosphate salt aqueous phase. Further studies are needed since it was found that the IL aqueous phase was very rapidly saturated by proteins. The CCC technique is mainly used to produce significant amounts of purified chemicals using as little solvent as possible. So the two liquid phases of the biphasic system selected must be able to dissolve significant amount of the material to be purified; otherwise it will not be possible to purify a large amount of material in one run.

Acknowledgments

Alain Berthod thanks the CNRS UMR 5180 (P. Lanteri) for financial support. Maria-Jose Ruiz-Angel thanks the Spanish Ministerio de Educación y Ciencia for a Ramón y Cajal contract. Samuel Carda-Broch thanks the Conselleria d'Empresa de la Universidad de la Generalitat Valenciana for a short stay financing.

References

1. Berthod, A., *Countercurrent Chromatography: The Support-Free Liquid Stationary Phase*, Comprehensive Analytical Chemistry, Vol. 38, Elsevier, Amsterdam, 2002.
2. Foucault, A., *Centrifugal Partition Chromatography*, Chromatographic Science Series, Vol. 68, Marcel Dekker, New York, 1995.
3. Seddon, K.R. cited in Freemantle, M., New frontiers for ionic liquids, *C & EN*, Jan. 1, 23–27, 2007.
4. Waichigo, M.M., Riechel, T.L., and Danielson, N.D., Ethylammonium acetate as a mobile phase modifier for RPLC, *Chromatographia*, 61, 17–23, 2005.
5. Carda-Broch, S., Berthod, A., and Armstrong, D.W., Solvent properties of 1-butyl-3-methylimidazolium hexafluorophosphate ionic liquid, *Anal. Bioanal. Chem.*, 375, 191–196, 2003.
6. Berthod, A. and Carda-Broch, S., Use of the ionic liquid 1-butyl-3-methylimidazolium hexafluorophosphate in CCC, *Anal. Bioanal. Chem.*, 380, 168–177, 2004.

7. Berthod, A. and Carda-Broch, S., A new class of solvents for CCC: The room temperature ionic liquids, *J. Liq. Chromr. Relat. Technol.*, 26, 1493–1508, 2003.
8. Albertson, P.A., *Partition of Cells and Macromolecules*, Wiley, New York, 1971.
9. Sutherland, I.A., Heywood-Waddington, D., and Ito, Y., Countercurrent chromatography: Applications to the separation of biopolymers, organelles and cells using either aqueous—organic or aqueous—aqueous phase systems, *J. Chromatogr.*, 384, 197–207, 1987.
10. Knight, M., in Mandava, N.B., and Ito, Y. (Eds), *Countercurrent Chromatography: Theory and Practice*, Chromatographic Science Series, Vol. 44, Ch. 10, Marcel Dekker, New York, 1988.
11. Gutowski, K.E., Broker, G.A., Willauer, H.D., Huddelston, R.P., Swatloski, J.D., Holbrey, J.D., Rogers, R.D., Controlling the aqueous miscibility of ionic liquids: Aqueous biphasic systems of water-miscible ionic liquids and water-structuring salts for recycle, metathesis and separations, *J. Am. Chem. Soc.*, 125, 6632–6633, 2003.
12. Li, S.H., He, C.Y., Liu, H.W., Li, K., and Liu, J., Ionic liquid-based aqueous two-phase system, a sample pretreatment procedure prior to high-performance liquid chromatography of opium alkaloids, *J. Chromatogr. B*, 826, 58–62, 2005.
13. Wood, P.L., Hawes, D., Janaway, L., and Sutherland, I.A., Stationary phase retention in CCC: Modelling the J-type centrifuge as a constant pressure drop pump, *J. Liq. Chromatogr. & Rel. Technol.*, 26, 1373–1396, 2003.
14. Ruiz-Angel, M.J., Pino, V., Carda-Broch, S., Berthod, A., Solvent systems for CCC: An aqueous two phase liquid system based on a room temperature ionic liquid, *J. Chromatogr., A*, 1151, 65–73, 2007.
15. Liu, Q., Yu, J., Li, W., Hu, X., Xia, H., and Yang, P., Partitioning behavior of penicillin G in aqueous two phase system formed by ionic liquid and phosphate, *Sep. Sci. Technol.*, 41, 2849–2858, 2006.
16. Sutherland, I.A., Mandava, N.B., and Ito, Y. (Eds), *Countercurrent Chromatography: In Theory and Practice*, Chromatographic Science Series, Vol. 44, Ch. 11, Marcel Dekker, New York, 1988.

chapter eight

Gas solubilities in ionic liquids and related measurement techniques

Joan Frances Brennecke, Zulema K. Lopez-Castillo, and Berlyn Rose Mellein

Contents

8.1   Introduction

There has been tremendous growth in interest in the solubility of various gases in ionic liquids (ILs), especially since the discovery that CO_2 has a relatively high solubility in these interesting materials [1]. However, even prior to that time there was a need for gas solubility measurements since a variety of reactions involving permanent gases had been performed in ILs. In fact, the use of ILs for gas separations and the possibility of doing reactions that involve permanent gases in ILs are the two main reasons why gas solubilities in ILs have been studied extensively since 2000.

For gas separation and storage applications, one needs to know gas solubilities to determine carrying capacities and selectivities. Since CO_2 is relatively soluble in ILs, they have been suggested for the removal of CO_2 from postcombustion flue gas, natural gas, and hydrogen [2]. They may also be suited for olefin/paraffin separations and drying of gases. Air Products has developed ILs for safe storage, transportation, and use of toxic gases, such as BF_3 and PF_3, used in the microelectronics industry [3].

Reactions that have been performed in ILs involving permanent reactions include oxidations, hydrogenations, and hydroformylations. In all cases, one must know the solubility in the IL solutions at the experimental partial pressure to determine the kinetics and obtain quantitative values of the reaction rate constants.

One of the main impetuses for using ILs for gas separations and as a solvent for reactions involving permanent gases is that most ILs have extremely low-vapor pressures at normal operating conditions. Thus, one will not lose any of the solvent in the purified gas stream or in the products. Another attractive feature is that ILs are highly tunable by varying the cation, anion, and substituents. Thus, they can be tailored for specific applications to optimize selectivity, capacities, reactant or product solubilities, and rates.

For real reaction and separation systems, one is frequently interested in the solubility of gases when they are present in a mixture. Thus, both pure gas and mixed gas solubilities are of importance. Here we will focus on the solubility of pure gases in ILs. We will describe various methods that have been used to measure the solubility of pure gases in ILs. Some of these methods are general techniques for measuring gas solubility in liquids. However, some are uniquely suited for determining gas solubilities in ILs, which frequently have negligible volatility at the temperatures under investigation. The techniques described include the stoichiometric technique, the pressure drop technique, and various gravimetric methods. Of course, one can also determine gas solubilities with various *in-situ* spectroscopic techniques, such as Fourier transform infrared spectroscopy (FTIR) for CO_2. However, quantitative measurements require knowledge of how extinction coefficients vary with temperature and pressure, which introduces additional complexity into the measurements and thus are not covered here.

8.2   Stoichiometric technique

The stoichiometric (or synthetic) method, and variations thereof, involves metering in known amounts of liquid and gas into a high-pressure viewcell and determining the gas solubility in one of several ways. One technique is to accurately measure the volume of the cell and the volume of the vapor and liquid phases. This information is used to determine the number of moles of gas remaining in the gas phase and, subsequently, the number of moles of gas dissolved in the liquid, by difference. A typical apparatus based on this technique is shown in Figure 8.1. The main components of the system are a feed pump to deliver the gas, temperature, and pressure indicators and controllers, a water bath to maintain the cell at constant temperature, a cathetometer to measure the volume of the liquid in the cell, an optically transparent cell and cell holder (a sapphire tube rated to several hundred bars can be obtained readily), and an agitation system to stir the sample. In a typical experiment, the dry IL is loaded into the cell in a glove box under a dry nitrogen atmosphere to avoid inadvertent sorption of water from the

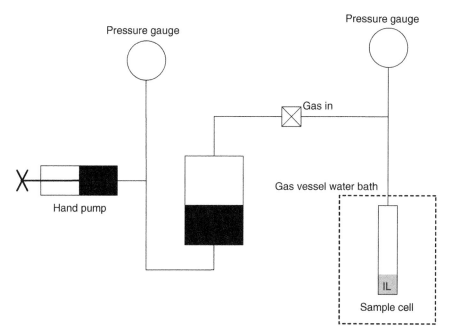

Figure 8.1   Simplified schematic of the stoichiometric gas solubility apparatus.

atmosphere. The cell is then attached to the apparatus and placed in the constant-temperature water bath to reach thermal equilibrium. A known amount of gas is added to the cell and the liquid is stirred until equilibrium is attained (i.e., no further change in pressure). An accurate equation of state for the pure gas is used to determine the number of moles of gas remaining in the gas phase. Since the IL is nonvolatile at test conditions, the gas phase is, in fact, pure gas. Knowing the number of moles of gas that have been introduced into the cell, the amount of gas dissolved in the liquid can be determined by difference.

Another variation of the stoichiometric method involves loading known amounts of gas and IL into the cell and then increasing the pressure (at constant temperature) until all the gas dissolves in the liquid and, consequently, the vapor phase disappears. Using different loadings of the gas, one can determine the solubility at various different pressures and temperatures. Mercury was used as the pressurization fluid by Peters and coworkers to determine gas solubilities in ILs [4]. Maurer and coworkers used a similar method, but they introduced and withdrew additional known amounts of the IL to pressurize or depressurize the mixture and observe the phase change [5].

Representative data for CO_2 solubility in various ILs obtained in a high-pressure stoichiometric apparatus is shown in Figure 8.2. We compare the solubility of CO_2 in four different imidazolium-based ILs [6,7]. While the

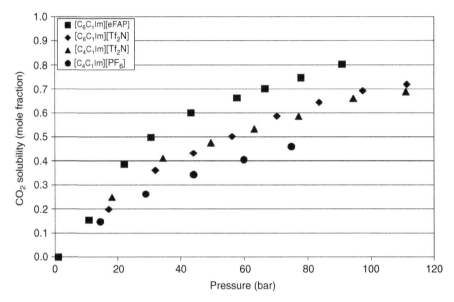

Figure 8.2 CO$_2$ solubility in [C$_4$C$_1$Im][PF$_6$], [C$_4$C$_1$Im][Tf$_2$N], [C$_6$C$_1$Im][Tf$_2$N], and [C$_6$C$_1$Im][eFAP] at 60°C. (From Aki, S. N. V. K. et al., *J. Phys. Chem. B*, 108, 20355, 2004; Muldoon, M. J. et al., *J. Phys. Chem. B*, 111, 9001, 2007.)

solubility is quite substantial in all the ILs, it is lowest in 1-butyl-3-methyl-imidazolium hexafluorophosphate ([C$_4$C$_1$Im][PF$_6$]). The solubility increases substantially when the [PF$_6$]$^-$ anion is replaced by the [Tf$_2$N]$^-$ anion in 1-butyl-3-methylimidazolium *bis*(trifluoromethylsulfonyl)imide ([C$_4$C$_1$Im][Tf$_2$N]). Increasing the length of one of the alkyl chains from butyl to hexyl in 1-hexyl-3-methylimidazolium *bis*(trifluoromethylsulfonyl)imide ([C$_6$C$_1$Im][Tf$_2$N]) increases the solubility slightly, likely due to the concomitant decrease in density and increase in free volume. Even further increases in CO$_2$ solubility can be achieved by replacing the anion with a fluoroalkyl phosphate. Specifically, CO$_2$ solubility in 1-hexyl-3-methylimidazolium tris(pentafluoroethyl)-trifluorophosphate ([C$_6$C$_1$Im][eFAP]) is shown in the Figure 8.2.

8.3   Pressure drop technique

In the pressure drop technique, a known number of moles of gas (determined by accurate measurements of temperature and pressure in a precalibrated volume) is secured in one section of the apparatus. The IL is metered into another section, whose volume is also known accurately. Then a valve is opened and the gas is allowed to expand into the entire apparatus, dissolving in the IL. Measurement of the full pressure drop when equilibrium is reached allows the number of moles of gas to be determined in the vapor phase and, subsequently, the number of moles in the liquid phase to be determined by difference.

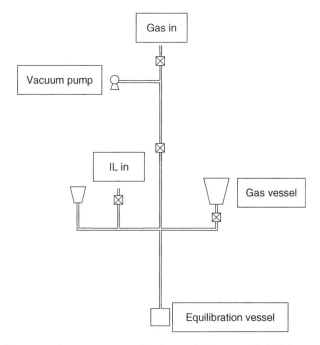

Figure 8.3   Pressure drop apparatus for determining gas solubilities.

While this technique can be used for gas solubility in volatile liquids, where the vapor pressure of the liquid is determined prior to the introduction of the gas, it is uniquely suited for the measurement of gas solubilities in ILs because the gas phase remains pure. This is the technique used by Costa Gomez and coworkers [8–10] to measure the solubility of various gases in ILs and a schematic of the apparatus is shown in Figure 8.3. These apparatuses are frequently made entirely from glass and, therefore, are limited to low-pressure operation. Nonetheless, this makes them ideal for determining Henry's law constants.

Data determined by the pressure drop technique [9] are shown in Figure 8.4. The solubility of CO_2, CH_4, and O_2 in 1-butyl-3-methylimidazolium tetrafluoroborate [C_4C_1Im][BF_4]) at 303 K shows the much greater solubility of CO_2 compared to CH_4 and O_2. Henry's law constants can be determined from the slopes of these graphs.

8.4   Gravimetric methods

The method most widely used to determine gas solubilities in ILs is gravimetric. As we first showed in 2001 [11] gravimetric techniques are unique, suited to measure the solubility of gases and vapors in ILs because the liquid does not evaporate into the vapor, complicating the measurements. Thus, apparatuses designed to measure gas sorption on solids can be used readily for ILs.

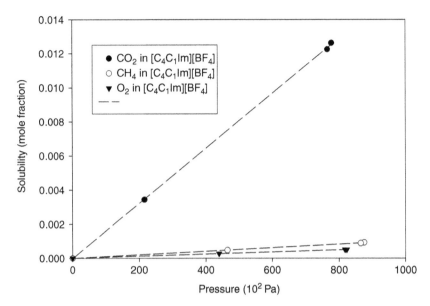

Figure 8.4 Solubility of CO_2, CH_4, and O_2 in $[C_4C_1Im][BF_4]$ at 303 K, as determined by the pressure drop technique. (From Jacquemin, J. et al., _J. Chem. Thermodyn._, 38, 490, 2006.)

The simplest form of gravimetric determination of gas solubility is to bubble the gas through a cylinder containing the IL and to weigh the cylinder before and after. While somewhat crude, this is one of the methods used by Davis and coworkers to determine large uptake of CO_2 by an IL that had been functionalized with a free amine [12].

More accurate gravimetric measurements can be made with gravimetric microbalances. For instance, a schematic of the Hiden Intelligent Gravimetric Analyzer (IGA003) is shown in Figure 8.5. A schematic of another gravimetric balance is shown in Figure 8.6.

The Hiden microbalance consists of a sample pan and counterweight symmetrically configured to minimize buoyancy effects. Since the balance has a 1 µg stable resolution, IL samples as small as 75 mg can be used. The IL samples should be predried, but they can be further dried _in-situ_ by evacuating the chamber (to a ca. 10^{-9} bar with a turbomolecular pump) and heating, if necessary. After a steady weight is obtained, gas is introduced into the chamber at a specified pressure and the chamber is maintained at the set point temperature. The weight is monitored as a function of time as the IL sample absorbs the gas. The weight change is monitored until the mass did not change significantly for a significant period, after which the sample is deemed to have reached equilibrium, thus yielding a single point on the absorption isotherm. This process can be repeated through a predetermined set of pressures until the maximum pressure is reached. Following this, it is

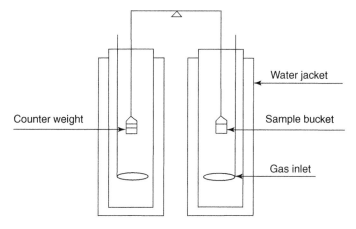

Figure 8.5 Schematic of intelligent gravimetric analyzer (Hiden Analytical Limited, England).

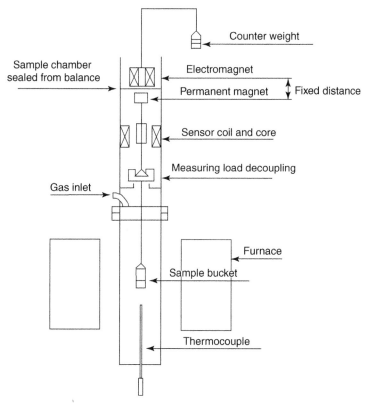

Figure 8.6 Schematic of Rubotherm magnetically coupled microbalance (Rubotherm, Germany).

advisable to reverse the process; lowering the gas pressure above the sample gradually in a series of small desorption steps, during which the decrease in sample mass is recorded. This yields a complete absorption/desorption isotherm. The degree of hysteresis between the two isotherm branches gives an indication of the accuracy of each value.

There are two factors critical in performing experiments with a gravimetric microbalance. Firstly, great care must be taken to account for buoyancy effects in the system, even when a symmetric balance is used. This requires accurate knowledge of the gas density, which can be determined from an accurate pure component equation of state. One must also know the density of the IL as a function of temperature and make the assumption that the sorption of gas into the IL does not dramatically change the density (unless independent measurements of IL/gas mixture densities are available). Secondly, sufficient time must be allowed for the system to reach equilibrium. The microbalance does not have any stirring and, therefore, relies on the diffusion of gas into the IL. Since ILs can be somewhat viscous, equilibrium time can be several hours per point, depending on the IL.

The Rubotherm magnetic suspension balance (Rubotherm, Germany) works on a similar principle, except that it is equipped for a high-pressure (to 500 bar) operation. In addition, the balance is magnetically coupled to the sorption chamber, so the balance is not contacted by the gas, allowing investigation of more toxic and corrosive gases, such as SO_2 [13]. To obtain accurate results typical sample sizes are on the order of a gram, which translates to longer diffusion times.

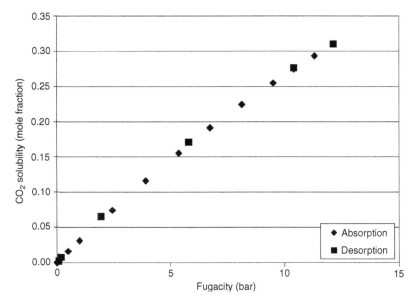

Figure 8.7 Absorption and desorption curves for CO_2 in $[C_6C_1Im][Tf_2N]$ at 25°C.

Absorption and desorption curves for CO_2 sorption into ($[C_6C_1Im][Tf_2N]$) at 25°C, for the Hiden IGA003 are shown in Figure 8.7. Clearly, the data points are at equilibrium, with no hysteresis apparent. Comparative data of the solubility of CO_2 in $[C_6C_1Im][Tf_2N]$ and in ($[C_4C_1Im][Tf_2N]$) at 25°C [7,14] from this apparatus are shown in Figure 8.8.

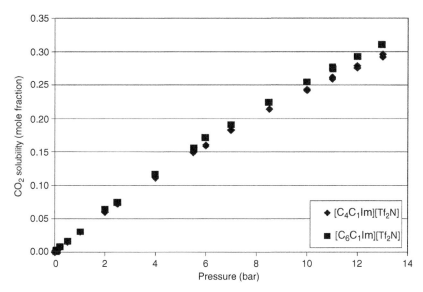

Figure 8.8 CO_2 solubility in $[C_4C_1Im][Tf_2N]$ and $[C_6C_1Im][Tf_2N]$ at 25°C. (From Muldoon, M. J. et al., *J. Phys. Chem. B,* 109, 6366, 2005.)

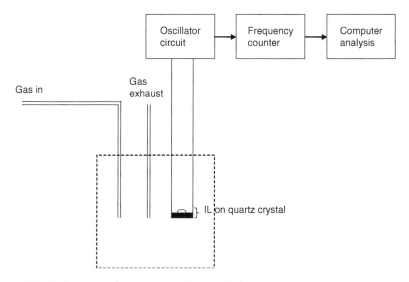

Figure 8.9   Schematic of quartz crystal microbalance apparatus.

Table 8.1 Examples of Henry's Law Constants for Gas Solubilities in Various Ionic Liquids

Cation	Anion	Gas	Henry's law constant (bar)	References
$[C_2C_1Im]$	$[Tf_2N]$	CO_2	35.6	17
$[C_2C_1C_1Im]$	$[Tf_2N]$	CO_2	39.6	17
$[C_3C_1Im]$	$[PF_6]$	CO_2	52.0	15
$[C_3C_1Im]$	$[Tf_2N]$	CO_2	37.0	15
$[C_4C_1Im]$	$[PF_6]$	CO_2	53.4	18
			57.1	19
$[C_4C_1Im]$	$[BF_4]$	CO_2	59.0	14
			60.2	19
$[C_4C_1Im]$	$[Tf_2N]$	CO_2	33.0	14
			37.0	15
$[C_4C_1C_1Im]$	$[PF_6]$	CO_2	61.8	17
$[C_4C_1C_1Im]$	$[BF_4]$	CO_2	61.0	17
$[C_5C_1Im]$	Tris(nonafluorobutyl)trifluoro-phosphate	CO_2	20.2	7
$[C_6C_1Im]$	Tris(pentafluoroethyl)trifluoro-phosphate	CO_2	25.2	7
$[C_6C_1Im]$	Tris(heptafluoropropyl)trifluoro-phosphate	CO_2	21.6	7
$[C_6C_1Im]$	$[Tf_2N]$	CO_2	35.0	15
			31.6	20
$[C_8C_1Im]$	$[Tf_2N]$	CO_2	30.0	15
1-Methyl-3-(3,3,4,4,5,5,6,6-nonafluorohexyl) imidazolium	$[Tf_2N]$	CO_2	28.4	7
1-Methyl-3-(3,3,4,4,5,5,6,6,7,7,8,8-tridecafluorooctyl) imidazolium	$[Tf_2N]$	CO_2	4.5	15
			6.0	16
			27.3	7

Cation	Anion	Gas	Value	Ref
1,4-Dibutyl-3-phenylimidazolium	[Tf₂N]	CO_2	63.0	15
1-*n*-Butyl-3-phenylimidazolium	[Tf₂N]	CO_2	180.0	15
Methyl-tri-butyl-ammonium	[Tf₂N]	CO_2	43.5	14
1-*n*-Butyl-methylpyrrolidinium	[Tf₂N]	CO_2	38.6	14
1-*n*-Hexyl-3-methylpyridinium	[Tf₂N]	CO_2	32.8	7
(1-Methylimidazole)(triethylamine)boronium	[Tf₂N]	CO_2	33.1	7
[C₄C₁Im]	[Tf₂N]	CO	950 (295 K)	21
N-Hexylpyridinium	[Tf₂N]	CO	1140 (295 K)	21
1-*n*-Hexyl-3-methylpyridinium	[Tf₂N]	CO	880 (295 K)	21
1-*n*-Butyl-methylpyrrolidinium	[Tf₂N]	CO	3260 (295 K)	21
(2-Hydroxyethyl)butyl dimethylammonium	[Tf₂N]	CO	1240 (295 K)	21
[C₄C₁Im]	[BF₄]	O_2	368 (314 K)	8
[C₄C₁Im]	[BF₄]	H_2	5800 (293 K)	22
			1630	23
[C₄C₁Im]	[PF₆]	H_2	6600 (293 K)	22
			5380	23
[C₄C₁Im]	[Tf₂N]	H_2	4500	22
N-Butylpyridinium	[Tf₂N]	H_2	3900	22
[C₆C₁Im]	[Tf₂N]	SO_2	1.64	13
1-*n*-Hexyl-3-methylpyridinium	[Tf₂N]	SO_2	1.54	13

Note: Values are at 298 K, unless otherwise noted.

Another type of microbalance that has been used to determine gas solubilities in ILs is a quartz crystal microbalance [15]. A schematic of the apparatus used by Baltus and coworkers is shown in Figure 8.9.

The principle of operation in a quartz crystal microbalance is based on the piezoelectric effect. When a voltage is applied to the crystal, it oscillates at its resonance frequency. If mass is added to the crystal, its resonance frequency changes in direct proportion to the mass, as described by the Sauerbrey equation. Care must be taken to ensure that the thickness of the film satisfies the thin film limit and it may be necessary to correct for damping. While data for the gas solubilities in some ILs obtained by this method agree with data from other sources (e.g., $[C_4C_1Im][Tf_2N]$ in Ref. 15), other data [15,16] deviate by a factor of more than four [7].

Examples of Henry's law constants measured by the various techniques described above are shown in Table 8.1.

Summary

Measurements of the solubility of gases in ILs are increasingly important as researchers explore the use of ILs for gas separation and gas storage, as well as a solvent media for reactions involving permanent gases. Here we present several different methods that have been used to obtain these measurements. These include traditional synthetic and pressure drop methods, as well as gravimetric methods that are particularly well-suited for the measurement of gases in nonvolatile liquids.

References

1. Blanchard, L. A. et al., Green processing using ionic liquids and CO_2, *Nature*, 399, 28, 1999.
2. Brennecke, J. F. and Maginn, E. J., Purification of gas with liquid ionic components, U.S. Patent 6579343, 2003.
3. Tempel, D. J., Henderson, P. B., and Brzozowski, J. R., Reactive liquid based gas storage and delivery systems, U.S. Patent 7172646, 2007.
4. Shariati, A. and Peters, C. J., High-pressure phase behavior of systems with ionic liquids: Measurements and modeling of the binary system fluoroform + 1-ethyl-3-methylimidazolium hexafluorophosphate, *J. Supercrit. Fluids*, 25, 109, 2003.
5. Kumelen, J. et al., Solubility of CO_2 in the ionic liquid [hmim][Tf$_2$N], *J. Chem. Thermodyn.*, 38, 1396, 2006.
6. Aki, S. N. V. K. et al., High-pressure phase behavior of carbon dioxide with imidazolium-based ionic liquids, *J. Phys. Chem. B*, 108, 20355, 2004.
7. Muldoon, M. J. et al., Improving carbon dioxide solubility in ionic liquids, *J. Phys. Chem. B*, 111, 9001, 2007.
8. Husson-Borg, P., Majer, V., and Gomes, M. F. C., Solubilities of oxygen and carbon dioxide in butyl methyl imidazolium tetrafluoroborate as a function of temperature and at pressures close to atmospheric pressure, *J. Chem. Eng. Data*, 48, 480, 2003.

9. Jacquemin, J. et al., Solubility of carbon dioxide, ethane, methane, oxygen, nitrogen, hydrogen, argon, and carbon monoxide in 1-butyl-3-methylimidazolium tetrafluoroborate between temperatures 283 K and 343 K and at pressures close to atmospheric, *J. Chem. Thermodyn.*, 38, 490, 2006.

10. Jacquemin, J. et al., Low-pressure solubilities and thermodynamics of solvation of eight gases in 1-butyl-3-methylimidazolium hexafluorophosphate, *Fluid Phase Equilib.*, 240, 87, 2006.

11. Anthony, J. L., Maginn, E. J., and Brennecke, J. F., Solution thermodynamics of imidazolium-based ionic liquids and water, *J. Phys. Chem. B*, 105, 10942, 2001.

12. Bates, E. D. et al., CO_2 capture by a task-specific ionic liquid, *J. Am. Chem. Soc.*, 124, 926, 2002.

13. Anderson, J. L. et al., Measurement of SO_2 solubility in ionic liquids, *J. Phys. Chem. B*, 110, 15059, 2006.

14. Anthony, J. L. et al., Anion effects on gas solubility in ionic liquids, *J. Phys. Chem. B*, 109, 6366, 2005.

15. Baltus, R. E. et al., Low-pressure solubility of carbon dioxide in room-temperature ionic liquids measured with a quartz crystal microbalance, *J. Phys. Chem. B*, 108, 721, 2004.

16. Baltus, R. E. et al., Examination of the potential of ionic liquids for gas separations, *Sep. Sci. Technol.*, 40, 525, 2005.

17. Cadena, C. et al., Why is CO_2 so soluble in imidazolium-based ionic liquids?, *J. Am. Chem. Soc.*, 126, 5300, 2004.

18. Anthony, J. L., Maginn, E. J., and Brennecke, J. F, Solubilities and thermodynamic properties of gases in the ionic liquid 1-*n*-butyl-3-methyl imidazolium hexafluorophosphate, *J. Phys. Chem. B*, 106, 7315, 2002.

19. Shiflett, M. B. and Yokozeki, A., Solubilities and diffusivities of carbon dioxide in ionic liquids: [bmim][PF_6] and [bmim][BF_4], *Ind. Eng. Chem. Res.*, 44, 4453, 2005.

20. Hert, D. G. et al., Enhancement of oxygen and methane solubility in 1-hexyl-3-methyl imidazolium bis(trifluoromethylsulfonyl) imide using carbon dioxide, *Chem. Commun.*, 2603, 2005.

21. Ohlin, C. A., Dyson, P. J., and Laurenczy, G., Carbon monoxide solubility in ionic liquids: Determination, prediction and relevance to hydroformylation, *Chem. Commun.*, 1070, 2004.

22. Dyson, P. J. et al., Determination of hydrogen concentration in ionic liquids and the effect (or lack of) on rates of hydrogenation, *Chem. Commun.*, 2418, 2003.

23. Berger, A. et al., Ionic liquid-phase asymmetric catalytic hydrogenation: Hydrogen concentration effects on enantioselectivity, *Tetrahedron: Asymmetry*, 12, 1825, 2001.

chapter nine

Liquid–liquid extraction of organic compounds

Igor V. Pletnev, Svetlana V. Smirnova,
and Vladimir M. Egorov

Contents

9.1 Introduction

Among the separation techniques, liquid–liquid (solvent) extraction is one of the best-known, well-established, versatile, and easy to use. However, traditional extraction employs conventional organic solvents immiscible with water, which are typically volatile, flammable, and health hazardous. This makes extraction inappropriate for modern and future environmental-friendly technologies and analysis processes. Another problem with conventional solvents is that their number is rather limited, so it may be difficult to find the solvent ideally suited for a particular application (even considering solvent mixtures).

Room-temperature ionic liquids (ILs) are extensively investigated as a replacement for organic solvents in chemistry, in general, and in chemical analysis, in particular. ILs have a considerable potential as solvents for separations. It is believed that, being typically nonflammable and nonvolatile compounds, they may help to design environmentally safe processes. Another quite attractive feature of IL solvents is that their polarity, hydrophobicity, viscosity, and other physical and chemical properties may be relatively easily altered by changing the nature of the cationic or anionic constituent. Also, high solvation ability and ionic nature of ILs make them promising candidates for solving a challenging problem of efficient extraction of ionic and zwitterionic compounds from aqueous solutions, including bio liquids. Evidently, an IL solvent may act as a provider of hydrophobic counterions for extraction, supplying them in high concentration.

Additionally, ILs may serve as a convenient medium for analysis after extraction. For example, their ionic nature opens an exciting opportunity of direct electrochemical analysis of the extract, impossible for most of the conventional water-immiscible solvents.

Given all these attractive features, it is easy to understand the increasing interest in the application of ILs in solvent extraction. The review gives an introduction into the rapidly growing area; it focuses on the extraction of organic compounds, metal ion extraction being considered in Chapter 10 of this book.

9.2   Extraction of organic compounds from aqueous solutions

Rogers et al. [1] were the first to report about the use of ILs as extractants of organic compounds. The authors investigated the distribution of aniline, aromatic carboxylic acids, and other aromatic compounds in the biphasic system water/1-butyl-3-methylimidazolium hexafluorophosphate. Still today, this IL, $[C_4C_1Im][PF_6]$, is the most popular IL for extraction (although it is now well-known that this solvent is not so *green*, mainly due to the slow hydrolysis of the anion). The distribution ratios were determined radiometrically, using labeled compounds. It was shown that the distribution ratios (D) in the water/IL system correlate with the corresponding water/1-octanol partition coefficients (P), but are about an order of magnitude lower. As with common solvents, distribution ratios of ionizable solutes highly depend on the pH of the aqueous phase, and maximal values correspond to the uncharged form of the extracting compounds.

Rogers et al. [2] demonstrated the reversible pH-dependent liquid–liquid partitioning of the well-known indicator thymol blue between water and $[C_4C_1Im][PF_6]$. Neutral (zwitterionic) form of the dye partitions mostly to the IL phase, while the charged forms prefer the water phase. To adjust pH, the authors used mineral acids and bases; no significant difference in the extraction dependent on the nature of that acid or base was observed (interestingly, it was also suggested to use gaseous CO_2 and NH_3 as pH adjusters).

Rogers and coworkers published the results concerning the extraction of aromatic compounds to ILs which consist of *N*-alkylisoquinolinium cation (Alkisoq$^+$; C_{8-} and C_{14-}derivatives were used) and *bis*(perfluoroethylsulfonyl) imide (BETI)$^-$, anion [3]. The extended aromatic system presented in the IL structure would presumably lead to specific interactions with other aromatics, thus enhancing the efficiency of extraction. Indeed, the distribution ratios for the substituted aromatic compounds were generally higher than those obtained with imidazolium liquids (1,2,4-trichlorobenzene: D = 1280, $[C_{14}isoq][BETI]$; D = 524, $[C_4C_1Im][PF_6]$). In general, distribution ratios increase in parallel with solute's hydrophobicity. The exception is carboxylic acids (probably, due to hydrogen bonding and partitioning of complex species).

The extraction of aliphatic alcohols with alkyl chain lengths from 1 to 5 is covered in the paper [4]. The distribution ratios of alcohols increased with increasing alkyl chain length but were lower than those of aromatic nonionizable compounds. This should be due to relative hydrophilicity of the studied alcohols. The authors of [5] investigated the recovery of butyl alcohol from fermentation broth to $[C_4C_1Im][PF_6]$ and $[C_8C_1Im][PF_6]$. The distribution ratios of butanol for both ILs were similar (ca. 0.9, room temperature). At the same time, in the case of $[C_8C_1Im][PF_6]$ the extraction selectivity ($DBuOH/DH_2O$) was higher than that for $[C_4C_1Im][PF_6]$. Higher temperatures improved the extraction selectivity. The authors also covered the use of pervaporation through polydimethylsiloxane membrane for the separation of liquid mixtures.

Armstrong et al. measured distribution ratios for 40 compounds, including organic acids, organic bases, amino acids, and neutral compounds, between water and $[C_4C_1Im][PF_6]$, as well as between hexane and $[C_4C_1Im][PF_6]$ [6] (measurements were performed by liquid chromatography). The distribution ratios for ionizable compounds were strongly dependent on pH of the aqueous phase (pH values 2, 5.1, and 10 were used), which allowed the determination of both molecular and ion distribution ratios for each substance. Probably, the data reported on charged forms need additional examination, as the used pH values do not always correspond to the predominance of the single ionic form in the aqueous solution.

The distribution ratios obtained were compared to the corresponding distribution ratios between water and 1-octanol. The authors also calculated solvent parameters of $[C_4C_1Im][PF_6]$ (Section 9.3). It was shown that phenolate-ion associates with $[C_4C_1Im][PF_6]$ more strongly than other ions. The authors also mention the possibility of extraction of amino acids into $[C_4C_1Im][PF_6]$ in the presence of crown ether dibenzo-18-crown-6, though at rather moderate efficiency.

MacFarlane et al. [7] have investigated the extraction of polar organic compounds (acids, alcohols, and functionalized aromatic compounds) into nine hydrophobic ILs, namely, $[C_4C_1Im][PF_6]$; $[C_4C_1Im][Tf_2N]$, $[C_6C_1Im][Tf_2N]$,

[C_8C_1Im][Tf_2N]; [$(C_6)_3C_{14}$P][Tf_2N] and [C_4C_1pyr][Tf_2N], trihexyltetradecylphosphonium and 1-butyl-1-methyl-pyrrolidinium *bis*(trifluoromethanesulfonyl) amides; [$(C_6)_3C_{14}$P][C_{12}PhSO$_3$] and [$(C_4)_3C_{14}$P][C_{12}PhSO$_3$], trihexyltetradecylphosphonium and tributyltetradecylphosphonium dodecylbenzenesulfonates; [$(C_6)_3C_{14}$P][C_1SO$_3$], trihexyltetradecylphosphonium methanesulfonate. The effect of salinity, temperature, concentration, and pH on the distribution ratios was studied.

Acetic acid did not significantly partition into the IL phase, except for the sulfonate ILs. The extraction of much more hydrophobic hexanoic acid is significant, molecular form being efficiently recovered by all the ILs studied. Other organic compounds also showed high distribution ratios, up to 500. The distribution ratios for toluene, 1-nonanol, cyclohexanone, and hexanoic acid were independent of the phase/volume ratio in the range 0.02–1.0. The authors studied the regeneration of the ILs by rinsing and heating. Regeneration did not prove successful in all the cases. The final conclusion was that, despite the high efficiency of extraction, practical use of ILs in the intended application, detection and removal of organic compounds from water, appears to be hard due to small but significant solubility of ILs in water and due to difficulty in solvent regeneration.

The distribution of several organic acids (acetic, glycolic, propionic, lactic, pyruvic, and butyric) between water and 1-alkyl-3-methylimidazolium hexafluorophosphate ILs—[C_4C_1Im][PF_6], [C_6C_1Im][PF_6], and [C_8C_1Im][PF_6] is described in the paper [8]. The distribution constants of all the organic acids were low, their values roughly correlating with the water/1-octanol partition coefficients for the acids (i.e., with solute hydrophobicity). The highest distribution constant was observed for butyric acid (1.06, extraction into [C_6C_1Im][PF_6]), the most hydrophobic one; the most hydrophilic glycolic acid has not been extracted yet. The general order of IL extraction ability, with minor exceptions, is as follows: [C_8C_1Im][PF_6] < [C_4C_1Im][PF_6] < [C_6C_1Im][PF_6] (however, the difference is rather small). In the case of lactic acid recovery, additional extractant, tri-*n*-butylphosphate, was used. This resulted in efficient extraction, comparable with that characteristic of the combination of conventional organic solvents and tri-*n*-butylphosphate.

Several studies are devoted to the extraction of phenolic compounds. These compounds are particularly interesting from a practical viewpoint, as phenol derivatives are toxic pollutants that have marked detrimental effects on living organisms in general; therefore, the development of effective methods of phenols recovery is a long-standing problem of analytical chemistry. To determine phenolic compounds at the trace level, typically preconcentration and separation from accompanying substances is required, but the extraction of phenolic compounds with conventional solvents is often not quantitative. From a more theoretical viewpoint, phenolic compounds exhibit a wide structural variability, thus, a study of their

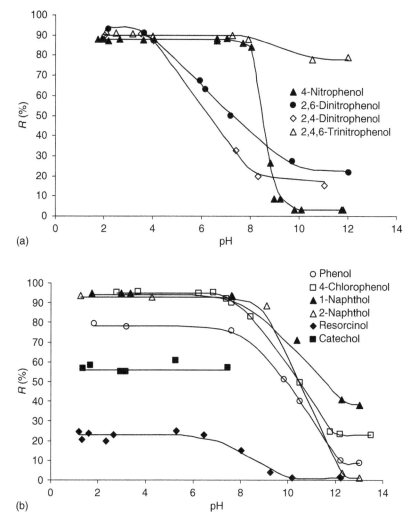

Figure 9.1 Extraction of phenols ($5 \cdot 10^{-5}$ to $2 \cdot 10^{-4}$ M) into [C$_4$C$_1$Im][PF$_6$] versus pH. (From Khachatryan, K.S., Smirnova, S.V., Torocheshnikova, I.I., Shvedene, N.V., Formanovsky, A.A., and Pletnev, I.V. *Anal. Bioanal. Chem.*, **381**, 464–470, 2005. With permission.)

extraction may give important highlights concerning the solvation/extraction ability of ILs.

The present authors studied the extraction of various phenol derivatives to a number of imidazolium-based ILs [9]. Extraction of most phenols is efficient at pH < pK_a, as illustrated in Figure 9.1 for the case of [C$_4$C$_1$Im][PF$_6$].

The pH values of efficient extraction correspond to the pH range where the molecular form of the respective phenol dominates. The recovery of 4-nitrophenol, 2,4-dinitrophenol, 2,6-dinitrophenol, 4-chlorophenol, 1-naphthol, and 2-naphthol is above 90% (the ratio of aqueous: organic phase volume is 3:1). The extraction of naphthol and 4-chlorophenol is significant even at pH > pK_a, more than 40 and 24% at pH > 10, respectively. Recovery of picric acid (2,4,6-trinitrophenol) is about 90% at pH 1.5–12.0, where the anionic form of picric acid dominates. Obviously, the high extraction is caused by high hydrophobicity of picrate anions. Recovery of the phenol itself and diatomic phenols, catechol and resorcinol is rather moderate (79, 58, and 20%, respectively; pH 1–7), which could be explained by relatively high hydrophilicity of these compounds.

To summarize, the more hydrophobic phenolic compounds are extracted better than the less hydrophobic ones, and the extraction is maximal at pH < pK_a. In general, for all the compounds distribution ratios are relatively high and comparable to those achieved with conventional active solvents like 1-octanol. This may be attributed to the ability of IL's imidazolic proton at C_2 to hydrogen bonding and specific solvation of the phenolic molecule.

In this work it was also shown that $[C_4C_1Im][PF_6]$ is suitable for extraction-voltammetric determination of phenols without back-extraction or adding a support electrolyte.

The paper [10] is devoted to the extraction of chlorophenols (2-chlorophenol, 2,4-dichlorophenol, 2,4,6-trichlorophenol, 2,3,4,5-tetrachlorophenol, pentachlorophenol; also, unsubstituted phenol) from an aqueous solution to two room-temperature ILs, $[C_4C_1Im][PF_6]$ and 1-ethyl-3-methylimidazolium *bis* (perfluoroethylsulfonyl)imide ($[C_2C_1Im][BETI]$). The partitioning was monitored by both aqueous and IL phases with high pressure liquid chromatography (HPLC). Extraction efficiency was found to be higher for $[C_4C_1Im][PF_6]$. Maximal recovery was attained if the pH of the aqueous solution was at least one unit below the pK_a value of the solute, once again indicating that neutral forms were extracted. Partitioning to both ILs increased as the number of chlorine atoms in the chlorophenol increased, in parallel with the hydrophobicity measured by log$P$(water/1-octanol). Notably, distribution ratios of chlorinated phenols to the studied ILs are nearly an order of magnitude lower than those for extraction to 1-octanol. It is noted that the ionic strength of the aqueous phase had no significant effect on the IL-water distribution ratios of chlorinated phenols but had a dramatic effect on the solubility of ILs in water.

The present authors studied the extraction of aromatic amines into ILs. As is seen from experimental data for $[C_4C_1Im][PF_6]$ (Figure 9.2), aniline, napthylamine, and *o*-toluidine are efficiently extracted from the alkaline aqueous solution. Thus, as in the case of phenols, neutral (molecular) forms of solutes were extracted. Another example of the same behavior is given by many polyfunctional compounds, for example, 8-hydroxyquinoline (Figure 9.3 presents a comparison of extraction pH-profile with the distribution diagram for ionic forms of the solute).

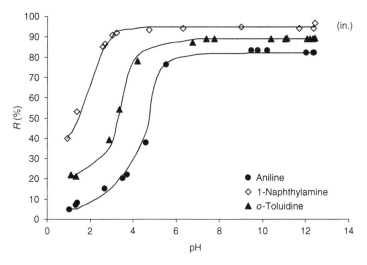

Figure 9.2   Extraction of aromatic amines ($5 \cdot 10^{-5}$ to $2 \cdot 10^{-4}$ M) into $[C_4C_1Im][PF_6]$ versus pH.

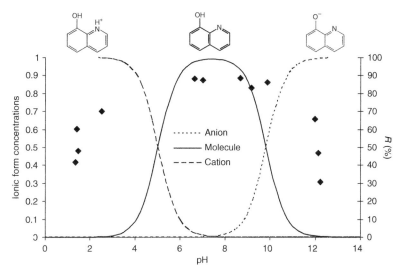

Figure 9.3   Distribution diagram for the ionic forms of 8-hydroxyquinoline in comparison with pH-profile of its extraction into $[C_4C_1Im][PF_6]$.

9.3   Comparison of $[C_4C_1Im][PF_6]$ with conventional extraction solvents

As the reader may notice, most of the described extraction systems are based on $[C_4C_1Im][PF_6]$. The amount of data available for this IL on the extraction of neutral organic compounds allows one to compare it with other solvents

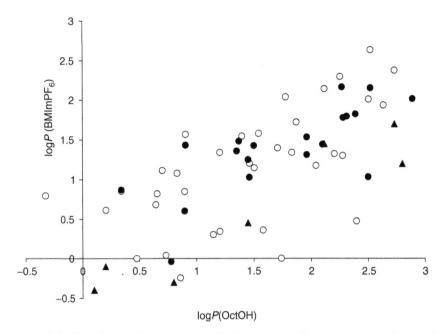

Figure 9.4 Correlation between distribution ratios of various neutral organic compounds from aqueous solution to 1-octanol (Korenman, Ya.I., *Distribution Ratios of Organic Compounds,* Voronezh University Press, Voronezh, Russia, 1992. p. 336) and to [C₄C₁Im][PF₆] (literature and our own data, [Huddleston, J.G., Visser, A.E., Reichert, V.M., Willauer, H.D., Broker, G.A., Rogers, R.D., *Green Chem.,* **3**, 156–164, 2001; Carda-Broch, S., Berthod, A., Armstrong, D.W., *Anal. Bioanal. Chem.,* **375**, 191–199, 2003; Pletnev, I.V., Smirnova, S.V., Khachatryan, K.S., Zernov, V.V., *Rus. Chem. J.–J. Russ. Chem. Soc.,* **48**, 51–58, 2004]).

by the extraction ability (naturally, a number of publications compare the solvents by other parameters, like polarity derived from solvatochromic data, and so on—for example, see Refs 11–13).

An opinion often quoted in the literature is that by its extraction ability [C₄C₁Im][PF₆] is mainly similar to 1-octanol, and typically the efficiency of extraction is somewhat lower. However, at closer examination, the correlation of the distribution ratio for the same compounds extracted from the aqueous solution to [C₄C₁Im][PF₆] and to 1-octanol seems to be not particularly impressive (Figure 9.4; see Refs 14 and 15).

By the use of the multiple linear regression, Abraham and Rogers [16] determined the parameters of the Abraham linear solvation free energy relationship (LSFER) equation [17,18]

$$\lg D = c + eE + sS + aA + bB + vV$$

for [C₄C₁Im][PF₆] and compared them with the parameters for the conventional extraction solvents.

Here E is the solute excess molar refractivity, S is the solute dipolarity/polarizability; A and B are the overall or summation hydrogen-bond acidity and basicity, respectively; and V is the McGowan characteristic volume; lower-case letters stand for respective coefficients which are characteristic of the solvent, c is the constant. By help of statistical methods like the principal component analysis and nonlinear mapping, the authors determined the mathematical distance (i.e., measure of dissimilarity) from an IL to seven conventional solvents immiscible with water. It appears that the closest to the IL conventional solvent is 1-octanol. Even more close to IL is an aqueous biphasic system based on PEG-200 and ammonium sulfate (and even closer are ethylene glycol and trifluoroethanol, as calculated for hypothetical water–solvent systems involving these solvents).

Notably, Armstrong et al. [6] performed a similarly targeted independent study and also reported on the LSFER parameters for [C₄C₁Im][PF₆]. In Table 9.1 are the data from both groups, as well as the known parameters for 1-octanol.

It is worth mentioning that the coefficients for the same IL in Table 9.1 significantly differ. Even the sign is different for the two coefficients.

One possible reason of such a discrepancy is that during regression fitting an experimental uncertainty may *spread* over various parameters, thus leading to somewhat distorted final picture. Probably, a more reliable way to measure similarity/dissimilarity of solvents would be to rely on the direct experimental measurements of the distribution ratios rather than on the derived quantities, that is, LSFER parameters. The present authors employed that approach [15].

We compiled literature and our own extraction data to compare the distribution of the same solutes from water to [C₄C₁Im][PF₆] and 48 various conventional solvents. As a measure of similarity of the extraction properties of any two solvents, we used the Pearson correlation coefficient between lg$D$ for the same solutes. Note that a high correlation coefficient does not mean that the distribution ratios determined with the two solvents are close by absolute value; rather, it means that the distribution ratios change in the same manner from one solute to another.

Table 9.1   The Abraham LSFER Parameters for [C₄C₁Im][PF₆] [6,16] Compared to the Parameters for 1-Octanol [19]

	Ref. [6]	Ref. [16]	Ref. [19]
c	1.79	−0.17	0.088
e	1.29	0.45	0.562
s	−0.73	0.23	−1.054
a	−0.76	−1.76	0.034
b	−2.39	−1.83	−3.46
v	0.64	2.150	3.814

In total, 4,777 solutes and 10,198 lgD values were analyzed. From these data, the correlation matrix of size 49 × 49 (48 conventional solvents + IL) was derived. It appeared that the solvents most close to $[C_4C_1Im][PF_6]$ (i.e., having the highest pairwise correlation coefficient solvent/IL) are esters with a short alkyl chain (ethyl acetate: correlation coefficient $r = 0.95$, as determined by the distribution ratios for $n = 11$ solutes; butyl acetate: $r = 0.92$, $n = 30$) and substituted aromatic hydrocarbons (m-xylene: $r = 0.92$, $n = 20$). The most *distant* from ILs are aliphatic hydrocarbons. Interestingly, the correlation with 1-octanol is moderate, $r = 0.76$, $n = 56$.

These data allow one to suggest simple correlation equations to predict the distribution ratios of neutral organic molecules between water and $[C_4C_1Im]$ $[PF_6]$ by the distribution ratios measured for conventional solvents:

$$\lg D_{[C_4C_1Im][PF_6]} = 0.05 + 0.72 \lg D \text{ butyl acetate} \quad (n = 30, r^2 = .84)$$

$$\lg D_{[C_4C_1Im][PF_6]} = 0.135 + 0.62 \lg D \text{ butyl acetate}$$
$$+ 0.11 \lg D \text{ m-xylene} \qquad (n = 18, r^2 = .95)$$

An overall picture of similarity between $[C_4C_1Im][PF_6]$ and conventional solvents is given by the so-called nonlinear map, or Sammon map [20] in Figure 9.5. The closer the points representing solvents on the map, the more similar the solvents are. As is seen, the position of the IL on the map is between esters and aromatic hydrocarbons.

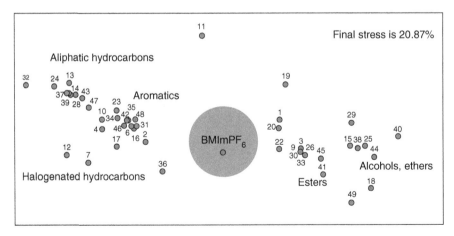

Figure 9.5 Sammon map representing similarity of extraction ability of solvents (with respect to extraction of neutral organic compounds). The closer the points are on the map, the more similar are the corresponding solvents. Conventional solvents are designated by numbers, some representative numbers are: m-xylene—35, butyl acetate—9, 1-octanol—40. The point marked *BMImPF₆* represents 1-butyl-3-methyl-imidazolium hexafluorophosphate ($[C_4C_1Im][PF_6]$).

9.4 Comparison of imidazolium-based ionic liquids as extraction solvents

One of the advantages often attributed to ILs as a class is that solvent properties may be easily modified, for example, through substitution of the structure of the cationic or anionic constituent. This opens an avenue to the fine tuning of the solvent. In this context, it is interesting to compare how the change of the cation or the anion influences the extraction ability of widespread imidazolium-based ILs.

Having this in mind, the present authors systematically studied the extraction of the same organic compounds into $[C_4C_1Im][PF_6]$, $[C_4C_1C_1Im][PF_6]$, $[C_4C_1Im][Tf_2N]$, and $[C_4C_1C_1Im][Tf_2N]$. The representative solutes were phenol, 4-nitrophenol, resorcinol, aniline, adrenaline, and dobutamine ($1 \cdot 10^{-5}$ to $1 \cdot 10^{-4}$M); extraction was performed at the aqueous organic phase ratio equal to 3:1 and phase contact time of 10–15 min.

It appears that the extraction mechanism of various compounds does not change on the variation of solvent composition/structure. As is $[C_4C_1Im][PF_6]$, the related ILs $[C_4C_1C_1Im][PF_6]$, $[C_4C_1Im][Tf_2N]$, and $[C_4C_1C_1Im][Tf_2N]$ efficiently recover phenolic compounds from an acidic aqueous solution at pH $< pK_a$, where the neutral form of the solute dominates. Aniline is quantitatively recovered into $[C_4C_1C_1Im][PF_6]$, $[C_4C_1Im][Tf_2N]$, and $[C_4C_1C_1Im][Tf_2N]$ from alkaline aqueous solutions at pH $> pK_a$, also the region where a molecular form of solute dominates. Recovery of catecholamines, adrenaline, dobutamine, and dopamine is nearly constant at pH 2–8, where catecholamines exist in the cationic form (they are extracted as cations). The more hydrophobic $[C_4C_1C_1Im]$ derivatives and, in particular, ILs containing *bis*-(trifluoromethanesulfonyl)imide anion, typically demonstrate higher distribution ratios for the same solutes (Figure 9.6). In general, the extraction power of ILs decreases in the following series: $[C_4C_1Im][PF_6] \approx [C_4C_1C_1Im][PF_6] < [C_4C_1Im][Tf_2N] < [C_4C_1C_1Im][Tf_2N]$.

Naturally, the consideration above concerns only that particular set of ILs. In general, the potential of variations in IL's nature is so high that one may expect much greater variations in extraction efficiency and selectivity, as well as in extraction mechanisms, particularly if novel classes of ILs are synthesized and used.

For example, in our recent study of the tetrahexylammonium dihexylsulfosuccinate IL (ILs comprising dihexylsulfosuccinate anion were first synthesized by Kakiuchi et al. [21]) we found that its properties are quite different from those of $[C_4C_1Im][PF_6]$ and other conventional imidazolium ILs. In particular, it is a much more efficient solvent for polar compounds, and it also demonstrates different pH-profiles of extraction [22].

Another example was given in the recent study [23] of the extraction of lactic acid into two phosphonium ILs incorporating trihexyltetradecylphosphonium cation and *bis*(2,4,4-trimethylpentyl)phosphinate (IL-104) and chloride (IL-101) anions. The behavior of the two ILs is quite different and

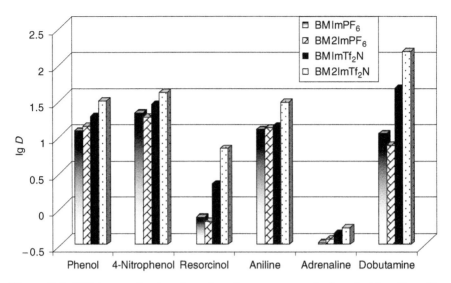

Figure 9.6 Efficiency of extraction of various compounds into imidazolium ILs (optimal conditions). Abbreviations: BMIm—1-butyl-3-methylimidazolium, C_4C_1Im; BM2Im—1-butyl-2,3-dimethylimidazolium, $C_4C_1C_1Im$.

complicated. Phosphinate IL effectively extracts only the dissociated form of lactic acid; the authors proposed a coordination mechanism of extraction via H-bonding solute-solvent. Interestingly, the solubility of water in hydrophobic IL-104 is surprisingly high, up to 14.4%, which is explained by the formation of reverse micelles. In the case of lactic acid extraction by IL-101 an ion-exchange mechanism (see the following sections for discussion) makes a remarkable contribution, especially at high pH values where the anionic form of lactic acid is dominant.

9.5   Ion-exchange extraction from aqueous solutions

The above consideration of the similarity and dissimilarity of ILs and conventional extraction solvents ignores one particularly striking feature of ILs. In sharp contrast to common solvents immiscible with water, ILs are capable of ion exchange. We exemplify this very important ability by considering the extraction of amino acids on the basis of our work [24].

Practically motivated, the aim was to develop methods for recovery and determination of amino acids in the context of analytical chemistry and biotechnology. Amino acids are hydrophilic compounds, which therefore are difficult targets for conventional solvent extraction. Extraction to an organic solvent may be enhanced by the addition of lipophilic cationic or anionic extractants, forming extractable complexes with amino acids, or by the use of macrocyclic compounds, which form stable hydrophobic *host-guest* complexes. The most popular reagents from the latter group are crown ethers, which form complexes via hydrogen bonding of protonated amino groups. However, even in the presence of crown ethers, extraction is not highly efficient, especially for the most hydrophilic amino acids like glycine.

Owing to their ionic nature, ILs seem promising candidates to facilitate the extraction of amino acids. As a solvent, we used $[C_4C_1Im][PF_6]$; dicyclohexano-18-crown-6, DC18C6, was also added. It was shown that the equilibrium is achieved within 15 min and the extraction of Trp, Gly, Ala, Leu, Lys, Arg $(2 \cdot 10^{-5}$ to $2 \cdot 10^{-1}$ mol $L^{-1})$ from acidic solutions is nearly quantitative (Figure 9.7).

Most interestingly, the pH of maximal extraction corresponds well to the range in which the cationic form of amino acids dominates. With increasing pH amino acids are transformed from cations to zwitterions and then to anions, which results in a progressive decrease of extraction. The range of efficient extraction of arginine and lysine is much broader than in the case of Trp, Gly, Ala, and Leu, as Arg and Lys can form both cationic and dicationic species; dicationic species dominate at low pH, whereas increasing pH changes the state to the single-charged cation.

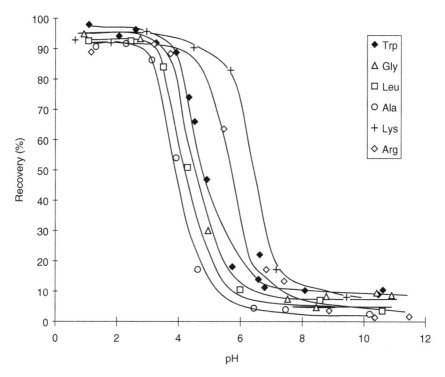

Figure 9.7   Extraction of amino acids $(1 \cdot 10^{-4}M)$ with dicyclohexano-18-crown-6 $(1 \cdot 10^{-1}M)$ into $[C_4C_1Im][PF_6]$. (From Smirnova, S.V., Torocheshnikova, I.I., Formanovsky, A.A., and Pletnev, I.V., *Anal. Bioanal. Chem.*, **378**, 1369–1375, 2004. With permission.)

The extraction becomes significant only in the presence of crown ether, which strongly indicates that crown ether has a critical role as a complexing reagent. The ratio of the amino acid to crown ether in extractable species for Trp, Leu, Gly, Ala is 1:1 and for Arg and Lys 1:2. At pH 2 the dicationic forms of Arg and Lys dominate. It means that the extraction in ILs proceeds as in traditional solvents, where each ammonium group interacts with one molecule of crown ether.

The most interesting point is that extraction of amino acids into IL occurs without addition of a counterion. Typically, for amino acid extraction into conventional solvents (including extraction with crown ether), a hydrophobic counterion is required. Moreover, in most cases, even the presence of such a counteranion does not provide an efficient recovery.

The following equilibrium equation can be used to describe the extraction of the cationic of amino acids (AmH_2^+):

$$AmH_2^+ + CE_o + C_4C_1Im_o^+ \rightleftarrows AmH_2^+ \cdot CE_o + C_4C_1Im^+$$

This expression corresponds to the cation-exchange reaction in which the amino acid cation goes into the organic phase and the dialkylimidazolium cation into the aqueous phase. This scheme was confirmed by experimental data. The significant leakage of the dialkylimidazolim cation into the aqueous phase on the extraction of amino acids is confirmed by monitoring the corresponding UV band. No influence of the foreign anions presented in the system was observed. The study of amino acids extraction from nitric and hydrochloric acids solutions reveals no influence of anions of the aqueous phase on the extraction, moreover, the partition coefficients of amino acids were practically the same at different aqueous concentrations of the acid used to create an acidic medium, for example, HNO_3, HCl (therefore, coextraction of aqueous phase anions does not occur).

High distribution ratios and almost quantitative recovery were observed for all amino acids. In contrast to conventional solvents, the extraction of the most hydrophilic amino acids such as Gly is quantitative. Recovery of Trp, Leu, Ala, Gly, Lys, and Arg is 96, 93, 92, 95, 94, and 92%, respectively. Also, amino acids, including highly hydrophilic, can be extracted with high efficiency from the mixture. For example, the recovery of Trp, Val, Gly from their equimolar mixture is equal to 99, 94, and 93%, respectively. Extraction was performed by adding 3 mL of IL with crown ether concentration 0.10 mol L with a 3 mL aqueous solution of amino acids ($5 \cdot 10^{-4}$ mol/L each; 1.8 pH) and shaking for 15 min.

Using the described extraction system, we developed methods of amino acid recovery from pharmaceutical samples and fermentation broth. Amino acids were extracted efficiently from the diluted solution of fermentation broth into $[C_4C_1Im][PF_6]$ in the presence of DC18C6 and may be well back-extracted by the alkaline aqueous solution (pH > 9). These methods served as a basis for the corresponding analytical procedures.

Interestingly, authors of a subsequent paper [25] studied the extraction of amino acids (Trp, Phe, Tyr, Leu, Val) into four imidazolium-based ILs ($[C_4C_1Im][PF_6]$, $[C_6C_1Im][PF_6]$, $[C_6C_1Im][BF_4]$, and $[C_8C_1Im][BF_4]$) in the absence of crown ether and observed the same pH-profiles of extraction (though, naturally, distribution ratios were lower).

The ion-exchange mechanism of extraction does not occur only for amino acids. We observed it also for catecholamines [26]. These compounds are efficiently extracted into ILs in the cationic form, at pH 1–8. At these pH, the primary (dopamine) or secondary (adrenaline and dobutamine) amino groups are protonated (catecholamines are oxidized in alkaline solutions at pH > 8). By analogy with amino acids, extraction may be described by the cation-exchange reaction:

$$R_2NH_2^+ + C_4C_1Im_o^+ \rightleftarrows R_2NH_{2o}^+ + C_4C_1Im^+$$

It should be noted that cation exchange is not the only possibility: as our results show, some anionic compounds are also extracted into ILs (at least, into $[C_4C_1Im][PF_6]$) through the ion-exchange mechanism. The most striking example is the picric acid which is efficiently extracted in the pH range from 1.5 to 12, where it exists in the anionic form. For some other phenols, significant extraction was also observed at high pH where anions dominate (e.g., are 1-naphthol and 4-chlorophenol at pH > 10 and 2,6-dinitrophenol at pH > 5). The experiments showed that an excess concentration of Na^+ or $[C_4C_1Im]^+$ does not influence extraction; in case of the ion-pair extraction mechanism, this excess of potential counterion should affect the recovery. Therefore, we concluded that the extraction of the mentioned phenolates proceeds via the anion-exchange path:

$$PhO^- + PF_{6o}^- \rightleftarrows PhO_o^- + PF_6^-$$

presuming that when the phenolate goes to the organic phase, an equal amount of IL's anion is transferred to water.

Evidently, the capability of cation-exchange and anion-exchange extraction, makes a great difference between ILs and conventional solvents. From a practical viewpoint, it may be both a benefit and a disadvantage, depending on the particular task.

9.6   Practical applications: extraction of biologically important compounds

This section covers some recent practical applications of solvent extraction utilizing ILs in one particularly important area. Many biologically important compounds can hardly be recovered by using conventional solvents due to their high polarity, polymeric nature, instability, and so on. Given also high structural diversity of biologically relevant molecules, it is not surprising that using novel *designer solvents* for their isolation attracts more and more attention. Some representative examples are given below.

Biphasic IL/water system containing water-soluble $[C_4C_1Im]Cl$, and K_2HPO_4 (salting-out agent) was used to isolate testosterone and epitestosterone from human urine with subsequent determination by reversed-phase HPLC [27]. The optimal salt amount for phase separation and extraction of steroids was found. Under the optimal conditions, the extraction efficiency for both steroids is ~80 to 90% (single-step extraction). The authors suggested a method of analysis for testosterone and epitestosterone, which requires 3.0 mL of urine, due sample preparation and single extraction step followed by the HPLC analysis. Detection limits were at the level of 1 ng/mL for both substances, and linear range was 10–500 ng/mL.

Soto et al. [28] investigated the extraction of two antibiotics, amoxicillin and ampicillin from water into the $[C_8C_1Im][BF_4]$, at pH 4 and 8. It was shown that for both antibiotics recovery is much higher in the alkaline region, where the compounds exist in the anionic form. Given that, the authors supposed that the extraction mechanism relies on the electrostatic interactions between the IL ions and charged antibiotic molecules (alternatively, ion exchange may take place). The partition coefficients of ampicillin were higher than those of amoxicillin at both pH values studied (explained by higher hydrophilicity of the latter caused by an extra OH group).

The partition of penicillin G in the aqueous biphasic system comprising $[C_4C_1Im]Cl$ and NaH_2PO_4 aqueous solution is described in Ref. 29. The effect of concentrations of NaH_2PO_4, penicillin G, and $[C_4C_1Im]Cl$ on the partition coefficient and extraction yield of penicillin G was determined. The recovery in optimal conditions reaches 93%. The procedure is applied to the isolation of penicillin G from the fermentation broth.

$[C_4C_1Im][PF_6]$, a widespread IL, was recently used for the direct extraction of double-stranded DNA (dsDNA) [30]. The authors demonstrated that DNA may be extracted with high efficiency, >95%, while proteins and metalloproteins do not interfere extraction. The back-extraction of DNA into the aqueous phase with the efficiency of 30% was performed in the presence of phosphate–citrate buffer solution (pH 4).

The extraction was shown to be endothermic with an enthalpy of 34.3 kJ/mol. The mechanism of extraction was proposed and partially verified by means of ^{31}P nuclear magnetic resonance (NMR) and FT-IR spectra. This mechanism involves interactions between the IL cation and P–O bonds of phosphate groups in the DNA molecule. The authors also suggested an extraction-based procedure for the quantitative determination of dsDNA.

Several ILs incorporating $[Tf_2N]$ anion were shown to be capable of quantitative partitioning of cytochrome *c* (cyt *c*) from the aqueous phase in the presence of crown ether DC18C6 [31]. The structures and abbreviations of the ILs employed in this work are as follows:

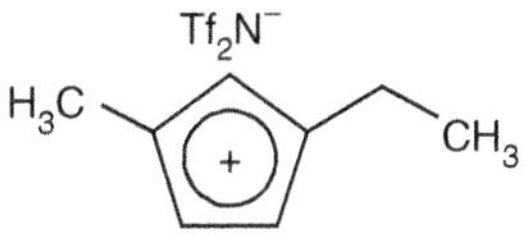

EMImTf$_2$N   CnOHMImTf$_2$N   C$_1$OC$_1$MImTf$_2$N

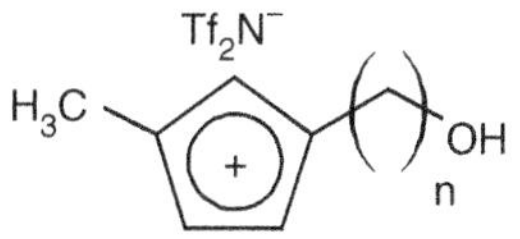

n = 2, 3, 6, 8

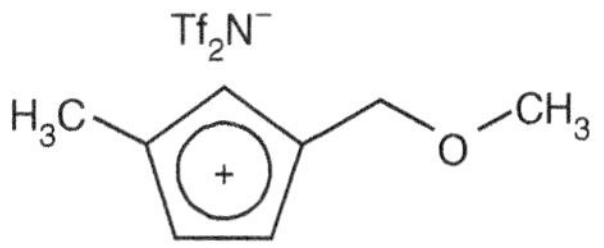

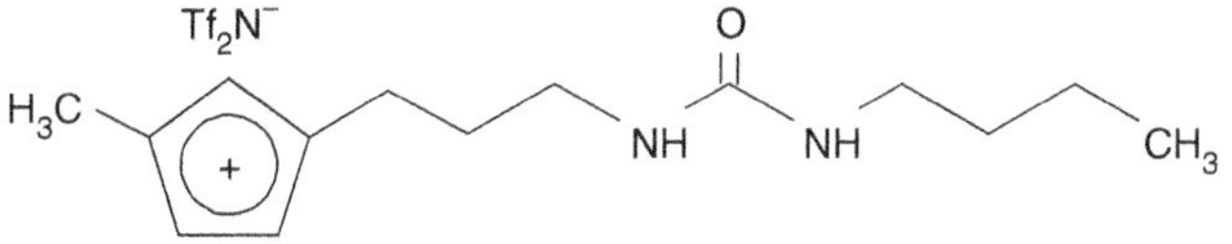

C$_3$UC$_4$MImTf$_2$N

In the absence of crown ether in the IL phase the extraction of cyt *c* was found to be negligibly small. However, the addition of crown ether to the extraction system improved the extractability. High recovery was observed for the ILs containing two hydroxyl groups, [C$_2$OHMIm][Tf$_2$N] and [C$_3$OHMIm][Tf$_2$N] (100.0 and 90.5%, respectively). The improvement of extraction in the presence of crown ether was explained by the complexation of crown ether molecule with positively charged lysine residues of cyt *c*. The extraction of cyt *c* into conventional organic solvents—chloroform, toluene, isooctane and 1-octanol—was also inspected; no transfer of cyt *c* to organic phase was observed. Interestingly, authors suggested a very practical way of cyt *c* isolation from the IL after extraction—the addition of potassium chloride, which forms a complex with crown ether.

An aqueous biphasic system based on the hydrophilic [C$_4$C$_1$Im]Cl and K$_2$HPO$_4$ was used for direct extraction of proteins from human body fluids [32]. Proteins present at low levels were quantitatively extracted into the [C$_4$C$_1$Im]Cl-rich upper phase with a distribution ratio of about 10 (between upper and lower phases) and an enrichment factor of 5. Addition of an appropriate amount of K$_2$HPO$_4$ to the separated upper phase again resulted in a phase separation, which improves the enrichment factor (up to 20). FT-IR and UV spectroscopies demonstrated that no chemical (bonding) interactions between the IL and the protein functional groups were identifiable. The authors suggested that the system could be useful for the quantification of proteins in human urine after on-line phase separation in a flow system.

9.7   Practical applications: extraction from nonaqueous solutions

The consideration above was limited to the extraction from aqueous solutions. However, nonaqueous systems are widely used in technology and analytical applications, so it is natural that ILs are now actively studied in the area. Distillation is a traditional and still very important technique of separation of the mixtures of organic compounds. Quite often, it is hampered by the formation of azeotropes. In many cases, ILs have been shown to be effective azeotrope breakers; alternatively, they may present a possibility of extractive separations.

The paper [33] describes an extraction technique for the separation of ethanol–heptane azeotropic mixture with the use of $[C_4C_1Im][PF_6]$. The authors studied the liquid–liquid equilibrium (LLE) for the ternary system heptane-ethanol-IL at 298.15 K. Tie-line compositions were determined by measuring the density and applying the corresponding fitting polynomials. The selectivity and the solute distribution ratio were determined from the tie-lines data, and their values were compared to the values obtained with $[C_8C_1Im]Cl$ as a solvent [34]. After extraction heptane was collected at the top of the column and ethanol was separated from the mixture with the IL by distillation. The ethanol-free IL was recycled to the extractive column.

The LLE for another ternary system, ethyl *tert*-butyl ether (ETBE) + ethanol + $[C_4C_1Im][TfO]$, at 298.15 K was studied by Arce et al. [35]. To determine the tie-line compositions, they used the NMR spectroscopy. The values of the solute distribution ratio ($\beta = x_{EtOH}^{II}/x_{EtOH}^{I}$, where II refers to an IL-rich phase) and selectivity ($S = \beta_{EtOH}/\beta_{ETBE}$) were calculated from tie-line data. In general, both the solute distribution ratio and the selectivity decreased as the molar fraction of ethanol in the organic-rich phase increased, the maximal values being ca. 3.5 and ca. 22, respectively. The ETBE + ethanol + IL system was compared to the ETBE + ethanol + water system.

Arce et al. also investigated the effect of anion fluorination in $[C_2C_1Im]$ ILs on the extraction of ethanol from ETBE [36]. For this purpose, two anions, methanesulfonate and trifluoromethanesulfonate, were selected. The corresponding phase diagrams were plotted for both ternary ETBE + ethanol + IL systems. The solute distribution ratios for the IL with the nonfluorinated anion were higher, especially at low concentrations. As for the selectivities, better results were obtained for methanesulfonate IL at low solute concentrations, while at higher solute concentrations the selectivity was better for the fluorinated analog.

The work reported in Ref. 37 is devoted to the extraction of terpenes from the model solution (a mixture of limonene and linalool) imitating citrus-essential oil. Arce et al. studied $[C_2C_1Im][TfO]$, as well as conventional solvents—2-butene-1,4-diol and ethylene glycol. The ability of the organic solvents and the IL to extract linalool from the essential oil mixture was evaluated in terms of solute distribution ratio and selectivity. The results were compared with available data obtained with the use of other solvents—diethylene glycol, 1,2-propanediol, 1,3-propanediol, and 2-aminoethanol. It was demonstrated that only the IL and 1,2-propanediol have a favorable solute distribution ratio (though not in the entire concentration range). The best values of selectivity were observed for the IL.

Several studies have analyzed the challenging problem of the isolation of aromatic hydrocarbons from their mixtures with alkanes. Their boiling points are often close and the formation of azeotropes of various compositions is a common difficulty.

Meindersma et al. [38,39] compared several ILs as solvents for the extraction of toluene from toluene–heptane mixtures. The following ILs were tested: 1-methylimidazolium hydrogensulfate; 1,3-dimethylimidazolium methylsulfate and dimethylphosphate; 1-ethyl-3-methylimidazolium hydrogensulfate, methylsulfonate, ethylsulfate, diethylphosphate, tosylate and tetrafluoroborate; 1-butyl-3-methylimidazolium tetrafluoroborate and methylsulfate; 1-octyl-3-methylimidazolium tetrafluoroborate; N-butylpyridinium methylsulfate; 1-butyl-4-methylpyridinium methylsulfate and tetrafluoroborate; ethylisoquinolinium ethylsulfate; tributylmethylammonium methylsulfate; tetrabutylphosphonium *bis*[oxalato(2-)]borate. The toluene/heptane selectivities at 40 and 75°C for the ILs were 1.5–2.5 times higher than those achieved with sulfolane, which is the most common industrial solvent for the extraction of aromatic hydrocarbons from the aromatic–aliphatic hydrocarbon stream. Of these ILs, 1-butyl-4-methylpyridinium tetrafluoroborate, $[4\text{-}C_1C_4py][BF_4]$ appeared the most suitable, because of a high toluene distribution ratio ($D = 0.44$) and a high toluene/heptane selectivity ($S = 53.6$). For this reason, the additional extraction experiments with $[4\text{-}C_1C_4py][BF_4]$ and other aromatic/aliphatic mixtures were carried out. The obtained arene/alkane selectivity was in the same range as for toluene/heptane mixture. The distribution ratios of aromatic or aliphatic compounds decreased with the lengthening of the alkyl chain in these compounds (for aromatics it means lengthening of the alkyl side-chain).

The separation of aliphatic and aromatic hydrocarbons, namely hexane and benzene, using $[C_2C_1Im][Tf_2N]$ was investigated by Arce et al. [40]. The LLE for the ternary system IL–benzene–hexane was studied at 25 and 40°C. The obtained values of distribution ratio and the selectivity were found better than those reached with sulfolane. Notably, the effect of temperature on the separation is very small, therefore the extraction process does not require high energy consumption.

Another and, probably, the most important area of application of ILs in extraction from nonaqueous media is the removal of sulfur-containing organic compounds from hydrocarbon fuels. Attention to this problem is growing rapidly. The reason is the increasingly strict regulatory requirements which are currently being placed on sulfur content in fuels. Naturally, these efforts are intended to decrease the emission of sulfur dioxide and to prolong car engine lifetime; moreover, lower sulfur content in fuel would make it possible to diversify the set of catalysts available to reduce emission of nitrogen oxides.

Common industrial procedures of desulfurization involve hydrogenation that is generally good only for aliphatic and alicyclic sulfur compounds. Therefore, alternative approaches are of great value.

Wasserscheid and coworkers were the first to attempt to use ILs for the desulfurization of model solutions (dibenzothiophene [DBT], in *n*-dodecane) and real diesel fuels [41]. For extraction, the authors used ILs with 1-alkyl-3-methylimidazolium cations ([C_nC_1Im], n = 2, 4, 6) and various anions. Also, binary mixtures of 1-alkyl-3-methylimidazolium chloride with $AlCl_3$ (Lewis-acidic ILs), the equimolar mixture of cyclohexyldiethylammonium and tri-butylammonium methanesulfonates (Brønsted-acidic IL) and the equimolar mixture of cyclohexyldiethylmethylammonium and tributylmethylammo-nium methanesulfonates were tested.

The best extraction of DBT was observed with the use of the binary mix-ture [C_4C_1Im]Cl/$AlCl_3$. For *neutral* imidazolium ILs the authors found that the bigger the cation or anion of IL, the better is the DBT extraction. It was shown that the extraction power of the ILs, at least for Brønsted-acidic ILs, does not uniquely depend on chemical interactions involving the acid proton. Multistage desulfurization of *real* predesulfurized diesel oil was carried out using several ILs.

Despite the fact that the proposed Lewis-acid IL systems show a very good extraction of DBT, their use on industrial scale is not desirable since $AlCl_3$ is hydrolytically unstable and can cause equipment corrosion. Many attempts have been made to find better IL systems for the desulfurization of oil.

Zhand and Zhang [42] tested three ILs ([C_4C_1Im][PF_6], [C_2C_1Im][BF_4], and [C_4C_1Im][BF_4]) for removing sulfur from fuels. They measured the absorption capacities of the ILs for eight model compounds (S-containing and S-free, aliphatic and aromatic). All the three ILs showed high selectivity for S-containing C_5 aromatics over C_6 aromatics, while S-containing non-aromatic compounds were absorbed poorly. To extract most of the aromatic compounds, ILs can be ordered by their absorption capacities as follows: [C_4C_1Im][PF_6] > [C_4C_1Im][BF_4] > [C_2C_1Im][BF_4]. Experiments with real gaso-lines showed quite satisfactory extraction of S-compounds. The sulfur-loaded ILs may be regenerated by distillation.

In their later paper [43], the same authors broadened the range of the tested ILs, adding [C_8C_1Im][BF_4], [C_6C_1Im][PF_6], and trimethylammonium chloroaluminate (binary mixtures of $AlCl_3$ and trimethylammonium chlo-ride with molar ratios 2:1 and 1.5:1). The extraction of nitrogen-containing compounds from fuels was also studied. It was found that 1-alkyl-3-methy-limidazolium ILs have a good ability to extract aromatic compounds from the main aliphatic fuels, the absorption capacity decreasing as the aromatics become decorated with the side-chain alkyls. For example, the absorption capacity decreased in the order of benzene > toluene > xylene > cumene. A similar phenomenon was observed for S- and N-containing aromatics. Alkanes, cycloalkanes, and alkenes along with their S-containing analogs were absorbed poorly. The preferable extraction of aromatic compounds over aliphatic ones was explained by the favorable electronic interactions between polarized aromatic molecules and the charged ion pairs of the IL. The comparison of gasoline and diesel fuel desulfurization showed that the removal of sulfur from the latter is more difficult than from the former.

Lo et al. [44] described a desulfurization procedure that involved a combination of chemical oxidation and liquid–liquid extraction using IL ([C_4C_1Im][PF_6] or [C_4C_1Im][BF_4]). A mixture of 30% H_2O_2 and acetic acid was used as an oxidizer. The desulfurization procedure was applied to both the model solution of DBT, in tetradecane and to the real light oil (sulfur content 8040 ppm). The extraction efficiency without oxidation step was 47 and 39% for [C_4C_1Im][PF_6] and [C_4C_1Im][BF_4], respectively. To combine extraction and oxidation, DBT was oxidized in a solvent phase as it was extracted, and a continuous decrease in the sulfur content in tetradecane was observed. Interestingly, the oxidation step drastically improved separation in the case of [C_4C_1Im][PF_6], but had lower effect with [C_4C_1Im][BF_4] used as a solvent.

The study of the desulfurization of model oils, predesulfurized diesel oil, and FCC-gasoline focused on extraction by halogen-free ILs (alkylsulfates of alkylmethylimidazoliums) was reported in Ref. 45. The highest DBT partition coefficient was observed for [C_4C_1Im][C_8SO_4]. After four extraction steps, sulfur concentration in the oil decreased from 500 to 10 ppm.

Huang et al. [46] synthesized metallated IL by mixing [C_4C_1Im]Cl with CuCl and used this IL to remove thiophene from model oil and gasoline. The extraction efficiency for model oil was ca. 23%, while for thiophene-loaded gasoline it was slightly lower (this is explained by the interference of other gasoline components). Sulfur removal improved as the initial sulfur content in the gasoline decreased. The author suggested that the mechanism of desulfurization is related to the complexation of thiophene and Cu(I). Notably, extractive desulfurization with the use of CuCl-based IL does not generate side reactions (polymerization of olefins, alkylation of benzene derivatives, and so on) that are common with chloroaluminate ILs. Another recent work concerns the use of ILs containing alkyl phosphate anions [47]. The authors report that, considering relatively high sulfur removal ability, low fuel dissolvability, and low influence on the gasoline treated, 1-ethyl-3-methylimidazolium diethylphosphate [C_2C_1Im][$2C_2P$] might be used as a promising solvent for the extractive desulfurization of gasoline.

9.8   Conclusions

As is seen from the presented data, ILs are true designer solvents. The modification of their constituents gives solvents with very different properties, namely, efficiency and selectivity of extraction of various solutes. In the near future, it should have important implications, as more and more finely tuned solvents are likely to appear for a variety of tasks.

For the chemists, this variability of ILs presents a challenge. As concerns extraction, it is associated with the possibility of different mechanisms of solute transfer, different mutual solubilities in biphasic systems, different solvation abilities, and probably, even different bulk phase structures. We should develop a deeper understanding of IL-based extraction systems; undoubtedly technological and analytical applications will benefit a great deal.

Acknowledgments

The authors are much indebted to their coworkers, Dr Kristine Khacha-tryan, Dr Nataliya Shvedene, Dr Irina Torocheshnikova, and Dr Vladimir Zernov, who contributed to the study of ILs as solvents for separations. We are also grateful to Russian Foundation for Basic Research (grant 05-03-32976) and INTAS (grant 05-1000008-8020) for the financial support of the work on ILs.

References

1. Huddleston, J.G., Willauer, H.D., Swatloski, R.P., Visser, A.E., Rogers, R.D., Room temperature ionic liquids as novel media for clean liquid-liquid extraction, *Chem. Commun.*, 1765–1766, 1998.
2. Visser, A.E., Swatloski, R.P., and Rogers, R.D., pH-dependent partitioning in room temperature ionic liquids provides a link to traditional solvent extraction behavior, *Green Chem.*, **2**, 1–4, 2000.
3. Visser, A.E., Holbrey, J.D., Rogers, R.D., Hydrophobic ionic liquids incorporating N-alkylisoquinolinium cations and their utilization in liquid-liquid separations, *Chem. Commun.*, 2484–2485, 2001.
4. Huddleston, J.G., Visser, A.E., Reichert, V.M., Willauer, H.D., Broker, G.A., Rogers, R.D., Characterization and comparison of hydrophilic and hydrophobic room temperature ionic liquids incorporating the imidazolium cation, *Green Chem.*, **3**, 156–164, 2001.
5. Fadeev, A., Meagher, M., Opportunities for ionic liquids in recovery of biofuels, *Chem. Commun.*, 295–296, 2001.
6. Carda-Broch, S., Berthod, A., Armstrong, D.W., Solvent properties of the 1-butyl-3-methylimidazolium hexafluorophosphate ionic liquid, *Anal. Bioanal. Chem.*, **375**, 191–199, 2003.
7. McFarlane, J., Ridenour, W.B., Luo, H., Hunt, R.D., DePaoli, D.W., Ren, R.X., Room temperature ionic liquids for separating organics from produced water, *Sep. Sci. Technol.*, **40**, 1245–1265, 2005.
8. Matsumoto, M., Mochiduki, K., Fukunishi, K., Kondo, K., Extraction of organic acids using imidazolium-based ionic liquids and their toxicity to Lactobacillus rhamnosus, *Sep. Purif. Tech.*, **40**, 97–101, 2004.
9. Khachatryan, K.S., Smirnova, S.V., Torocheshnikova, I.I., Shvedene, N.V., Formanovsky, A.A., Pletnev, I.V., Solvent extraction and extraction-voltammetric determination of phenols using room temperature ionic liquid, *Anal. Bioanal. Chem.*, **381**, 464–470, 2005.

10. Bekou, E., Dionysiou, D.D., Qian, R.-Y., Botsaris, G.D., Extraction of chlorophenols from water using room temperature ionic liquids, *ACS Sympos. Ser.*, **856**, 544–560, 2003.
11. Fletcher, K.A., Storey, I., Hendricks, A.E., Pandey, S., Pandey, S., Behavior of the solvatochromic probes Reichardt's dye, pyrene, dansylamide, Nile Red and 1-pyrenecarbaldehyde within the room-temperature ionic liquid bmimPF(6), *Green Chem.*, **3**, 210–215, 2001.
12. Aki, S.N.V.K., Brennecke J.F., Samanta A., How polar are room-temperature ionic liquids? *Chem. Commun.*, 413–414, 2001.
13. Baker, S.N., Baker, G.A., Bright, F.V., Temperature-dependent microscopic solvent properties of dry and wet 1-butyl-3-methylimidazolium hexafluorophosphate: Correlation with ET(30) and Kamlet-Taft polarity scales, *Green Chem.*, **4**, 165–169, 2002.
14. Korenman, Ya.I., *Distribution Ratios of Organic Compounds*, Voronezh University Press, Voronezh, Russia, 1992, p. 336.
15. Pletnev, I.V., Smirnova, S.V., Khachatryan, K.S., Zernov, V.V., Application of ionic liquids in solvent extraction, *Rus. Chem. J.–J. Russ. Chem. Soc.*, **48**, 51–58, 2004.
16. Abraham, M.H., Zissimos, A.M., Huddleston, J.G., Willauer, H.D., Rogers, R.D., Acree, W.E., Some novel liquid partitioning systems: Water-ionic liquids and aqueous biphasic systems, *Ind. Eng. Chem. Res.*, **42**, 413–418, 2003.
17. Kamlet, M.J., Doherty, R.M., Abraham, M.H., Marcus, Y., Taft, R.W., Linear solvation energy relationship. 46. An improved equation for correlation and prediction of octanol/water partition coefficients of organic nonelectrolytes (including strong hydrogen bond donor solutes), *J. Phys. Chem.*, **92**, 5244–5255, 1988.
18. Abraham, M.H., Scales of solute hydrogen-bonding: Their construction and application to physicochemical and biochemical processes, *Chem. Soc. Rev.*, **22**, 73–83, 1993.
19. Platts, J.A., Butina, D., Abraham, M.H., Hersey, A., Estimation of molecular linear free energy relation descriptors using a group contribution approach, *J. Chem. Inf. Comput. Sci.*, **39**, 835–845, 1999.
20. Sammon, J.W. Jr., A non-linear mapping for data structure analysis, *IEEE Trans. Comp.*, **C-18**, 401–409, 1969.
21. Nishi, N., Kawakami, T., Shigematsu, F., Yamamoto, M., Kakiuchi, T., Fluorine-free and hydrophobic room-temperature ionic liquids, tetraalkylammonium bis(2-ethylhexyl)sulfosuccinates, and their ionic liquid-water two-phase properties, *Green Chem.*, **8**, 349–355, 2006.
22. Egorov, V.M., Samoylov, V.Yu., Smirnova, S.V., Petrov, S.I., Pletnev, I.V., Extraction of phenols and amines into new ionic liquid tetrahexylammonium dihexylsulfosuccinate, *Proc. Int. Conf., Green Solvents for Processes*, October 8–11, 2006. Friedrichshafen, Germany, p. 123.
23. Marták, J., Schlosser, Š., Phosphonium ionic liquids as new, reactive extractants of lactic acid, *Chem. Pap.*, **60**, 395–398, 2006.
24. Smirnova, S.V., Torocheshnikova, I.I., Formanovsky, A.A., Pletnev, I.V., Solvent extraction of amino acids into a room temperature ionic liquid with dicyclohexano-18-crown-6, *Anal. Bioanal. Chem.*, **378**, 1369–1375, 2004.
25. Wang, J., Pei, Y., Zhao, Y., Hu, Zh., Recovery of amino acids by imidazolium based ionic liquids from aqueous media, *Green Chem.*, **7**, 196–202, 2005.
26. Shvedene, N.V., Nemilova, M.Yu., Khachatryan, K.S., Mamonov, N.A., Shukhaev, A.V., Formanovsky, A.A., Pletnev, I.V., Extraction-voltammetric determination of catecholamines with the use of new-class solvents, ionic liquids, *Moscow Univ. Chem. Bull.*, **45**, 324–332, 2004.

27. He, C., Li, S., Liu, H., Li, K., Liu, F., Extraction of testosterone and epitestosterone in human urine using aqueous two-phase systems of ionic liquid and salt, *J. Chromatogr. A*, **1082**, 143–149, 2005.

28. Soto, A., Arce, A., Khoshkbarchi, M.K., Partitioning of antibiotics in a two-liquid phase system formed by water and a room temperature ionic liquid, *Sep. Purif. Tech.*, **44**, 242–246, 2005.

29. Liu, Q., Yu, J., Li, W., Hu, X., Xia, H., Liu, H., Yang, P., Partitioning behavior of penicillin G in aqueous two phase system formed by ionic liquids and phosphate, *Sep. Sci. Technol.*, **41**, 2849–2858, 2006.

30. Wang, J.-H., Cheng, D.-H., Chen, X.-W., Du, Z., Fang, Z.-L., Direct extraction of double-stranded DNA into ionic liquid 1-butyl-3-methylimidazolium hexafluorophosphate and its quantification, *Anal. Chem.*, **79**, 620–625, 2007.

31. Shimojo, K., Nakashima, K., Kamiya, N., Goto, M., Crown ether-mediated extraction and functional conversion of cytochrome c in ionic liquids, *Biomacromol.*, **7**, 2–5, 2006.

32. Du, Z., Yu, Y.-L., Wang, J.-H., Extraction of proteins from biological fluids by use of an ionic liquid/aqueous two-phase system, *Chem. Eur. J.*, **13**, 2130–2137, 2007.

33. Pereiro, A.B., Tojo, E., Rodríguez, A., Canosa, J., Tojo, J., HMImPF(6) ionic liquid that separates the azeotropic mixture ethanol plus heptane, *Green Chem.*, **8**, 307–310, 2006.

34. Letcher, T.M., Deenadayalu, N., Soko, B., Ramjugernath, D., Naicker, P.K., Ternary liquid-liquid equilibria for mixtures of 1-methyl-3-octylimidazolium chloride plus an alkanol plus an alkane at 298.2 K and 1 bar, *J. Chem. Eng. Data*, **48**, 904–907, 2003.

35. Arce, A., Rodríguez, H., Soto, A., Purification of ethyl tert-butyl ether from its mixtures with ethanol by using an ionic liquid, *Chem. Eng. J.*, **115**, 219–223, 2006.

36. Arce, A., Rodríguez, H., Soto, A., Effect of anion fluorination in 1-ethyl-3-methylimidazolium as solvent for the liquid extraction of ethanol from ethyl tert-butyl ether, *Fluid Phase Equilib.*, **242**, 164–168, 2006.

37. Arce, A., Marchiaro, A., Rodríguez, O., Soto, A., Essential oil terpenless by extraction using organic solvents or ionic liquids. *AIChE J*, **52**, 2089–2097, 2006.

38. Meindersma, G.W., Podt, A., de Haan, A.B., Selection of ionic liquids for the extraction of aromatic hydrocarbons from aromatic/aliphatic mixtures, *Fuel Proc. Tech.*, **87**, 59–70, 2005.

39. Meindersma, G.W., Podt, A., de Haan, A.B., Ternary liquid-liquid equilibria for mixtures of toluene plus n-heptane plus an ionic liquid, *Fluid Phase Equilib.*, **247**, 158–168, 2006.

40. Arce, A., Earle, M.J., Rodríguez, H., Seddon, K.R., Separation of aromatic hydrocarbons from alkanes using the ionic liquid 1-ethyl-3-methylimidazolium bis{(trifluoromethyl) sulfonyl} amide, *Green Chem.*, **9**, 70–74, 2007.

41. Bösmann, A., Datsevich, L., Jess, A., Lauter, A., Schmitz, C., Wasserscheid, P., Deep desulfurization of diesel fuel by extraction with ionic liquids, *Chem. Commun.*, 2494–2495, 2001.

42. Zhang, S., Zhang, Z.C., Novel properties of ionic liquids in selective sulfur removal from fuels at room temperature, *Green Chem.*, **4**, 376–379, 2002.

43. Zhang, S., Zhang, Q., Zhang, Z.C., Extractive desulfurization and denitrogenation of fuels using ionic liquids, *Ind. Eng. Chem. Res.*, **43**, 614–622, 2004.

44. Lo, W.-H., Yang, H.-Y., Wei, G.-T., One-pot desulfurization of light oils by chemical oxidation and solvent extraction with room temperature ionic liquids, *Green Chem.*, **5**, 639–642, 2003.

45. Eßer, J., Wasserscheid, P., Jess, A., Deep desulfurization of oil refinery streams by extraction with ionic liquids, *Green Chem.,* **6**, 316–322, 2004.
46. Huang, C., Chen, B., Zhang, J., Liu, Z., Li, Y., Desulfurization of gasoline by extraction with new ionic liquids, *Energy & Fuels,* **18**, 1862–1864, 2004.
47. Nie, Y., Li, C., Sun, A., Meng, H., Wang, Z., Extractive desulfurization of gasoline using imidazolium-based phosphoric ionic liquids, *Energy & Fuels,* **20**, 2083–2087, 2006.

chapter ten

Separation of metal ions based on ionic liquids

Huimin Luo and Sheng Dai

Contents

10.1 Introduction

Separation of metal ions based on solvent extraction is widely used in hydrometallurgy, environmental cleanup, and advanced fuel cycles. The focus of this chapter will be on the use of ionic liquids (ILs) as advanced and functional solvents for separation of metal ions related to fission products. For example, the separation of radioactive cesium-137 and strontium-90 from high-level nuclear wastes and spent nuclear fuels is a serious priority environmental problem related to contaminated nuclear sites and a requirement in advanced fuel cycles. Since the primary chemical components of nuclear wastes in contaminated facilities are sodium nitrate and sodium hydroxide, the majority of the wastes could be disposed of as low-level wastes and it would cost significantly less to treat the rest of the wastes if cesium-137 and some strontium-90 could be selectively removed. The selective removal of cesium and strontium metal ions from nuclear wastes will also be essential to the safe and cost-effective production of associated waste forms for superior postclosure performance in a repository.

When compared to conventional solvent-extraction processes, which typically employ a somewhat volatile hydrocarbon diluent and a polar phase modifier to improve solvation, using polar ILs as carriers for macrocyclic agents offers the promise of improved economics. Higher distribution coefficients (D_M's) are attainable for a given extractant concentration, so that less of the expensive macrocyclic agents are used to obtain the same level of decontamination. Greater flexibility in process flow-sheet design can be achieved if the IL-based process can be made *tunable* to allow for a larger ratio of D_M between extraction and stripping. The general simplification of the process with a smaller number of components is possible because a single IL could replace a two-component modified diluent. In addition, the low volatility of ILs means that they would not need to be periodically replaced on account of evaporative losses, as would occur with hydrocarbon diluents typically used in conventional solvent extraction.

Crown ethers and other macrocycles have been used to extract Sr^{2+} and Cs^+ from various types of aqueous media, including actual acidic and alkaline nuclear wastes. For Sr^{2+}, a great deal of work has been reported by both McDowell and coworkers [1–4] and Horwitz and coworkers [5,6]. Horwitz and coworkers showed that dicyclohexano-18-crown-6 (DCH18C6) and its di-*tert*-butyl derivative could successfully extract strontium ion from nitric acid media and they developed a process (the strontium extraction [SREX] process) suitable for the extraction of Sr^{2+} from actual acidic nuclear tank waste. The Sr^{2+} distribution ratios (D_{Sr}) using the SREX process are typically in the range of 5–30, depending on the nitric acid concentration, the crown-ether concentration, and the makeup of the diluent. McDowell found that the extraction of Sr^{2+} could be enhanced by the addition of organophilic alcohols [1], phenols [1], phosphoric acids, or (especially) carboxylic acids [4] to solutions containing the organophilic crown ethers. The extraction of Sr^{2+} from 4.0 M sodium nitrate solution by a toluene solution of dicyclohexano-18-crown-6 (at 0.05 M) and Versatic acid 1519 (a branched carboxylic acid, at 0.10 M) gave Sr^{2+}-distribution ratios of as high as 100 at pH 8, reaching a maximum of slightly over 1000 at pH 9.4, after which the D_{Sr} value began to drop [4]. At pH 9.4, the acid alone affords a D_{Sr} of about 100, through simple ion exchange. D_{Sr} values on the order of 10^3 appear to be the highest distribution ratios conventionally obtainable by solvent extraction that we are aware of. Despite the high D_{Sr} values obtainable from the combined crown ether–carboxylic acid solvent system, carboxylic acids have not been utilized in actual processes, because they tend to coextract large quantities of other metal ions such as sodium ion.

Cesium ion extraction using crown ethers has also been investigated by McDowell [7], Horwitz [8,9], and their coworkers, as well as many others [10–13]. McDowell showed that the distribution ratio of cesium ion (D_{Cs}) from 0.1 M nitric acid solutions could be on the order of 10^2 when the crown ether *bis(tert*-butylbenzo)-21-crown-7 was used in combination with the organophilic anion didodecylnaphthalenesulfonic acid (in toluene solution). Recently, a new class of extremely powerful macrocyles for the extraction of cesium ion, the calixarene-crown ethers, in particular mono- and bis-crown derivatives of calix[4]arenes in the 1,3-alternate conformation, have been shown to possess both extremely high extractive strength for cesium and excellent selectivity (generally exceeding 10^4) for cesium ion over sodium ion [14,15]. Without the aid of an organophilic anion, these calixarene crown ethers are capable of extracting cesium from both acidic and alkaline media with distribution ratios more than unity and the D values can be as high as 100, depending on the concentration of competitive cations such as potassium ion in the aqueous solution, the concentration of the calixarene crown, and the polarity of the diluent. By comparison, the crown ether *bis(tert*-butylbenzo)-21-crown-7 would give D_{Cs} values under the same conditions for at least two orders of lower magnitude [15]. More lipophilic derivatives of the calixarene crown ethers, calix[4]-*bis*-crown-6, and calix[4]-*bis*-1,2-benzo-crown-6, are currently being investigated in solvent-extraction solvents for the removal of cesium ion from alkaline nuclear wastes [15]. Use of crown ethers for strontium ion extraction, and calixarene crown ethers for cesium ion extraction, in ionic-liquid media, can greatly increase the strontium and cesium distribution ratios to 10^4 or more.

10.2 Rationale for Ionic liquid-based extraction systems

The distribution ratios (D_M's) for crown-ether-based extraction processes using conventional solvents depend on two major factors: (1) the thermodynamic driving force for cation complexation by a crown ether and (2) the solvation of the cation and counter anion by the organic solvent [1,4]; the former factor is usually thermodynamically favored (see Equation 10.3). Difficulties in increasing the solvent-extraction efficiency of conventional solvent-extraction systems using crown ethers as extractants lie in the unfavorable transport of cations and counter anions from aqueous phases to organic phases [16–19]. Limited solubilities of inorganic metal ions in nonionic organic solvents (Equations 10.1 and 10.2) are the main problem associated with conventional solvent extraction. The inability of organic solvents to solubilize ionic crown-ether complexes and their counter anions is one of the main obstacles in improving the separation efficiency of fission products based on conventional solvent extraction. Because the solvation of ion-pair species is more favorable than the solvation of individual ions in organic solvents, the crown-ether complexes and counter anions are compelled to form ion pairs (Equation 10.4), despite the entropic cost [16–23]. Conventional organic solvent extractions using crown ethers fundamentally involve the transport of ion pairs from the aqueous phase to the organic phase. Although the formation of ion pairs helps to reduce the total thermodynamic cost of transporting ionic species into the organic phase, the free-energy gain in this process is *not* enough to make up its loss in solubilizing individual ions in the organic phase (Equations 10.1 and 10.2). Accordingly, this leads generally to a very small distribution ratio for the overall extraction process (Equation 10.5). For example, the distribution ratio for the extraction of strontium nitrate into nonpolar organic phases from aqueous solutions by crown ethers is generally less than one [1,4,16–19], even though the thermodynamic driving force for the complexation of Sr^{2+} with a number of crown ethers is very favorable [16–19]. This distribution ratio is low because the solvation free energies for simple inorganic anions such as nitrate by organic solvents are *extremely* unfavorable [24].

A number of strategies have been proposed to address this problem, including the addition of hydrophobic anions to the aqueous solution or the addition of hydrophilic solvents to the organic phase. For example, McDowell et al. [1–4] made use of lipophic carboxylic acids to enhance the distribution ratio of Sr^{2+}. The solvation of the carboxylic anion in organic phases is more favorable than those of simple inorganic anions. The obvious drawback to these approaches is that more chemicals would have to be added to the system thereby increasing the complexity of the aqueous solution. Another drawback to this approach is the loss of some selectivity due to the nonselectivity of the ion-exchange capability of the ion exchanger (carboxylate anion).

$$M^{n+}(aq) = M^{n+}(org) \text{ Thermodynamiclly unfavorable} \qquad (10.1)$$

$$nA^-(aq) = nA^-(org) \text{ Thermodynamically unfavorable} \qquad (10.2)$$

$$crown(org) + M^{n+}(org) = \{crown\,M^{n+}\}(org)$$
$$\text{Thermodynamically favorable} \qquad (10.3)$$

$$\{crown\,M^{n+}\}(org) + n\,A^-(org) = \{crown\,M^{n+}n\,A^-\}(org)$$
$$\text{Formation of ion pair} \tag{10.4}$$

$$crown\,(org) + M^{n+}(aq) = \{crown\,M^{n+}n\,A^-\}(org)$$
$$\text{Thermodynamically unfavorable} \tag{10.5}$$

These key deficiencies associated with current extraction processes based on crown ethers prompted us to search for alternative extraction media that could convert the solvation of ionic species into a *more favorable* thermodynamic process. We and others [25] have demonstrated the basis for the use of hydrophobic ILs containing macrocyclic ligands as extraction solvents to selectively remove the fission products (Sr^{2+} and Cs^+) [26–28]. An ionic liquid [29] is a fluid phase, comprised only of cations and anions (Figure 10.1), which is formed by melting an organic salt, and is conveniently defined as an organic salt with a melting point below ~100°C. ILs are liquid at 25°C. ILs are proving to be increasingly interesting fluids for application in soft-matter systems from electrochemistry to energetic materials, and are rapidly becoming established as viable media for synthesis and separations operations [30–33]. As a class of liquids, they are fundamentally different from molecular solvents and from salt solutions. They are electrically conducting, nonvolatile fluids and have, by definition, low melting points and wide liquid ranges. These unique properties have encouraged researchers around the world to explore the uses of ILs in place of volatile organic solvents. The most promising current developments, however, appear to be emerging not from the use of ILs as new solvents, but in utilizing the unique combinations of chemical and physical properties and characteristics of ILs. The key chemical, structural, and physical features of ILs have been widely described, most comprehensively in ILs in synthesis [29] and the general trends in the physical

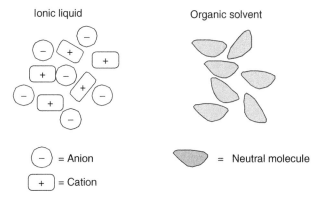

Figure 10.1 Comparison of the structures of an ionic liquid and an organic solvent. In the ionic liquid, cations are solvated by anions and vice versa.

properties of ILs are reasonably well established. For example, variation in the thermal-decomposition temperatures with anion type and melting points of alkyl-methylimidazolium salts with alkyl-chain length [34]. The intrinsic nonvolatile nature of these fluids provides an opportunity to reduce, or even completely eliminate, hazardous and toxic emissions associated with the loss of volatile compounds to the atmosphere, thus providing the promise for significant environmental benefits. A second consideration is that solvation and dissolution characteristics of ILs vary significantly when compared to conventional solvents such as water, dichloromethane, or hexane, with a number of different solvent characteristics, all contributing to the overall solvent properties. These solvent properties have been investigated using a range of techniques including the use of solvatochromic probes, linear solvent free-energy relationships, and partitioning of organic solutes.

From a thermodynamic perspective, the solvation of ionic species (see Equations 10.6 and 10.7), such as crown-ether complexes, NO_3^-, and SO_4^{2-}, in the ILs, should be much more favored thermodynamically than those of conventional solvent extractions (Equations 10.1 and 10.2). This is one of the key advantages of using ILs in separations involving ionic species. In this case, cationic crown-ether complexes and their counter anions are *not* expected to form ion pairs, but to be solvated separately by ionic species from the ILs. Therefore, the extraction process using crown ethers in ILs may *not* be an ion-pair extraction process.

$$M^{n+}(aq) = M^{n+}(IL) \text{ Thermodynamically much more}$$
$$\text{favorable than Equation 10.1} \tag{10.6}$$

$$A^{n-}(aq) = A^{n-}(IL) \text{ Thermodynamically much more}$$
$$\text{favorable than Equation 10.2} \tag{10.7}$$

In contrast to molecular solvents used as extracting phases, ILs are comprised of two ions, rather than a single neutral molecule, and the ions can participate in the extraction process. Understanding, control, and utilization of the different solvent extraction, and cation- and anion-exchange mechanisms possible in any one IL solvent system are both the challenge and the key to exploiting ILs in separation schemes. By adjusting the nature of the extracting phase through choice or modification of the cationic or anionic components of ILs, the roles of solvent extraction and ion-exchange mechanisms in the metal-ion separation process can be controlled.

Rhinebarger et al. [35] and Eyring et al. [36,37] have used lithium-7 nuclear magnetic resonance (NMR) chemical shift data to determine the stability constants for crown-ether complexes of Li^+ in two IL systems consisting of 55/45 mole% N-butylpyridinium chloride-aluminum chloride and 1-ethyl-3-methyl-imidazolium chloride-aluminum chloride. The stability constants for

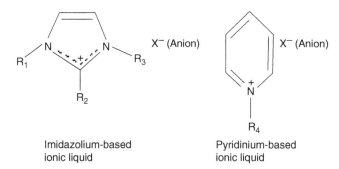

Imidazolium-based
ionic liquid

Pyridinium-based
ionic liquid

Figure 10.2 Structures of two basic kinds of ionic liquids. (Reproduced from the American Chemical Society from Luo, H. M., Dai, S., Bonnesen, P. V., *Anal. Chem.*, 76, 2773–2779, 2004. With permission.)

1:1 complexes between crown ethers and lithium ions in the ILs are greater than those in organic solvents. The unique solvation environment of those ILs enhances the stability of the crown-ether complexes. Accordingly, the extraction efficiency and selectivity using ILs would be expected to be very high. This is an important rationale that compels us to develop an IL-based solvent-extraction process using macrocyclic ligands.

Figure 10.2 gives the structure templates of the two most common classes of ILs. The cation is usually a heterocyclic cation, such as a dialkyl imidazolium ion or an *N*-alkylpyridinium ion. These organic cations, which are relatively large compared with simple inorganic cations, account for the low melting points of the salts. Anions that yield ILs range from simple inorganic anions such as Cl^- or $[NO_3]^-$ to large complex anions such as the carborane $[CB_{12}H_{12}]^-$ [38]. Introducing large, perfluorinated anions, such as $[PF_6]^-$, or $[Tf_2N]^-$ [25], can control the solvent's reactivity with water, coordinating ability, and hydrophobicity; the $[PF_6]^-$ and $[Tf_2N]^-$ anions produce hydrophobic solvents due to their lack of hydrogen-bond accepting ability [39], though not all ILs containing perfluorinated anions are hydrophobic [34]. Control over hydrophobicity and other physical properties is governed by cation–anion pair interactions, enabling the formation of two-phase in contact with aqueous phases, thereby making these systems available for aqueous liquid–liquid separations [40].

In comparison with conventional molecular solvents, ILs exhibit enhanced distribution coefficients (D_M's) for a number of complexing neutral ligands in extraction of metal ions from aqueous solutions. Such enhancements were initially attributed to two major fundamental factors: (1) ion-recognition capabilities of complexing ligands and (2) unique ionic solvation environments provided by ILs for ionic species. Bartsch and coworkers [28] and Dietz and coworkers [41] reported very interesting results, in which the effect of varying the alkyl group in $[C_nC_1Im][PF_6]$ on the efficiency and selectivity of competitive extraction of five alkali-metal cations by DCH18C6 was observed.

The distribution coefficients have been found to decrease with the alkyl-chain length of the IL cations. This observation led Dietz [41] and his coworkers to propose that the extraction process also involves the exchange of the IL cations with metal ions (Equation 10.10). The longer the alkyl-chain lengths of the IL cations are, the more hydrophobic are the IL cations and more difficult to be transported into aqueous phases via ion exchange. Most recently, Dietz and coworkers have shown how the extracting mechanisms in ionic-liquid systems can be controlled and changes from solvent extraction to ion exchange selectively by changing the composition of the IL extracting phase [42]. The two direct effects of the ion-exchange process are: (1) enhanced extraction via synergism of complexation and ion exchange and (2) some loss of ILs. Accordingly, one of the key challenges associated with IL-based extraction is to minimize the loss of ILs via ion-exchange processes and direct dissolutions. The partitioning of solutes in IL/aqueous systems [31,43] has been shown to follow traditional octanol/water distributions [40], which is useful for applying hydrophobic ILs as direct replacements for solvents, such as benzene, toluene, dichloromethane, or chloroform in two-phase system separation schemes [44,45]. The overall analysis indicates that ILs are less polar than water (i.e., in terms of hydrogen-bond acidity and basicity), which is a reasonable conclusion, but has important implications for separations. Ions and ionic compounds will not partition to an IL from water in the absence of a complexing extractant.

$$[M^+]_{aq} + [\text{Crown-Ether}]_{IL} + [C_n min^+]_{IL} \rightleftharpoons [M\text{-Crown-Ether}^+]_{IL}$$
$$+ [C_n min^+]_{aq} \tag{10.8}$$

In following sections, the selected key recent results from our research group on the separation of fission products (Sr^{2+} and Cs^+) are outlined. For readers who are interested in IL-based extraction systems for other metal ions, several excellent reviews are very helpful. Rogers and coworkers [44] have written an excellent review on use of ILs for extraction of lanthanides and actinides.

10.2.1 Development of highly selective extraction process for Cs+ based on calixarenes

Extraction. We extended the original investigation of our IL-based extraction for Sr^{2+} to the extraction of Cs^+. A lipophilic derivative of a calix[4]-*bis*-crown-6, calix[4]arene-*bis*(*tert*-octylbenzo-crown-6) (BOBCalixC6), first prepared for the caustic side solvent-extraction process for Cs removal from alkaline nuclear wastes [15,46,47] as shown in Figure 10.3 [48] was utilized as a cation receptor. These studies were directed primarily toward acquiring an understanding of the mechanism of extraction in the complex IL solvents, and secondarily toward determining if a Cs-removal technology utilizing ILs is possible. The cations of the ILs used in this work were $[C_nC_1Im]^+$, where $C_n = C_2$(ethyl), C_4(butyl), C_6(hexyl), and C_8(octyl). The anion is $[Tf_2N]^-$.

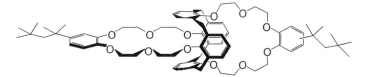

Figure 10.3 Structure of BOBCalixC6. (Reproduced from the American Chemical Society from Luo, H. M., Dai, S., Bonnesen, P. V., Buchanan, A. C., Holbrey, J. D., Bridges, N. J., Rogers, R. D., *Anal. Chem.*, 76, 3078–3083, 2004. With permission.)

Table 10.1 Distribution Ratios (D_{Cs}) between the Ionic Liquids and 2.5 mM CsNO$_3$

BOBCalixC6 (mM)	D_{Cs} as a function of C_n in $[C_nC_1Im][Tf_2N]$						
	Ethyl, C_2[a]	Propyl, C_3	Butyl, C_4	Butyl, C_4[b]	Hexyl, C_6[c]	Octyl, C_8[c]	CHCl$_3$[c]
0	0.084	0.055	0.024	0.19	ud	ud	ud
1.00	1.61	1.36	0.77	0.95	0.56	0.45	ud
3.40	4.68	3.90	3.67	2.92	3.20	2.50	ud
7.71	15.3	14.6	13.8	11.7	11.5	8.10	ud
13.6	nm	137	131	63.8	57.4	17.9	0.034

[a] nm: Not measured because concentration of BOBCalixC6 could not be reached.
[b] Work conducted using radiotracer technique.
[c] ud: Undetectable via ion chromatography.

Source: Reproduced from the American Chemical Society from Luo, H. M., Dai, S., Bonnesen, P. V., Buchanan, A. C., Holbrey, J. D., Bridges, N. J., Rogers, R. D., *Anal. Chem.*, 76, 3078–3083, 2004. With permission.

The D_{Cs} values of the plain ILs in the absence of an extractant are very low. The small amount of Cs$^+$ which does partition into the ILs correlates with the observed hydrophobicities of the IL organic cations. The hydrophobicity is also inversely related to the solubility of the corresponding cation $[C_nC_1Im]^+$ in the aqueous phase, and thus its ion-exchange capability. Accordingly, the selectivity of these plain IL solvents toward Cs$^+$ is dominated by the hydrophobicity of the metal ions extracted (i.e., their Hofmeister selectivity) [49,50]. The solubilities of BOBCalixC6 in the ILs are quite high, even though this complexant was designed specifically for use in an organic extracting phase. As seen in Table 10.1 [48], the distribution coefficients (D_{Cs}) of Cs$^+$ from water to ILs phase with BOBCalixC6 strongly depend on the concentration of BOB-CalixC6 in the ILs and increase with the concentration of the extractant. This observation is very similar to that noted in the extraction of Sr^{2+} with *cis*-dicyclohexano-18-crown-6 (DCH18C6) in ILs, indicating that the complexation of Cs$^+$ plays a key role in the partitioning processes [26,51,52].

Selectivity. The selectivity for Cs$^+$ was investigated using aqueous solutions containing competitive Na$^+$, K$^+$, and Sr^{2+} ions in contact with BOBCalixC6-loaded $[C_4C_1Im][Tf_2N]$. The selectivity coefficient for Cs$^+$ in the presence of competitor cations can be obtained from the ratios of the D_{Cs} values to the corresponding distribution coefficients of the competitive ions. The distribution coefficients for Na^+ and Sr^{2+} using BOBCalixC6 in $[C_4C_1Im][Tf_2N]$ were too small to be detected by ion chromatography. This observation is consistent with what has previously been observed for BOBCalixC6 and related calixarene-crown ethers in conventional solvents [15,42]. Coextraction of K^+ along with Cs^+ has been observed, however, using BOBCalixC6 in conventional solvents [15]. The concomitant extraction of K^+ along with Cs^+ is also observed in our studies. The selectivity coefficients determined under various conditions range from ~3.94 to ~68.7, which are lower than the selectivities of ~220 seen in organic solvents such as 1,2-dichloroethane [53]. The lower selectivity of cesium over potassium may be attributable to the greater polarity of ILs than 1,2-dichloroethane (dielectric constant 10.19 at 25°C). The distribution of Cs^+ to the IL phase was observed to decrease with increasing chain length of the substituted alkyl group (R) in the organic cation of the IL. This observation is consistent with a mechanism of ion exchange during the extraction. The longer the alkyl chain is, the more hydrophobic the organic cation becomes, and accordingly the less the IL will partition to the aqueous phase [42].

10.2.2   Demonstration of facilitated sacrificial ion-exchange extraction processes to reduce the loss of ionic liquids and to increase extractive strength of ionic liquids

Based on the ion-exchange model, the imidazolium cations of ILs would be partitioned from the IL phase to the aqueous phase during extraction of metal cations from the aqueous phase to the IL phase (Equation 10.8). Thus, the addition of a sacrificial cationic species (e.g., Na^+ and H^+) to the IL phase that will preferentially transfer to the aqueous phase should reduce the loss of imidazolium cations. A sacrificial cation in the IL which has no affinity toward BOBCalixC6 and which is more hydrophilic than the IL's imidazolium cation should result in an enhancement of D_{Cs}. Sodium tetraphenylborate ($NaBPh_4$) was chosen in our experiments, as such a sacrificial hydrophilic cationic exchanger, to demonstrate the principle. The release of the $[BPh_4]^-$ anion from the imidazolium-based ILs to aqueous solutions should be negligible in the presence of the organic cations in aqueous phases.

Table 10.2 [48] compares the extraction results obtained using $[C_4C_1Im][Tf_2N]$ containing various concentrations of BOBCalixC6 without and with 0.12 M $NaBPh_4$. As seen in Table 10.2, no significant changes of the corresponding D_{Cs} values with the addition of $NaBPh_4$ were observed. The small changes that were observed in D_{Cs} with the addition of $NaBPh_4$ imply that high concentrations of Na^+ may compete with Cs^+ for the coordination with BOBCalixC6 and that the extraction of Cs^+ is not totally through the ion-exchange process. Support for the substitution of $[C_4C_1Im]^+$ by Na^+ in the ion-exchange process proposed earlier comes from the analysis of the concentration of $[C_4C_1Im]^+$ released to the corresponding aqueous phases during extraction. As seen in Figure 10.4 [48],

Table 10.2   Effects of Extractable Counter-Anion on the Distribution
Ratios (D_{Cs}) from [C$_4$C$_1$Im][Tf$_2$N] Phases Containing BOBCalixC6
with and without NaBPh4[a]

| | [BOBCalixC6] (mM) | | | | | | | |
| | 7.71 | | 10.0 | | 13.8 | | 20.0 | |
Anion	0.0 M NaBPh$_4$	0.12 M NaBPh$_4$	0.0 M NaBPh$_4$	0.12 M NaBPh$_4$	0.0 M NaBPh$_4$	0.12 M NaBPh$_4$	0.0 M NaBPh$_4$	0.12 M NaBPh$_4$
NO$_3^-$	10.9	9.76	22.4	21.9	97.6	92.7	629	608
Cl$^-$	12.2	11.4	22.6	22.6	75.3	66.6	956	940
OAc$^-$	12.8	13.4	24.9	22.6	120	85.4	925	**798**

[a]   The initial aqueous concentration of Cs$^+$ was 2.5 mM with the specified anion.

Source: Reproduced from the American Chemical Society from Luo, H. M., Dai, S., Bonnesen, P. V.,
Buchanan, A. C., Holbrey, J. D., Bridges, N. J., Rogers, R. D., *Anal. Chem.*, 76, 3078–3083,
2004. With permission.

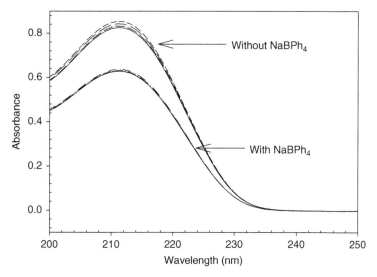

Figure 10.4   Comparison of UV absorption spectra of aqueous solutions equilibrated
with Cs$^+$-containing ionic liquids with and without NaBPh$_4$. (Reproduced from the
American Chemical Society from Luo, H. M., Dai, S., Bonnesen, P. V., Buchanan,
A. C., Holbrey, J. D., Bridges, N. J., Rogers, R. D., *Anal. Chem.*, 76, 3078–3083, 2004.
With permission.)

the addition of NaBPh$_4$ decreases the loss of ILs by about 24%, as determined by
measuring the UV spectra of [C$_4$C$_1$Im]$^+$ in the corresponding aqueous phases.

As originally revealed by Dietz and Dzielawa [41], the mechanism for
the extraction of metal ions by extractants in ILs involves the ion-exchange
process (Equation 10.8). We have explored the concept of using sacrificial

metal ions (Na$^+$) to replace alkylimidazolium cations in the ion-exchange process. The rationale behind this assertion is that sodium ions are more hydrophilic than imidazolium cations and therefore should replace the imidazolium cations in the ion-exchange process (Equation 10.8). As seen from Figure 10.4, we can significantly cut the loss of ILs through the ion-exchange process via this sacrificial ion-exchange method. Another environmentally benign cation for the sacrificial ion-exchanging process is proton. In fact, organic acids have been extensively explored for the synergistic effect on conventional solvent extractions via proton transfer [54]. The preliminary experiments have been conducted to evaluate weak organic acids to test this concept. A significant enhancement has been observed using the proton ions from oleic acid in [C$_4$C$_1$Im][Tf$_2$N]. The pH dependence of this enhancement is correlated with the pK_a value of oleic acid. The basic condition of aqueous phases can enhance this sacrificial ion-exchange process based on proton ions. Accordingly, the sacrificial ion-exchange process should be directly applicable to high pH waste compositions.

10.2.3 Development of recyclable ionic liquid-based extraction systems

A series of *N*-alkyl aza-18-crown-6 ethers were synthesized and characterized. These monoaza-substituted crown ethers in ILs were investigated as recyclable extractants for separation of Sr^{2+} and Cs$^+$ from aqueous solutions. The pH-sensitive complexation capability of these ligands allows for a facile stripping process to be developed so that both macrocyclic ligands and ILs can be reused.

Syntheses of *N*-alkyl aza crown ethers have been based on a modified protocol previously investigated for similar compounds [5,55,56]. The reaction used for synthesizing recyclable monoaza crown ethers in this study is illustrated in Scheme 10.1 [57]. Eight monoaza crown ethers were successfully synthesized

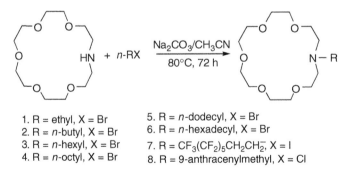

1. R = ethyl, X = Br
2. R = *n*-butyl, X = Br
3. R = *n*-hexyl, X = Br
4. R = *n*-octyl, X = Br
5. R = *n*-dodecyl, X = Br
6. R = *n*-hexadecyl, X = Br
7. R = CF$_3$(CF$_2$)$_5$CH$_2$CH$_2$, X = I
8. R = 9-anthracenylmethyl, X = Cl

Scheme 10.1 The reaction used for synthesis of recyclable monoaza-substituted crown ethers. (Reproduced by permission of the American Chemical Society from Luo, H. M., Dai, S., and Bonnesen, P. V., *Anal. Chem.*, 76, 2773–2779, 2004.)

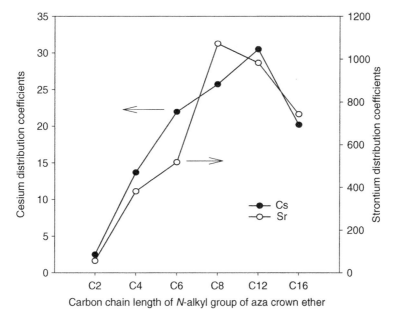

Figure 10.5   Effect of N-alkyl group variation of monoaza crown ethers on extraction efficiency of $[C_4C_1Im][Tf_2N]$ containing N-alkyl aza-18-crown-6 (0.1 M). (Reproduced from the American Chemical Society from Luo, H. M., Dai, S., Bonnesen, P. V., *Anal. Chem.*, 76, 2773–2779, 2004. With permission.)

in good yield except for the fluorinated alkyl crown ether (7). The basic reaction involves the alkylation of the monoaza group of 1-aza-18-crown-6. The hydrophobicity of the longer chain alkyl-group substituents should significantly reduce the solubilities of the aza-substituted crown ethers in water.

Figure 10.5 shows the variation of D_{Sr} and D_{Cs} as a function of the alkylchain length on the aza group of the crown ethers. As seen from Figure 10.5, the value of D_{Sr} and D_{Cs} increases as the alkyl chain length on the aza group of crown ethers increases peaking with octyl substitution. This observation can be rationalized according to the hydrophobicity of the different alkylsubstituted crown ethers. The longer the alkyl chain is, the less partitioning of the crown ethers to the aqueous phases. Therefore, the enhanced D_{Sr} values could be expected for the crown ethers with longer alkyl chains. D_{Sr} values as high as 1000 can be achieved with a concentration of 0.1 M of N-octyl aza-18-crown-6 in $[C_4C_1Im][Tf_2N]$, which is comparable to the extraction efficiency of DCH18C6 at the same condition [26,41]. However, the D_{Sr} values decrease slightly as the alkyl chain length on the aza group of crown ethers increases from C_8 to C_{12} and C_{16}. This may be due to solubility problems associated with the latter compounds.

The cation-binding capabilities of monoaza-substituted crown ethers are extremely pH sensitive [58]. At low pH, the aza crown will be protonated, and

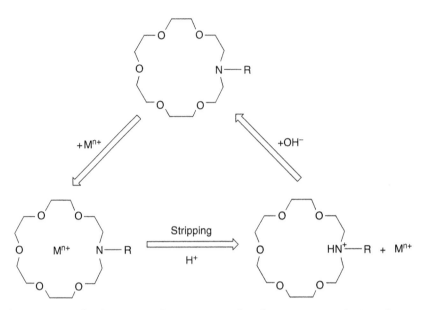

Scheme 10.2 The basic recycling strategy for the monoaza-substituted crown ethers. (Reproduced by permission of the American Chemical Society from Luo, H. M., Dai, S., and Bonnesen, P. V., *Anal. Chem.*, 76, 2773–2779, 2004.)

the binding constant for metal cations will be severely reduced, relative to that of the neutral aza crown, due to charge repulsion. The binding affinity can accordingly be switched by varying the pH, with metal ion binding taking place at neutral or high pH, and metal ion release taking place at low pH. This switchable binding property forms the main rationale for us to develop IL-based separation processes using monoaza-substituted crown ethers. The basic recycling strategy for the monoaza-substituted crown ethers is summarized in Scheme 10.2 [57]. The recycling experiments with N-octyl aza-18-crown-6 (4) or N-(9-anthracenylmethyl) aza-18-crown-6 (8) in [C$_4$C$_1$Im][Tf$_2$N] for extraction of metal ions from the aqueous solution containing CsNO$_3$ or Sr(NO$_3$)$_2$ were conducted. In each case, more than 98% Cs or Sr cations were recovered. The IL phases containing N-alkyl aza crown ethers could be reused for solvent extractions after deprotonation with a base. No significant changes of extraction efficiencies were observed, indicating that a stripping process based on monoaza crown ethers and ILs is feasible.

10.2.4   Effect of 1-alkyl group of ionic liquids on extraction efficiency of N-alkyl aza-18-crown-6 and comparison with DCH18C6

Table 10.3 shows the comparison of D_{Sr} and D_{Cs} for DCH18C6 and five N-alkyl aza-18-crown-6 in four different ILs. The distribution coefficients for both Sr^{2+} and Cs$^+$ increase with the decrease of the alkyl chain length in ILs. Such dependence can be attributed to the ion-exchange capability of

Table 10.3 Effects of Different Ionic Liquids on Extraction Efficiency of N-alkyl aza-18-Crown-6 and Comparison with DCH18C6[a]

Concentration of crown ethers in ionic liquids	Aqueous phase (1.5 mM)	D_M as a function of C_n in $[C_nC_1Im][Tf_2N]$			
		Ethyl, C_2	Butyl, C_4	Hexyl, C_6	Octyl, C_8
0.10 M DCH18C6	CsCl	589	380	66.0	8.35
	SrCl$_2$	10700[b]	935	82.2	3.95
0.10 M N-dodecyl aza-18-crown-6	CsCl	45.5	30.5	19.4	6.66
	SrCl$_2$	3840[b]	982	209	17.5
0.10 M N-octyl aza-18-crown-6	CsCl	25.2	25.7	14.3	3.22
	SrCl$_2$	8430[b]	1070	169	5.29
0.10 M N-hexyl aza-18-crown-6	CsCl	24.4	22.0	14.3	4.31
	SrCl$_2$	2900	518	135	4.47
0.10 M N-butyl aza-18-crown-6	CsCl	17.4	13.7	6.59	2.74
	SrCl$_2$	1000	381	26.0	2.47
0.10 M N-ethyl aza-18-crown-6	CsCl	6.02	2.48	3.10	1.74
	SrCl$_2$	799	55.8	5.62	0.38

[a] Single species extraction.
[b] Determined by ICP-AE.

Source: Reproduced from the American Chemical Society from Luo, H. M., Dai, S., Bonnesen, P. V., *Anal. Chem.*, 76, 2773–2779, 2004. With permission.

the corresponding cations of a specific IL. The D_{Sr} values obtained using N-alkyl aza-18-crown-6 in different ILs exhibit the same trend. In fact, the extraction efficiencies of Sr^{2+} for the IL systems containing DCH18C6 and N-alkyl aza-18-crown-6 are very similar. For instance, as can be seen from Table 10.3, N-octyl aza-18-crown-6 gave D_{Sr} values of 8430 in $[Tf_2N]$ and 1070 in $[C_4C_1Im][Tf_2N]$, which are comparable to 10,700 in $[C_2C_1Im][Tf_2N]$ and 935 in $[C_4C_1Im][Tf_2N]$ for the corresponding DCH18C6-based system. Accordingly, the extractive strength of N-octyl aza-18-crown-6 for Sr^{2+} in ILs is very similar to that of DCH18C6.

The extraction results for the competitive solvent extraction of the aqueous solution with DCH18C6 in four ILs are illustrated in Figure 10.6. (The data are corrected for the levels of metal cations that are extracted by the ILs alone.) As can be seen from Figure 10.6, there is a smooth decrease in the extraction efficiency for each metal cation as the carbon chain length in ILs increases. Figure 10.7 shows the influence of varying the carbon chain length of ILs on the selectivities of Sr^{2+}/Na^+, Sr^{2+}/Cs^+, and Sr^{2+}/K^+ for competitive extraction by DCH18C6. With the exception of Sr^{2+}/Na^+ in $[C_2C_1Im][Tf_2N]$, the Sr^{2+}/Na^+, Sr^{2+}/Cs^+, and Sr^{2+}/K^+ selectivities decrease as the 1-alkyl group of ILs is lengthened. The order of extraction efficiency is $K^+ \gg Sr^{2+} > Cs^+ > Na^+$ in all four ILs. This observation is consistent with the reported results in Ref. 28 for $[C_nC_1Im][PF_6]$. The high extraction efficiency for K^+ significantly hampers the applications of these systems for extraction of Sr^{2+} in the presence of K^+. The results for competitive extraction of the same aqueous solutions

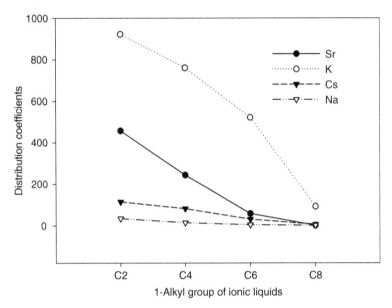

Figure 10.6 Effect of 1-alkyl group variation of ionic liquids on efficiency of competitive metal cation extraction from aqueous solutions by DCH18C6 (0.1 M) in $[C_nC_1Im][Tf_2N]$. (Reproduced from the American Chemical Society from Luo, H. M., Dai, S., Bonnesen, P.V., *Anal. Chem.*, 76, 2773–2779, 2004. With permission.)

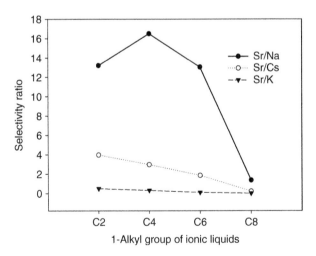

Figure 10.7 Effect of 1-alkyl group variation of ionic liquids on Sr^{2+}/Na^+, Sr^{2+}/Cs^+, and Sr^{2+}/K^+ selectivities in competitive metal cation extraction from aqueous solutions by DCH18C6 (0.1 M) in $[C_nC_1Im][Tf_2N]$. (Reproduced from the American Chemical Society from Luo, H. M., Dai, S., Bonnesen, P. V., *Anal. Chem.*, 76, 2773–2779, 2004. With permission.)

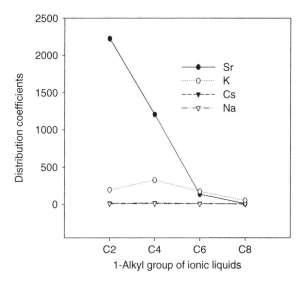

Figure 10.8 Effect of 1-alkyl group variation of ionic liquids on efficiency of competitive metal cation extraction from aqueous solutions by *N*-octyl aza-18-crown-6 (0.1 M) in $[C_nC_1Im][Tf_2N]$. (Reproduced from the American Chemical Society from Luo, H. M., Dai, S., Bonnesen, P. V., *Anal. Chem.*, 76, 2773–2779, 2004. With permission.)

with *N*-octyl aza-18-crown-6 in the same four ILs are shown in Figure 10.8. With the exception of the D_K value for $[C_4C_1Im][Tf_2N]$, extraction efficiency for each metal cation decreases as the carbon chain length in ILs increases. The key change from the selectivity order of the DCH18C6 system is that the order of extraction efficiency is $Sr^{2+} \gg K^+ > Cs^+ > Na^+$ for the *N*-octyl aza-18-crown-6 extraction systems based on $[C_2C_1Im][Tf_2N]$ and $[C_4C_1Im][Tf_2N]$. This observation clearly demonstrates the advantage of using *N*-alkyl aza-18-crown-6 for extraction of Sr^{2+} based on ILs in the presence of other competitive cations. Interestingly, the order of extraction efficiency ($K^+ > Sr^{2+} > Cs^+ > Na^+$) is the same as that of DCH18C6 for the *N*-octyl aza-18-crown-6 extraction systems based on $[C_6C_1Im][Tf_2N]$ and $[C_8C_1Im][Tf_2N]$. The reversal of the extraction efficiency for Sr^{2+} and K^+ induced by the variation of the IL cations further demonstrates the strong role played by ILs in determining not only distribution coefficients but also orders of the selectivities. Such an effect is unusual in solvent extractions based on conventional solvents.

Figure 10.9 shows the influence of varying the carbon chain length of ILs on the Sr^{2+}/Na^+, Sr^{2+}/Cs^+, and Sr^{2+}/K^+ for competitive metal cation extraction by *N*-octyl aza-18-crown-6. The Sr^{2+}/Na^+, Sr^{2+}/Cs^+, and Sr^{2+}/K^+ selectivities increase as the carbon chain length in ILs decreases. For $[C_2C_1Im][Tf_2N]$, a Sr^{2+}/Na^+ selectivity of more than 300 can be achieved. Similar trends were observed for other four *N*-alkyl aza-18-crown-6, where the alkyl groups are *n*-dodecyl, *n*-hexyl, *n*-butyl, and ethyl. For the most part, D_{Sr}/D_K is relatively constant in a given IL as the alkyl chain on the aza crown is varied.

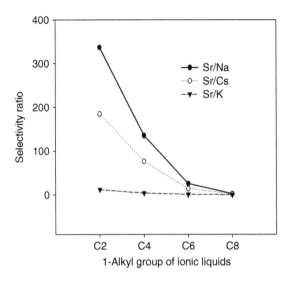

Figure 10.9 Effect of 1-alkyl group variation of ionic liquids on Sr^{2+}/Na^+, Sr^{2+}/Cs^+, and Sr^{2+}/K^+ selectivities in competitive metal cation extraction from aqueous solutions by *N*-octyl aza-18-crown-6 (0.1 M) in $[C_nC_1Im][Tf_2N]$. (Reproduced from the American Chemical Society from Luo, H. M., Dai, S., Bonnesen, P. V., *Anal. Chem.*, 76, 2773–2779, 2004. With permission.)

10.2.5 Effects of anions of ionic liquids on extraction of fission products

The extraction efficiency of crown ethers in ILs has been demonstrated to be highly correlated to the hydrophilicity of IL cations. Dietz and coworkers [41] demonstrated that this dependence is controlled by the ion-exchange capabilities of the corresponding IL cations. Our recent experiments [59] have showed that the distribution ratios of the ILs with *bis*[(trifluoromethyl) sulfonyl] imide anion are much larger than those with hexafluorophosphate anion (see Figure 10.10). This indicates that the anion also plays a key role in solvent extraction involving ILs. This dependence is opposite to that found for IL cations, indicating a significant advantage of using ILs with hydrophobic anions for cation extraction [59]. Furthermore, the extraction selectivity for Sr^{2+} over Na^+, K^+, and Cs^+ can be significantly improved through the use of hydrophobic anions for the ILs containing 1-ethyl-3-methylimidazolium or 1-butyl-3-methylimidazolium cations. The application of Le Chatelier's principle [60] to Equation 10.8 indicates that as the concentration of $[C_nC_1Im]^+$ in aqueous solution increases, M^{n+} partitions less into ILs. The concentration of $[C_nC_1Im]^+$ in aqueous solution results mainly from the solubilization of ILs into aqueous phases. Therefore, the organic cations dissolved in aqueous phases through solubilization have an adverse effect on the transport of Sr^{2+} and Cs^+ from aqueous phases into ILs. The enhanced hydrophobicity

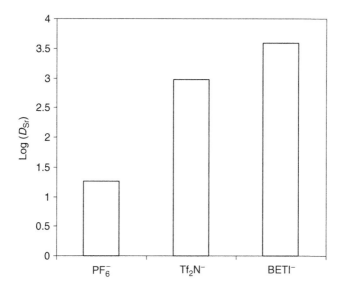

Figure 10.10 Dependence of Log (D_{Sr}) on ionic liquid anions of $[C_4C_1Im]^+$-based ionic liquids.

of the conjugate anions in ILs can give rise not only to greater distribution coefficients but also to decreased losses of ILs, as revealed by the ion-exchange model.

To further confirm the validity of the proposed rationalization of the anion effect via the ion-exchange model, the organic cations ($[C_nC_1Im]^+$) were deliberately added to the aqueous phases in the form of the corresponding bromide salts ($[C_nC_1Im]Br$) to shift the equilibrium of Equation 10.8) to the left side [61]. The concomitant significant reduction of Sr^{2+} extraction was observed consistent with Le Chatelier's principle based on the ion-exchange model. For example, the addition of $[C_4C_1Im]Br$ (19.0 mM) into the aqueous solution in the extraction system based on $[C_4C_1Im][BETI]$ (*bis*(perfluoroethanesulfonyl)imide [BETI]) reduces D_{Sr} from 3950 to 357. The value of D_{Sr} decreased to 23.1 on further addition of $[C_4C_1Im]Br$ (85.9 mM). Accordingly, the intrinsic solubilities of ILs in aqueous solutions have an adverse effect on the extraction of cationic species.

10.2.6 *Electrochemical method for recycling ionic liquid–ionophore extractant phase in ionic liquid-based extraction of Cs^+ and Sr^{2+}*

To develop an efficient and economical method for the extraction of radioactive Cs^+ and Sr^{2+} from nuclear waste with an IL, a nondestructive process must be developed for removing the captured Cs^+ and Sr^{2+} so that the IL can be recycled. This method must also concentrate the recovered Cs^+ and Sr^{2+}

into the smallest volume possible. The electrochemical reduction of these ions into a liquid electrode material, such as Hg, is an attractive candidate for this process. However, the reduction of Cs^+ and Sr^{2+} in both native form and complexed with the ionophores of interest, BOBCalixC6 and DCH18C6, respectively, is observed close to the negative potential limit of the di- and trialkylimidazolium $[Tf_2N]^-$ or $[PF_6]^-$ salts that have been used as extraction solvents for these ions [26]. Thus, the prospects for the development of a simple, efficient electrochemical method for the removal of these captured ions from imidazolium ILs is not good. However, it was recently demonstrated that ILs based on $[Tf_2N]^-$ anions and quaternary tetraalkylammonium cations such as tri-1-butylmethyammonium $[(C_4)_3C_1N]^+$ are hydrophobic liquids at room temperature and exhibit much greater cathodic stability than the corresponding imidazolium salts [62–65]. Thus, at the University of Mississippi, Hussey and his coworkers [66–68] have investigated mixtures of this IL with the ionophores, BOBCalixC6 and DCH18C6, to determine if these extraction mixtures can be used to selectively extract Cs^+ and Sr^{2+} from simulated nuclear wastes and to assess the prospects of using electrochemical methods to selectively remove the extracted Cs^+ and Sr^{2+} so that the IL–ionophore extraction solvent can be recycled. They have found that the $[(C_4)_3C_1N][Tf_2N]$ containing BOBCalixC6 and DCH18C6 is an excellent solvent for the extraction of Cs^+ and Sr^{2+} from simulated aqueous wastes. The extraction efficiency is comparable to the corresponding imidazolium-based ILs under active investigation at our research institute. Furthermore, it is possible to efficiently remove the Cs^+ and Sr^{2+} complexed by BOBCalixC6 and DCH18C6, respectively, from the $[(C_4)_3C_1N][Tf_2N]$ IL by electrochemical reduction at a mercury film electrode (MFE). In addition, because of the difference in the reduction potentials of Cs^+ complexed by BOBCalixC6 and Sr^{2+} complexed by DCH18C6, it is also possible to selectively remove either Cs^+ or both Cs^+ and Sr^{2+} from the extraction mixture by careful control of the reduction potential during controlled-potential electrolysis, leaving the unaltered ionophores in the IL. Thus, they have demonstrated the feasibility of an IL-based process for removing Cs^+ and Sr^{2+} from aqueous simulated nuclear wastes by extraction and electrochemical concentration of the extracted Cs and Sr into a liquid–metal electrode. This process permits recycling of the IL–ionophore mixture and provides a route for transporting the recovered radioactive Cs and Sr from the extraction facility.

10.2.7   Ionic liquids based on metal complexes of crown ethers

We have developed a new strategy for synthesizing a new class of ILs [69]. The basic concept and proof-of-principle experiments have been summarized in a patent application, which was filed on December 2, 2003 by UT-Battelle, LLC. The conventional formation of ILs can be regarded as the complexation reactions of simple anionic species and neutral compounds.

$$AlCl_3 + EMI^+Cl^- \rightleftharpoons EMI^+AlCl_4^- \qquad (10.9)$$

The most classic example is the formation of the ILs [C_2C_1Im]Cl/$AlCl_3$ [32]. When two solids are mixed stoichiometrically, the complexation of Cl^- by $AlCl_3$ gives rise to an IL composed of [C_2C_1Im][$AlCl_4$] (Equation 10.9). The reaction is thermodynamically favored, resulting in the release of heat. The essence of our new strategy for synthesizing a new class of ILs is through the complexation reactions of metal cations by neutral ligands.

Crown ethers are well-known neutral ligands for their strong complexation of metal ions to form cationic coordination complexes. The preliminary experiments have been conducted to test the aforementioned new strategy for formation of ILs based on asymmetric neutral crown ethers. The systems we chose to demonstrate the basic principle of our strategy consist of a neutral asymmetric crown ether and a metal-organic salt. The crown ether we used in our preliminary experiments is cyclohexano-15-crown-5, which is very asymmetric and known to complex strongly with Li^+. *N*-lithiotri-fluoromethanesulfonimide (Li[Tf_2N]) was used as an alkaline organic salt to provide cations (Li^+) to be complexed by the crown ether and conjugate anions [Tf_2N]$^-$. When the above crown ether and Li[Tf_2N] were mixed, a release of heat was observed, resulting in a clear solution. This observation is reminiscent of [C_2C_1Im]Cl/$AlCl_3$ reaction. The basic reaction is given by Equation 10.10.

$$(10.10)$$

This new IL also has negligible vapor pressure at ambient conditions. No reduction of mass has been observed under vacuum rotary evaporation at 100°C. The miscibilities of this IL with a number of organic solvents (e.g., acetonitrile and acetone) are very good. However, it is immiscible with aqueous solutions. Accordingly, it can be utilized as a potential new solvent for solvent extraction of ionic species. Our preliminary experiments have shown that Cs^+ can be quantitatively extracted to this new IL.

10.2.8   Task-specific ionic liquids containing an aza crown-ether fragment

Davis first introduced the term task-specific ILs (TSILs) to describe ILs prepared using the concept of increasing the ILs affinity of the extractants by incorporating the complexing functionality as an integral part of the IL [44].

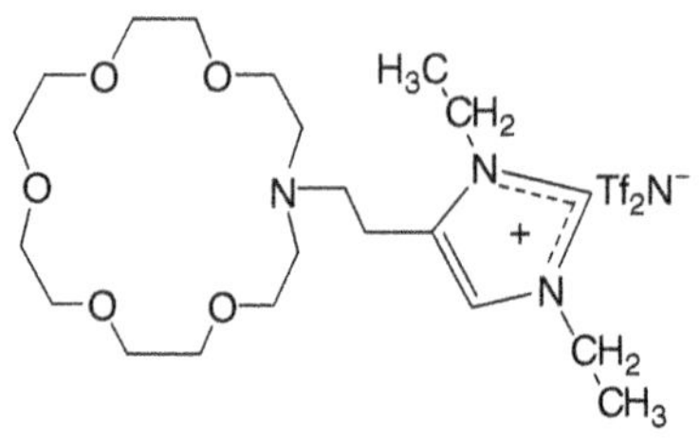

Figure 10.11   Structure of a new ionic liquid containing the aza-crown fragment.

Thus, ILs with built-in extracting capability, and differentiation from both aqueous and organic phases, through modification of the IL comiscibility can be prepared. Based on our current success in development of a stripping process with *N*-alkyl aza crown ether [56], we have synthesized several new ILs covalently attached with an aza crown-ether fragment [70]. These new ILs can function as both extractant and solvent. The solvation properties of these ILs can be systematically influenced by both the carbon chain length between the aza crown ether and IL fragment and alkyl substituents (R) on organic cations [70]. The ILs containing the aza crown ether with various combinations of substituent groups R and different carbon chain length (*n*) were synthesized and characterized for solvent-extraction applications. As our preliminary results, we have successfully synthesized an IL covalently attached to an aza crown (Figure 10.11). The extraction result was very promising.

Conclusions

In this chapter, we provide an overview of our recent research on solvent extraction of fission products based on ILs. Since the inception of the IL-based extraction system for metal ions in 1999, this approach has proven to be highly efficient for the extraction of metal ions. The success of the IL-based extraction systems lies in its ionicity, tunability, ion exchangeability, and nonvolatility. Although this review is focused on the recent research activities of our research groups, we hope to convince readers that the separation of metal ions based on IL is a fascinating research arena. Existing achievements and many anticipated future advances in this area will have fundamental and practical impacts on separation sciences.

Acknowledgment

This research was supported by the Environmental Management Sciences Program, Office of Biological and Environmental Research, U.S. Department of Energy and DOE NA-22 program under Contract DE-AC05-00OR22725 with Oak Ridge National Laboratory (ORNL), managed and operated by UT-Battelle, LLC. The authors thank T. G. Keever for running some tracer experiments at ORNL.

References

1. McDowell, W. J., Shoun, R. R., An evaluation of crown compounds as solvent extraction reagents. *Proc. ISEC '77, 1,* 95–97, 1977.
2. McDowell, W. J., Case, G. N., Aldrup, D. W., Investigation of ion-size-selective synergism in solvent extraction, *Sep. Sci. Techol., 18,* 1483–1507, 1983.
3. McDowell, W. J., Moyer, B. A., Bryan, S. A., Chadwick, R. B., Case, G. N., Selectivity and equilibria in extraction with organophilic acids, *Proc. ISEC '86,* Munchen, FRG, 1986.
4. McDowell, W. J., Crown ethers as solvent extraction reagents: Where do we stand, *Sep. Sci. Technol., 23,* 1251–1268, 1988.
5. Horwitz, E. P., Dietz, M. L., Fisher, D. E., Extraction of strontium from nitric acid solutions using dicyclohexano-18-crown-6 and its derivatives, *Solv. Extr. Ion Exch., 8,* 557–572, 1990.
6. Horwitz, E. P., Dietz, M. L., Fisher, D. E. I., Srex: A new process for the extraction and recovery of strontium from acidic nuclear waste streams, *Solv. Extr. Ion Exch., 9,* 1–25, 1991.
7. McDowell, W. J., Case, G. N., McDonough, J. A., Bartsch, R. A., Selective extraction of cesium from acidic nitrate solutions with didodecylnaphthalenesulfonic acid synergized with bis(tert-butylbenzo)-21-crown-7, *Anal. Chem., 64,* 3013–3017, 1992.
8. Dietz, M. L., Horwitz, E. P., Jensen, M. P., Rhoads, S., Bartsch, R. A., Palka, A., Krzykawski, J., Nam, J., Substituent effects in the extraction of cesium from acidic nitrate media with macrocyclic polyethers, *Solv. Extr. Ion Exch., 14,* 357–384, 1996.
9. Dietz, M. L., Horwitz, E. P., Rhoads, S., Bartsch, R. A., Krzykawski, J., Extraction of cesium from acidic nitrate media using macrocyclic polyethers: The role of organic phase water, *Solv. Extr. Ion Exch., 14,* 1–12, 1996.
10. Gerow, I. H., Davis, M. V., Use of 24-crown-8s in the solvent extraction of $CsNO_3$ and $Sr(NO_3)_2$, *Sep. Sci. Technol., 14,* 395–414, 1979.
11. Gerow, I. H., Smith, J. E., Davis, M. W., Extraction of Cs^+ and Sr^{2+} from HNO_3 solution using macrocyclic polyethers, *Sep. Sci. Technol., 16,* 519–548, 1981.
12. Blasius, E., Nilles, K.-H., The removal of cesium from medium-active waste solutions. 1. Evaluation of crown ethers and special crown-ether adducts in the solvent extraction of cesium, *Radiochim Acta, 35,* 173–182, 1984.
13. Schulz, W. W., Bray, L. A., Solvent-extraction recovery of by-product Cs-137 and Sr-90 from HNO_3 solutions: A technology review and assessment, *Sep. Sci. Technol., 22,* 191–214, 1987.
14. Dozol, J.-F., Casas, J., Sastre, A. M., Transport of cesium from reprocessing concentrate solutions through flat-sheet-supported liquid membranes: Influence of the extractant, *Sep. Sci. Technol., 30,* 435–448, 1995.
15. Haverlock, T. J., Bonnesen, P. V., Sachleben, R. A., Moyer, B. A., Applicability of a calixarene-crown compound for the removal of cesium from alkaline tank waste, *Radiochim Acta, 76,* 103–108, 1997.
16. Hiraoka, M., *Crown Compounds: Their Characteristics and Applications,* Elsevier, Amsterdam, 1982.
17. Lindoy, L. F., *The Chemistry of Macrocyclic Ligand Complexes,* Cambridge University Press, Cambridge, 1992.
18. Moyer, B. A., In Gokel, G. W. E., (Ed.), *Molecular Recognition: Receptors for Cationic Guests,* Vol. 1, Pergamon, Elsevier, Oxford, 1996.

19. Namor, A. F. D., Cleverley, R. M., Zapata-Ormachea, M. L., Thermodynamics of calixarence chemistry, *Chem. Rev., 98*, 2495, 1998.
20. Pedersen, C. J., Cyclic polyethers and their complexes with metal salts, *J. Am. Chem. Soc., 89*, 7017, 1967.
21. Pedersen, C. J., The discovery of crown ethers (noble lecture), *Angew. Chem. Int. Ed., 27*, 1021–1027, 1988.
22. Cram, D. J., The design of molecular hosts, guests, and their complexes (nobel lecture), *Angew. Chem. Int. Ed., 27*, 1009–1020, 1988.
23. Lehn, J. M., Supramolecular chemistry: Scope and perspectives molecules, supermolecules, and molecular devices, *Angew. Chem. Int. Ed., 27*, 89–112, 1988.
24. Dietz, M. L., Horwitz, E. P., Rogers, R. D., Extraction of strontium from acidic nitrate media using a modified purex solvent, *Solv. Extr. Ion Exch., 13*, 1–17, 1995.
25. Bonhote, P., Dias, A. P., Papageorgiou, N., Kalyanasundaram, K., Gratzel, M., Hydrophobic, highly conductive ambient-temperature molten salts, *Inorg. Chem., 35*, 1168–1178, 1996.
26. Dai, S., Ju, Y. H., Barnes, C. E., Solvent extraction of strontium nitrate by a crown ether using room-temperature ionic liquids, *J. Chem. Soc., Dalton Trans.,* 1201–1202, 1999.
27. Visser, A. E., Swatloski, R. P., Reichert, W. M., Griffin, S. T., Rogers, R. D., Traditional extractants in nontraditional solvents: Groups 1 and 2 extraction by crown ethers in room-temperature ionic liquids, *Ind. Eng. Chem. Res., 39*, 3596–3604, 2000.
28. Chun, S., Dzyuba, S. V., Bartsch, R. A., Influence of structural variation in room-temperature ionic liquids on the selectivity and efficiency of competitive alkali metal salt extraction by a crown ether, *Anal. Chem., 73*, 3737–3741, 2001.
29. Wasserscheid, P., Keim, W., Ionic liquids: New solutions for transition metal catalysis, *Angew. Chem. Int. Ed., 39*, 3773–3789, 2000.
30. Visser, A. E., Swatloski, R. P., Reichert, W. M., Mayton, R., Sheff, S., Wierzbicki, A., Davis, J. H., Rogers, R. D., Task-specific ionic liquids incorporating novel cations for the coordination and extraction of Hg^{2+} and Cd^{2+}: Synthesis, characterization, and extraction studies, *Environ. Sci. Tech., 36*, 2523–2529, 2002.
31. Abraham, M. H., Zissimos, A. M., Huddleston, J. G., Willauer, H. D., Rogers, R. D., Acree, W. E., Some novel liquid partitioning systems: Water-ionic liquids and aqueous biphasic systems, *Ind. Eng. Chem. Res., 42*, 413–418, 2003.
32. Hussey, C. L., Room-temperature molten salt systems, *Adv. Molten Salt Chem., 5*, 185, 1983.
33. Hussey, C. L., In Popov, A., Mamantov, G. (Eds), *Chemistry of Nonaqueous Solvents*, VXH Publishers: New York, 1994.
34. Holbrey, J. D., Seddon, K. R., The phase behaviour of 1-alkyl-3-methylimidazolium tetrafluoroborates: Ionic liquids and ionic liquid crystals, *J. Chem. Soc., Dalton Trans.,* 2133–2139, 1999.
35. Rhinebarger, R. R., Rovang, J. W., Popov, A. I., Multinuclear magnetic- resonance studies of macrocyclic complexation in room-temperature molten salts, *Inorg. Chem., 23*, 2558, 1984.
36. Eyring, E. M., Cobranchi, D. P., Garland, B. A., Gerhard, A., Highley, A. M., Huang, Y. H., Konya, G., Petrucci, S., Eldik, R. V., Lithium ion–crown-ether complexes in a molten salt, *Pure and Appl. Chem., 65*, 451, 1993.
37. Gerhard, A., Cobranchi, D. P., Highley, A. M., Huang, Y. H., Konya, G., Zahl, A., Eldik, R. V., Petrucci, S., Eyring, E. M., Li-7-NMR determination of stability constants as a function of temperature for lithium crown-ether complexes in a molten salt mixture, *J. Phys. Chem., 98*, 7923, 1994.

38. Larsen, A. S., Holbrey, J. D., Tham F. S., Reed, C. A., Designing ionic liquids: Imidazolium melts with inert carborane anions, *J. Am. Chem. Soc., 122*, 7264–7272, 2000.
39. Wilkes, J. S., Zaworotko, M. J., Air and water stable 1-ethyl-3-methylimidazolium based ionic liquids, *J. Chem. Soc. Chem. Commun.*, 965–967, 1992.
40. Huddleston, J. G., Willauer, H. D., Swatloski, R. P., Visser, A. E., Rogers, R. D., Room temperature ionic liquids as novel media for "clean" liquid-liquid extraction, *Chem. Commun.*, 1765–1766, 1998.
41. Dietz, M. L., Dzielawa, J. A., Ion-exchange as a mode of cation transfer into room-temperature ionic liquids containing crown ethers: Implications for the "greenness" of ionic liquids as diluents in liquid-liquid extraction, *Chem. Commun.*, 2124–2125, 2001.
42. Dietz, M. L., Dzielawa, J. A., Laszak, I., Young, B. A., Jensen, M. P., Influence of solvent structural variations on the mechanism of facilitated ion transfer into room-temperature ionic liquids, *Green Chem., 5*, 682–685, 2003.
43. Huddleston, J. G., Broker, G. A., Willauer, H. D., Rogers, R. D., Free-energy relationships and solvatochromatic properties of 1-alkyl-3-methylimidazolium ionic liquids. In *Ionic Liquids*, ACS Symposium Series 818, 270–288, 2002.
44. Visser, A. E., Swatloski, R. P., Griffin, S. T., Hartman, D. H., Rogers, R. D., Liquid/liquid extraction of metal ions in room temperature ionic liquids, *Sep. Sci. Techn., 36*, 785–804, 2001.
45. Cull, S. G., Holbrey, J. D., Vargas-Mora, V., Seddon, K. R., Lye, G. J., Room-temperature ionic liquids as replacements for organic solvents in multiphase bioprocess operations, *Biotech. Bioeng., 69*, 227–233, 2000.
46. Bonnesen, P. V., Delmau, L. H., Moyer, B. A., Lumetta, G. J., Development of effective solvent modifiers for the solvent extraction of cesium from alkaline high-level tank waste, *Solv. Extr. Ion Exch., 21*, 141–170, 2003.
47. Bonnesen, P. V., Delmau, L. H., Moyer, B. A., Leonard, R. A., A robust alkaline-side CSEX solvent suitable for removing cesium from Savannah river high level waste, *Solv. Extr. Ion Exch., 18*, 1079–1107, 2000.
48. Luo, H. M., Dai, S., Bonnesen, P. V., Buchanan, A. C., Holbrey, J. D., Bridges, N. J., Rogers, R. D., *Anal. Chem., 76*, 3078–3083, 2004.
49. Kavallieratos, K., Moyer, B. A., Attenuation of Hofmeister bias in ion-pair extraction by a disulfonamide anion host used in strikingly effective synergistic combination with a calix-crown Cs+ host, *Chem. Commun.*, 1620–1621, 2001.
50. Lee, B., Bao, L. L., Im, H. J., Dai, S., Hagaman, E. W., Lin, J. S., Synthesis and characterization of organic-inorganic hybrid mesoporous anion-exchange resins for perrhenate (ReO_4^-) anion adsorption, *Langmuir, 19*, 4246–4252, 2003.
51. Makote, R. D., Luo, H. M., Dai, S., Synthesis of ionic liquid and silica composites doped with dicyclohexyl-18-crown-6 for sequestration of metal ions. In *Clean Solvents*, ACS Symposium Series 819, 26–33, 2002.
52. Dai, S., Shin, Y. S., Toth, L. M., Barnes, C. E., Comparative UV-Vis studies of uranyl chloride complex in two basic ambient-temperature melt systems: The observation of spectral and thermodynamic variations induced via hydrogen bonding, *Inorg. Chem., 36*, 4900–4902, 1997.
53. Sachleben, R. A., Bonnesen, P. V., Descazeaud, T., Haverlock, T. J., Urvoas, A., Moyer, B. A., Surveying the extraction of cesium nitrate by 1,3-alternate calix 4-arene crown-6 ethers in 1,2-dichloroethane, *Solv. Extr. Ion Exch., 17*, 1445–1459, 1999.

54. Levitskaia, T. G., Bonnesen, P. V., Chambliss, C. K., Moyer, B. A., Synergistic pseudo-hydroxide extraction: synergism and anion selectivity in sodium extraction using a crown ether and a series of weak lipophilic acids, *Anal. Chem.*, 75, 405–412, 2003.

55. Kimura, K., Sakamoto, H., Koseki, Y., Shono, T., Excellent, ph-sensitive cation-selectivities of bis(monoazacrown ether) derivatives in membrane cation-transport, *Chem. Lett.*, 1241–1244, 1985.

56. Nakatsuji, Y., Sunagawa, T., Masuyama, A., Kida, T., Ikeda, I., Synthesis, structure, and complexation properties of monoazacryptands, *Chem. Lett.*, 445–446, 1995.

57. Luo, H., Dai, S., Bonnesen, P. V., Solvent extraction of Sr^{2+} and Cs^+ based on room-temperature ionic liquids containing monoaza-substituted crown ethers, *Anal. Chem.*, 76, 2773–2779, 2004.

58. Matsushima, K., Kobayashi, H., Nakatsuji, Y., Okahara, M., Proton driven active-transport of alkali-metal cations by using alkyl monoaza crown ether derivatives, *Chem. Lett.*, 701–704, 1983.

59. Luo, H., Dai, S., Bonnesen, P. V., Haverlock, T. J., Moyer, B. A., Buchanan, A. C., A striking effect of ionic-liquid anions in the extraction of Sr^{2+} and Cs^+ by dicyclohexano-18-crown-6, *Solv. Extr. Ion Exch.*, 24, 19–31, 2006.

60. Pauling, L., *General Chemistry*, Dover Publications, New York, 1970.

61. Shimojo, K., Goto, M., Solvent extraction and stripping of silver ions in room-temperature ionic liquids containing calixarenes, *Anal. Chem.*, 76, 5039–5044, 2004.

62. Matsumoto, H., Yanagida, M., Tanimoto, K., Nomura, M., Kitagawa, Y., Miyazaki, Y., Highly conductive room temperature molten salts based on small trimethylalkylammonium cations and bis(trifluoromethylsulfonyl)imide, *Chem. Lett.*, 922–923, 2000.

63. Quinn, B. M., Ding, Z., Moulton, R., Bard, A. J., Novel electrochemical studies of ionic liquids, *Langmuir*, 18, 1734–1742, 2002.

64. Sun, J., Forsyth, M., MacFarlane, D. R., Room-temperature molten salts based on the quaternary ammonium ion, *J. Phys. Chem. B.*, 102, 8858–8864, 1998.

65. Murase, K., Nitta, K., Hirato, T., Awakura, Y., Electrochemical behaviour of copper in trimethyl-n-hexylammonium bis((trifluoromethyl)sulfonyl)amide, an ammonium imide-type room temperature molten salt, *J. Appl. Electrochem.*, 31, 1089, 2001.

66. Chen, P. Y., Hussey, C. L., Electrochemistry of ionophore-coordinated Cs and Sr ions in the tri-1-butylmethylammonium bis((trifluoromethyl)sulfonyl)imide ionic liquid, *Electrochim. Acta*, 50, 2533–2540, 2005.

67. Chen, P. Y., Hussey, C. L., Electrodeposition of cesium at mercury electrodes in the tri-1-butylmethylammonium bis((trifluoromethyl)sulfonyl)imide room-temperature ionic liquid, *Electrochim. Acta*, 49, 5125–5138, 2004.

68. Tsuda, T., Hussey, C. L., Luo, H. M., Dai, S., Recovery of cesium extracted from simulated tank waste with an ionic liquid: Water and oxygen effects, *J. Electrochem. Soc.*, 153, D171–D176, 2006.

69. Huang, J. F., Luo, H. M., Dai, S., A new strategy for synthesis of novel classes of room-temperature ionic liquids based on complexation reaction of cations, *J. Electrochem. Soc.*, 153, J9–J13, 2006.

70. Luo, H. M., Dai, S., Bonnesen, P. V., Buchanan, A. C., Separation of fission products based on ionic liquids: Task-specific ionic liquids containing an aza-crown ether fragment, *J. Alloy. Compd.*, 418, 195–199, 2006.

chapter eleven

Molecular spectroscopy and ionic liquids

Mihkel Koel

Contents

11.1   Introduction

Molecular systems can be characterized by using a wide range of spectroscopic methods that describe their energy term schemes in the form of spectra. Usually ultraviolet–visible (UV–Vis) spectra are obtained by measuring the intensity of the absorption of the monochromatic radiation with a wavelength of 190–1100 nm passing through a solution in the cuvette. It means that usually for UV–Vis measurements the sample or chromophore is in a liquid state or solvated in certain solvent. On typical chromophore, transitions in molecular orbitals and in the ligand fields of metal chelates and charge-transfer (CT) bands can be observed. The solvents can actively interact with certain solutes and thus change the UV–Vis spectra formed by either removing vibrational fine structures or shifting absorption band maxima, or both. In case of the molecular photoluminescence (fluorescence, phosphorescence) and chemiluminescence, the emission spectra are obtained in the above-mentioned wavelength range when the excited molecules decay to their ground states.

Pure ionic liquids (ILs) are solvents that remain colorless and transparent *throughout almost* the whole visible and near-infrared (NIR) spectral regions. This property coupled with excellent stability makes ILs very attractive optical solvents that may be used for the absorption and fluorescence studies of dissolved substances, as well as for monitoring the reactions performed in these solvents. Also, the UV–Vis spectroscopy can be and is widely used to analyze the properties/behavior of ILs, as well as to conduct theoretical studies to get a better understanding of the nature of ILs.

This chapter will briefly treat some of these aspects. An attempt is made to give an idea about the possibilities of applying ILs to molecular spectroscopic studies.

11.2 Solvent for spectroscopy

Already in the era of haloaluminate ILs it was reported that ILs are able to solvate a wide range of species, including organic, inorganic, and organometallic compounds. Notably a variety of transition metal complexes, which are unstable in other media, may be studied in ambient temperature ILs [1]. The use of ILs as solvents circumvents the problems of solvation and solvolysis of the halide complexes of transition metals and permits reliable solution spectra of these species to be recorded. In some case the spectra exhibit quite a remarkable resolution. In above-mentioned work, simple halide complexes of titanium, vanadium, chromium, manganese, iron, cobalt, nickel, copper, molybdenum, tungsten, rhenium, ruthenium, osmium, rhodium, and iridium were subjected to study. In all cases it was observed enhanced resolution and significant spectral shifts compared with the standard solution spectra published. However, the results obtained were similar to those obtained with solvents having a low dielectric constant (such as dichloromethane) [2]. Other investigators have confirmed that haloaluminate ILs are excellent solvents to be applied to absorption spectroscopic and spectroelectrochemical studies. Advantageous is that the methods used to study the properties or behavior of complex species in room-temperature haloaluminates are mostly the same as those used to carry out similar investigations in conventional solvents [3].

Haloaluminate ILs are tunable solvents that allow the dependence of the oxidation state of a compound on the solvent acidity to be studied. An example [4] is the characterization of iodine and the iodine species in different oxidation states at widely varied acidities of $[C_4py]Cl/AlCl_3$. The variety of the species formed at different solvent acidities reflects the features of the characteristic acid–base interactions in chloroaluminate molten salts as well. Combining electrochemical and spectroscopic data, the electro-active species can be identified as it was done in Ref. 5.

In Ref. 6, the Brønsted superacidity of HCl in liquid chloroaluminate IL ($[C_2C_1Im]Cl/AlCl_3$) was studied by the protonation of arens during which the degree of protonation was measured by using absorption spectroscopy. The arens were stable in the liquid chloroaluminate for many hours and their protonated forms (arenium ions) were stable for 1 h or more.

The easy handling of water- and air-stable ILs also enables their application to spectroscopic measurements. ILs have good solvating properties and a high spectral transparency (the UV cut off wavelength of alkylimidazolium-based

ILs is in the range of 230–250 nm) which make them suitable to be used as solvents for spectroscopic measurements especially in the visible region. Because of their ionic origin, ILs allow the coordination of a complex compound in a liquid state to be studied. An additional advantage of ILs is that their solvating properties can be designed in such a way that differently coordinating solvents are obtained. A lot of examples can be presented on spectroscopic studies with lanthanides and actinides.

The solubility and coordination properties of uranium(VI) in ILs strongly depend on the chemical form of the uranium(VI) induced in solution, the composition of ILs, and the presence of any other anions [7]. The coordination chemistry of iron(III) tetraarylporphyrins in ILs studied by UV–Vis spectroscopy could be mentioned [8].

ILs are becoming useful solvents to investigate the spectroscopic behavior of both organic and inorganic lanthanide complexes in solution, especially of complexes with weakly binding ligands, which otherwise would be unable to compete with the solvent molecules for a binding site on the lanthanide ion [9].

Researchers have observed an unusually stable behavior of Eu(II). In Ref. 10, the solvation of Eu(II) and its complexation effects in the presence of the crown ether DC15C5 solubilized in $[C_4C_1Im][PF_6]$ were studied. This extra stabilization can be explained by a local order of ions at short distances existing in ILs.

It was demonstrated that the visible luminescence from Eu(III) in ILs can provide an insight into the role of the catalyst and its interactions with ionic surrounding and traces of water. The spectroscopically active lanthanide ion can interact with the light through an induced electric dipole. In spectra, the transition having an absorption band maximum at 586 nm shows *hypersensitivity* and already small coordination changes shift the band. The spectra exhibit the structure of lanthanide complexes to be much finer than is commonly the case with the complexes in solution. Molecular dynamics simulations indicate that the first coordination shell is occupied by anions whose number depends on the ionic radius of the lanthanide ion. The imidazolium cations are situated in the second coordination sphere. Using noncoordinating anions the spectra have a less fine structure.

The luminescence of lanthanides can be improved by using an intramolecular energy transfer, the antenna effect. If the ligand (β-diketonates and noncharged adducts such as 1,10-phenanthroline or 2,2'-bipyridine) contains a light-absorbing group, the energy absorbed by the chromophore can be transferred to the lanthanide ion. The antenna effect improves the intensity just as in other solvents [11]. The ILs doped with luminescent lanthanide complexes could find potential applications as active components in, for instance, chelate lasers, which operate in a similar way as dye lasers do.

ILs are effective solvents also for fullerenes C_{60}, the solubility of it is up to 0.1 mg/mL. The solubility of C_{60} in ILs depends very much on the structure of the component ions [12]. The varying degrees of aggregation of C_{60}

molecules in ILs were also observed from the concentration dependence of the UV–Vis spectra of solutions of C_{60}.

Using ILs for the preparation of different nanoparticles is becoming popular. The latter can be easily characterized by using UV–Vis absorption spectroscopy, as it was the case with the Au nanoparticles obtained under different preparation conditions [13].

As already mentioned, UV–Vis spectroscopy is an effective tool to study metal complexes. For different actinide(IV) compounds in ILs (Np(IV), Pu(IV), U(IV) as $[C_4C_1Im]_2[AnCl_6]$ complexes in $[C_4C_1Im][Tf_2N]$) a similarity with solid complexes having an octahedral An(IV) environment was estimated [14,15]. In other ILs, for example, uranyl ions dissolved in $[C_4C_1Im]$ [NfO], $[UO_2]^{2+}$ may be present as a bare cation [16].

In case of ions, it is important to know the coordination environment to extract metals. UV–Vis spectroscopic measurements provide the respective information. There are systems in which both the biphasic extraction equilibrium and the metal coordination environment in an IL and a molecular organic solvent are the same [17].

Spectroscopic, that is, fluorescence and circular dichroism (CD), fourier transform infared (FTIR) measurements have been widely applied to analyzing changes in enzyme structures in an attempt to explain the stabilization or denaturation phenomena associated with enzyme environments, for example, temperature, solvents, and so on. All these measurements can also be performed in ILs where fluorescence and CD spectroscopy demonstrated the conformational changes in the native structure of calcium binding proteins (CALB) which resulted in the higher synthetic activity and stability in ILs as compared to those obtained in classical organic solvents [18].The stabilization of enzymes by ILs may be related to the associated structural changes of proteins [19].

11.3   Photochemistry and solvatochromism in ionic liquids

The influence of the solvent in molecular organic photochemistry is well established and it is natural to expect the same from ILs when they are used as a medium in photochemistry. Photochemical reactions include energy transfer, hydrogen abstraction, oxygen quenching, and electron transfer. From the screening of photochemical reactions in ILs the following are estimated: (i) a remarkable low oxygen solubility (can be measured by a pyrene fluorescence decay); (ii) the slow molecular diffusion rendering the diffusion-controlled processes about two orders of magnitude slower than those in common organic solvents (can be estimated from the diffusion-controlled energy transfer between xantone and naphthalene); (iii) a long lifetime of the triplet excited states (a xantone triplet excited state as a polarity probe) and radical ions; and (vi) a weaker CT interaction (competitive interactions with the ions of ILs) decreasing the association constant of the CT complexes and shifting λ_{max} to a longer wavelength (a CT complex between anthracene and methylviologen) compared to acetonitrile. ILs can be used to study electron transfer processes but they are also useful media to follow for energy and hydrogen transfer although not for the singlet oxygen generation. Photochemistry is strongly influenced by viscosity and diffusion rate in the media [20]. Finally, the slowdown of the diffusion rate could be useful for performing fast processes (sub ns) in more widely accessible timescales (ns systems).

The photochromism in ILs can be monitored in the same way as in organic solvents. The process parameters in ILs were reported to be similar to those reported in a dimethylsulfoxide (dmso) solution [21].

All ILs showed a rich photochemistry after UV photolysis leading to the buildup of various long-lived intermediate products as evidenced from the observation that ILs turn yellow upon continuous irradiation. On the other hand, exposing ILs to short excitation pulse (a rapid-scan method) significantly suppressed the formation of halides [22].

Spectroscopic measurements of solvatochromic and fluorescent probe molecules in room temperature ILs provide an insight into solvent intermolecular interactions, although the interpretation of the different and generally uncorrelated *polarity* scales is sometimes ambiguous [23]. It appears that the same solvatochromic probes work in ILs as well [24], but up to now only limited data are available on the behavior of electronic absorption and fluorescence solvatochromic probes within ILs and IL–organic solvent mixtures.

It was demonstrated that the addition of a cosolvent may drastically alter the physicochemical properties of ILs and hence increase their overall analytical utility toward solute solvation [25].

On the basis of the response of solvatochromic probes it is possible to investigate the aggregation behavior of common anionic, cationic, and nonionic surfactants when solubilized in low-viscosity ILs. The investigations have demonstrated that several common nonionic surfactants display an aggregation behavior in ILs as judged from the changes observed in the solvatochromic probe response. This method may be indicative of that aggregation that may to some extent occur in an IL solution. However, the method of using solvatochromic probes has some limitations. Even at fairly low concentrations, the probe may interact with such aggregates, distorting them during the process. Alternatively, the self-association of the probe or the interaction of the latter with the surfactant monomer may also lead to changes in the solvatochromic probe response [26].

The time-resolved spectroscopy is a sensitive tool to study the solute–solvent interactions. The technique has been used to characterize the solvating environment in the solvent. By measuring the time-dependent changes of the fluorescence signals in solvents, the solvation, rotation, photoisomerization, or excimer formation processes of a probe molecule can be examined. In conventional molecular solutions, many solute–solvent complexes, including excimers and exciplexes, show characteristic fluorescence bands indicating the formation of these complexes, in addition to the monomer fluorescence bands.

There are few similar measurements performed in ILs. The results of measurement of time-resolved fluorescence spectra and the fluorescence anisotropy decay of 2-aminoquinoline probe molecules in ILs, including imidazolium-based aromatic and nonaromatic ones, have shown to differ from measurements in ordinary molecular liquids. The time-resolved fluorescence spectra observed are indicative of the formation of π–π aromatic complexes in some aromatic ILs but not in nonaromatic ILs. It is argued that the aromaticity of the imidazolium cation plays a key role in the formation of a local structure of ions in imidazolium-based ILs [27]. The results obtained provide a further support for a hypothesis that specific local structures are formed in imidazolium-based ILs and the latter should be considered as *nanostructured fluids*. The investigations have also demonstrated that the aromaticity of the imidazolium cation plays a crucial role in the formation of a local structure, most probably through the π–π interaction.

As it was mentioned, the properties of ILs can be successfully altered by the addition of cosolvents. However, the binary and ternary solvent systems reveal a nonideal behavior depending on the cosolvent concentration and solvatochromic probe nature [28]. It has also been observed in case of IL mixtures with CO_2 where a local enhancement of ILs formation around a chromophore takes place, maintaining the solvent strength on initial level even at fairly high loadings of CO_2, whereas the microviscosity in the vicinity of the solute is dramatically reduced, leading to an enhanced mass transport and facilitated separation [29]. ILs and CO_2 can be used together with organic cosolvents which *solvate* the constituent ions of an IL, resulting in a decrease in the aggregation of these ions (lower viscosity and higher conductivity).

UV–Vis spectroscopy is a widely used technique to monitor the acid–base equilibrium in common organic solvents. This is also true for ILs. However, considering ionic nature of this solvent, the interpretation of the results is not such simple. The work is going on to establish the Brønsted acidity scale in ILs [30] which can later be used to predict various interactions and structure–property relationships.

As many ILs form transparent, good-quality glasses on freezing at 77 K, they can be used as matrixes over a wide range of temperatures in combination with the UV–Vis–NIR spectroscopy for the detection of the reactive species generated by an ionizing radiation. The first study by I. Dunkin et al. has shown ILs to be excellent media for the generation and spectral characterization of radical ions of solute molecules [31]. It has been demonstrated that the radical ions of interest can be generated under cryogenic conditions and their reactivity can be monitored during the thermal annealing of the solvent up to ambient temperature. Most notably, there exists a possibility of generating both radical cations and anions in the same experiment. This novel application of ILs could be made use of especially in characterizing biologically important compounds with low solubility in conventional organic matrixes. As the properties of ILs can be to a large extent controlled by the variation of both the cation and the anion, the desired properties of the liquids can be obtained by an appropriate design and synthesis.

11.4   The purity of ionic liquids

One aspect that limits the use of ILs in optical applications is that many synthetic methodologies lead to yellowish or even brown ILs. Specifically, during the preparation of 1-alkyl-3-methylimidazolium halides that are frequently used as starting materials in the synthesis of other ILs, discoloration is difficult to avoid. Thus, most commercially available ILs contain colored impurities. Apparently these impurities do not affect organic or catalytic reactions which are carried out in ILs.

The impurities that cause this discoloration are reported to be below the detection limit of nuclear magnetic resonance (NMR) spectroscopy. Although the chemical nature of colored impurities is not known yet, it is obvious that the quaternization step in the synthesis of ILs is the critical part in the procedure. The conditions of the quaternization step have to be optimized (mild reaction conditions and low-reaction temperatures). It is suggested to apply a synthetic procedure which involves slight modifications of commonly used synthetic routes, which means the lower reaction temperature (room temperature could be used), starting with ice-cooled reactants which were allowed to warm up very slowly and, respectively, longer reaction time (3–4 days) when the reaction proceeds under inert (Argon) atmosphere. The best purity of ILs is obtained when rather small quantities of reactants were used. The higher quantities of the reaction mixture resulted in a decrease of the transparency of ILs, probably due to the temporary higher local temperatures in the reaction mixture.

Taking into account these findings, a microreaction system has been developed by IoLiTec Ionic Liquid Technologies GmbH & Co to improve mass and heat transfer. The production capacity of the currently running system is about 0.5kg/h with a purity >99%.

In articles on spectroscopy and electrochemistry one can find discussions about the purification of ILs. Thoroughgoing approach to that problem is presented by Gordon in Ref. 32 where he is discussing about the synthesis of ILs and the purification methods, including the purification by charcoal, alumina columns, and the precautions on purity to be taken for the starting materials. It is evident that to obtain pure ILs, carefully purified reactants have to be used and the reagents should be stored in the dark. One latest article completely devoted to purification of ILs is given in Ref. 33.

Owing to the low volatility of ILs, only a few purification methods are suitable so that the formation of colored impurities should be prevented as far as possible during the preparation of an IL. The most commonly used methods for purifying ILs utilize an activated carbon and alumina (or sometimes silica) as a stationary phase in column liquid chromatography [34,35].

The treatment of ILs with active charcoal may enhance the transparency of ILs in the visible spectral region, but the method does not work well when it is aimed at obtaining ILs with a good transparency in the UV region. It appears that after ILs have passed through an alumina or a silica column, respectively, they exhibit a slight Tyndall effect (a light scattering was observed). This is indicative of the presence of small particles in the liquid, probably colloidal particles of alumina or silica [36]. Al_2O_3 seems to be dissolved in ILs at low concentrations, and it can be seen in the form of deposits in electrochemical experiments [37].

A conclusion can be drawn that it is difficult to remove traces of disturbing impurities by charcoal or alumina. A slight decolorization may occur, but not without a considerable loss of the product due to the dilution and evaporation of the solvent.

Studies on the properties/behavior of ILs may be complicated because the latter easily absorb small amounts of water (up to 3–5% w/w). The absorbed water affects the bulk solvent properties (i.e., viscosity and density) and may therefore alter the local solvent microenvironment surrounding a solute. Water can quench the luminescence and the resulting luminescence properties of ions can be strongly influenced by water. The dissolved organic solvent and water in ILs can be removed by keeping the IL under vacuum (10^{-2}–10^{-3} bar) at 50–60°C for at least 7–10 h [38].

Salts are purified by recrystallization and this procedure can be applied for ILs also. The recrystallization is performed by solving the compound in a small amount of acetonitrile and by adding a small amount of toluene to the solution. The solution was cooled in a freezer (–18°C). White crystals of the IL were formed. After filtration, the crystals were washed with an ice-cold toluene and the remaining solvent was removed by evaporation on a rotavap under reduced pressure [39]. This method is not universal as many ILs are oily liquids at room temperature or even at temperatures well below room temperature and are quite difficult to crystallize. There is of course always a loss of IL due to the washing of the samples. It is always necessary to avoid using environmentally unfriendly solvents, especially halogen-containing ones that are not suitable for industrial applications.

Some more exotic procedures can be suggested as well: The combination of ILs with the use of supercritical carbon dioxide ($scCO_2$) as an extractant represents a potential combination for the reaction of synthesis and downstream separation, as well as for purification.

Zone melting is a possibly generic approach to IL purification. The solidification of ILs often results in the formation of glass. However, it is possible to determine/choose conditions under which single crystals of ILs with a melting point down to –25°C (but not all) can be grown [39]. Where crystallization is seen, then separation of impurities can be demonstrated.

Recently it has been demonstrated that it is possible to distill ILs. Unfortunately, this is a slow process and is not currently suitable to be used for the purification of large amounts of material.

From the other side it is clear that UV–Vis spectrophotometry is an efficient method to control the quality of ILs and assess the purity. For this purpose UV–Vis spectroscopy is the simplest and relatively efficient method. In the pure material there is no absorbance above 300 nm. If significant absorption occurs above 350 nm, the IL is impure.

Fluorescence spectroscopy is probably the most sensitive method for determining the purity of an IL. It has been observed that the fluorescence of ILs can be effectively reduced to zero by a systematic purification of precursor salts. For the lowest fluorescence response, particular care should be taken while carrying out synthesis. For convenience it is possible to judge the degree of fluorescence impurity in the IL by exposing a sample to the UV light from a hand scanner at 380 nm. If the IL has significant impurities, it will fluoresce visibly [40].

It is important to emphasize again that considerable care should be taken when handling compounds that may possibly contain hydrogen fluoride (HF). The evolution of acidic HF fumes and formation of a colorless solid as a crystalline decomposition product from a hydrolytic degradation have frequently been observed to take place around the opening to vessels containing $[C_4C_1Im][PF_6]$ [41]. This kind of degradation of ILs indicates that in many cases the use of $[PF_6]$ containing ILs is neither suitable nor desirable.

11.5   Outlook

Molecular spectroscopy is a key method in almost all fields of ILs research. Starting with the assessment of the purity of ILs and study of their properties using different spectroscopic probes and their absorption and emission spectra, the reactions taking place in ILs are almost impossible to be studied without using molecular spectroscopy. Recording the UV–Vis or luminescence spectra is a commonly used technique for the detection of compounds by chromatography and electrophoresis, and ILs are more widely used in the respective studies. So, it is important to further investigate the applicability of ILs to molecular spectroscopy.

References

1. Appleby, D., Hussey, C.L., Seddon, K.R., Turp, J.E., Room-temperature ionic liquids as solvents for electronic absorption spectroscopy of halide complexes, *Nature*, 323, 614–615, 1986.
2. Seddon, K.R., Electronic absorption spectroscopy in room-temperature ionic liquids, *NATO ASI Series, Ser. C: Math. Phys. Sci.*, 202(*Molten Salt Chemistry*), 365–381, 1987.

3. Hussey, C.L., Room temperature haloaluminate ionic liquids. Novel solvents for transition metal solution chemistry, *Pure & Appl. Chem.*, 60, 1763–1772, 1988.
4. Karpinski, Z.J., Osteryoung, R.A., Spectrophotometric studies of iodine complexes in an aluminum chloride—butylpyridinium chloride ionic liquid, *Inorg. Chem.*, 23, 4561–4565, 1984.
5. Sinha, S.P., Spectrochemical and electrochemical properties of some lanthanides and actinides in room temperature melt, In Mamantov, G., Mamantov, C.B., and Brausnstein, J. (Eds), *Advances in Molten Salt Chemistry*, 6, Elsevier, *Proc. Electrochem. Sci. (Proc. Int. Symp. Molten Salts)* 87-7, 458–468, 1987.
6. Smith, G.P., Dworkin, A.S., Pagni, R.M., Zingg, S.P., Brønsted superacidity of HCl in liquid chloroaluminate, AlCl3—1-ethyl-3-methyl-1H-imidazolium chloride, *J. Am. Chem. Soc.*, 111, 525–529, 1989.
7. Gaillard, C., Chaumont, A., Billard, I., Henning, C., Ouadi, A., Wipff, G., Uranyl coordination in ionic liquids, *Inorg. Chem.*, 46, 4815–4826, 2007.
8. Campagna, S.R., Koronaios, P., Osteryoung, R.A., Cornman, C.R., Spectroscopy and coordination chemistry of iron(III) tetraarylporphyrins in room-temperature ionic liquids, Book of Abstracts INOR-121, 217th ACS National Meeting, Anaheim, California, 1999.
9. Arenz, S., Babai, A., Binnemans, K., Driesen, K., Giernoth, R., Mudring, A.-V., Nockemann, P., Intense near-infrared luminescence of anhydrous lanthanide(III) iodides in an imidazolium ionic liquid, *Chem. Phys. Let.*, 402, 75–79, 2005.
10. Billard, I., Moutiers, G., Labet, A., El Azzi, A., Gaillard, C., Mariet, C., Luetzenkirchen, K., Stability of divalent europium in an ionic liquid: Spectroscopic investigations in 1-methyl-3-butylimidazolium hexafluorophosphate, *Inorg. Chem.*, 42, 1726–1733, 2003.
11. Driesen, K., Nockemann, P., Binnemans, K., Ionic liquids as solvents for near-infrared emitting lanthanide complexes, *Chem. Phys. Let.*, 395, 306–310, 2004.
12. Liu, H., Tao, G., Evans, D.G., Kou, Y., Solubility of C-60 in ionic liquids, *Carbon*, 43, 1782–1785, 2005.
13. Guo, S., Shi, F., Gu, Y., Yang, J., Deng, Y., Size-controllable synthesis of gold nanoparticles via carbonylation and reduction of hydrochloroauric acid with CO and H_2O in ionic liquids, *Chem. Lett.*, 34, 830–831, 2005.
14. Nikitenko, S.I., Moisy, Ph., Formation of higher chloride complexes of Np(IV) and Pu(IV) in water-stable room-temperature ionic liquid [BuMeIm][Tf2N], *Inorg. Chem.*, 45, 1235–1242, 2006.
15. Nikitenko, S.I., Cannes, C., Le Leaour, C., Moisy, P., Trubert, D., Spectroscopic and electrochemical studies of U(IV)-hexachloro complexes in hydrophobic room-temperature ionic liquids [BuMeIm][Tf2N] and [MeBu3N][Tf2N], *Inorg. Chem.*, 44, 9497–9505, 2005.
16. Mizuoka, K., Ikeda, Y., Structural study on uranyl ion in 1-butyl-3-methylimidazolium nonafluorobutanesulfonate ionic liquid, *Prog. Nucl. Ener.*, 47, 426–433, 2005.
17. Cocalia, V.A., Jensen, M.P., Holbrey, J.D., Spear, S.K., Stepinski, D.C., Rogers, R.D., Identical extraction behavior and coordination of trivalent or hexavalent f-element cations using ionic liquid and molecular solvents, *J. Chem. Soc., Dalton Trans.*, 11, 1966–1971, 2005.
18. De Diego, T., Lozano, P., Gmouh, S., Vaultier, M., Iborra, J.L., Understanding structure-stability relationships of candida antartica lipase B in ionic liquids, *Biomacromol.*, 6, 1457–1464, 2005.

19. Lozano, P., De Diego, T., Gmouh, S., Vaultier, M., Iborra, J.L., Dynamic structure-function relationships in enzyme stabilization by ionic liquids, *Biocat. Biotransf.*, 23, 169–176, 2005.

20. Alvaro, M., Ferrer, B., Garcia, H., Narayana, M., Screening of an ionic liquid as medium for photochemical reactions, *Chem. Phys. Let.*, 362, 435–440, 2002.

21. Rack, J.J., Mockus, N.V., Room-temperature photochromism in cis- and trans-[Ru(bpy)2(dmso)2]2+. *Inorg. Chem.*, 42, 5792–5794, 2003.

22. Brands, H., Chandrasekhar, N., Unterreiner, A.-N., Ultrafast dynamics of room temperature ionic liquids after ultraviolet femtosecond excitation, *J. Phys. Chem. B*, 111, 4830–4836, 2007.

23. Baker, S.N., Baker, G.A., Bright, F.V., Temperature-dependent microscopic solvent properties of 'dry' and 'wet' 1-butyl-3-methylimidazolium hexafluorophosphate; corelation with ET(30) and kamlet-taft polarity scales, *Green Chem.*, 4, 165–169, 2002.

24. Reichardt, C., Pyridinium N-phenoxide betaine dyes and their application to the determination of solvent polarities, part 29—Polarity of ionic liquids determined empirically by means of solvatochromic pyridinium N-phenolate betaine dyes, *Green Chem.*, 7, 339–351, 2005.

25. Fletcher, K.A., Storey, I.A., Hendricks, A.E., Pandey, S., and Pandey, S., Behavior of the solvatochromic probes Reichardt's dye, pyrene, dansylamide, Nile Red and 1-pyrenecarbaldehyde within the room-temperature ionic liquid bmim PF(6), *Green Chem.*, 3, 210–215, 2001.

26. Fletcher, K.A., Pandey, S., Surfactant aggregation within room temperature ionic liquid 1-ethyl-3-methylimidazolium bis(trifluoromethylsulfonyl)imide, *Langmuir*, 20, 33–36, 2004.

27. Iwata, K., Kakita, M., Hamaguchi, H., Picosecond time-resolved fluorescence study on solute-solvent interaction of 2-aminoquinoline in room-temperature ionic liquids: Aromaticity of imidazolium-based ionic liquids, *J. Phys. Chem. B*, 111, 4914–4919, 2007.

28. Koel, M., Solvatochromic probes within ionic liquids, *Proc. Est. Acad. Sci. Chem.*, 54, 3–11, 2005.

29. Lu, J., Liotta, C.L., Eckert, C.A., Spectroscopically probing microscopic solvent properties of room-temperature ionic liquids with the addition of carbon dioxide, *J. Phys. Chem. A*, 107, 3995–4000, 2003.

30. Thomazeau, C., Olivier-Bourbigou, H., Magna, L., Luts, S., Gilbert, B., Determination of an acidic scale in room temperature ionic liquids, *J. Am. Chem. Soc.*, 125, 5264–5265, 2003.

31. Marcinek, A., Zielonka, J., Geübicki, J., Gordon C.M., and Dunkin, I.R., Ionic liquids: Novel media for characterization of radical ions, *J. Phys. Chem. A*, 105, 9305–9309, 2001.

32. Gordon, C.M., Chap. 2.1 In Wasserscheid, P. and Welton, T. (Eds), *Ionic Liquids in Synthesis*, Wiley-VCH Verlag, Weinheim, 2003.

33. Earle, M.J., Gordon, C.M., Plechkova, N.V., Seddon, K.R., Welton, T., Decolorization of ionic liquids for spectroscopy, *Anal. Chem.*, 79, 758–764, 2007.

34. Karmakar, R., Samanta, A., Intramolecular excimer formation kinetics in room temperature ionic liquids, *Chem. Phys. Let.*, 376, 638–645, 2003.

35. Dahl, K., Sando, G.M., Fox, D.M., Sutto, T.E., Owrutsky, J.C., Vibrational spectroscopy and dynamics of small anions in ionic liquid solutions, *J. Chem. Phys.*, 123, 084504, 2005.

36. Billard, I., Moutiers, G., Labet, A., El Azzi, A., Gaillard, C., Mariet, C., Lutzenkirchen, K., Stability of divalent Europium in an ionic liquid; Spectroscopic investigations in 1-methyl-3-butylimidazolium hexafluorophosphate, *Inorg. Chem.*, 42, 1726–1733, 2003.
37. Nockemann, P., Binnemans, K., Driesen, K., Purification of imidazolium ionic liquids for spectroscopic applications, *Chem. Phys. Lett.*, 415, 131–136, 2005.
38. Endres, F., El Abedin, S.Z., Borissenko, N., Probing lithium and alumina impurities in air- and water stable ionic liquids by cyclic voltammetry and *in situ* scanning tunneling microscopy, *Zeits. Phys. Chem.—Int. J. Res. Phys. Chem. Chem. Phys.*, 220, 1377–1394, 2006.
39. Choudhury, A.R., Winterton, N., Steiner, A., Cooper, A.I., Johnson, K.A., In situ crystallization of low-melting ionic liquids, *J. Am. Chem. Soc.*, 127, 16792–16793, 2005.
40. Burrell, A.K., Del Sesto, R.E., Baker, S.N., McCleskey, T.M., Baker, G.A., The large scale synthesis of pure imidazolium and pyrrolidinium ionic liquids, *Green Chem.*, 9, 449–454, 2007.
41. Swatloski, R.P., Holbrey, J.D., and Rogers, R.D., Ionic liquids are not always green: hydrolysis of 1-butyl-3-methylimidazolium hexafluorophosphate, *Green Chem.*, 5, 361–363, 2003.

chapter twelve

Raman spectroscopy, ab initio *model calculations and conformational equilibria in ionic liquids*

Rolf W. Berg

Contents

12.1 Introduction

Generally room-temperature ionic liquids (ILs) consist mostly of ions [1]. In these liquids the Coulomb interaction plays a major role, in contrast to the situation in ordinary molecular liquids where only dipolar and higher-order multipolar electrostatic interactions occur. The long-range nature of the Coulomb force tends to make the melting points of ionic crystals much higher than for molecular crystals. The ILs are extraordinary with their low melting points and reasons for this will be considered later.

The intermolecular interactions in ILs are of great importance for their general use, and the basic knowledge in the physical chemistry of these solvent systems is under intensive growth. The number of publications is rising steeply and it is difficult to get an overview. Early studies probed the nature of interactions in so-called first-generation chloroaluminate ionic liquids [2], and now also information becomes available for second generation, air-stable systems. Many IL studies have covered theoretical aspects [3–7] and x-ray crystallography of the frozen melts [8–12], and solubilities of gases in ILs [13–15].

Hydrogen-bonding is known to occur between the cations and the anions in most ionic liquids. As demonstrated by several investigators, by for example, IR and Raman spectroscopies on systems containing 1-alkyl-3-methylimidazolium, all three ring protons at C2, C4, and C5 (see Figure 12.1) form strong hydrogen bonds to halide ions, and the C2 proton is hydrogen bonded to the $[BF_4]^-$ anion in neat $[C_2C_1Im][BF_4]$ [16].

The degree of hydrogen bonding between the ring-bound hydrogen atoms and the anion seems to change significantly when going, from, for example, a neat chloride to a hexafluorophosphate. IR spectroscopy has provided detailed information on the hydrogen-bonded interaction between water molecules and ILs, in for example $[C_4C_1Im][PF_6]$ and $[C_4C_1Im][BF_4]$ (see Refs. 15,17–19).

Raman spertroscopy of $[C_4C_1Im]Cl$—$EtAlCl_2$ iILs (Et = ethyl) has shown that the distribution of the ethylchloroaluminate(III) species follows a chlorobasicity pattern similar to that found in alkali chloroaluminate(III) ILs [20,21]. Hence, in these ILs Cl^- and $[EtAlCl_3]^-$ ions are found when the liquid is chlorobasic. In moderately acidic ILs, $EtAlCl_3^-$ and $[Et_2Al_2Cl_5]^-$ ions are present, and in highly acidic compositions $[Et_3Al_3Cl_7]^-$ and $Et_2Al_2Cl_4$ are important components. Similar results are found for $[C_2C_1Im]Cl$–Et_2AlCl IL systems [22]. Closer inspection of the Raman spectra of the acidic $[C_2C_1Im]Cl$–$EtAlCl_2$ ILs has revealed that the species $[AlCl_4]^-$, $[EtAl_2Cl_6]^-$, Et_2AlCl and $Et_3Al_2Cl_3$ are present [22]. Exchange of ethyl and chloride ligands must be taking place; a quite likely behavior for such complex mixtures.

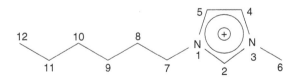

Figure 12.1 Numbering scheme in the 1-hexyl-3-methylimidazolium cation, $[C_6C_1Im]^+$, showing the three ring protons H2, H4, and H5.

The ILs interact with surfaces and electrodes [23–25], and many more studies have been done that what we can cite. As one example, *in situ* Fourier-transform infrared reflection absorption spectroscopy (FT-IRAS) has been utilized to study the molecular structure of the electrified interphase between a 1-ethyl-3-methylimidazolium tetrafluoroborate $[C_2C_1Im][BF_4]$ liquid and gold substrates [26]. Similar results have been obtained by surface-enhanced Raman scattering (SERS) for $[C_4C_1Im][PF_6]$ adsorbed on silver [24,27] and quartz [28].

Mixed systems of organic molecules and ILs that form separate phases (by thermomorphic phase separation) have been also studied by Raman spectroscopy [29].

12.2 Brief introduction to Raman spectroscopy

12.2.1 Basics

The effect of Raman scattering may be defined as instantaneous inelastic scattering of electromagnetic radiation (light) and was discovered in 1928 by Indian physicists Raman and Krishnan [30–33]. When a photon collides with a sample, it may be elastically scattered (called Rayleigh scattering) or an amount of energy may be exchanged with the sample (Stokes or anti-Stokes Raman processes) as shown schematically in the quantum energy level diagram in Figure 12.2.

During the IR absorption process, a quantum of radiation (a photon) of a particular energy E and with a frequency $v(E = hv)$ is absorbed (h is Planck's constant). During the absorption the molecular system undergoes a transition from the ground state (quantum number $v = 0$) to an excited state ($v = 1$),

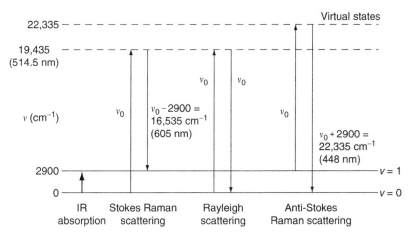

Figure 12.2 The relationships between IR absorption, Rayleigh, and Raman scattering.

in the present case, for example corresponding to a CH_3 group C–H bond stretching with a wave number shift of 2900 cm^{-1}. In contrast to this, during Rayleigh and Raman scattering, an exiting photon of much higher energy hits the molecular system and raises it to a virtual state, from where it *immediately* falls back. There are two possibilities, here illustrated with green Ar^+ light of 514.5 nm wavelength corresponding to 19,435 cm^{-1}. In so-called Stokes Raman scattering (not so likely), the system falls back to the $v = 1$ state (emitting a 16,535 cm^{-1} photon), or in Rayleigh scattering (more likely) to the $v = 0$ ground state (emitting light at ~19,435 cm^{-1}), producing the so-called Rayleigh wing. If the system is starting from the $v = 1$ state (not so likely at room temperature because of the Boltzman distribution), similar transitions can happen. Now also a so-called anti-Stokes Raman process is possible producing photons at 22,335 cm^{-1}. Most Raman spectroscopy studies report data corresponding to Stokes Raman transitions. Samples (or impurities therein) having energy states near the *virtual* ones (here at e.g., ~19,435 cm^{-1}) may absorb photons from the incident light and later reemit the light as a broad intensive background called *fluorescence*.

Accordingly, the outgoing photon has less or more energy than the incoming one. A Raman process corresponds to a (fundamental) transition among certain group vibrational states. For the Raman spectral band to occur with a significant intensity, the molecular bond stretching or angle deformation vibration must cause a *change* in the polarizability of the molecule. The ensemble of light scattering bands constitute the Raman spectrum. Stokes-shift Raman spectra are most often measured; that is, the scattered photons have lower frequency than the incident radiation.

Dramatic improvements in instrumentation (lasers, detectors, optics, computers, and so on) have during recent years raised the Raman spectroscopy technique to a level where it can be used for *species specific'* quantitative chemical analysis. Although not as sensitive as, for example IR absorption, the Raman technique has the advantage that it can directly measure samples inside ampoules and other kinds of closed vials because of the transparency of many window materials. Furthermore, with the use of polarization techniques, one can derive molecular information that cannot be obtained from IR spectra. Good starting references dealing with Raman spectroscopy instruments and lasers are perhaps [34–38].

Raman and IR spectroscopies are closely interrelated in that they both depend on characteristic *group* molecular motions in the sample that give rise to the vibrational bands in the spectra. As an example, bands occurring near 2950 cm^{-1} can often be assigned to aliphatic C–H stretching transitions (although sometimes in *Fermi-resonance* with overtones and other nonfundamental transitions). So-called empirical group frequency charts are available, specifying *fingerprint* bands, that may be used to identify pure materials or the presence of a particular component in a mixture (see Refs. 39–42).

Although similar transitional energy ranges occur in IR and Raman spectroscopies, different selection rules govern the intensities in Raman scattering and IR absorption spectra. Hence both types of spectra are often required to fully characterize a substance: A necessary requirement for a molecular motion (such as a vibration, rotation, rotation/vibration, or lattice normal mode) to be measurable in IR spectra it is needed that an oscillating dipole moment is produced during the vibration (in Raman the motion within the molecular system should vary the polarizability). Combinations, differences, or overtones of these transitions can occur, but normally only weakly. The selection rules of the transitions are described in quantum mechanics and group theory (see, for example, Refs. 31–33,43,44).

12.2.2 *Experimental fluorescence- and Fourier-transform-Raman spectroscopy instrumentation*

Applications of Raman spectroscopy in analytical chemistry have been limited by the presence of fluorescence from some samples (or from some impurities in the samples). In case of strong fluorescence the use of less-energetic near-IR lasers for the excitation is often a requirement. FT-Raman instruments have been developed, that successfully apply, for example, ~1064 nm laser excitation (from solid state Nd-YAG or Nd-YVO$_4$ lasers) to avoid fluorescence [45]. The advantage of Raman spectroscopy over IR and other analytical techniques (when the fluorescence problems can be circumvented) stems from the ability of Raman to identify discrete species *in situ*. Raman spectra can be obtained directly from samples of any phase, in for example, glass cells. With a minimum effort, temperature and pressure limitations can be overcome. The polarization properties of the Raman scattered light may be employed to select only the isotropic intensity of the symmetric vibrational modes, thereby helping conclusive assignment of the spectra.

The Raman effect is weak, perhaps only 10^{-8} of the photons hitting the sample are scattered in Raman. The use of high-power lasers (to circumvent the low scattering efficiency) often results in sample decomposition, and fluorescence interference from impurities must be considered a likely problem for visible light. Recent development of charge coupled detectors (CCDs) and notch filters have revolutionized the Raman technique for samples that do not emit much fluorescence. Sampling through a microscope, under high magnification, is an effective way to collect Raman light over a large solid angle, and then only minute sample quantities are necessary.

Room temperature ILs have been the object of several Raman spectroscopy studies but often ILs emit intensive broad fluorescence. In our own experiments, the use of visible laser light (green 514.5 nm or red 784 nm) resulted in strong fluorescence [29,46]. Similar observations have been reported for many IL systems. Our experimental spectra needed to be obtained by use of a 1064 nm near-IR exciting source (Nd-YAG laser at 100 mW of power). The scattered light was filtered and collected in a Bruker

IFS66-FRA-106 FT-Raman spectrometer equipped with a liquid-N_2 cooled Ge-diode detector. Samples were in small glass capillary tubes at ~23°C. The spectra were calculated by averaging ~200 scans followed by apodization and fast-Fourier-transformation to obtain a resolution of ~2 cm^{-1} and a precision better than 1 cm^{-1}. The spectra were not corrected for (small) intensity changes in detector response versus wavelength.

12.3   Brief introduction to ab initio *model calculations*

Ab initio and semiempirical molecular orbital (MO) model calculations have become an efficient way to predict chemical structures and vibrational (i.e., Raman scattering and IR emission) spectra. We and others have used such approaches to better understand certain features of the spectra, as explained in the following. The basic principles underlying *ab initio* model calculations have been described in many textbooks and papers (see for example Refs. 44,47,48). Applications in relation to ILs and similar systems have also been reported, as discussed later.

The MO calculations may nowadays be performed with the GAUSSIAN 03W program package [49]. A guessed molecular geometry (conformation) is used as input together with some kind of approximation to the atomic orbitals (so-called basis sets, often sums of Gaussian functions). The total energy is minimized by restricted Hartree–Fock (RHF), Møller–Plesset (MP2) or density functional theory (DFT) principles, using, for example, third-order Becke–Lee–Yang–Parr (B3LYP) procedures [44,48,49]. Common basis sets are the split valence 6-31+G(d,p) sets including diffuse orbitals (d) augmented with Pople's polarization functions (*p*) [49]. The molecular ions are commonly assumed to be in a hypothetical *gaseous free state* and without any preassumed symmetry, but some calculations also involve better approximations to real systems. After optimization, giving a geometry with a minimum energy, perhaps not a global one, vibrational frequencies and intensities (spectra) and eigenvectors for the normal modes are calculated and displayed on a computer screen, to identify the dominating motions. The frequencies (wave numbers) are correlated with the Raman and IR bands.

The calculated and experimental vibrational spectra are in more or less good agreement. The wave number scale is often calculated as slightly too high, due to the lack of good modeling of the orbitals and interactions with the surroundings. In the gas phase empirical scale factors of ~0.95 are sometimes used to get fairly accurate wave number fits. A scaling factor of 1 was used in this work.

12.4   *Case study on Raman spectroscopy and structure of imidazolium-based ionic liquids*

As mentioned above, vibrational spectroscopy is known to be a very powerful tool in the study of molecular stuctures and interionic interactions in

ILs [18,19,50]. This is especially so when done in combination with crystal structure studies, as explained in the following:

To illustrate the situation, we start our discussion with the example of the alkyl-methylimidazolium liquids, from $[C_2C_1Im]^+$ to $[C_{18}C_1Im]^+$, and a number of different anions. Although other techniques such as IR spectroscopy, x-ray and neutron diffraction studies have been used to study these ions in the liquid or solid state or at surfaces [12,51–61], a real gain in our understanding came with the combination of crystal structure solution, Raman spectroscopy and *ab initio* DFT calculations [50,62]. We concentrate the story on the instructive example of the l-butyl-3-methylimidazolium cation, $[C_4C_1Im]^+$ (see Figure 12.1, without carbon atoms 11 and 12), that makes a number of ILs with varying properties, depending on the different anions. The two prototype ILs $[C_4C_1Im][BF_4]$ and $[C_4C_1Im][PF_6]$ have already been used extensively in fundamental investigations as well as in practical applications. Therefore, the elucidation of their crystal and liquid structures was an important first step for the understanding of ILs in general [50].

The most fundamental question about ILs to be discussed is: Why are ILs liquids at the ambient temperature, despite the fact that they are composed solely of ions? This question can be answered as described in the following:

$[C_4C_1Im]Cl$ and $[C_4C_1Im]Br$ are crystals at room temperature, while $[C_4C_1Im]I$ is an IL (melting point −72°C [63]). A typical ionic crystal such as NaI only melts at ~660°C. By cooling $[C_4C_1Im]Cl$ and $[C_4C_1Im]Br$ liquids below their melting points, supercooled liquids are easily obtained. Crystals could be grown of the $[C_4C_1Im]Cl$ and $[C_4C_1Im]Br$ salts and x-ray diffraction could be used to determine the crystal structures. These systems thus comprised unique systems for studying the structure of the $[C_4C_1Im]^+$ cation in the liquid and crystalline states.

A crystal polymorphism was discovered in the $[C_4C_1Im]Cl$. It adopts a monoclinic (melting point ~41°C) and an orthorhombic (melting point ~66 °C) crystal structure [57,58,64,65]. In the following, we use the Hamaguchi notation *Crystal (1)* and *Crystal (2)*. The crystal structures were determined by x-ray diffraction of $[C_4C_1Im]Cl$ *Crystal (1)* $[C_4C_1Im]Br$ at room temperature [57] and independently, of $[C_4C_1Im]Cl$ *Crystal (2)*, as well as that of $[C_4C_1Im]Br$ at −100°C [58]. The structures determined by different authors agreed quite well with each other, taking into account that the lattice constants vary with temperature. The molecular structure of the $[C_4C_1Im]^+$ cation in $[C_4C_1Im]Cl$ *Crystal (2)* is different from that in (1) but it was the same as that in $[C_4C_1Im]Br$, as also proved later by the Raman spectra.

The $[C_4C_1Im]^+$ cations in the two polymorphs were found predominantly to differ with respect to conformation: The structural results showed that the polymorphism is due to a rotational isomerism of the butyl group of the $[C_4C_1Im]^+$ cation around C7-C8, as defined in Figure 12.1. In the monoclinic polymorph, the butyl chain is in *anti* (or trans) conformation around C7-C8, and in the orthorhombic polymorph it is *gauche* around C7-C8. The conformational difference reveals itself in the rotation of the butyl chain around the C7-C8 bond, that differed by 106.16° between the two conformers [58]. The C8-C9 conformation was found to be *anti* in both polymorphs. In a convenient and obvious notation, these two conformers of the $[C_4C_1Im]^+$ cation are *here* referred to as the anti-anti (AA) and the gauche-anti (GA) forms (Hamaguchi et al. denote them as TT and GT). Also the crystal structure of $[C_4C_1Im]Br$ has been reported [58].

The structure of the monoclinic $[C_4C_1Im]Cl$ *Crystal (1)* is shown in Figure 12.3.

Details of structural data are available from the Cambridge Crystallographic Data Centre [66]. The crystal belongs to space group P21/n with

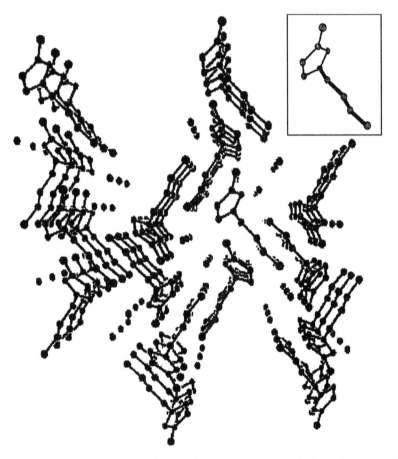

Figure 12.3 Crystal structure of $[C_4C_1Im]Cl$ *Crystal (1)* viewed along the a axis. Only carbon atoms, nitrogen atoms, and chloride anions are shown. The AA conformation of the $[C_4C_1Im]^+$ cation is shown in the inset. The butyl group C–C bonds are shown as thick bars. Note that the cations and chloride anions form characteristic columns along the crystal a axis. (Adapted from Hamaguchi, H., and Ozawa, R., *Adv. Chem. Phys.*, 131, 85–104, 2005. With permission.)

a = 9.982(10)Å, b = 11.590(12)Å, c = 10.077(11) Å, and β = 121.80(2)°. Both the $[C_4C_1Im]^+$ cations and the chloride anions form separate columns extending along the crystal **a** axis. The imidazolium rings are all planar pentagons. The stretched n-butyl group of the $[C_4C_1Im]+$ cation takes an AA conformation with respect to the C7-C8 and C8-C9 bonds, as shown in the inset of Figure 12.3. The butyl groups stack together (aliphatic interaction) and form columns extending along the **a** axis, in which all the imidazolium ring planes are parallel with one another. Two types of cation columns with different orientations exist, the planes of the imidazolium rings in the two different columns making an angle of 69.5°. Zig-zag chains of Cl⁻ anions directed in the a direction are accommodated in channels formed by four cation columns, of which two opposite columns have the same orientation. The three shortest distances between Cl⁻ anions in the zig-zag chain were 4.84 Å, 6.06 Å, and 6.36 Å and these distances are much larger than the sum of the van der Waals radii of Cl⁻ (3.5 Å). There seems to be no specific interaction among the Cl⁻ anions, and they are likely to be aligned under the effect of Coulombic forces. The chloride ion is very close to the hydrogen H2 in the ring (2.55 Å), and to the two methylene protons on C7 (2.72 and 2.73 Å) [57,58], meeting the criteria for relatively strong hydrogen bonds [67,68]. Similarly strong hydrogen bonds are observed in the orthorhombic form [58]. Also other crystal structures, for example, of the 1-ethyl-3-methylimidazolium chloride ($[C_2C_1Im]C_1$) [15], the tetrafluoroborate ($[C_2C_1Im][BF4]$) and other salts [69] have been reported.

The crystal structure of orthorhombic $[C_4C_1Im]Br$ (melting point 77.6°C [65]) is shown in Figure 12.4.

The detailed structural data are available from the Cambridge Crystallographic Data Center [66]. The $[C_4C_1Im]Br$ crystal belongs to the space group ***Pna2₁*** with a = 10.0149(14) Å, b = 12.0047(15) Å, c = 8.5319(11) Å. As for the $[C_4C_1Im]Cl$ *Crystal (1)*, the cations and anions form separate columns extending along the *a* axis. In $[C_4C_1Im]Br$ the *n*-butyl group takes a GA conformation with respect to the C7–C8 and C8–C9 bonds (see inset of Figure 12.4). Only one kind of cation column is found. The imidazolium rings are stacked so that the N–C–N moiety of one ring interacts with the C=C portion of the adjacent ring. The adjacent ring plane can be obtained by rotation of the ring by about 73° around an axis involving the two N atoms. The zig-zag chain of Br⁻ anions resides in the channel produced by four cation columns, extending in the *a* direction. The shortest three Br⁻–Br⁻ distances (4.77, 6.55, and 8.30 Å) are all longer than the sum of the van der Waals radii (3.7 Å). This indicates that there is no specific interaction among the Br⁻ anions and that the zig-zag molecular arrangement is a result of Coulombic interactions.

12.5 Raman spectra and structure of $[C_4C_1Im]^+$ liquids

The information obtained from the study of the $[C_4C_1Im]Cl$ crystals can be used as a basis to better understand the liquid structure of the $[C_4C_1Im]X$

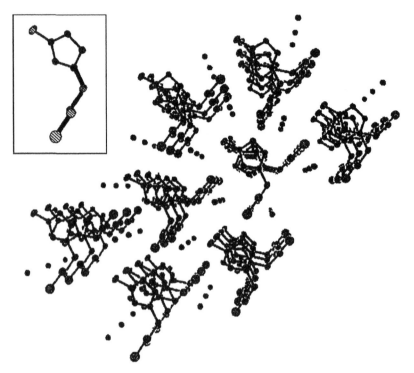

Figure 12.4 Crystal structure of [C$_4$C$_1$Im]Br viewed in the direction of the a axis. Only carbon atoms, nitrogen atoms, and bromide anions are shown. The gauche-anti (GA) conformation of the [C$_4$C$_1$Im]$^+$ cation is shown in the inset. The butyl group C-C bonds are shown as thick bars. (Adapted from Hamaguchi, H., and Ozawa, R., *Adv. Chem. Phys.,* 131, 85–104, 2005. With permission.)

ILs (X is an anion). It is well-known that Raman spectroscopy facilitates comparative studies of the structures in crystals and liquids. Raman spectra of [C$_4$C$_1$Im]Cl *Crystals (1) and (2)*, and [C$_4$C$_1$Im]Br by Hamaguchi et al. [50,57,59,64,70] are shown in Figure 12.5.

As seen, the two polymorphs of [C$_4$C$_1$Im]Cl gave distinct Raman spectra differing considerably, while those of [C$_4$C$_1$Im]Cl *Crystal (2)* and [C$_4$C$_1$Im]Br were almost identical. These findings are consistent with the x-ray diffraction experimental results. The halogen anions are inactive in Raman scattering—except for the lattice vibrations, that are observed in the wave number region < 200 cm^{-1} [71]. Therefore, all the Raman bands seen in Figure 12.5 can be ascribed to the [C$_4$C$_1$Im]$^+$ cation. Figure 12.6 was accordingly interpreted to indicate that the [C$_4$C$_1$Im]$^+$ cation takes two different conformations in those salts. To be in accordance with the x-rays results, at least the cation must adopt the same molecular conformation in [C$_4$C$_1$Im]Cl *Crystal (2)* and [C$_4$C$_1$Im]Br, and a different one in [C$_4$C$_1$Im]Cl *Crystal (1)*. In this way it emerged that the Raman spectral differences in Figure 12.5 most likely

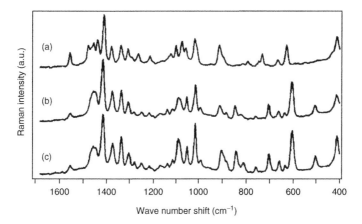

Figure 12.5 Raman spectra of (a) [C₄C₁Im]Cl *Crystal (1)*, (b) [C₄C₁Im]Cl *Crystal (2)*, and (c) [C₄C₁Im]Br crystals. (a) differs from (b) and (c) (Adapted from Hamaguchi, H., and Ozawa, R., *Adv. Chem. Phys.*, 131, 85–104, 2005. With permission.)

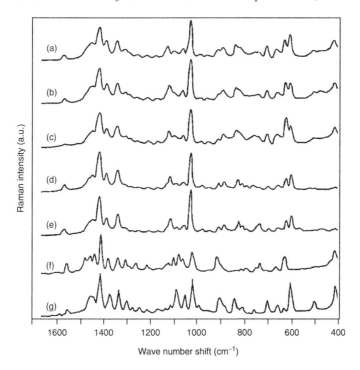

Figure 12.6 Raman spectra of liquid [C₄C₁Im]X, where X = Cl (a), Br (b), I (c), [BF₄] (d), and [PF₆] (e). The anion bands in (d) and (e) have been deleted (Hamaguchi, H., and Ozawa, R., *Adv. Chem. Phys.*, 131, 85–104, 2005). Spectra of [C₄C₁Im]Cl *Crystal (1)* and crystalline [C₄C₁Im]Br, respectively, are included as (f) and (g), for reference purposes (Adapted from Hamaguchi, H., and Ozawa, R., *Adv. Chem. Phys.*, 131, 85–104, 2005. With permission.)

originated from the rotational isomerism around the C7-C8 (AA and GA isomerism) of the butyl chain of the $[C_4C_1Im]^+$ cation [64,70].

Raman spectra of liquid $[C_4C_1Im]X$ (X = Cl, Br, I, $[BF_4]$, $[PF_6]$) measured at room temperature are shown in Figure 12.6.

The spectra of $[C_4C_1Im]Cl$ *Crystal (1)* and $[C_4C_1Im]Br$ are also included for reference purposes. Spectra for fluids $[C_4C_1Im]Cl$ and $[C_4C_1Im]Br$ were taken from supercooled liquids. The Raman spectra of the $[C_4C_1Im]X$ liquids were surprisingly alike. One should note that the Raman spectral bands of the separate $[BF_4]^-$ and $[PF_6]^-$ anions that are already well-known [41] have been *deleted in Figure 12.6*. However, from the similarity of the spectra it seems that the structural properties of the $[C_4C_1Im]^+$ cation in these liquids are very similar. But what else can be deduced from the spectra?

12.6   Normal mode analysis and rotational isomerism of the $[C_4C_1Im]^+$ cation

To pursue this question further, Ozawa et al. [70] have performed DFT calculations with Gaussian98 at the B3LYP/6-31G+** level. In the calculation, the structures of AA and GA forms of $[C_4C_1Im]^+$ were optimized in the vicinity of the determined x-ray crystal structures for $[C_4C_1Im]Cl$ *Crystal (1)* and $[C_4C_1Im]Br$, respectively. The structures of the optimized $[C_4C_1Im]^+$ cations in the two crystals are depicted in Figure 12.7, together with the experimental spectra (in a limited wave number region of 1000–400 cm^{-1}).

The calculated fundamental frequencies and intensities were included in Figure 12.7 as thick vertical bars. As seen the calculated *bar*-spectra reproduced the observed spectra quite well.

The normal mode calculation was used to elucidate the rotational isomerization equilibrium of the $[C_4C_1Im]X$ liquids. In the wave number region near 800–500 cm^{-1}, where ring deformation bands are expected, two Raman bands appeared at ~730 cm^{-1} and ~625 cm^{-1} in the $[C_4C_1Im]Cl$ *Crystal (1)*. In the $[C_4C_1Im]Br$ these bands were not found. Here instead, another couple of bands appeared at ~701 and ~603 cm^{-1}. To assist the interpretation of the spectra, the normal modes of vibrations calculated by Hamaguchi et al. [50] are shown in Figure 12.8.

It shows modes for the $[C_4C_1Im]^+$ ion of the geometry of $[C_4C_1Im]Cl$ *Crystal (1)* at 735 and 626 cm^{-1}; and similarly the modes for $[C_4C_1Im]Br$ occurring at 696 cm^{-1} and 596 cm^{-1}. Obviously the 626 cm^{-1} band of $[C_4C_1Im]Cl$ *Crystal (1)* and the 596 cm^{-1} band of $[C_4C_1Im]Br$ originate from similar kind of *ring* deformation vibrations, but they have different magnitudes of the coupling with the CH_2 rocking motion of the C8 carbon (encircled in Figure 12.8). It thus seems that more intensive coupling occurs between (i) the CH_2 rocking motion and (ii) the ring deformation vibrations in the GA form (596 cm^{-1}) than in AA form (626 cm^{-1}), resulting in an overall lower frequency of the mode and therefore a lower–wave number position of the Raman band [50].

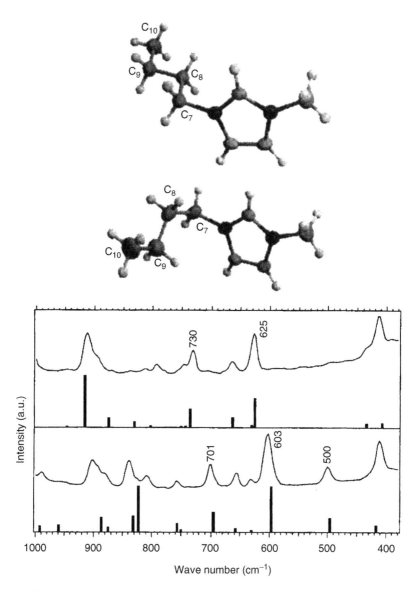

Figure 12.7 Optimized structures of the $[C_4C_1Im]+$ cation in the two crystals. Experimental (continuous lines) and calculated Raman spectra (solid vertical bars) of $[C_4C_1Im]Cl$ *Crystal (1)* (above) and $[C_4C_1Im]Br$ (below) are shown (Adapted from Hamaguchi, H., and Ozawa, R., *Adv. Chem. Phys.,* 131, 85–104, 2005; Ozawa, R., Hayashi, S., Saha, S., Kobayashi, A., and Hamaguchi, H., *Chem. Lett.* 32, 948–949, 2003. With permission.)

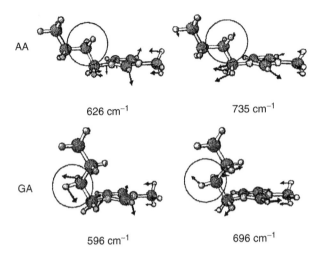

AA

626 cm⁻¹   735 cm⁻¹

GA

596 cm⁻¹   696 cm⁻¹

Figure 12.8 Calculated normal modes of key bands of the AA and GA forms of the $[C_4C_1Im]^+$ cation. The arrows indicate vibrational amplitudes of atoms. The C8 methylene group is surrounded by a circle. Obviously it appears that the CH_2 rocking vibration is coupled to the ring modes only for the GA conformer, thereby lowering the frequencies. (Adapted from Hamaguchi, H., and Ozawa, R., *Adv. Chem. Phys.*, 131, 85–104, 2005. With permission.)

By comparing their normal coordinate analysis results and their *liquid* experimental Raman spectra in Figure 12.6, Hamaguchi et al. [50,57,59,64,70] concluded that the two rotational isomers AA and GA must coexist in the IL state. According to the Raman spectra of all the liquids in Figure 12.6, both the key bands for the AA conformer (625 cm⁻¹ and 730 cm⁻¹ bands), and for the GA conformer (603 cm⁻¹ and 701 cm⁻¹ bands), respectively, appeared in the spectra. Therefore, the two isomers of rotational freedom around the C7-C8 and C8-C9 bonds, AA and GA, must coexist in these $[C_4C_1Im]X$ liquids.

Furthermore, the observed relative intensity of the 625 cm⁻¹ band to that of the 603 cm⁻¹ band should be correlatable with the AA/GA population ratio of the conformation equilibrium. The observed ratios depended slightly on the anion: For the halides, it seems to increase in the order $[BF_4]^- \approx [PF_6]^- \approx Cl^- < Br^- < I^-$ [50].

During our work on $[C_4C_1Im]^+$ and $[C_6C_1Im]^+$, we have repeated the experiments and calculations for the $[C_4C_1Im]^+$ cation and found the results of Hamaguchi et al. to be essentially reproducible (details explained in Ref. 46). Our calculated Raman spectra in the whole range for the AA and GA conformers of $[C_4C_1Im]^+$ are shown in Figures 12.9 and 12.10.

Our assignments (approximate descriptions of the modes giving origin to the Raman bands) are listed in Table 12.1, based on the calculated vibrational frequencies and we communicate the intensities of the IR and Raman bands.

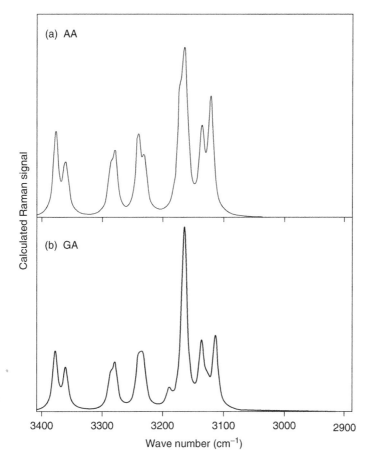

Figure 12.9 Calculated Raman spectra of two conformers of the [C$_4$C$_1$Im]$^+$ cation in the range between 3400 and 2900 cm^{-1}.(a) the AA conformer and (b) the GA conformer. (Data from Berg, R. W., Unpublished results, 2006.

The movements were depicted on a PC screen and assignments were derived using the Gaussian03W software.

Our recalculated modes of the [C$_4$C$_1$Im]$^+$ cation were obtained with somewhat higher frequencies: the modes at 626, 735, 596, and 696 cm^{-1} by Hamaguchi et al. in Figure 12.8 became 636 and 749 cm^{-1} for AA mode and 622 and 713 cm^{-1} for GA mode (see Figure 12.11).

According to the calculated minimum energy E_e of the conformers, the GA was more stable than the AA conformer, but at 298.15 K the *Gibbs* energy of the AA conformer was 0.168 kJ mol^{-1} less than that of the GA conformer, indicating 52% of AA versus 48% of GA or almost equal amounts of the two conformers at equilibrium at room temperature [46]. A higher difference

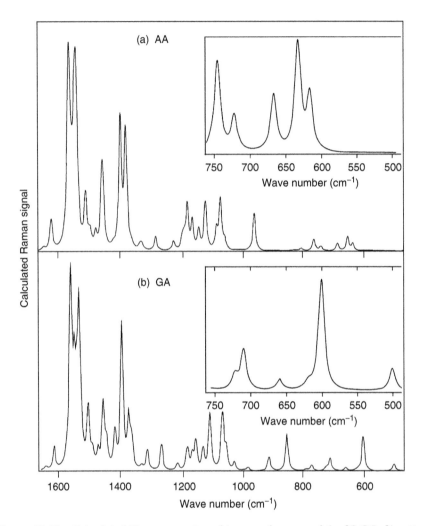

Figure 12.10 Calculated Raman spectra of two conformers of the $[C_4C_1Im]^+$ cation in the range between 1650 and 400 cm^{-1}: (a) the AA conformer and (b) the GA conformer. (Data from Berg, R. W., Deetlefs, M., Seddon, K. R., Shim, I., and Thompson, J. M., *J. Phys. Chem. B*, 109, 19018–19025, 2005; Berg, R. W., Unpublished results, 2006. With permission.)

between the AA and the GA energies was found in other calculations [72]. These results are consistent with the observation of both conformers being simultaneously present in the spectra of the $[C_4C_1Im]^+$ liquids as observed by Ozawa et al. [70] (see Figure 12.6).

Our obtained experimental Raman signals for the $[C_4C_1Im]^+$ cation liquids are given in Figure 12.12 and the assignments are specified in Table 12.2, based on the calculations, some of which are summarized in Table 12.3.

Table 12.1 Approximate Descriptions[a] of Vibrational Frequencies (IR and Raman Bands) as Determined in MP2 Calculations for the [C$_4$C$_1$lm]$^+$ Cation with the Butyl Group in the AA or GA Conformations

Mode	Butyl group in the anti-anti (AA) conformation		Butyl group in the gauche-anti (GA) conformations	
	ν/(cm^{-1})	Approximate description	ν/(cm^{-1})	Approximate description
1	30.4	N-C7 tor	27.5	N-C7 tor
2	58.3	N-C6 tor	58.3	N-C6 tor
3	74.3	C7-C8 tor	76.6	N-C6 tor + C7-C8 tor
4	81.8	N-C6 tor + N-C7-C8 bend	82.5	C7-C8 tor
5	116.5	N-C7 tor	156.7	N-C6 oopl bend + C8H$_2$ rock
6	203.6	Chain def + N-C6 oopl bend	205.1	Chain def + N-C6 oopl bend
7	250.9	N-C6 + N-C7H$_2$ oopl ooph bend	251.1	N-C6 + N-C7H$_2$ oopl ooph bend
8	252.8	C9-C10 tor	258.4	C9-C10 tor
9	278.8	C7H$_2$ rock + C6H$_3$ ipl iph bend	296.1	C7H$_2$ rock + C6H$_3$ ipl iph bend
10	327.1	Ring wag + chain def	333.3	Ring wag + chain def
11	407.0	Ring rot + C7H$_2$ rock + C6H$_3$ rock	418.0	Ring rot + C7H$_2$ rock + N-C6H$_3$ ipl rock
12	441.0	N-C7 oopl bend + N-C6 oopl ooph bend + chain bend + CH$_2$ wag + ring rot	503.7	N-C7 oopl bend + N-C7-C8-C9 angles bend
13	619.6	Ring oopl def + C7H$_2$ rock + C7-C8-C9 bend	603.9	N-C6 N-C7 iph str + ring oopl def + C8H$_2$ rock + N-C7-C8 bend.
14	636.2	Ring def (C2-H oopl bend) + N-C6 N-C7 iph str + C7H$_2$ rock + C7-C8-C9 bend	622.8	Ring def (C2-H C4-H iph oopl bend) + C8H$_2$ rock + N-C7-C8 bend.
15	670.2	N-C6 str + ring def (N1 and H on C2 oopl ooph departure) + C8H$_2$ wag + N-C7-C8 bend	662.8	Ring def (bend around line NN) + C8H$_2$ rock + N-C7-C8 bend
16	725.5	Ring C-H oopl bend (bend around NN line)	712.8	N-C6 N-C7 ooph str + ring ipl def + C8H$_2$ rock + C7-C8 tor
17	748.8	N-C6 N-C7 ooph str + ring ipl def + N-C7-C8 and C7-C8-C9 bend	725.1	Ring C-H oopl bend (bend around NN line)
18	754.6	Chain CH$_2$ sci + rock	773.5	Chain CH$_2$ rock + ring C-H oopl iph bend (umbrella)
19	791.7	Ring C-H oopl iph bend (umbrella)	790.4	Ring C-H oopl iph bend (umbrella)

(continued)

Table 12.1 (Continued)

Mode	Butyl group in the anti-anti (AA) conformation		Butyl group in the gauche-anti (GA) conformation	
	v/(cm^{-1})	Approximate description	v/(cm^{-1})	Approximate description
20	813.7	C4-H C5-H oopl ooph bend (twi)	811.5	C4-H C5-H oopl ooph bend (twi)
21	819.3	Chain CH$_2$ sci	853.9	Chain CH$_2$ sci
22	945.3	Chain def + C10H$_2$ rock	912.4	Chain def + C10H$_3$ rock
23	969.8	Chain def (CH$_2$ twi + rock)	981.5	Chain def (CH$_2$ sci + rock)
24	1043.6	Ring def + chain def	1025.7	Ring def + chain def
25	1057.3	Ring def + N-C6 str + C4-H C5-H ipl iph bend	1051.9	Ring def + N-C6 str + N-C7 str + C7-C8 str
26	1070.3	C4-H ipl bend + C7-C8 str	1064.0	C4-H C5-H ipl bend (sci) + N-C6 str + N-C7 str + C7-C8 str
27	1106.6	Chain def	1104.9	Chain def
28	1128.4	C6H$_3$ ipl rock + ring def	1127.4	C6H$_3$ ipl rock + ring def
29	1149.9	C4-H C5-H ipl bend (sci)	1151.0	C4-H C5-H ipl bend (sci)
30	1167.0	Chain def (C-C str)	1162.6	Chain def (C-C str)
31	1176.9	C6H$_3$ def (oopl rock)	1176.4	C6H$_3$ def (oopl rock)
32	1182.7	C2-H ipl bend, C6H$_3$ def, chain CH$_2$ def	1180.5	C2-H ipl bend, C6H$_3$ def, chain CH$_2$ def
33	1212.3	C6-N C7-N ooph str + ring C-H ipl bend	1211.1	C6-N C7-N ooph str + ring C-H ipl bend
34	1272.5	Ring CH iph ipl bend + chain CH$_2$ def	1263.6	Ring CH iph ipl bend + chain CH$_2$ def
35	1318.4	Ring CH iph ipl bend + chain CH$_2$ wag	1311.0	Ring CH iph ipl bend + chain CH$_2$ def
36	1324.3	Ring CH iph ipl bend + chain CH$_2$ def	1329.2	Ring CH iph ipl bend + chain CH$_2$ def
37	1353.0	Chain CH$_2$ def + ring CH iph ipl bend	1362.5	Chain CH$_2$ def
38	1371.8	C8H$_2$ C9H$_2$ twi	1371.7	Chain CH$_2$ def
39	1388.9	Ring breathing + C7H$_2$ twi	1393.3	Ring breathing + C7H$_2$ twi
40	1413.6	C7H$_2$ C9H$_2$ wag	1416.7	C7H$_2$ wag
41	1443.7	Chain CH$_2$ wag	1445.1	Chain CH$_2$ wag
42	1450.6	Ring asym str + C7H$_2$ twi + C6H$_3$ def	1456.1	Ring asym str + C7H$_2$ twi + C6H$_3$ def
43	1472.7	C10H$_3$ def (umbrella)	1473.0	C10H$_3$ def (umbrella)
44	1490.9	Ring asym str + C6H$_3$ def (umbrella)	1491.1	Ring asym str + C6H$_3$ def (umbrella)

45	1505.4	$C6H_3$ def (umbrella)	1504.8	$C6H_3$ def (umbrella)
46	1534.6	$C6H_3$ def	1531.8	$C8H_2$ def (sci)
47	1539.1	$C7H_2 + C8H_2$ bend (sci)	1534.5	$C6H_3$ def
48	1545.1	Chain CH_2 bend (sci)	1541.3	$C7H_2$ bend (sci)
49	1554.8	$C7H_2 + C8H_2 + C10H_2$ bend (sci)	1549.0	$C8H_2 + C9H_2 + C10H_2$ bend (sci)
50	1559.8	$C10H_3$ def	1559.5	$C10H_3$ def
51	1562.7	$C6H_3$ def	1562.2	$C6H_3$ def
52	1566.6	Chain CH_2 bend (sci)	1565.4	$C8H_2 + C9H_2 + C10H_2$ bend (sci)
53	1620.5	C4-C5 ring str + C4-H + C5-H ipl sym bend	1618.2	C4-C5 ring str + C4-H + C5-H ipl sym bend
54	1645.7	C2-N ring asym str + C2-H ipl bend	1644.3	C2-N ring asym str + C2-H ipl bend
55	3120.7	$C8H_2 + C9H_2$ iph str (sym)	3113.1	$C8H_2 + C9H_2$ iph str (sym)
56	3128.6	$C8H_2 + C9H_2$ ooph str (sym)	3126.5	$C8H_2 + C9H_2$ ooph str (sym)
57	3135.6	$C10H_3$ iph str (sym)	3136.1	$C10H_3$ iph str (sym)
58	3160.9	$C7H_2$ iph str (sym)	3162.8	$C7H_2$ iph str (sym) + $C8H_2 + C9H_2$ ooph str
59	3164.3	$C6H_3$ iph str (sym)	3164.3	$C6H_3$ iph str (sym)
60	3171.6	$C8H_2 + C9H_2$ ooph str (asym)	3165.5	$C7H_2$ sym str + $C8H_2 + C9H_2$ str (asym)
61	3190.6	$C8H_2 + C9H_2$ ooph str (asym)	3189.4	$C8H_2 + C9H_2$ ooph str (asym)
62	3229.5	$C10H_3$ ooph str (asym)	3230.7	$C10H_3$ ooph str (asym)
63	3232.0	$C7H_2$ ooph str (asym)	3234.2	$C7H_2$ ooph str (asym)
64	3240.4	$C10H_3$ ooph str (sym)	3240.1	$C10H_3$ ooph str (sym)
65	3278.6	$C6H_3$ ooph str (asym)	3278.6	$C6H_3$ ooph str (asym)
66	3286.3	$C6H_3$ ooph str (sym)	3286.4	$C6H_3$ ooph str (sym)
67	3359.7	C3-H C4-H ooph str (asym)	3360.7	C3-H C4-H ooph str (asym)
68	3363.0	C2-H str	3361.9	C2-H str
69	3377.4	C3-H C4-H iph str (sym)	3378.5	C3-H C4-H iph str (sym)

Note: Data derived from movements as depicted on a PC screen.

[a] Key of approximate group vibrations: asym = asymmetric, bend = angle bending (scissoring), breathing = all ring bonds iph, def = more complicated deformation of skeleton, ipl = in plane, iph = in phase (symmetric), oopl = out of ring plane, ooph = opposite motion, out of phase (asymmetric), ring = imidazole core, rot = ring rotation, as a wheel, with carbon H atoms, rock = rocking (like V to V by rotation around an axis out of the paper), sci = non-connected scissoring, str = bond stretching, sym = symmetric, tor = torsion around specified bond, twi = twisting of CH_2 group or chain, wag = wagging (like V to v by rotation around an axis in the paper, →).

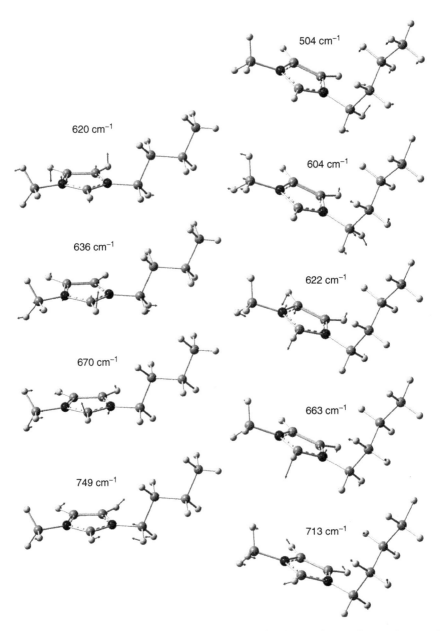

Figure 12.11 **(See color insert following page 224.)**   Some of our calculated normal modes of certain bands of the AA and GA forms of the $[C_4C_1Im]^+$ cation. The arrows indicate vibrational amplitudes of atoms. As found by Hamaguchi et al. (Hamaguchi, H., and Ozawa, R., *Adv. Chem. Phys.*, 131, 85–104, 2005) also our C8 methylene CH_2 rocking vibration was coupled to the ring modes only for the gauche-anti conformer (Berg, R. W., Deetlefs, M., Seddon, K. R., Shim, I., and Thompson, J. M., *J. Phys. Chem. B*, 109, 19018–19025, 2005) and (Berg, R. W., Unpublished results, 2006. With permission.)

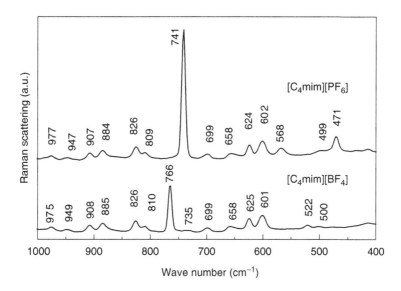

Figure 12.12 Details of FT-Raman spectra of the $[C_4C_1Im][PF_6]$ and $[C_4C_1Im][BF_4]$ ionic liquids at ~25°C (Berg, R. W., Unpublished results, 2006. With permission.) Note that the characteristic bands of the AA and GA forms of the $[C_4C_1Im]^+$ cation are present in both melts, as also found, for example, by Hamaguchi et al. [50].

Note in Figure 12.12 that bands from the $[PF_6]^-$ and $[BF_4]^-$ anions are visible, $\nu_1([PF_6]^-)$ symmetric stretching at 741 cm^{-1}, $\nu_2([PF_6]^-)$ stretching at 568 cm^{-1}, $\nu_5([PF_6]^-)$ symmetric bending at 471 cm^{-1}, $\nu_1([BF_4]^-)$ symmetric stretching at 766 cm^{-1} and $\nu_4([BF_4]^-)$ bending at 522 cm^{-1} [73]. Similar kind of experimental and calculational results for $[C_nC_1Im]X$ liquids were obtained by Carper and others [74–78].

To conclude the situation for $[C_4C_1Im]^+$, Ozawa et al. [50,70] have discovered, by the combined use of x-ray crystallography, Raman spectroscopy, and DFT calculations, that one can use certain Raman bands as key bands to probe the conformation around the C7-C8 bond of the $[C_4C_1Im]^+$ cation.

The calculated bands at ~626–636 cm^{-1} and ~735–749 cm^{-1} are characteristics of the AA conformer (AA conformation around the C7-C8 bond), as compared to the experimental values of ~624 and ~730 cm^{-1} (Table 12.2), whereas the ~596–604 cm^{-1} and ~696–713 cm^{-1} calculated bands are characteristic of the GA conformer (GA conformation). Furthermore a characteristic frequency was calculated as ~504 cm^{-1}, (Table 12.1), as compared to the experimental values of ~500, ~602, and ~699 cm^{-1} (Table 12.2). Experimentally the Raman spectral bands are occurring at easily recognizable peak positions and with intensities obtainable from *ab initio* DFT calculations. Bands measured by Ozawa and Hamaguchi at 701, 625, 603, and 500 cm^{-1} correspond within experimental error to our $[C_4C_1Im][PF_6]$ liquid bands at 698, 624, 601, and 498 cm^{-1} (Table 12.2).

Table 12.2 Experimentally Observed Raman Spectral Bands for Two Common
[C$_4$C$_1$Im]$^+$ Ionic Liquids, Given in cm^{-1}, and Approximate Assignments

ν/(cm^{-1})		Assignments[a]
[C$_4$C$_1$Im] [PF$_6$]	[C$_4$C$_1$Im] [BF$_4$]	
498	500	N-C7 opl bend + N-C7-C8-C9 angles bend (GA12)
568		ν_2(PF$_6^-$) stretching
601	601	N-C6 N-C7 iph str + ring oopl def + C8H$_2$ rock + N-C7-C8 bend. (GA13)
624	625	Ring def (C2-H oopl bend) + N-C6 N-C7 iph str + C7H$_2$ rock + C7-C8-C9 bend (AA14)
656	658	N-C6 str + ring def (N1 and H on C2 oopl ooph departure) + C8H$_2$ wag + N-C7-C8 bend (AA15) + ring def (bend around line NN) + C8H$_2$ rock + N-C7-C8 bend (GA15)
698	699	N-C6 N-C7 ooph str + ring ipl def + C8H$_2$ rock + C7-C8 tor (GA16)
730 ? hidden	~735	N-C6 N-C7 ooph str + ring ipl def + N-C7-C8 and C7-C8-C9 bend (AA17)
741		ν_1(PF$_6^-$) stretching

[a] Key for descriptions of approximate group vibrations, see Table 12.1; AA and GA mode numbers given in parentheses.

In a more refined gas phase-ion pair model aimed at understanding the interaction in the [C$_4$C$_1$Im][PF$_6$] liquid, Meng et al. [5] in a combined spectroscopic, semiempirical and *ab initio* study, observed hydrogen bonding between the [PF$_6$]$^-$ ion and the hydrogen atom at C2 in the aromatic ring of the [C$_4$C$_1$Im]$^+$ cation. Virtually identical theoretical results were obtained using both HF and DFT. The DFT minimized structure is shown in Figure 12.13.

Obviously Meng et al. [5] have reached an *anti-gauche* (AG) conformation that probably just is local but not a global minimum. The hydrogen bonding has previously been detected by ^{13}C NMR relaxation studies on [C$_4$C$_1$Im] [PF$_6$] and related systems in the liquid state [5,6,75,76].

It is well known that hydrogen bonding significantly supports the formation of ion pairs (and even higher aggregates) in electrolyte solutions when compared to systems without specific interactions. Hanke et al. [4] in another simulation study found that the largest probability for finding an anion is near C2 below and above the ring. Most likely dimeric and tetrameric ion pairs and higher aggregates are formed with a type of layer structure, in which the anions are located mainly above and below the aromatic ring near C2 [5,6]. The occurrence of hydrogen bonding in addition to the Coulombic interactions was put forward to explain the quite high viscosity and other specific macroscopic properties of the [C$_4$C$_1$Im][PF$_6$] IL.

Table 12.3   Selected Vibrational Modes as Determined in MP2 Calculations
for the $[C_4C_1Im]^+$ Cation in AA and GA Conformation

Mode	$v\,(cm^{-1})$	IR intensity (km mole^{-1})	Raman activity ($Å^4\,amu^{-1}$)	Approximate description[a]
In AA conformation				
13	620	0.52	1.34	Ring oopl def + C7H$_2$ rock + C7-C8-C9 bend
14	636	9.40	2.59	Ring def (C2-H oopl bend) + N-C6 N-C7 iph str + C7H$_2$ rock + C7-C8-C9 bend
15	670	12.23	1.36	N-C6 str + ring def (N1 and H on C2 oopl ooph departure) + C8H$_2$ wag + N-C7-C8 bend
17	749	13.06	2.13	N-C6 N-C7 ooph str + ring ipl def + N-C7-C8 and C7-C8-C9 bend
In GA conformation				
12	504	0.65	1.19	N-C7 opl bend + N-C7-C8-C9 angles bend
13	604	1.46	5.99	Ring ipl def + C8H$_2$ rock + N-C7-C8 bend
14	623	2.82	0.36	Ring def (C2-H C4-H iph oopl bend) + C8H$_2$ rock + N-C7-C8 bend
15	663	15.62	0.56	Ring def (bend around line NN) + C8H$_2$ rock + N-C7-C8 bend
16	713	6.65	2.15	N-C6 N-C7 ooph str + ring ipl def + C8H$_2$ rock + C7-C8 tor

[a] For key of approximate group vibrations, see Table 12.1.
Source: Berg, R. W., Deetlefs, M., Seddon, K. R., Shim, I., and Thompson, J. M., *J. Phys. Chem. B*, 109, 19018–19025, 2005.

12.7   Other studies on $[C_nC_1Im]^+$ liquids

Measured IR and Raman spectra of a series of 1-*alkyl*-3-methylimidazolium halides and hexafluorophosphate ILs ($[C_2C_1Im][PF_6]$ to $[C_4C_1Im][PF_6]$) were correlated with results of *ab initio* DFT calculations at the 6-31+G* and 6-311+G(2d,p) levels [72,74]. It was suggested that common Raman C–H stretching frequencies in these liquids may serve as possible probes in studies of ionic interactions. Hydrogen-bonding interactions were observed between the fluorine atoms of the $[PF_6]^-$ anion and the C2 hydrogen on the imidazolium ring, and between $[PF_6]^-$ anion and the H atoms on the adjacent alkyl side chains. There are at least four discernible strong vibrations in the 2878–2970 cm^{-1} region of the $[C_2C_1Im][PF_6]$ Raman spectrum [74].

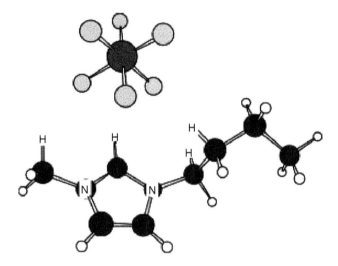

Figure 12.13 Minimized molecular structure of [C$_4$C$_1$Im][PF$_6$] (B3LYP/6-31G*) {5}. Found hydrogen bonds included: F2-H25 = 2.279 Å, F2-H18 = 2.042 Å, Fl-H18 = 2.441 Å, F5-H18 = 2.070 Å, F5-H22 = 2.419 Å, and Fl-H26 = 2.377 Å. (Adapted from Meng, Z., Dölle, A., and Carper, W. R., *J. Molec. Struct. (THEOCHEM)* 585, 119–128, 2002. With permission.)

These Raman vibrations are represented by the calculated vibrations in the 3153–3220 cm^{-1} region and represent a complex combination of multiple stretching and bending vibrations. The weak Raman bands observed at 3116 and 3179 cm^{-1} were assigned to vibrations associated with the imidazolium ring (H–C–C–H and N–(C–H)–N) C–H stretches.

Similar studies were done on [C$_2$C$_1$Im][BF$_4$] and other 1-ethyl-3-methyl-imidazolium liquid salts [16,69,79,80]. In the Raman spectral range 200–500 cm^{-1}, Umebayashi et al. [80] for liquids containing [BF$_4$]$^-$, [PF$_6$]$^-$, [TfO]$^-$, and [Tf$_2$N]$^-$ found bands at 241, 297, 387, 430, and 448 cm^{-1} that must originate from the [C$_2$C$_1$Im]$^+$ ion. However, the 448 cm^{-1} band could not be reproduced by theoretical calculations in terms of a single given [C$_2$C$_1$Im]$^+$ conformer. The ethyl group bound to one N atom of the imidazolium ring is able to rotate around the C–N bond to yield two different torsion conformers (see Figure 12.14).

The energy barrier of this rotation was calculated with an energy amplitude of ~2 kJ mol^{-1} [80]. Two local minima were found, suggesting that the two conformers can be present in equilibrium. Full geometry optimizations followed by normal frequency analyses indicated that the two conformers of minimum energy were those with planar and nonplanar ethyl groups against the imidazolium ring plane, and that the nonplanar conformer was the most favorable. The Raman bands at 241, 297, 387, and 430 cm^{-1} were found to

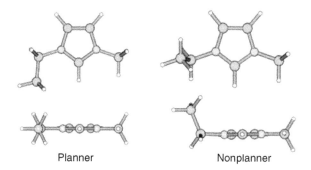

Planner   Nonplanner

Figure 12.14 Calculated structures of the $[C_2C_1Im]^+$ cation, showing the two different torsion conformers obtainable by rotation of the ethyl group around the C–N bond relative to the imidazolium ring. Planar (*left*) and nonplanar (*right*) rotamers are viewed perpendicular to and along the ring plane. The nonplanar form is known from x-ray structures {69}. (Adapted from Umebayashi, Y., Fujimori, T, Sukizaki, T., Asada, M., Fujii, K., Kanzaki, R., and Ishiguro, S.-i., *J. Phys. Chem. A,* 109, 8976–8982, 2005. With permission.)

mainly originate from the nonplanar conformer, whereas the 448 cm^{-1} band originated from the planar conformer. Indeed, the enthalpy for conformational change from nonplanar to planar $[C_2C_1Im]^+$, obtained experimentally by analyzing band intensities of the conformers at various temperatures, was practically the same as the enthalpy evaluated by theoretical calculations. We thus conclude that the $[C_2C_1Im]^+$ ion exists as planar or nonplanar conformers in equilibrium in its liquid salts [80], and this was confirmed by x-ray diffraction [69]. For the longer-chain $[C_nC_1Im]^+$ systems also nonplanar forms are most stable, for example, compare with AA and GA conformers of $[C_4C_1Im]^+$ in Figure 12.8.

We have shown [46] that the same situation as for $[C_4C_1Im]^+$ exists for longer alkyl chain systems, at least for the 1-hexyl-3-methylimidazolium cation. Raman spectra for $[C_6C_1Im]^+$ cation systems have bands at 698, 623, and 601 cm^{-1} (but no distinct band at ~498 cm^{-1}). A comparison between typical experimental spectra is shown in Figure 12.15.

Also the calculations came out in much the same way, as can be seen by comparing the results in Figure 12.16 with those from Figure 12.10 (insets).

All in all, it was recognized, both from spectra and calculations that the characteristic frequencies do not change significantly when the butyl group was exchanged for a hexyl group, and we conclude that the AA-GA isomerism phenomenon probably is general, and not specific to the $[C_4C_1Im]X$ ILs. For a discussion of the hexyl systems, we refer to our comprehensive report [46].

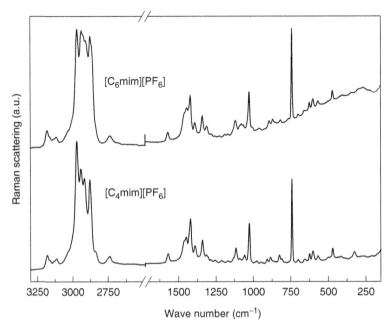

Figure 12.15 FT-Raman spectra of the $[C_6C_1Im][PF_6]$ and $[C_4C_1Im][PF_6]$ ionic liquids at ~25°C. Note that the characteristic bands of the AA and GA forms of the $[C_4C_1Im]+$ cation are very much like the AAAA and GAAA bands from the $[C_6C_1Im]+$ cation. (Berg, R. W., Deetlefs, M., Seddon, K. R., Shim, I., and Thompson, J. M., *J. Phys. Chem. B*, 109, 19018–19025, 2005; Berg, R. W., Unpublished results, 2006. With permission.)

12.8   Conformations equilibria in liquids versus temperature

The Raman spectra of $[C_nC_1Im][BF_4]$ from $n = 10$ to $n = 2$ in the liquid state at room temperature are shown in Figure 12.17.

Clearly the rotational isomerism around C7–C8 interconvert AA and GA conformers so as to change the AA/GA ratio with the cation. For $[C_2C_1Im]$ $[BF_4]$ there can be no rotational isomerism around C7-C8 because C8 is the end methyl group [16] and only one Raman band is observed at 596 cm^{-1}, corresponding to the band from GA conformation of $[C_4C_1Im][BF_4]$. This observation was rationalized by Hamaguchi et al. [50] who noted that the methyl rocking motion of the C8 carbon is strongly coupled to the ring deformation vibration and pushes down the frequency in $[C_2C_1Im]^+$ exactly as in the case of GA conformation of the $[C_4C_1Im]^+$ cation.

For longer side chains, the 625/603 cm^{-1} Raman intensity ratio increases with increasing n because the *all*-anti AA... kind of band at 625 cm^{-1} grows in intensity. Since the vibrational modes are very similar—being localized within the imidazolium ring and around C7 and C8 carbons

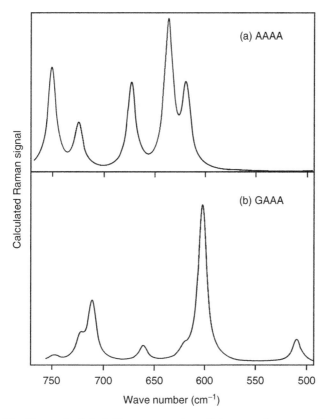

Figure 12.16 Our calculated Raman spectra of two conformers of the hexyl $[C_6C_1Im]^+$ cation between 750 and 500 cm^{-1} : (a) AAAA and (b) GAAA conformers. (Adapted from Berg, R. W., Deetlefs, M., Seddon, K. R., Shim, I., and Thompson, J. M., *J. Phys. Chem. B,* 109, 19018–19025, 2005. With permission.)

(see Figure 12.11), their Raman cross sections are thought to be quite in dependent of the chain length [46]. Therefore, the 625/603 cm^{-1} Raman intensity ratio can be regarded as a direct measure of the AA.../GA... isomer ratio. The observed *increase* of the 625/603 cm^{-1} intensity *ratio* with n means that the AAA.../GAA... isomer ratio increases as the chain becomes longer. The AAA... structure stabilization relatively to the GAA... one for longer alkyl chains is understandable only if we assume interactions among the cations. Otherwise, the relative stability would be determined alone by the energy difference between the *anti* and the *gauche* conformations around the C7-C8 bond and would be likely to be independent of the chain length. Instead, the AA.../GA... ratio dependence of the chain length therefore must depend on aliphatic interactions between the alkyl chains. In Figure 12.17, the broad CH_2 rocking and bending features observed for longer chain $[C_nC_1Im][BF_4]$ liquids (n = 7–10) in the region near 800–950 cm^{-1},

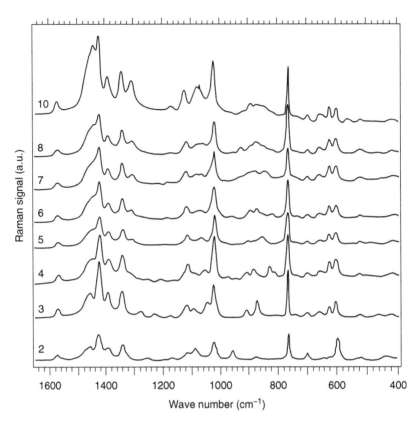

Figure 12.17 Raman spectra of 1-alkyl-3-methylimidazolium tetrafluoroborate liquids, $[C_nC_1Im][BF_4]$ for $n = 10, 8, 7, 6, 5, 4, 3,$ and 2. (Adapted from Hamaguchi, H., and Ozawa, R., *Adv. Chem. Phys.*, 131, 85–104, 2005. With permission.)

are indicative of such interactions between the alkyl chains [50], similar to the interactions found for certain mesophase liquid crystals when $n > 12$, (see Refs. 12,55,81).

An unusually long equilibration time has been observed upon melting of a single crystal of the AA polymorph of $[C_4C_1Im]Cl$ *Crystal (1)* heated rapidly from room temperature to 72°C to form a droplet of liquid in a non-equilibrium state [50]. Raman versus time spectra are reproduced in Figure 12.18.

Before melting and for some time after only the band at 625 cm^{-1} of the AA $[C_4C_1Im]^+$ cation was observed in the 600–630 cm^{-1} region. Gradually 603 cm^{-1} band due to the GA conformer became stronger. After about 10 min the AA/GA intensity ratio became constant. The interpretation [50] is that the rotational isomers do not interconvert momentarily at the molecular level. Most probably it involves a conversion of a larger *local structure* as a whole. The existence of such local structures of different rotamers has been found by optical heterodyne-detected Raman-induced Kerr-effect spectroscopy (OHD-RIKES) [82], Coherent anti-Stokes Raman scattering (CARS) [83],

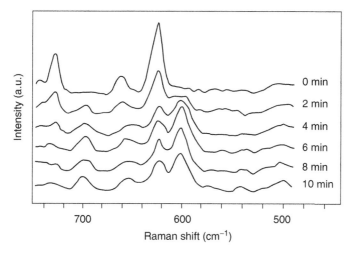

Figure 12.18 Time-resolved Raman spectra of the melting and thermal equilibration process at 72°C for a [C₄C₁Im]Cl *Crystal (1)* sample. (Adapted from Hamaguchi, H., and Ozawa, R., *Adv. Chem. Phys.*, 131, 85–104, 2005.)

neutron scattering and diffraction experiments [53,54] and by theoretical molecular dynamics calculations [4,7,84–86].

The enthalpy difference between the AA and the GA conformers in the [C₄C₁Im][BF₄] is much smaller than the corresponding enthalpy difference between the conformers of a free butane chain. This indicates that the 1-butyl-3-methylimidazolium cations most likely form *local liquid structures* specific to each rotational isomers [50]. Coexistence of these local structures, incorporating different rotational isomers, seems to hinder crystallization. This is probably the reason for the low melting points of such ILs.

These local structures most probably distinguish IL from conventional molecular liquids and may explain why IL phases are between a liquid and a crystal.

12.9   Local structures in ionic liquids

From NMR spectroscopy it is known that conformational isomers of alkane chains give coalesced peaks indicating transformation between the conformers taking place much faster than a second. Accordingly, one should expect single [C₄C₁Im]⁺ cations undergoing AA to GA transformation almost instantaneously [87]. The observed ~10 min long equilibration time in liquid [C₄C₁Im]Cl (Figure 12.18) therefore has been taken to indicate that the conformers do not transform *at the single molecular level* but only interconvert through slow collective transformations of *ensembles* of [C₄C₁Im]⁺ cations (analogous to a phase transition) [50,87]. Most probably the two rotational isomers are incorporated in specific local structures tending to interconvert

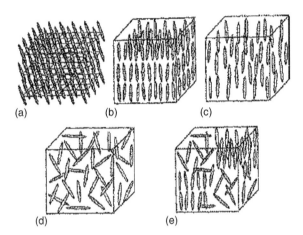

Figure 12.19 Conceptual structure of ionic salt crystals and liquids: a = crystal, b and c = liquid crystals, d = liquid, e = ionic liquid, according to the model of Hamaguchi and Ozawa. (Adapted from Hamaguchi, H., and Ozawa, R., *Adv. Chem. Phys.*, 131, 85–104, 2005. With permission.)

only through conversion of the local structures *as a whole* and giving rise to wide premelting ranges and other features [65].

The ordering of the anions in ILs has, for the case of $[C_4C_1Im]I$, been confirmed by large-angle x-ray scattering experiments [59]. In this way it seems that the zig-zag chains found in the $[C_4C_1Im]X$ crystals do exist in the IL state as well, at least partially. Thus, by combining Raman spectroscopy with several other experimental and theoretical techniques, Hamaguchi and coworkers have come to mean that both the cations and the anions in $[C_4C_1Im]X$ ILs might have local ordering of their structures. Their conceptual structure of ILs is reproduced in Figure 12.19.

According to the model, the supposed local structures are positioned and oriented randomly, and there seems to be no translational and orientational order at the macroscopic level. The local structure modeling of Ozawa and Hamaguchi in Figure 12.19 is shown for the $[C_4C_1Im]X$ ILs (e) in comparison with the structures of a crystal (a), liquid crystals (b and c), a conventional liquid (d). In the crystal (a), component molecules or ions are arranged to form a periodic lattice with long-range order. In a standard liquid state (d), the molecules or ions take random positions and random orientations and there is no order. In liquid crystals (b and c), different kinds of long-range orders exist, with for example, only partial orientational order (b) or random position (c).

In $[C_4C_1Im]X$ ILs, the supposed *local structures* are positioned and oriented randomly, and there seems to be no translational and orientational orders at the macroscopic level. Taking into account that the $[C_4C_1Im]X$ ILs are all transparent (not opaque), the dimension of those *local structures* must be much smaller than the wavelength of visible light (<100 Å) [50,87].

Microphase segregation in imidazolium-based ILs has also been discussed, and the existence of polar and nonpolar microsegregated domains in ILs has been predicted in molecular simulation dynamics [88]. The structural analysis helps the understanding of solvation of nonpolar, polar, and associating solutes in these media [53,88]. The existence of an extended hydrogen-bonded network structure was suggested by Abdul-Sada et al. [89] for 1-alkyl-3-methylimidazolium halides based on results from fast-atom bombardment mass spectroscopy. Charge ordering in ILs was discussed by Hardacre et al. [53]. They obtained the radial distribution curves of dimethylimidazolium chloride and hexafluorophosphate liquids using neutron diffraction and argued for charge ordering of ions in ILs resembling what is found in the solid state. Charge ordering has also been discussed in a number of molecular dynamics computer simulation studies on 1-alkyl-3-methylimidazolium-based ILs in recent few years [7,84,85,90–100]. The radial distribution functions calculated in these papers all have suggested long-range charge ordering, giving support to the idea that ILs are unique in having more structure ordering than do conventional molecular liquids.

Many unique properties may be expected to arise from these local structures in ILs. One example is the *unusually high viscosity* of certain ILs arising from the hindering of the translational motion of the ions. Magnetic behavior is another most unusual and interesting property that arise when magnetic ions (strongly interacting with one another) are locally aligned in a liquid. Recently it was demonstrated [101,102] that *magnetic* ILs can be made by mixing imidazolium chlorides ($[C_4C_1Im]Cl$ or $[nC_4C_1Im]Cl$) and $FeCl_3$, forming 1-butyl-3-methylimidazolium tetrachloroferrate $[C_4C_1Im]$ $[FeCl_4]$ or 1-butyronitrile-3-methylimidazolium tetrachloroferrate $[nC_4C_1Im]$ $[FeCl_4]$ (IUPAC name of $[nC_4C_1Im]^+$ cation: 1-(3-cyanopropyl)-3-methyl-1*H*-imidazol-3-ium). These nearly paramagnetic liquids show strong responses to magnetic fields, probably because of local ordering of the magnetic high-spin $[FeCl_4]^-$ anions. The surfaces of the liquids bend (deviate from being horizontal) when they are approached by a magnet, an interesting property that might find applications. FT-Raman spectroscopy indicated that the magnetic liquids contained $[C_4C_1Im]^+$ and $[nC_4C_1Im]^+$ cations. The constitution of the liquids thus were confirmed by their Raman spectra. By combining many different cations and magnetic anions it might be possible to prepare superparamagnetic or even ferromagnetic ILs [101,102].

Another interesting behavior of an IL has been observed: The molecular arrangements of 1-butyronitrile-3-methylimidazolium halides, in the presence and absence of intruded water molecules, form a new kind of ice that has been studied by Raman spectroscopy [103,104]. Single crystals of the ice were isolated and the structure was elucidated by single-crystal x-ray crystallography. Apparently the water changes the physical properties of the IL at the molecular level and this was found to change the conformation of the *n*-butyronitrile chain of the cation. The hydrogen-bonding interaction between the anion and the water molecule seems to lead to loose molecular packing arrangements of the IL. As the unique properties are related to the structures and molecular arrangements of the ILs, the presence of water, wanted or unwanted, must be carefully examined in any kind of IL research and applications [104].

12.10   Other systems

Raman spectra of N-butylpyridinium chloride, aluminum trichloride liquid systems (e.g., [C_4py][$AlCl_4$]) were obtained at ambient temperatures already in 1978 [105]. The [C_4py][$FeCl_4$] also is an IL, and the phase diagram of the binary system [C_4py]Cl–$FeCl_3$ liquids are formed in a wide mole fraction composition range from 0.26 to 0.58 [106]. Unrestricted HF calculations were performed with 6–31G* basis sets in order to predict the structures, energies, bond lengths, and vibrational (Raman) frequencies. Both the Raman scattering experiments and the *ab initio* calculations indicate that [$FeCl_4$]$^-$ is the predominant anion in the IL at a mole fraction of 0.50 [106].

The structure of the *bis*(trifluoromethylsulfonyl)imide ([Tf_2N]$^-$) anion in the liquid state has been investigated by means of IR and Raman spectroscopy and *ab initio* self-consistent HF and DFT calculations on the free ion, aiming at a determination of the equilibrium geometry and understanding of the vibrational spectrum [80,90,91, 107–109]. A pronounced delocalization of the negative charge on the nitrogen and oxygen atoms was found, and a marked double-bond character of the S-N-S moiety for the anion [107,108]. A tentative assignment of some characteristic vibrations of the [Tf_2N]$^-$ anion was performed using the spectra of aqueous solutions for comparison in order to analyze the conformational isomerism and ion-pairing effects [107,108]. Also, it was found that dihedral S-N-S-C torsion angle in the bistriflylimide anion has a complex energy profile, which was precisely reproduced when tested by confrontation against liquid-phase Raman spectroscopic data [90–92].

The Raman spectra of the [C_2C_1Im][Tf_2N] show relatively strong bands arising from the [Tf_2N]$^-$ ion at ~398 and ~407 cm^{-1}, see Figure 12.20 [108].

Interestingly, the ~407 cm^{-1} band, relative to the ~398 cm^{-1} one, grows appreciably in intensity with raising temperature. This feature is suggesting that an equilibrium is established between [Tf_2N]$^-$ conformers in the liquid state (see Figure 12.21).

According to the DFT calculations (followed by normal frequency analyses), two conformers (of C_2 or C_1 point group symmetry; two-fold rotational axis or no symmetry, respectively) constitute global and local minima, and have an energy difference of 2.2–3.3 kJ mol^{-1} [108]. The observed spectra over the range 380–440 cm^{-1} could be understood from the calculation of [Tf_2N]$^-$ conformers: The omega-SO_2 wagging vibration appeared at 396 and 430 cm^{-1} for the C_1 conformer and at 387 and 402 cm^{-1} for the C_2 one. The enthalpy of

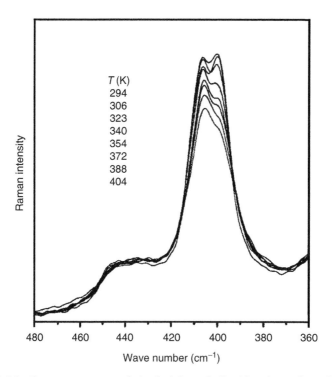

Wave number (cm⁻¹)

Figure 12.20 Raman spectra of 1-ethyl-3-methylimidazolium liquid [C_2C_1Im] [Tf_2N] showing the temperature-dependent SO_2 wagging bands at ~398 and ~407 cm⁻¹. According to Fujii et al. (Fujii, K., Fujimori, T., Takamuku, T., Kanzaki, R., Umebayashi, Y., and Ishiguro, S., *J. Phys. Chem. B.* 110, 8179–8183, 2006) the bands arise from different conformers of the [Tf_2N]- ion, known also from crystal structures (Matsumoto, K., Hagiwara, R., and Tamada, O., *Solid State Sci.*, 8, 1103–1107, 2006.) (Fujii, K., Fujimori, T., Takamuku, T., Kanzaki, R., Umebayashi, Y., and Ishiguro, S., *J. Phys. Chem. B.* 110, 8179–8183, 2006. With Permission.)

the conformational change from the most stable C_2 to C_1 came out in good agreement with the theoretical calculation. It was concluded that a conformational equilibrium indeed must exist between the C_1 and the C_2 conformers of [Tf_2N]⁻ in liquid [C_2C_1Im][Tf_2N] [109]. Three different conformers (named cis and trans by the authors) have recently been found in the x-ray crystal structure of the salt Li_2[C_2C_1Im][Tf_2N]$_3$ [110].

We were able to obtain the same wagging omega-SO_2 vibrational bands in our Raman spectrum of liquid [C_4C_1Im][Tf_2N] (see Figure 12.22).

Apparently the splitting between the two bands (at 411 and 403 cm⁻¹) is not so large for the case of the 1-butyl-3-methylimidazolium *bis*(trifluoromethylsulfonyl)imide liquid. The symmetric CF_3 stretching and deformation bands are seen very strongly at ~1242 and ~742 cm⁻¹ in our spectra, as found also by Rey et al. [107,108]. The bands at 500–750 cm⁻¹ discussed in relation with Figure 12.6 can be faintly seen, showing that the AA/GA

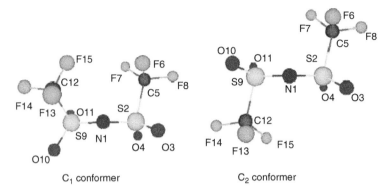

Figure 12.21 Different conformers of symmetry C_1 and C_2 of the [Tf$_2$N]- ion, as determined by Fujii et al. [109] by means of DFT calculations for the 1-ethyl-3-methylimidazolium liquid [C$_2$C$_1$Im][Tf$_2$N]. (Adapted from Fujii, K., Fujimori, T., Takamuku, T., Kanzaki, R., Umebayashi, Y., and Ishiguro, S., *J. Phys. Chem. B.* 110, 8179–8183, 2006. With permission.)

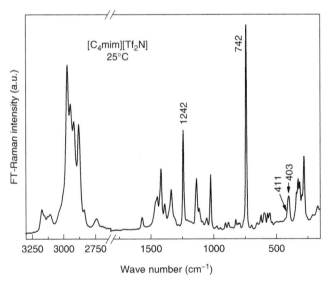

Figure 12.22 Our Raman spectrum of liquid [C$_4$C$_1$Im][Tf$_2$N]. Apparently the splitting between the two conformation sensitive bands for the [Tf$_2$N]– ion, near ~400 cm^{-1}, is not so large in this liquid as for the 1-ethyl [C$_2$C$_1$Im][Tf$_2$N] case [109]. The CF$_3$ symmetric stretching and deformation bands are seen at 1242 and 742 cm^{-1}. The AA/GA conformational equilibrium bands at 500–700 cm^{-1} discussed in relation with Figure 12.6 can also be weakly seen.

conformational equilibrium of the butyl group in $[C_4C_1Im]^+$ is established, as discovered by Hamaguchi et al. [50].

The $[Tf_2N]^-$ anion was further studied and discussed in Raman investigations on the IL N-propyl-N-methylpyrrolidinium *bis*(trifluoromethylsulfonyl)imide ($[C_1C_3pyr][Tf_2N]$) and its 2/1 mixture with Li[Tf_2N] [111], as well as on the N-butyl-N-methylpyrrolidinium *bis*(trifluoromethanesulfonyl)imide ($[P14][Tf_2N]$ or $[C_1C_4pyr][Tf_2N]$) and other useful reaction media such as $[C_1C_4pyr][Tf_2N]$ [112,113]. Raman results have also shown that $[Tf_2N]^-$ anions only interact weakly with $[P13]^+$ cations, but coordinated strongly to Li^+ cations over a temperature range extending from −100 to +60 °C, that is, in the crystalline and melt states [114].

By means of Raman spectroscopy and DFT calculations on the $[C_1C_4pyr]$ $[Tf_2N]$ and $[C_1C_4pyr]Br$ systems, various types of conformations with respect to the pyrrolidinium ring and N-butyl group were found [114]. The calculations indicated that, among others, the so-called *eq-* and *ax-envelope* conformers with the butyl group at equatorial and axial positions, respectively, against the plane of four atoms of the *envelope* pyrrolidinium ring (see Figure 12.23) were relatively stable, and the former gave the global minimum [114].

By comparing observed and calculated Raman spectra it was found that the $[C_1C_4pyr]^+$ ion was present mainly as the *ax-envelope* conformer in the $[C_1C_4pyr]Br$ crystal, while the *eq-* and *ax-envelope* conformers were present in equilibrium in the $[C_1C_4pyr][Tf_2N]$ IL (called [P14][TFSI] in [114]).

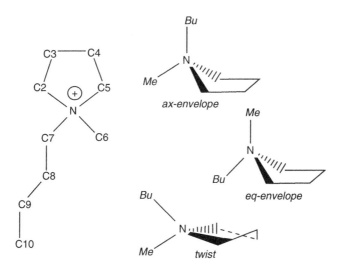

Figure 12.23 Structure and conformations of N-butyl-N-methylpyrrolidinium cation $[C_1C_4pyr]^+$. In [P14]+, P denotes pyrrolidinium and the digits denote the number of carbon atoms in radicals R^1 and R^2. Also, the ion is commonly called [bmpy]+ (N is atom number 1). The ring of the $(CH_2)4NR^1R^2$ pyrimidinium ion is not planar and has two stable (twist and envelope) forms.

The presence of conformational equilibria was further experimentally supported by Raman spectra measured at different temperatures. It was established that the conformation of the butyl group was restricted to a so-called trans-TT conformation, in which the butyl group is located trans against a ring carbon atom (C2 or C5), and all carbon atoms in the butyl chain are located trans to each other [113].

We have briefly investigated some $[C_1C_4pyr]^+$ room temperature liquids, namely $[C_1C_4pyr][Tf_2N]$ and $[C_1C_4pyr][TfO]$ [73]. The experimental Raman spectra of our liquids looked much like *sums* of the spectrum of the $[C_1C_4pyr]^+$ ion (as measured from the chloride salt) and the spectra of the $[Tf_2N]^-$ or $[TfO]^-$ ions (as measured from the lithium salts). An example of our results on $[C_1C_4pyr][Tf_2N]$ is shown in Figure 12.24.

The spectra in Figure 12.24 clearly show that the constituent ions in the liquid and in the respective solid salts vibrate *rather* independent of the surroundings. Therefore the liquid spectrum looks much like the sum of the solid salts. This conclusion is of course not new, but nevertheless it is still *quite* applicable in the evaluation of many IL Raman (and IR) spectra. However, the presence of conformational equilibria for both of the IL ions makes a closer study worth while. We therefore recommend the interested reader to study the work by Umebayashi et al. [114] in which subtle spectral band shape details, for example, around 930–880 cm^{-1} are evaluated to show information on the *eq-envelope*:trans-TT and *ax-envelope*:trans-TT interconversion of the $[C_4C_1Im]^+$ ion in the liquid. Also note that the crystal structure of the $[C_1C_4pyr][Tf_2N]$ salt was recently solved; it contained the *eq-envelope*:trans-TT conformer of the cation [115]. Also conformers of symmetry C_1 and C_2 of the $[Tf_2N]^-$ ion show their presence burried in the band at 400–440 cm^{-1} [109].

Prior to the publication of the work by Umebayashi et al. [114] on the 1-butyl-1-methyl-pyrrolidinium *bis*(trifluoromethylsulfonyl)imide, some preliminary calculations were done aiming at a better understanding of the spectra on that system (shown in Figure 12.24). To illustrate how useful such procedures are, we depict two examples of our own results in Figure 12.25.

The shown *ax-envelope*:trans-TG and -TT confomations are just some of the many possible conformations. The more likely ones, such as the *ax-envelope*:trans-TT *and* *eq-envelope*:trans-TT were included in the study by Umebayashi et al. [114]. The calculated spectra of the different conformers looked rather much like each other. As seen in Figure 12.25 the *gauche* form of the C8-C9 bond did not change much relative to the *ax-envelope*:trans-TT form. Quite satisfactory one-to-one correspondences between calculated and observed bands are found in Figure 12.25 and also in the work of Umebayashi et al. [114]. One should not expect perfect fits (frequencies are calculated too high and intensities are perturbed, because of the simplicity of the modeling).

When the theoretical spectra of the $[C_1C_4pyr]^+$ ion (e.g., those in Figure 12.25) and similarly calculated spectra of conformations of the $[Tf_2N]^-$

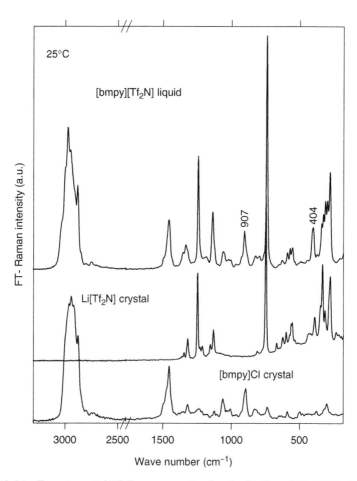

Figure 12.24 Experimental FT-Raman spectra for the [C$_1$C$_4$pyr][Tf$_2$N] liquid (in the figure [bmpy][Tf$_2$N]) (Fujimori, T., Fujii, K., Kanzaki, R., Chiba, K., Yamamoto, H., Umebayashi, Y., and Ishiguro, S.-i., *J. Mol. Liquids* 131-132, 216-224, 2007), showing that the spectum (top) at room temperature essentially consists of bands from both the [Tf$_2$N]$^-$ anion (*middle*) and the [C$_1$C$_4$pyr]$^+$ cation (bottom) (shifted conveniently). The Li$^+$ and Cl$^-$ do not contribute bands directly in the liquid but have influence on the structures of the salts and are interactive with the ions and influence the conformational equilibria in the IL (Berg, R. W., Unpublished results, 2006.)

ion (in Figure 12.26, see later) were added, we obtained sums (not shown) that essentially corresponded to the spectrum of the [C$_1$C$_4$pyr][Tf$_2$N] liquid (in Figure 12.24, top) [72].

New 2-hydroxypropyl-functionalized imidazolium cation ionic liquids (containing an appended hydroxyl functionality) have been made [116] by use of an *atom efficient one-pot reaction* between 1-methylimidazole and acid with propylene oxide. Unfortunately the systems were not studied

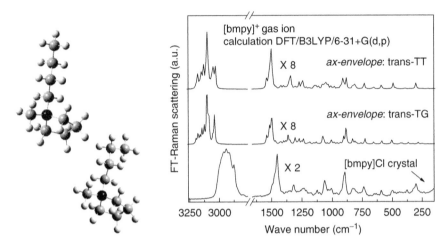

Figure 12.25 Minimized structures of the so-called ax-envelope:trans-TT and -TG confomations of the $[C_1C_4pyr]^+$ ion in assumed gaseous state calculated (Berg, R. W., Unpublished results, 2006) at the DFT/B3LYP/6-31+ G (p,d) level. The minimized bond distances and angles had standard magnitudes. The model spectra shown compare well to the experimental FT-Raman spectrum of the $[C_1C_4pyr]^+$ ion in the solid chloride salt (Berg, R. W., Unpublished results, 2006.)

by FT-Raman spectroscopy so far. We have shown in a study on 1-hexanol in [1,3-*bis*-[2-(methoxyethoxy)ethyl]imidazolium] *bis*-trifluoromethylsulfonylimide [29] that Raman spectroscopy has a potential for finding clues to what goes on in ILs that contain hydroxyl groups (alcohols) and where hydrogen bonding between the IL and the hydroxyl group is of importance.

A rather new class of room temperature ILs is based on the N,N,N',N'-tetramethylguanidinium $[((CH_3)_4)_2CNH_2]^+$ or $[TMGH]^+$ cation. A dedicated force field was developed [117] to fit the experimental bonds and angles and the vibration frequencies, for five kinds of $[TMGH]^+$ ILs, where the anion was formate, lactate, *perchlorate*, trifluoroacetate, and trifluoromethylsulfonate, respectively. *Ab initio* calculations were performed and predictions in good agreement with the experimental data were obtained. Radial and spatial distribution functions (RDFs and SDFs) were investigated to depict the microscopic structures of the ILs [13].

An illustrative FT-Raman spectrum was recorded on the $[TMGH][Tf_2N]$ liquid to be compared with calculated spectra of the constituent isolated ions in their equlibrium geometries obtained at the RHF/6-31G(d) level using DFT/ B3LYP methods [49]. Some data obtained [73] are shown in Figure 12.26. Unfortunately the highest-frequent N–H stretchings were out of our instrumental range (limited to 3500–100 cm^{-1}). However, many experimental details are being convincingly accounted for by the modeling, just by summation of the calculated spectra, even when other conformers are left out.

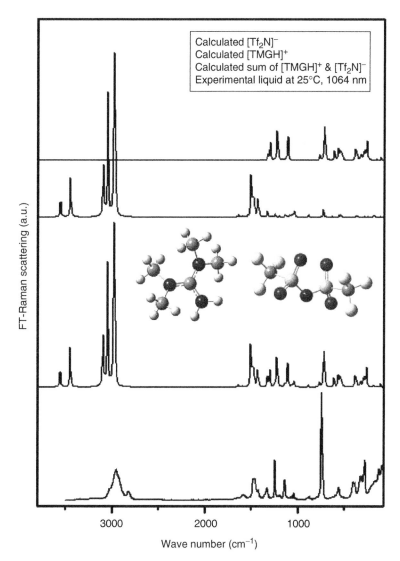

Figure 12.26 Illustrative example of the power of *ab initio* methods combined with Raman spectroscopy, applied on the [TMGH][Tf₂N] liquid. The two top spectral curves show calculated Raman spectra of minimized conformers of $[Tf_2N]^-$ and $[TMGH]^+$ at the DFT/B3LYP/6-31G(d) level. The geometries of the ions are also depicted together with the sum of the spectra, constituting a hypothetical [TMGH][Tf₂N] liquid of non-interacting ions (shifted conveniently). Bottom: the experimental FT-Raman spectrum. Unfortunately measurements exist only below 3500 cm^{-1} (Berg, R. W., Unpublished results, 2006.)

12.11   Other applications of Raman spectroscopy

Raman spectroscopy has up to now mainly been applied to elucidate conformational forms and associated conformational equilibria of the IL components. Yet other applications are appearing in these years. One example is the characterization of metal ions like Mn^{2+}, Ni^{2+}, Cu^{2+}, and Zn^{2+} in coordinating solvent mixtures by means of titration Raman Spectroscopy [118]. Another issue is the study of solvation of probe molecules in ILs. In such a study [118], for example, acceptor numbers (AN) of ILs in diphenylcyclopropenone (DPCP) were estimated by an empirical equation associated with a C=C / C=O stretching mode Raman band of DPCP. According to the dependence of AN on cation and anion species, the Lewis acidity of ILs was considered to come mainly from the cation charge [119].

Finally, Raman spectroscopy has a potential of being used for qualitative and quantitative analysis. We have used Raman spectroscopy to verify the presence of components in a two-phase system [29].

12.12   Conclusions

Recently Raman spectroscopy has been applied, in combination with other methods, to show that certain characteristic spectral bands can be identified that are characteristic for conformational forms (conformers) of the IL components, and that the associated conformational equilibria might be partly responsible for the salts to have such low melting points.

In this review we have discussed in detail some examples of the conformational equilibria, for example, those discovered by Hamaguchi et al. [70] in liquids containing 1-butyl-3-methylimidazolium cation. Also, we have examined in some detail liquids containing the *bis*(trifluoromethanesulfonyl) imide anion, as described above. We have extended the knowledge on the characteristic Raman bands to include conformers of the 1-hexyl-3-methylimidazolium cation [46]. Vibrational analysis has been made of the components of the systems to improve our understanding of what goes on in the liquids. These results, although not surprising, add weight to our understanding of the existence of mixtures of low-symmetry conformers that disturb the crystallization process. Arguments have been presented for the belief that this is the reason for the low melting points of the ILs relative to *normal molecular salts* with much higher melting points.

We have seen that the *ab initio* self-consistent quantum mechanical functional methods such as DFT/B3LYP with the chosen 6-31+G(d,p) basis sets are well suited to calculate reasonable molecular ion structures and vibrational spectra of these ions. The results obtained by us or others have indicated that the neglect of the presence of cation-anion interactions is a reasonable approximation for a rather successful prediction of the Raman spectra. Based on such calculations, detailed and reliable assignments of the spectra can be given and information on conformational equilibria can be obtained.

Acknowledgments

I would like to thank Niels J. Bjerrum, Anders Riisager, Rasmus Fehrmann, and Irene Shim from Department of Chemistry, DTU, Denmark, Cory C. Pye from Department of Chemistry, Saint Mary's University, Halifax, Nova Scotia, Canada, N. Llewellyn Lancaster from Department of Chemistry, King's College, London University, UK, Ken Seddon, QUILL Research Centre, Queen's University Belfast, Northern Ireland, Mihkel Koel from Department of Chemistry, Tallin University of Technology, Estonia, and Susanne Brunsgaard Hansen (formerly from Department of Chemistry, DTU) for help with finishing the manuscript. Lykke Ryelund and Ole Faurskov Nielsen of the Department of Chemistry, H.C. Ørsted Institute, University of Copenhagen are thanked for much measurement assistance. The work was supported by the Technical University of Denmark.

References

1. Wilkes, J. S., Properties of ionic liquid solvents for catalysis, *J. Mol. Catal. A: Chem.*, 214, 11–17, 2004.
2. Tait, S., and Osteryoung, R. A., Infrared study of ambient-temperature chloroaluminates as a function of melt acidity, *Inorg. Chem.*, 23, 4352–4360, 1984.
3. Dymek, C. J., Jr., and Stewart, J. J. P., Calculation of hydrogen-bonding interactions between ions in room-temperature molten salts, *Inorg. Chem.*, 28, 1472–1476, 1989.
4. Hanke, C. G., Price, S. L., and Lynden-Bell, R. M., Intermolecular potentials for simulations of liquid imidazolium salts, *Molec. Phys.*, 99, 801–809, 2001.
5. Meng, Z., Dölle, A., and Carper, W. R, Gas phase model of an ionic liquid: Semi-empirical and ab initio bonding and molecular structure, *J. Molec. Struct. (THEOCHEM)*, 585, 119–128, 2002.
6. Carper, W. R., Meng, Z., Wasserscheid, P., and Dölle, A., NMR relaxation studies and molecular modeling of 1-butyl-3-methylimidazolium PF_6, [BMIM][PF_6], *International Symposium on Molten Salts*, Trulove, P. C., DeLong, H. C., and Mantz, R. A. (Eds), *Electrochem. Soc. Proceedings 2002–19 (Molten Salts XIII)*, 973–982, 2002.
7. Shah, J. K., Brennecke, J. F., and Maginn, E. J., Thermodynamic properties of the ionic liquid 1-n-butyl-3 methylimidazolium hexafluorophosphate from Monte Carlo simulations, *Green Chem.*, 4, 112–118, 2004.
8. Dymek, C. J., Grossie, D. A., Fratini, A. V., and Adams, W. W., Evidence for the presence of hydrogen bonded ion-ion interaction in the molten salt precursor, 1-methyl-3-ethylimidazolium chloride, *J. Molec. Struct.*, 213, 25–34, 1989.
9. Wilkes, J. S., and Zaworotko, M. J., Manifestations of noncovalent interactions in the solid state. Dimeric and polymeric self-assembly in imidazolium salts via face-to-face cation-cation Π-stacking, *Supramolec. Chem.*, 1, 191–193, 1993.
10. Fuller, J., Carlin, R. T., DeLong, H. C., and Haworth, D., Structure of 1-ethyl-3-methylimidazolium hexafluorophosphate: Model for room temperature molten salts, *Chem. Commun.*, 299–300, 1994.
11. Carmichael, A. J., Hardacre, C., Holbrey, J. D., Nieuwenhuyzen, M., and Seddon, K. R., Molecular layering and local order in thin films of 1-alkyl-3-methylimidazolium ionic liquids using x-ray reflectivity, *Molec. Phys.*, 99, 795–800, 2001.

12. Bradley, A. E., Hardacre, C., Holbrey, J. D., Johnston, S., McMath, S. E. J., and Nieuwenhuyzen, M., Small-angle x-ray scattering studies of liquid crystalline 1-alkyl-3-methylimidazolium salts, *Chem. Mater.*, 14, 629–635, 2002.

13. Huang, J., Riisager, A., Wasserscheid, P., and Fehrmann, R., Reversible physical absorption of SO_2 by ionic liquids, *Chem. Commun.*, 4027–4029, 2006.

14. Huang, J., Riisager, A., Berg, R. W., and Fehrmann, R., Tuning Ionic Liquids for High Gas Solubility and Reversible Gas Sorption, *J. Mol. Catalysis A (Chemical)*, 279, 170–176, 2007.

15. Elaiwi, A., Hitchcock, P. B., Seddon, K. R., Srinivasan, N., Tan, Y. M., Welton, T., and Zora, J. A., Hydrogen bonding in imidazolium salts and its implications for ambient-temperature halogenoaluminate(III) ionic liquids, *J. Chem. Soc., Dalton Trans.*, 1995, 3467–3472, 1995.

16. Katsyuba, S. A., Dyson, P. J., Vandyukova, E. E., Chernova, A. V., and Vidis, A., Molecular structure, vibrational spectra, and hydrogen bonding of the ionic liquid 1-ethyl-3-methyl-1H-imidazolium tetrafluoroborate, *Helv. Chim. Acta*, 87, 2556–2565, 2004.

17. Cammarata, L., Kazarian, S. G., Salter, P. A., and Welton, T., Molecular states of water in room temperature ionic liquids, *Phys. Chem. Chem. Phys.*, 3, 5192–5200, 2001.

18. Storhaug, V. J., and Carper, W. R., Ab initio molecular structure and vibrational spectra of ionic liquids, *Trends in Phys. Chem.*, 9, 173–177, 2003.

19. Endres, F., and Zein El Abedin, S., Air and water stable ionic liquids in physical chemistry, *Phys. Chem. Chem. Phys.*, 8, 2101–2116, 2006.

20. Gilbert, B., Chauvin, Y., and Guibard, I., Investigation by Raman spectrometry of a new room-temperature organochloroaluminate molten salt, *Vib. Spectrosc.*, 1, 299–304, 1991.

21. Gilbert, B., Pauly, J. P., Chauvin, Y., and Di Marco-Van Tiggelen, F., Raman spectroscopy of room temperature organochloroaluminate molten salts, *Proc. 9th Int. Symposium on Molten Salts*, Hussey, C. L. et al., (Eds), *Electrochem. Soc.*, Pennington, NJ., Vol. 94–13, 218–226, 1994.

22. Chauvin, Y., Di Marco-Van Tiggelen, F., and Olivier, H., Determination of aluminum electronegativity in new ambient-temperature acidic molten salts, *J. Chem. Soc., Dalton Trans.*, 1009–1011, 1993.

23. Howlett, P. C., Brack, N., Hollenkamp, A. F., Forsyth, M., and MacFarlane, D. R., Characterization of the lithium surface in N-methyl-N-alkylpyrrolidinium bis(trifluoromethanesulfonyl)imide room-temperature ionic liquid electrolytes, *J. Electrochem. Soc.*, 153, A595–A606, 2006.

24. Ping He, Hongtao Liu, Zhiying Li, Yang Liu, Xiudong Xu, and Jinghong Li, Electrochemical deposition of silver in room-temperature ionic liquids and its surface-enhanced Raman scattering effect, *Langmuir*, 20, 10260–10267, 2004.

25. Schafer, T., Di Paolo, R. E., Franco R.,and Crespo, J. G., Elucidating interactions of ionic liquids with polymer films using confocal Raman spectroscopy, *Chem. Commun.*, 2594–2596, 2005.

26. Nanbu, N., Sasaki, Y., and Kitamura, F., In situ FT-IR spectroscopic observation of a room-temperature molten salt gold electrode interphase, *Electrochem. Commun.*, 5, 383–387, 2003.

27. Santos, V. O., Jr., Alves, M. B., Carvalho, M. S., Suarez, P. A. Z., and Rubim, J. C., Surface-enhanced Raman scattering at the silver electrode/ionic liquid (BMIPF$_6$) interface, *J. Phys. Chem. B*, 110, 20379–20385, 2006.

28. Romero, C., and Baldelli, S., Sum frequency generation study of the room-temperature ionic liquids/quartz interface, *J. Phys. Chem. B,* 110, 6213–6223, 2006.

29. Riisager, A., Fehrmann, R., Berg, R. W., van Hal, R., and Wasserscheid, P., Thermomorphic phase separation in ionic liquid-organic liquid systems-conductivity and spectroscopic characterization, *Phys. Chem. Chem. Phys.,* 7, 3052–3058, 2005.

30. Raman, C. V., and Krishnan, K. S., A new type of secondary radiation, *Nature,* 121, 501, 1928.

31. Herzberg, G., *Molecular Spectra and Molecular Structure, Vol. II, Infrared and Raman Spectra of Polyatomic Molecules,* van Nostrand, New York, 1945.

32. Wilson, E. B., Jr., Decius, J. C., and Cross, P. C., *Molecular Vibrations, The Theory of Infrared and Raman Vibrational Spectra,* McGraw-Hill Book Co., New York, 1955.

33. Long, D. A., *The Raman Effect: A Unified Treatment of the Theory of Raman Scattering by Molecules,* J. Wiley & Sons, New York, 2002.

34. Diem, M., *Introduction to Modern Vibrational Spectroscopy,* J. Wiley & Sons, New York, 1993.

35. Laserna, J. J. (Ed.), *Modern Techniques in Raman Spectroscopy,* John Wiley & Sons, Chichester, U.K., 1996.

36. Coates, J., Vibrational spectroscopy: Instrumentation for infrared and Raman spectroscopy, *Appl. Spectrosc. Rev.,* 33, 267–425, 1998.

37. McCreery, R. L., *Raman Spectroscopy for Chemical Analysis,* John Wiley & Sons, Chichester, U.K., 2001, 1–415.

38. Smith, E., and Dent, G. *Modern Raman Spectroscopy—A Practical Approach,* John Wiley & Sons, New York, 2005, 1–210.

39. Lin-Vien, D., Colthup, N. B., Fateley, W. G., and Grasselli, J. G., *Handbook of Infrared and Raman Characteristic Frequencies of Organic Molecules,* Academic Press, New York, 1991.

40. Nyquist, R. A., Kagel, R. O., Putzig, C. L, Leugers, M. A., *Handbook of Infrared and Raman Spectra of Inorganic Compounds and Organic Salts* (4 vols.), Academic Press, New York and London, 1996.

41. Nakamoto, K., Infrared and Raman spectra of inorganic and coordination compounds, 5th Ed., Part A: *Theory and Applications in Inorganic Chemistry,* and Part B, *Applications in Coordination, Organometallic and Bioinorganic Chemistry,* J. Wiley & Sons, New York, 1997.

42. Socrates, G., *Infrared and Raman Characteristic Group Frequencies,* Tables and Charts, 3rd Ed., J. Wiley & Sons, Chichester U.K., 2001, 1–249.

43. Carter, R. L., *Molecular Symmetry and Group Theory,* J. Wiley & Sons, New York, 1998.

44. Atkins, P. W., and Friedman, R. S, *Molecular Quantum Mechanics,* 4th Ed., Oxford University Press, Oxford, UK, 2004, 1–608.

45. Chase, D. B., and Rabolt, J. F., *Fourier Transform Raman Spectroscopy,* Academic Press, New York, 1994.

46. Berg, R. W., Deetlefs, M., Seddon, K. R., Shim, I., and Thompson, J. M., Raman and ab initio studies of simple and binary 1-alkyl-3-methylimidazolium ionic liquids, *J. Phys. Chem. B,* 109, 19018–19025, 2005. See also supplementary information at http://pubs.acs.org/subscribe/journals/jpcbfk/suppinfo/jp050691r/jp050691rsi20050208_105000.pdf .

47. Atkins, P., and de Paula, J., *Atkins' Physical Chemistry,* 8th Ed., Oxford University Press, Oxford, UK, 2006.

48. Ratner, M. A., and Schatz, G. C., *Introduction to Quantum Mechanics in Chemistry*, Prentice-Hall Inc., N.J., p.305, 2001.

49. Frisch, M. J., Trucks, G. W., Schlegel, H. B., Scuseria, G. E., Robb, M. A., Cheeseman, J. R., Montgomery, J. A., Jr., Vreven, T., Kudin, K. N., Burant, J. C., Millam, J. M., Iyengar, S. S., Tomasi, J., Barone, V., Mennucci, B., Cossi, M., Scalmani, G., Rega, N., Petersson, G. A., Nakatsuji, H., Hada, M., Ehara, M., Toyota, K., Fukuda, R., Hasegawa, J., Ishida, M., Nakajima, T., Honda, Y., Kitao, O., Nakai, H., Klene, M., Li, X., Knox, J. E., Hratchian, H. P., Cross, J. B., Adamo, C., Jaramillo, J., Gomperts, R., Stratmann, R. E., Yazyev, O., Austin, A. J., Cammi, R., Pomelli, C., Ochterski, J. W., Ayala, P. Y., Morokuma, K., Voth, G. A., Salvador, P., Dannenberg, J. J., Zakrzewski, V. G., Dapprich, S., Daniels, A. D., Strain, M. C., Farkas, O., Malick, D. K., Rabuck, A. D., Raghavachari, K., Foresman, J. B., Ortiz, J. V., Cui, Q., Baboul, A. G., Clifford, S., Cioslowski, J., Stefanov, B. B., Liu, G., Liashenko, A., Piskorz, P., Komaromi, I., Martin, R. L., Fox, D. J., Keith, T., Al-Laham, M. A., Peng, C. Y., Nanayakkara, A., Challacombe, M., Gill, P. M. W., Johnson, B., Chen, W., Wong, M. W., Gonzalez, C., and Pople, J. A., Gaussian 03, Revision C.02, Gaussian, Inc.: Wallingford CT, 2004.

50. Hamaguchi, H., and Ozawa, R., Structure of ionic liquids and ionic liquid compounds: Are ionic liquids genuine liquids in the conventional sense?, *Adv. Chem. Phys.*, 131, 85–104, 2005.

51. Gordon, C. M., Holbrey, J. D., Kennedy, A. R., and Seddon, K. R., Ionic liquid crystals: Hexafluorophosphate salts, *J. Mater. Chem.*, 8, 2627–2636, 1998.

52. Holbrey, J. D., and Seddon, K. R., The phase behaviour of 1-alkyl-3-methylimi-dazolium tetrafluoroborates; ionic liquids and ionic liquid crystals, *J. Chem. Soc., Dalton Trans.*, 2133–2139, 1999.

53. Triolo, A., Mandanici, A., Russina, O., Rodriguez-Mora, V., Cutroni, M., Hardacre, C., Nieuwenhuyzen, M., Bleif, H.-J., Keller, L., and Ramos, M. A., Thermodynamics, structure, and dynamics in room temperature ionic liquids: The case of 1-butyl-3-methyl Imidazolium hexafluorophosphate ([bmim][PF$_6$]), *J. Phys. Chem. B*, 110, 21357–21364, 2006.

54. Triolo, A., Russina, O., Arrighi, V., Juranyi, F., Janssen, S., and Gordon, C. M., Quasielastic neutron scattering characterization of the relaxation processes in a room temperature ionic liquid, *J. Chem. Phys.*, 119, 8549–8557, 2003.

55. Roche, J. D., Gordon, C. M., Imrie, C. T., Ingram, M. D., Kennedy, A. R., Lo Celso, F., and Triolo, A., Application of complementary experimental techniques to characterization of the phase behavior of [C$_{16}$mim][PF$_6$] and [C$_{14}$mim][PF$_6$], *Chem. Mater.*, 15, 3089–3097, 2003.

56. Firestone, M. A., Dzielawa, J. A., Zapol, P., Curtiss, L. A., Seifert, S., and Dietz, M. L., Lyotropic liquid-crystalline gel formation in a room-temperature ionic liquid, *Langmuir*, 18, 7258–7260, 2002.

57. Saha, S., Hayashi, S., Kobayashi, A., and Hamaguchi, H., Crystal structure of 1-butyl-3-methylimidazolium chloride. A clue to the elucidation of the ionic liquid structure, *Chem. Let., (Japan)*, 32, 740–741, 2003.

58. Holbrey, J. D., Reichert, W. M., Nieuwenhuyzen, M., Johnston, S., Seddon, K. R., and Rogers, R. D., Crystal polymorphism in 1-butyl-3-methylimidazolium halides: Supporting ionic liquid formation by inhibition of crystallization, *Chem. Commun.*, 1636–1637, 2003.

59. Katayanagi, H., Hayashi, S., Hamaguchi, H., and Nishikawa, K., Structure of an ionic liquid, 1-n-butyl-3-methylimidazolium iodide, studied by wide-angle x-ray scattering and Raman spectroscopy, *Chem. Phys Lett.*, 392, 460–464, 2004.

60. Bowers, J., Vergara-Gutierrez, M. C., and Webster, J. R. P., Surface ordering of amphiphilic ionic liquids, *Langmuir,* 20, 309–312, 2004.

61. Bowers, J., Butts, C. P., Martin, P. J., Vergara-Gutierrez, M. C., and Heenan, R. K., Aggregation behavior of aqueous solutions of ionic liquids, *Langmuir,* 20, 2191–2198, 2004.

62. Takahashi, S., Curtiss, L. A., Gosztola, D., Koura, N., and Saboungi, M.-L., Molecular orbital calculations and raman measurements for 1-ethyl-3-methyl-imidazolium chloroaluminates, *Inorg. Chem.,* 34, 2990–2993, 1995.

63. Huddleston, J. G., Visser, A. E., Reichert, W. M., Willauer, H. D., Broker, G. A., and Rogers, R. D., Characterization and comparison of hydrophilic and hydrophobic room temperature ionic liquids incorporating the imidazolium cation, *Green Chem.,* 3, 156–164, 2001.

64. Hayashi, S., Ozawa, R., and Hamaguchi, H., Raman spectra, crystal polymorphism, and structure of a prototype ionic-liquid [bmim]Cl, *Chem. Lett.,* 32, 498–499, 2003.

65. Nishikawa, K., Wang, S., Katayanagi, H., Hayashi, S., Hamaguchi, H., Koga, Y., and Tozaki, K., Melting and freezing behaviors of prototype ionic liquids, 1-butyl-3-methylimidazolium bromide and its chloride, studied by using a nano-watt differential scanning calorimeter, *J. Phys. Chem. B,* 111, 4894–4900, 2007.

66. Cambridge Crystallographic Data Centre, [C4mim]Cl Crystal (1) and [C4mim]Br data are registered as CCDC deposition number 213959 and 213960, respectively, see http://www.ccdc.cam.ac.uk.

67. Aakeröy, C. B., Evans, T. A., Seddon, K. R., and Pálinkó, I., The C-H., Cl hydrogen bond: Does it exist?, *New J. Chem.,* 23, 145–152, 1999.

68. van den Berg, J.-A., and Seddon, K. R., Critical Evaluation of C-H··· X hydrogen bonding in the crystalline state, *Cryst. Growth Des.,* 3, 643–661, 2003.

69. Matsumoto, K., Hagiwara, R., Mazej, Z., Benki˘c, P., and Žemva, B., Crystal structures of frozen room temperature ionic liquids, 1-ethyl-3-methylimi-dazolium tetrafluoroborate (EMImBF$_4$), hexafluoroniobate (EMImNbF$_6$) and hexafluorotantalate (EMImTaF$_6$), determined by low-temperature x-ray diffraction, *Solid State Sci.,* 8, 1250–1257, 2006.

70. Ozawa, R., Hayashi, S., Saha, S., Kobayashi, A., and Hamaguchi, H., Rotational isomerism and structure of the 1-butyl-3-methylimidazolium cation in the ionic liquid state, *Chem. Lett.,* 32, 948–949, 2003.

71. Okajima, H., and Hamaguchi, H., Low frequency Raman spectroscopy and liquid structure of imidazolium based ionic liquids. Abstracts of Papers, 231st ACS National Meeting, Atlanta, U.S., March 26–30, 2006, IEC-015. American Chemical Society, Washington, D.C., 2006.

72. Turner, E. A., Pye, C. C., and Singer, R. D., Use of ab initio calculations toward the rational design of room temperature ionic liquids, *J. Phys. Chem. A,* 107, 2277–2288, 2003.

73. Berg, R. W., Unpublished results, 2006.

74. Talaty, E. R., Raja, S., Storhaug, V. J., Dölle, A., and Carper, W. R., Raman and infrared spectra and ab initio calculations of C$_{2-4}$MIM imidazolium hexafluorophosphate ionic liquids, *J. Phys. Chem. B,* 108, 13177–13184, 2004.

75. Antony, J. H., Mertens, D., Dölle, A., Wasserscheid, P., and Carper, W. R., Molecular reorientational dynamics of the neat ionic liquid 1-butyl-3-methyl-imidazolium hexafluorophosphate by measurement of ^{13}C nuclear magnetic relaxation data., *Chem. Phys. Chem.,* 4, 588–594, 2003.

76. Antony, J. H., Mertens, D., Breitenstein, T., Dölle, A., Wasserscheid, P., and Carper, W. R., Molecular structure, reorientational dynamics, and

intermolecular interactions in the neat ionic liquid 1-butyl-3-methylimidazo-
lium hexafluorophosphate, *Pure Appl. Chem.*, 76, 255–261, 2004.

77. Heimer, N. E., Del Sesto, R. E., and Carper, W. R., Evidence for spin diffusion
    in a H,H-NOESY study of imidazolium tetrafluoroborate ionic liquids, *Magn.
    Reson. Chem.*, 42, 71–75, 2004.

78. Heimer, N. E., Del Sesto, R. E., Meng, Z., Wilkes, J. S., and Carper, W. R.,
    Vibrational spectra of imidazolium tetrafluoroborate ionic liquids, *J. Mol.
    Liquids*, 124, 84–95, 2006.

79. Matsumoto, K., Hagiwara, R., Yoshida, R., Ito, Y., Mazej, Z., Benkic, P., Zemva,
    B., Tamada, O., Yoshino, H., and Matsubara, S., Syntheses, structures and prop-
    erties of 1-ethyl-3-methylimidazolium salts of fluorocomplex anions, *J. Chem.
    Soc. Dalton Trans.*, 144–149, 2004.

80. Umebayashi, Y., Fujimori, T., Sukizaki, T., Asada, M., Fujii, K., Kanzaki, R., and
    Ishiguro, S.-I., Evidence of conformational equilibrium of 1-ethyl-3-methylimi-
    dazolium in its ionic liquid salts: Raman spectroscopic study and quantum
    chemical calculations, *J. Phys. Chem. A*, 109, 8976–8982, 2005.

81. Downard, A., Earle, M. J., Hardacre, C., McMath, S. E. J., Nieuwenhuyzen, M.,
    and Teat, S. J., Structural studies of crystalline 1-alkyl-3-methylimidazolium
    chloride salts, *Chem. Mater.*, 16, 43–48, 2004.

82. Giraud, G., Gordon, C. M., Dunkin, I. R., and Wynne, K., The effects of
    anion and cation substitution on the ultrafast solvent dynamics of ionic
    liquids: A time-resolved optical Kerr-effect spectroscopic study, *J. Chem. Phys.*,
    119, 464–477, 2003.

83. Shigeto, S., and Hamaguchi, H., Evidence for mesoscopic local structures in
    ionic liquids: CARS signal spatial distribution of $C_n$mim[PF_6] (n = 4, 6, 8), *Chem.
    Phys. Lett.*, 427, 329–332, 2006.

84. Morrow, T. I., and Maginn, E. J., Molecular dynamics study of the Ionic Liquid
    1-n-Butyl-3-methylimidazolium hexafluorophosphate, *J. Phys. Chem. B*, 106,
    12807–12813 (2002).

85. Morrow, T. I., and Maginn, E. J., Erratum to document, *J. Phys. Chem. B*, 107,
    9160, 2003.

86. Shah, J. K., and Maginn, E. J., A Monte Carlo simulation study of the ionic liquid
    1-n-butyl-3-methylimidazolium hexafluorophosphate: Liquid structure, volu-
    metric properties and infinite dilution solution thermodynamics of CO_2, *Fluid
    Phase Equilb.*, 222–223, 195–203, 2004.

87. Hamaguchi, H., Saha, S., Ozawa, R., and Hayashi, S., Raman and x-ray stud-
    ies on the structure of [bmim]X (X = Cl, Br, I, [BF_4], [PF_6]): Rotational isomer-
    ism of the [bmim]$^+$ cation, in ACS Symposium Series 902 (Ionic Liquids IIIA:
    Fundamentals, Progress, Challenges, and Opportunities Roger, R. D. (Ed.)),
    68–78, 2005.

88. Lopes, J. N. C., Gomes, M. F. C., and Pádua, A. A. H., Nonpolar, polar, and asso-
    ciating solutes in ionic liquids, *J. Phys. Chem. B*, 110, 16816–16818, 2006.

89. Abdul-Sada, A. K., Elaiwi, A. E., Greenway, A. M., and Seddon, K. R., Evidence
    for the clustering of substituted imidazolium salts via hydrogen bonding
    under the conditions of fast atom bombardment mass spectrometry, *Eur.
    Mass Spectrom*, 3, 245–247, 1997.

90. Lopes, J. N. C., Deschamps, J., and Pádua, A. A. H., Modeling ionic liquids using
    a systematic all-atom force field, *J. Phys. Chem. B.*, 108, 2038–2047 and correction
    11250, 2004.

91. Lopes, J. N. C., and Pádua, A. A. H., Molecular force field for ionic liquids com-
    posed of triflate or bistriflylimid anions, *J. Phys. Chem. B*, 108, 16893–16898, 2004.

92. Lopes, J. N. A. C., and Pádua, A. A. H., Using spectroscopic data on imidazolium cation conformations to test a molecular force field for ionic liquids, *J. Phys. Chem. B*, 110, 7485–7489, 2006.

93. Lopes, J. N. A. C., and Pádua, A. A. H., Nanostructural organization in ionic liquids, *J. Phys. Chem. B*, 110, 3330–3335, 2006.

94. De Andrade, J., Böes, E. S., and Stassen, H., A force field for liquid state simulations on room temperature molten salts: 1-ethyl-3methylimidazolium tetrachloroaluminate, *J. Phys. Chem. B*, 106, 3546–3548, 2002.

95. De Andrade, J., Boes, E. S., and Stassen, H., Computational study of room temperature molten salts composed by 1-alkyl-3methylimidazolium cations-force field proposal and validation, *J. Phys. Chem., B*, 106, 13344–13351, 2002.

96. Margulis, C. J., Stern, H. A., and Berne, B. J., Computer simulation of a green chemistry room-temperature ionic solvent, *J. Phys. Chem. B.*, 106, 12017–12021, 2002.

97. Del Pópolo, M. G., and Voth, G. A., On the structure and dynamics of ionic liquids, *J. Phys. Chem. B*, 108, 1744–1752, 2004.

98. Yan, T., Burnham, C. J., Del Pópolo, M. G., and Voth, G. A., Molecular dynamics simulation of ionic liquids: The effect of electronic polarizability, *J. Phys. Chem. B*, 108, 11877–11881, 2004.

99. Liu, Z., Huang, S., and Wang, W., A refined force field for molecular simulation of imidazolium-based ionic liquids, *J. Phys. Chem. B*, 108, 12978–12989, 2004.

100. Urahata, S. M., and Ribeiro, M. C. C., Structure of ionic liquids of 1-alkyl-3-methylimidazolium cations: A systematic computer simulation study, *J. Chem. Phys.*, 120, 1855–1863, 2004.

101. Hayashi, S., and Hamaguchi, H., Discovery of a magnetic ionic liquid [bmim]$FeCl_4$, *Chem. Lett.*, 33, 1590–1591, 2004.

102. Hayashi, S., Saha, S., and Hamaguchi, H., A new class of magnetic fluids: bmim[$FeCl_4$] and nbmim[$FeCl_4$] ionic liquids, *IEEE Trans. Magn.*, 42, 12–14, 2006.

103. Miki, H., Hayashi, S., Kikura, H., and Hamaguchi, H., Raman spectra indicative of unusual water structure in crystals formed from a room-temperature ionic liquid, *J. Raman Spectros.*, 37, 1242–1243, 2006.

104. Saha, S., and Hamaguchi, H., Effect of water on the molecular structure and arrangement of nitrile-functionalized ionic liquids, *J. Phys. Chem. B*, 110, 2777–2781, 2006.

105. Gale, R. J., Gilbert, B., and Osteryoung, R. A., Raman spectra of molten aluminum chloride: 1-butylpyridinium chloride systems at ambient temperatures, *Inorg. Chem.*, 17, 2728–2729, 1978.

106. Tian, P., Song, X., Li, Y., Duan, J., Liang, Z., and Zhang, H., Studies on room temperature ionic liquid $FeCl_3$ - n-butylpyridinium chloride (BPC) system. *Huaxue Xuebao*, 64(23), 2305–2309, 2006. Journal in Chinese, cited from Chemical Abstracts.

107. Rey, I., Johansson, P., Lindgren, J., Lassegues, J. C., Grondin, J., and Servant, L., Spectroscopic and theoretical study of $(CF_3SO_2)_2N^-$ (TFSI$^-$) and $(CF_3SO_2)_2NH$ (HTFSI), *J. Phys. Chem. A*, 102, 3249–3258, 1998.

108. Rey, I., Lassegues, J. C., Grondin, J., and Servant, L., Infrared and Raman study of the PEO-LiTFSI polymer electrolyte, *Electrochim. Acta*, 43, 1505–1510, 1998.

109. Fujii, K., Fujimori, T., Takamuku, T., Kanzaki, R., Umebayashi, Y., and Ishiguro, S., Conformational equilibrium of bis(trifluoromethanesulfonyl) imide anion of a room-temperature ionic liquid: Raman spectroscopic study and DFT calculations, *J. Phys. Chem. B*, 110, 8179–8183, 2006.

110. Matsumoto, K., Hagiwara, R., and Tamada, O., Coordination environment around the lithium cation in solid Li$_2$(EMIm)(N(SO$_2$CF$_3$)$_2$)$_3$ (EMIm = 1-ethyl-3-methylimidazolium): Structural clue of ionic liquid electrolytes for lithium batteries, *Solid State Sci.*, 8, 1103–1107, 2006.
111. Castriota, M., Caruso, T., Agostino, R. G., Cazzanelli, E., Henderson, W. A., and Passerini, S., Raman investigation of the ionic liquid N-methyl-N-propylpyrrolidinium bis(trifluoromethanesulfonyl)imide and its mixture with LiN (SO$_2$CF$_3$)$_2$, *J. Phys. Chem. A*, 109, 92–96, 2005.
112. Lancaster, N. L., Salter, P. A., Welton, T., and Young, G., Brent, Nucleophilicity in ionic liquids. 2. Cation effects on halide nucleophilicity in a series of bis [tri-fluoromethylsulfonyl]imide ionic liquids, *J. Org. Chem.*, 67, 8855–8861, 2002.
113. Dal, E., and Lancaster, N. L., Acetyl nitrate nitrations in [bmpy][N(Tf)$_2$] and [bmpy][OTf], and the recycling of ionic liquids, *Org. Biomol. Chem.*, 3, 682–686, 2005.
114. Fujimori, T., Fujii, K., Kanzaki, R., Chiba, K., Yamamoto, H., Umebayashi, Y., and Ishiguro, S.-I., Conformational structure of room temperature ionic liquid N-butyl-N-methyl-pyrrolidinium bis(trifluoromethanesulfonyl)imide - Raman spectroscopic study and DFT calculations, *J. Mol. Liquids*, 131–132, 216–224, 2007.
115. Choudhury, A. R., Winterton, N., Steiner, A., Cooper, A. I., and Johnson, K. A., In situ crystallization of low-melting ionic liquids, *J. Am. Chem. Soc.*, 127, 16792–16793, 2005.
116. Holbrey, J. D., Turner, M. B., Reichert, W. M., and Rogers, D. R., New ionic liquids containing an appended hydroxyl functionality from the atomefficient, one-pot reaction of 1-methylimidazole and acid with propylene oxide, *Green Chem.*, 5, 731–736, 2003.
117. Liu, X., Zhang, S., Zhou, G., Wu, G., Yuan, X., and Yao, X., New force field for molecular simulation of guanidinium-based ionic liquids, *J. Phys. Chem. B.*, 110, 12062–12071, 2006.
118. Ishiguro, S.-I., Umebayashi, Y., and Kanzaki, R., Characterization of metal ions in coordinating solvent mixtures by means of raman spectroscopy, *Anal. Sci. (Japan)*, 20, 415–421, 2004.
119. Fujisawa, T., Fukuda, M., Terazima, M., and Kimura, Y., Raman spectroscopy study on solvation of diphenylcyclopropenone and phenol blue in room temperature ionic liquids, *J. Phys. Chem. A*, 110, 6164–6172, 2006.

chapter thirteen

Nuclear magnetic resonance spectroscopy in ionic liquids

Ralf Giernoth

Contents

13.1   Introduction

Nuclear magnetic resonance (NMR) spectroscopy is the most widely used spectroscopic technique in synthetic chemistry [1]. One main reason for the dominance of NMR is its versatility—by variation of only a few experimental parameters, a vast number of different NMR experiments can easily be performed, giving access to very different sets of information on the substance or the reaction under investigation. Today, NMR is dominant in structure elucidation, and *in situ* NMR spectroscopy can conveniently give insight into chemical reactions under real turnover conditions (in contrast to, e.g., x-ray crystallography, which can only provide a solid-state *snapshot* of a molecular conformation).

Keeping these facts in mind, there is an obvious interest in being able to apply NMR to the chemistry of and in ionic liquids (ILs). This chapter focuses on the main achievements in the field, giving a tutorial-like approach to NMR spectroscopy in ILs. If you prefer a more review-like source of information that aims at completeness and a more historical style, please refer to the recent literature [2].

13.2   Ionic liquids: general remarks

Of course, ILs are no invention of recent years—in fact, they are around for quite some time already [3]. But the use of the term *ionic liquid* as opposed to the more general term *molten salt* implies that the point of view has changed over the last decades. From an organic chemist's point of view, ILs are the molten salts that can be used as solvents for synthesis.

In this respect, it has become quite common to talk of two generations if ILs, although, strictly speaking, there are in fact three: the so-called first-generation ILs are the chloroaluminates while the second-generation ILs are the modern air- and moisture-stable ILs that can easily be handled even by untrained chemists. The long period of time before is not taken into account, since ILs did not play any significant role as solvents for chemical synthesis in that period of time.

There are many differences between first- and second-generation ILs. Therefore, it seems sensible to discuss them seperately. But since the first-generation (chloroaluminate) ILs have substantially lost in significance these days, this part will be kept as short as possible—even though the major advancements in IL-NMR has been achieved with chloroaluminates.

13.3   Nuclear magnetic resonance
with salt-containing samples

The challenge in NMR spectroscopy is to obtain well-resolved spectra with high sensitivity. NMR spectroscopy generally is a quite unsensitive method, for normally it relys solely on population differences of energy levels due to the Boltzmann distribution [1]. The resolution of NMR signals is directly correlated with sensitivity and line shape (line width versus signal height). The width of an NMR signal is influenced by many parameters, such as field homogenity, sample homogenity, sample viscosity, relaxation, and so on.

For NMR in ILs, three major effects play a key role (a) they often are relatively viscous, (b) they are normally not available in deuterated form, and (c) they comprise entirely of ions; they are liquid salts. As for viscosity (a), not much can be done about it (apart from applying MAS techniques, cf. Section 13.5.7). Problem (b) will be discussed in Section 13.5.5.

Salt-containing samples (c) give rise to broad lines because of high relaxation rates. In addition, some of the radiation used for the NMR pulse sequence may be absorbed by the sample, which leads to elongated NMR pulses (up to three times the lengths of pulses in standard NMR solvents, due to our own observations) and sometimes heating of the sample. Problems of this type may be overcome by using specialized cryo-probes with a special NMR tube geometry [4], although, apparently, this technique has not yet been used in conjunction with ILs. Nonetheless, after careful setup, routine NMR in and with ILs is possible (more on this topic in Section 13.5.1).

13.4   First-generation ionic liquids: chloroaluminates

Chloroaluminate ILs, the *first-generation ILs*, consist of a mixture of an -*onium* chloride (typically, of course, imidazolium or ammonium) and varying proportions of aluminum trichloride ($AlCl_3$). (The structures of all cations and anions that are discussed in this chapter are depicted in Figure 13.1.)

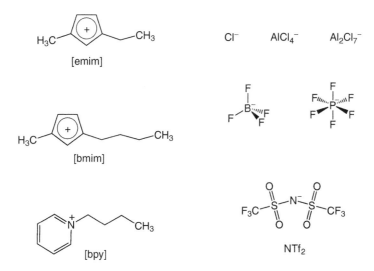

Figure 13.1   Ionic liquid cations and anions discussed in this chapter.

The $Cl/AlCl_3$ ratio defines the properties of an IL. If $AlCl_3$ is added substoichiometrically ($Cl/AlCl_3 > 1$), the resulting liquid will contain a mixture of $AlCl_4^-$ and Cl^-. Therefore, it is basic in nature. Addition of more than one equivalent of $AlCl_3$ ($Cl/AlCl_3 < 1$) leads to a liquid containing a mixture of $AlCl_4^-$, $Al_2Cl_7^-$ and higher-aggregated complexes, making these ILs acidic in nature (since the chloroaluminate anions are strong Lewis acids). Only a mixture of perfect one-to-one molar ratio ($Cl/AlCl_3 = 1$) gives a defined IL *[-onium]AlCl_4* like we would expect from second-generation ILs. (It has to be noted that, although to much smaller extent, bromoaluminate ILs are also described in the literature.)

13.4.1   Structure and speciation

Due to the fact that most chloroaluminate ILs comprise a mixture of species, much of the early NMR work of and in ILs was focused on speciation and on investigating the liquid-phase structure of these liquids.

13.4.1.1 ^1H and ^{13}C NMR: melt composition and H-bonding

Comparably few well-resolved NMR spectra of neat ILs have appeared in the literature. As pointed out above, the ionic nature of these media can result in very broad lines if the experiment (and, at the same time, the spectrometer) are not carefully optimized for these systems. Thus, it is easily understandable that frequently only the chemical shift information and/or signal integrals have been taken into account, whereas time-consuming spectrum optimization has been omitted.

Nonetheless, the chemical shift information of selected nuclei can be used advantageously for the determination of the composition of an unknown chloroaluminate melt. Wilkes et al. have recorded ^1H spectra of 1-ethyl-3-methyl imidazolium ($[C_2C_1Im]$) and butylpyridinium ($[C_4py]$) chloroaluminates and found that the 2-position in $[C_2C_1Im]$ as well as the 6-position in $[C_4py]$ are strongly composition-dependent [5,6]. With the help of this method, the melt composition can be estimated to $\pm0.001\chi(AlCl_3)$. Interestingly, the chemical shift dependence is far more pronounced in the basic than in the acidic region.

In the latter study [6], Wilkes was among the first to question ion pairing in ILs. By fitting ion aggregation models to the differences in chemical shifts with melt compositions, binary aggregates could not be explained. Instead, a much better fit to the experimental data was found for alternating cation–anion chains. Based upon these very early NMR investigations (and, of course, lots of subsequent studies), we know today that many ILs indeed form a characteristically ordered internal structure in the liquid phase. Much later, Johnson et al. have studied hydrogen bonding in $[C_2C_1Im]$ chloroaluminates in a similar fashion via chemical shift differences in variable-temperature ^1H NMR experiments [7]. They concluded that in fact the ability of the anions to form H bonds is responsible for supramolecular aggregation in these ILs.

More information on supramolecular aggregation phenomena has been produced with the help of relaxation data. Standard inversion recovery experiments on [C_4py] chloroaluminates gave T1 relaxation times that were different for basic and acidic melts, and also different for the α- and the δ-protons in the alkyl side chain of the cation [8]. Obviously, basic melts revealed strong cation–anion interactions, while in acidic melts the rotation of the terminal methyl group was largely uncoupled from the rest of the molecule, thus showing no participation in aggregation.

Finally, ^{1}H ROESY NMR experiments have been applied to study hydrogen bonding via intermolecular nuclear Overhauser effects (NOEs) in basic imidazolium chloroaluminates [9]. Interestingly, many NOEs could be detected—more than one would initially expect, since several of the nuclei were separated well beyond 4 Å, which is the natural "limit" for this kind of experiment.

13.4.1.2 ^{27}Al and ^{17}O NMR: anion speciation

Investigations on speciation and equilibria in chloroaluminate melts primarily concern the anions, since the cations are not subject to change. Therefore, ^{27}Al NMR is an obvious tool to investigate these ILs. With the help of one-dimensional NMR, the number of species not participating in rapid exchange can be studied, while two-dimensional exchange spectroscopy can reveal the number and the nature of the exchanging species [10–13]. Additionally, the fact that the ^{27}Al nucleus is quadrupolar can be used for investigations on the symmetry of the species involved, by means of line width and relaxation rate measurements: a high symmetry around Al decreases the quadrupolar relaxation, resulting in narrower signals [10,11,13–17].

Oxygen is normally introduced into chloroaluminate melts as a trace impurity. Since chloroaluminate anions are easily prone to oxidation, new species are formed this way. These are, of course, accessible by ^{17}O and ^{27}Al NMR spectroscopies. By adding ^{17}O-enriched water to [C_2C_1Im]Cl/AlCl$_3$, the group of Osteryoung was able to identify five different oxygen-containing species—three in acidic and two in basic melts [18,19] (Figure 13.2).

13.4.1.3   Dual spin probe experiments: quadrupolar coupling constants

Dual spin probe (DSP) techniques [20] are based on the correlation of relaxation data of two different nuclei, allowing for the determination of quadrupolar coupling constants. In cases where these coupling constants exist, the nuclei under investigation are within the same molecule or aggregate.

With this method, mixtures of [C_2C_1Im]Cl/AlCl$_3$ and of [C_2C_1Im]Cl/AlCl$_3$/NaCl have been studied via ^{13}C/^{27}Al dipolar relaxation measurements [21]. The quadrupolar coupling constants were larger in the latter NaCl-containing case, so strong coordination of Cl$^-$ by Na$^+$ was concluded, leading to a distortion of the tetrahedral symmetry of AlCl$_4^-$. With the help of additional ^{23}Na relaxation measurements [22], it could be shown that a hypothetic complex of the type [C_2C_1Im] (AlCl$_4$)$_n$Na$_m$, containing all three nuclei (^{13}C, ^{27}Al, and ^{23}Na), must in fact exist.

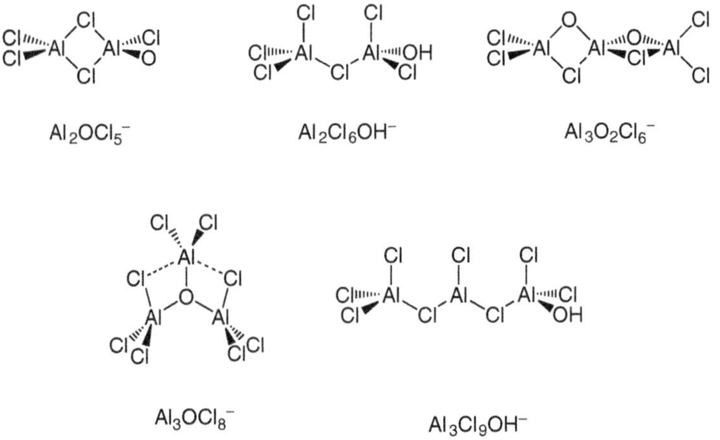

Figure 13.2 Oxygen-containing anion species in chloroaluminate ILs, as proposed by Osteryoung and Zawodzinsky (Zawodzinski, T. A., Jr. and Osteryoung, R. A., *Inorg. Chem.*, 29, 2842, 1990.).

13.4.2 Determination of physical properties

NMR spectroscopy has been successfully applied to determine physical properties of chloroaluminate ILs that are difficult to achieve by other means. These data include electronegativities, self-diffusion coefficients and Gutmann numbers.

Aluminum electronegativities for various mixtures have been determined by Chauvin et al. [23]. The group used a combination of ^{1}H NMR and the application of modified Dailey-Shoolery equations. In this fashion, composition-independent electronegativities of 1.43 could only be determined for basic melts. The electronegativities in acidic melts were strongly dependent on the melt composition, ranging from 1.2 to 1.6.

Self-diffusion coefficients have been obtained by diffusion-ordered spectroscopy (DOSY [24,25]). For the [C_2C_1Im]Cl/AlCl$_3$ system, the diffusion constants for all [C_2C_1Im] proton resonances were identical. A linear relationship between diffusion constants and conductivities of the melts demonstrated the fact that the transport properties are determined by the molar quantities (and not by the properties of the individual ions).

Gutmann acceptor numbers were determined in the "usual" way via the ^{31}P chemical shift variation of triphenylphosphine oxide by Osteryoung et al. [26]. While, again, the donor numbers were concentration- and composition-independent for basic melts, the acidic melts showed a strong composition dependence. Nonetheless, the acidity range was comparably small and was found around 100 (which compares to the acidity of trifluoroacetic acid). The donor number for basic melts was found to be 98, which was, of course, unrealistically high. As an explanation for this, the authors proposed that Et_3PO is a stronger base than Cl^-, so that the following (for the determination of acceptor numbers unwanted) side reaction takes place:

$$AlCl_4^- + Et_3PO \rightleftharpoons Et_3PO \cdot AlCl_3 + Cl^-$$

In this way, the acceptor number of $AlCl_3$ (and not the one of $AlCl_4^-$) was determined.

13.4.3   Chloroaluminate solvents

As mentioned earlier, the reactivity of chloroaluminate ILs toward oxygen (and hydrogen) as well as the fact that most of the times no defined single anion is present renders these melts unsuitable as solvents for many applications. Nonetheless, they are a few reports of reactions in chloroaluminate ILs that were studied via NMR.

Homogeneous Ziegler–Natta catalysis has been studied in $[C_2C_1Im]Cl/AlCl_3$ melts [27]. Interestingly, catalysis only took place in acidic melts. Strong complex formation with Cp_2TiCl_2 could be seen via the variation of the Cp proton shift as a function of melt composition. For Cp_2HfCl_2 and Cp_2ZrCl_2, only weak complexation was found.

The catalysis of Friedel–Crafts sulfonylation reactions by $[C_4C_1Im]Cl/AlCl_3$ has been studied by Salunkhe et al. [28]. The authors were able to distinguish between $AlCl_4^-$ and $Al_2Cl_7^-$ simply by looking at the line widths (as discussed earlier already): the higher symmetry of $AlCl_4^-$ reduces the line width of this ^{27}Al resonance considerably as compared to the other species. While in an acidic 1:2 mixture initially only $Al_2Cl_7^-$ was detectable, the population of $AlCl_4^-$ increased rapidly after addition of the substrate. As an explanation, the authors proposed the liberation of HCl during the reaction. Subsequently, an acid–base reaction between HCl and $Al_2Cl_7^-$ produces $AlCl_4^-$, with the sulfonylating agent tosyl chloride being the base (B):

$$Al_2Cl_7^- + HCl + B \rightleftharpoons 2AlCl_4^- + BH^-$$

13.5   Second-generation (modern) ionic liquids

13.5.1   Structure and speciation: ion pairing and hydrogen bonding

13.5.1.1   Ionic liquids in molecular solvents

"Second-generation" or "modern" ILs are well defined cation–anion combinations that are liquid at room (or reaction) temperature. They normally consist of exactly one cation and one anion, and they are often air- and moisture-stable. Therefore, investigations on structure and speciation of this class of ILs have a different focus. In addition, the higher stability of these

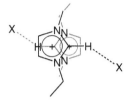

Figure 13.3 π-stacking of imidazolium cations (Avent, A. G., Chaloner, P. A., Day, M. P. et al., *J. Chem. Soc., Dalton Trans.*, 3405, 1994.)

ILs renders it much easier to investigate them in common NMR solvents, like any other chemical substance. In this fashion, the application of all common NMR techniques is possible.

The main focus in studies of this type lies on ion pairing and the presence of hydrogen bonds in these ILs. To this end, the variation of chemical shifts can be used as an indicator: isolated ions show different chemical shifts as opposed to ions with tight contact to other ions or molecules.

Many groups have investigated the changes in chemical shifts of various imidazolium-based ILs in different organic solvents at different concentrations [29–33], using, among others, ^1H, ^{13}C, ^{35}Cl, and ^{127}I NMR. From these studies, it could be shown that ILs form ion pairs in non-polar solvents, promoting π-stacking of the aromatic cations (Figure 13.3).

Hoffman et al. were in fact able to detect two different resonances for the cation of $[C_2C_1Im][Tf_2N]$ in $CDCl_3$, visualizing the $[C_2C_1Im][Tf_2N]$ ion pair and "free" $[C_2C_1Im]^+$ cations [34]. Dupont applied NOE techniques to detect inter- and intramolecular cation–anion interactions of $[C_4C_1Im][BPh_4]$ in $CDCl_3$ [35]. Unfortunately, the spectra in this publication show the presence of a considerable amount of water that may change the properties of these solutions severely. Polar solvents, like water in particular, lead to neglible cation–anion interactions and promote the formation of hydrogen bonds, especially with halide anions. This way, three-dimensional networks of anions and cations, linked by hydrogen bonds, were proposed.

Finally, DOSY [24,25] has been used to study ion aggregation, since the ionic radii are directly correlated to the respective diffusion coefficients. In this fashion, it could be demonstrated for $[C_4C_1Im]Br$ in D_2O that this IL is only completely dissolved for very dilute solutions [36]. Above a mole fraction of 0.015, ion pairing is dominant. A similar effect has been reported by Pregosin et al. who used a combination of ^1H, $\{^{19}F\}$ HOESY, ^1H and ^{19}F DOSY techniques [37].

13.5.1.2 Neat ionic liquids

Apart from (sometimes substantial) parameter adjustment, most standard NMR techniques can be performed in neat ILs as well [38]. The only *complications* are the high viscosities of these solvents (resulting in longer relaxation times and broader line widths) and the high number of residual solvent signals, especially in ^1H NMR.

Diffusion measurements in neat ILs can reveal information on the internal structure of these media and on ionic association. But due to the high viscosities of most ILs, strong gradients or quite long diffusion times are necessary. With this technique, it was found that most IL cations diffuse faster than their corresponding anions [39], but the molecular size of anions and cations did not correlate well with their diffusion coefficients. In addition to this, the diffusion coefficients of the cations were strongly anion-dependent [40].

Hydrogen bonding could be studied with the help of ^{13}C relaxation measurements in combination with NOE factor determination [41]. For $[C_4C_1Im][PF_6]$ it was found that the 2-proton on the imidazolium ring as well as the hydrogens at the butyl side chain are forming hydrogen bonds to the anion. For the same IL, the similar conclusions resulted from a reorientational dynamics study and led to the assumption of higher aggregates in a layer structure [42]. NOESY experiments on Tf_2N^- and BF_4^- salts, interestingly, detected no intermolecular contacts for the Tf_2N^- species [43]. In the BF_4^- case, the preferred relative orientations of the cations did not include parallel-aligned side chains, as one would expect. Thus, the formation of nonpolar domains in the liquid seems unlikely.

13.5.2   Charge distribution in imidazolium cations

A question of some (theoretical) debate for imidazolium cations is whether the cationic charge is equally distributed over the aromatic ring [44–48]. This can be addressed via nitrogen NMR. The group of Lyčka have applied ^{15}N NMR at natural abundance on 12 ILs [49]. For symmetrical ILs, only one NMR signal was found. Unsymmetrically substituted cations revealed two individual signals, hinting at a slight asymmetric charge distribution. In contrast, Blümel et al. used ^{14}N NMR measurements [50] to detect two individual signals for the two nitrogen nuclei with very similar line widths, proving a rather balanced charge distribution.

13.5.3   Impurities and residual solvents

Impurities and residual solvents can have dramatic effects on the physical properties of ILs [3]. Since many ILs are highly hygroscopic, water is especially dominant in this respect. By combination of H,H ROESY, and $^1H\{^{19}F\}$ NOE spectroscopy, further evidence was found for the fact that water breaks up ionic aggregates and minimizes ion–ion interactions [51]. At low water concentrations, selective interaction with the aromatic ring was found. The group of Seddon has studied the effects of water, chloride ions, and organic solvents on ILs [52]. High concentrations of chloride ions led to downfield shifts in the ^{1}H NMR, which was attributed to the formation of stronger hydrogen bonds. The change in physical properties of ILs upon addition of small contaminations has been studied by Kanakubo et al. [53]. Interestingly, diffusion coefficients varied up to 30% between a high-purity sample and commercially available [C$_4$C$_1$Im][PF$_6$] of 97% purity. This demonstrates how the presence of small amounts of cosolvents can weaken interionic Coulomb interactions.

13.5.4   Binary mixtures of modern ionic liquids

For the investigation of binary mixtures of "second-generation ILs", DOSY is the method of choice. In this fashion, Watanabe et al. have studied mixtures of imidazole with different molar ratios of bistriflic amide [54]. For amide-rich (acidic) mixtures, separate signals for each of the imidazole ring protons were found, whereas imidazolium-rich (basic) mixtures showed fast equilibria for both nitrogen-bound protons of the cation and the one in imidazole. In the same fashion, [C$_2$C$_1$Im][BF$_4$]/LiBF$_4$ mixtures have been studied by Hayamitsu et al. [55]. [C$_2$C$_1$Im]$^+$ was found to diffuse faster than BF$_4^-$, and Li$^+$ diffused the most slowly. Therefore, it needs to be noted that cation size here did not correlate with the respective diffusion coefficients at all.

13.5.5   Reaction monitoring

Since it was finally clear that ILs can be used like any other NMR solvent, work has been done on reaction monitoring and the detection of reactive intermediates in these solvents. The "easiest" approach, namely direct reaction monitoring via ^{1}H NMR, is often not applicable, for this normally requires the availability of perdeuterated ILs. Although the synthesis of perdeuterated ILs has been reported in the literature [56–58], this is often not the preferred option, since this obviously involves various additional synthetic steps, which are quite costly at the same time. Abu-Omar et al. therefore have applied reaction monitoring on the ^{2}D NMR channel by using deuterated substrates in nondeuterated ILs [59,60]. This way, they were able to monitor a variety of reactions in ILs and IL/molecular solvent mixtures and detect reactive intermediates.

Our own group has demonstrated a more accessible and straightforward method of solvent signal suppression [61]. By using simple and readily available DOSY techniques, the solvent signals can easily be separated from the signals stemming from the solute molecules (DOSY-editing; Figure 13.4). On the downside, this technique seems to be limited to mixtures without signal overlap in the direct dimension.

For a more specialized example, the reaction of CO$_2$ with various epoxides in different tetrahaloindate(III) ILs has been studied via high-pressure ^{1}H and ^{13}C NMR [62]. Here, it was possible to follow the reaction *in situ* and observe an indium chloroalkyl carbonate as a reactive intermediate. Rencurosi et al.

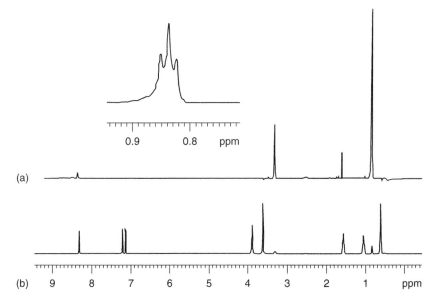

Figure 13.4 Solvent suppression via *DOSY-editing* (Giernoth, R. and Bankmann, D., *Eur. J. Org. Chem.*, 4529, 2005.) *Bottom*: ¹H NMR of one droplet of EtOH in neat [bmim]NTf₂. *Top*: DOSY-edited spectrum of the same sample.

were able to prove the participation of the IL ([C₂C₁Im][TfO]) in a glycolization reaction [63]. The triflate anion formed a transient α-glycosyl triflate, as demonstrated via ¹H and ¹⁹F NMR.

The dissolution of cellulose and other plant-based material in ILs has been studied by Moyna and Rogers [64–66]. Here, the application of ¹³C NMR excluded the need for solvent suppression techniques. In combination with ¹³C and ³⁵Cl/³⁷Cl relaxation measurements, it was found that dissolution of cellulose in [C₄C₁Im]Cl was promoted by the nonhydrated chloride ions, breaking up the hydrogen bonding networks via the cellulose hydroxyl groups.

Platinum-diphosphine complexes in [C₄C₁Im][PF₆] have been studied via ³¹P NMR [67,68]. Here, the insertion reaction of SnCl₂ into the Pt–Cl bond of PtCl₂(bdpp) was observed, as well as the formation of an additional cationic complex. The structure of the catalyst could be elucidated via characteristic ¹J$_{P,P}$ and ¹J$_{Pt,P}$ coupling constants.

13.5.6 Solvent reactivity

It is well-known that many ILs are not really inert solvents, but can be quite reactive. Dialkylimidazolium salts in particular are prone to deprotonation in the 2-position [69–71]. In addition to this, exchange with D₂O was found for all three imidazolium ring protons for halide salts [29,32], but not for BF₄⁻ and PF₆⁻ anions [32]. Our own work demonstrates that highly pure, neat imidazolium ILs do not exchange, but small amounts of basic impurities (like methylimidazole) give rise to immediate exchange [74].

In addition to these acidity/exchange effects, side reactions with supercritical CO_2 have also been studied [72]. The authors reported experiments in a high-pressure NMR cell withstanding pressures up to 1000 bar. For methanol/IL mixtures, ^{13}C NMRs showed signals of bicarbonate and methyl carbonate; ^{19}F investigations additionally revealed hydrated fluoride and methylfluoride, which was supposed to be a dehydration product of methanol with HF.

13.5.7   Ionic liquids on solid support

Blümel et al. have investigated ILs that were immobilized on silica particles [50]. Systems of this type can be used as catalysts for synthetic applications with the advantage of easy separation from the reaction mixture and possibility of recycling [73]. By using ^1H and ^{13}C HR-MAS NMR, well-resolved spectra of the silica surface were obtained, although $^{13}C/^1H$ cross polarization was found not to be very effective. The spectra could be improved substantially by suspending the particles in DMSO, leading to almost liquid state resolution. In this fashion, a number of experimental techniques, such as NOESY, TOCSY, HMQC, and so on could be performed successfully.

13.6   Outlook

Very much in contrast to common belief, the use of ILs as solvents does not preclude the application of NMR techniques. After careful parameter adjustment, virtually all standard and advanced NMR techniques can be applied, as demonstrated by the many examples in this chapter. Therefore, I would like to encourage everyone to pick up the NMR toolbox for applications on IL-based projects. I am convinced that in the near future many more advanced and even some more exotic techniques (like PHIP, CIDNP, ENDOR, TROSY, and so on) will be readily available for use in ILs.

References

1. Claridge, T. D. W., *High-Resolution NMR Techniques in Organic Chemistry*, Elsevier Science Ltd., Oxford, 1999.
2. Bankmann, D. and Giernoth, R., Magnetic resonance spectroscopy in ionic liquids, *Progr. Nucl. Magnet. Res. Spectr.*, 51, 63–90, 2007.
3. Wasserscheid, P. and Welton, T., (Eds), *Ionic Liquids in Synthesis*, 2nd Ed., Wiley-VCH, Weinheim, 2008.
4. Voehler, M. W., Collier, G., Young, J. K. et al., Performance of cryogenic probes as a function of ionic strength and sample tube geometry, *J. Magnet. Res.*, 183, 102, 2006.
5. Wilkes, J. S., Levisky, J. A., Pflug, J. L. et al., Composition determinations of liquid chloroaluminate molten-salts by nuclear magnetic-resonance spectrometry, *Anal. Chem.*, 54, 2378, 1982.

6. Fannin, A. A., Jr., King, L. A., Levisky, J. A. et al., Properties of 1,3-dialkyl-imidazolium chloride-aluminum chloride ionic liquids. 1. Ion interactions by nuclear magnetic resonance spectroscopy, *J. Phys. Chem.*, 88, 2609, 1984.

7. Campbell, J. L. E., Johnson, K. E., and Torkelson, J. R., Infrared and variable-temperature ^{1}H-NMR investigations of ambient-temperature ionic liquids prepared by reaction of HCl with 1-ethyl-3-methyl-1H-imidazolium chloride, *Inorg. Chem.*, 33, 3340, 1994.

8. Zawodzinski, T. A., Jr., Kurland, R., and Osteryoung, R. A., Relaxation time measurements in N-(1-butyl)pyridinium-aluminum chloride ambient temperature ionic liquids, *J. Phys. Chem.*, 91, 962, 1987.

9. Mantz, R. A., Trulove, P. C., Carlin, R. T. et al., ROESY NMR of basic ambient-temperature chloroaluminate ionic liquids, *Inorg. Chem.*, 34, 3846, 1995.

10. Wilkes, J. S., Frye, J. S., and Reynolds, G. F., ^{27}Al and ^{13}C NMR-studies of aluminum-chloride dialkylimidazolium chloride molten salts, *Inorg. Chem.*, 22, 3870, 1983.

11. Gray, J. L. and Maciel, G. E., ^{27}Al nuclear magnetic resonance study of the room-temperature melt $AlCl_3$/N-butylpyridinium chloride, *J. Am. Chem. Soc.*, 103, 7147, 1981.

12. Keller, C. E., Carper, W. R., and Piersma, B. J., Characterization of species in ethylaluminum dichloride molten-salts by ^{27}Al NMR, *Inorg. Chim. Acta*, 209, 239, 1993.

13. Keller, C. E. and Carper, W. R., ^{27}Al NMR relaxation studies of molten-salts containing ethylaluminum dichloride, *Inorg. Chim. Acta*, 210, 203, 1993.

14. Matsumoto, T. and Ichikawa, K., Determination of the ^{27}Al spin-lattice relaxation rate and the relative number of each chloroaluminate species in the molten 1-n-butylpyridinium chloride $AlCl_3$ system, *J. Am. Chem. Soc.*, 106, 4316, 1984.

15. Ichikawa, K. and Matsumoto, T., An aluminium-27 NMR study of chemical exchange and NMR line broadening in molten butylpyridinium chloride + $AlCl_3$ II, *J. Magn. Reson.*, 63, 445, 1985.

16. Takahashi, S., Saboungi, M.-L., Klingler, R. J. et al., Dynamics of room-temperature melts: Nuclear magnetic resonance measurements of dialkyl-imidazolium haloaluminates, *Faraday Trans.*, 89, 3591, 1993.

17. Keller, C. E. and Carper, W. R., ^{13}C and ^{27}Al NMR relaxation studies of ethyl-aluminum dichloride, *J. Magn. Reson. A*, 110, 125, 1994.

18. Zawodzinski, T. A., Jr. and Osteryoung, R. A., Aspects of the chemistry of water in ambient-temperature chloroaluminate ionic liquids: Oxygen-17 NMR studies., *Inorg. Chem.*, 26, 2920, 1987.

19. Zawodzinski, T. A., Jr. and Osteryoung, R. A., Oxide and hydroxide species formed on addition of water in ambient-temperature chloroaluminate melts. An ^{17}O NMR study, *Inorg. Chem.*, 29, 2842, 1990.

20. Dechter, J. J., Henriksson, U., Kowalewski, J. et al., Metal nucleus quadrupole coupling-constants in aluminum, gallium, and indium acetylacetonates, *J. Magn. Reson.*, 48, 503, 1982.

21. Carper, W. R., Pflug, J. L., and Wilkes, J. S., Dual spin probe NMR relaxation studies of ionic structure in 1-ethyl-3-methylimidazolium chloride-$AlCl_3$ molten-salts, *Inorg. Chim. Acta*, 202, 89, 1992.

22. Carper, W. R., Pflug, J. L., and Wilkes, J. S., Multiple spin probe NMR-studies of ionic structure in 1-methyl-3-ethylimidazolium chloride $AlCl_3$ molten-salts, *Inorg. Chim. Acta*, 193, 201, 1992.

23. Chauvin, Y., Di Marco-Van Tiggelen, F., and Olivier, H., Determination of aluminum electronegativity in new ambient-temperature acidic molten-salts based on 3-butyl-1-methylimidazolium chloride and $AlEt_{3-x}Cl_x$ (X = 0-3) by ^{1}H nuclear-magnetic-resonance spectroscopy, *J. Chem. Soc., Dalton Trans.*, 1009, 1993.

24. Gibbs, S. J. and Johnson, C. S., A PFG NMR experiment for accurate diffusion and flow studies in the presence of Eddy currents, *J. Magn. Reson.*, 93, 395, 1991.

25. Wu, D. H., Chen, A. D., and Johnson, C. S., An improved diffusion-ordered spectroscopy experiment incorporating bipolar-gradient pulses, *J. Magn. Reson., Ser A*, 115, 260, 1995.

26. Zawodzinski, T. A., Jr. and Osteryoung, R. A., Donor-acceptor properties of ambient-temperature chloroaluminate melts, *Inorg. Chem.*, 28, 1710, 1989.

27. Carlin, R. T. and Wilkes, J. S., Complexation of Cp_2MCl_2 in a chloroaluminate molten-salt. Relevance to homogeneous Ziegler-Natta catalysis, *J. Mol. Catal.*, 63, 125, 1990.

28. Nara, S. J., Harjani, J. R., and Salunkhe, M. R., Friedel-Crafts sulfonylation in 1-butyl-3-methylimidazolium chloroaluminate ionic liquids, *J. Org. Chem.*, 66, 8616, 2001.

29. Avent, A. G., Chaloner, P. A., Day, M. P. et al., Evidence for hydrogen bonding in solutions of 1-ethyl-3-methylimidazolium halides, and its implications for room-temperature halogenoaluminate(III) ionic liquids, *J. Chem. Soc., Dalton Trans.*, 3405, 1994.

30. Bonhôte, P., Dias, A.-P., Papageorgiou, N. et al., Hydrophobic, highly conductive ambient-temperature molten salts, *Inorg. Chem.*, 35, 1168, 1996.

31. Headley, A. D. and Jackson, N. M., The effect of the anion on the chemical shifts of the aromatic hydrogen atoms of liquid 1-butyl-3-methylimidazolium salts, *J. Phys. Org. Chem.*, 15, 52, 2002.

32. Lin, S.-T., Ding, M.-F., Chang, C.-W. et al., Nuclear magnetic resonance spectroscopic study on ionic liquids of 1-alkyl-3-methylimidazolium salts, *Tetrahedron*, 60, 9441, 2004.

33. Consorti, C. S., Suarez, P. A. Z., De Souza, R. F. et al., Identification of 1,3-dialkylimidazolium salt supramolecular aggregates in solution, *J. Phys. Chem. B*, 109, 4341, 2005.

34. Tubbs, J. D. and Hoffmann, M. M., Ion-pair formation of the ionic liquid 1-ethyl-3-methylimidazolium bis(trifyl)imide in low dielectric media, *J. Solution Chem.*, 33, 381, 2004.

35. Dupont, J., Suarez, P. A. Z., De Souza, R. F. et al., C-H-P interactions in 1-n-butyl-3-methylimidazolium tetraphenylborate molten salt: Solid and solution structures., *Chem. Eur. J.*, 6, 2377, 2000.

36. Nakakoshi, M., Ishihara, S., Utsumi, H. et al., Anomalous dynamic behavior of ions and water molecules in dilute aqueous solution of 1-butyl-3-methylimidazolium bromide studied by NMR, *Chem. Phys. Lett.*, 427, 87, 2006.

37. Nama, D., Kumar, P. G. A., Pregosin, P. S. et al., H-1, F-19-HOESY and PGSE diffusion studies on ionic liquids: The effect of co-solvent on structure, *Inorg. Chim. Acta*, 359, 1907, 2006.

38. Giernoth, R., Bankmann, D., and Schloerer, N., High performance NMR in ionic liquids., *Green Chem.*, 7, 279, 2005.

39. Noda, A., Hayamizu, K., and Watanabe, M., Pulsed-gradient spin-echo ^{1}H and ^{19}F NMR ionic diffusion coefficient, viscosity, and ionic conductivity of non-chloroaluminate room-temperature ionic liquids, *J. Phys. Chem. B*, 105, 4603, 2001.

40. Tokuda, H., Hayamizu, K., Ishii, K. et al., Physicochemical properties and structures of room temperature ionic liquids. 1. variation of anionic species, *J. Phys. Chem. B*, 108, 16593, 2004.

41. Carper, W. R., Wahlbeck, P. G., Antony, J. H. et al., A Bloembergen-Purcell-Pound ^{13}C NMR relaxation study of the ionic liquid 1-butyl-3-methylimidazolium hexafluorophosphate., *Anal. Bioanal. Chem.*, 378, 1548, 2004.

42. Antony, J. H., Mertens, D., Dölle, A. et al., Molecular reorientational dynamics of the neat ionic liquid 1-butyl-3-methylimidazolium hexafluorophosphate by measurement of ^{13}C nuclear magnetic relaxation data., *ChemPhysChem.*, 4, 588, 2003.

43. Mele, A., Romanò, G., Giannone, M. et al., The local structure of ionic liquids: Cation-cation NOE interactions and internuclear distances in neat [bmim][BF$_4$] and [bdmim][BF$_4$], *Angew. Chem. Int. Ed.*, 45, 1123, 2006.

44. Lopez-Martin, I., Burello, E., Davey, P. N. et al., Anion and cation effects on imidazolium salt melting points: A descriptor modelling study, *ChemPhysChem.*, 8, 690, 2007.

45. Lopes, J. N. C. and Padua, A. A. H., Molecular force field for ionic liquids III: Imidazolium, pyridinium, and phosphonium cations; chloride, bromide, and dicyanamide anions, *J. Phys. Chem. B*, 110, 19586, 2006.

46. Malek, K., Skubel, M., Schroeder, G. et al., Theoretical and experimental studies on selected 1,3-diazolium salts, *Vibrational Spectr.*, 42, 317, 2006.

47. Lopes, J. N. C., Deschamps, J., and Padua, A. A. H., Modeling ionic liquids using a systematic all-atom force field, *J. Phys. Chem. B*, 108, 2038, 2004.

48. Heinemann, C., Mueller, T., Apeloig, Y. et al., On the question of stability, conjugation, and aromaticity in imidazol-2-ylidenes and their silicon analogs, *J. Am. Chem. Soc.*, 118, 2023, 1996.

49. Lyčka, A., Doleček, R., Simonek, P. et al., ^{15}N NMR spectra of some ionic liquids based on 1,3-disubstituted imidazolium cations, *Magn. Reson. Chem.*, 44, 521, 2006.

50. Brenna, S., Posset, T., Furrer, J. et al., ^{14}N NMR and two-dimensional suspension ^{1}H and ^{13}C HRMAS NMR spectroscopy of ionic liquids immobilized on silica, *Chem. Eur. J.*, 12, 2880, 2006.

51. Mele, A., Tran, C. D., and De Paoli Lacerda, S. H., The structure of a room-temperature ionic liquid with and without trace amounts of water: The role of C-H...O and C-H...F interactions in 1-n-butyl-3-methylimidazolium tetrafluoroborate, *Angew. Chem., Int. Ed.*, 42, 4364, 2003.

52. Seddon, K. R., Stark, A., and Torres, M.-J., Influence of chloride, water, and organic solvents on the physical properties of ionic liquids, *Pure Appl. Chem.*, 72, 2275, 2000.

53. Umecky, T., Kanakubo, M., and Ikushima, Y., Self-diffusion coefficients of 1-butyl-3-methylimidazolium hexafluorophosphate with pulsed-field gradient spin-echo NMR technique, *Fluid Phase Equilib.*, 228–229, 329, 2005.

54. Noda, A., Susan, M. A. B. H., Kudo, K. et al., Brønsted acid-base ionic liquids as proton-conducting nonaqueous electrolytes, *J. Phys. Chem. B*, 107, 4024, 2003.

55. Hayamizu, K., Aihara, Y., Nakagawa, H. et al., Ionic conduction and ion diffusion in binary room-temperature ionic liquids composed of [emim][BF$_4$] and LiBF$_4$, *J. Phys. Chem. B*, 108, 19527, 2004.

56. Dieter, K. M., Dymek, C. J., Heimer, N. E. et al., Ionic structure and interactions in 1-methyl-3-ethylimidazolium chloride-AlCl$_3$ molten-salts, *J. Am. Chem. Soc.*, 110, 2722, 1988.

57. Hardacre, C., Holbrey, J. D., and McMath, S. E. J., A highly efficient synthetic procedure for deuteriating imidazoles and imidazolium salts, *Chem. Commun.*, 367, 2001.

58. Giernoth, R. and Bankmann, D., Transition-metal free ring deuteration of imidazolium ionic liquid cations, *Tetrahedron Lett.*, 47, 4293, 2006.

59. Durazo, A. and Abu-Omar, M. M., Deuterium NMR spectroscopy is a versatile and economical tool for monitoring reaction kinetics in ionic liquids, *Chem. Commun.*, 66, 2002.

60. Owens, G. S., Durazo, A., and Abu-Omar, M. M., Kinetics of MTO-catalyzed olefin epoxidation in ambient temperature ionic liquids: UV/VIS and ^{2}H NMR study, *Chem. Eur. J.*, 8, 3053, 2002.

61. Giernoth, R. and Bankmann, D., Application of diffusion-ordered spectroscopy (DOSY) as a solvent signal filter for NMR in neat ionic liquids., *Eur. J. Org. Chem.*, 4529, 2005.

62. Kim, Y. J. and Varma, R. S., Tetrahaloindate(III)-based ionic liquids in the coupling reaction of carbon dioxide and epoxides to generate cyclic carbonates: H-bonding and mechanistic studies, *J. Org. Chem.*, 70, 7882, 2005.

63. Rencurosi, A., Lay, L., Russo, G. et al., NMR evidence for the participation of triflated ionic liquids in glycosylation reaction mechanisms, *Carbohydrate Res.*, 341, 903, 2006.

64. Moulthrop, J. S., Swatloski, R. P., Moyna, G. et al., High-resolution ^{13}C NMR studies of cellulose and cellulose oligomers in ionic liquid solutions, *Chem. Commun.*, 1557, 2005.

65. Remsing, R. C., Swatloski, R. P., Rogers, R. D. et al., Mechanism of cellulose dissolution in the ionic liquid 1-n-butyl-3-methylimidazolium chloride: A ^{13}C and $^{35/37}$Cl NMR relaxation study on model systems, *Chem. Commun.*, 1271, 2006.

66. Fort, D. A., Swatloski, R. P., Moyna, P. et al., Use of ionic liquids in the study of fruit ripening by high-resolution ^{13}C NMR spectroscopy: Green solvents meet green bananas, *Chem. Commun.*, 714, 2006.

67. Rangits, G., Petocz, G., Berente, Z. et al., NMR investigation of platinum-diphosphine complexes in [bmim][PF$_6$] ionic liquid, *Inorg. Chim. Acta*, 353, 301, 2003.

68. Rangits, G., Berente, Z., Kégl, T. et al., The formation of [PtCl(diphosphine-I) (η1-diphosphine-II)]$^+$ species in the N-butyl-N'-methylimidazolium hexafluorophosphate ionic liquid: An NMR study, *J. Coord. Chem.*, 58, 869, 2005.

69. Wanzlick, H.-W., Beiträge zur nucleophilen Carben-Chemie, *Angew. Chem.*, 74, 129, 1962.

70. Arduengo, A. J., Harlow, R. L., and Kline, M., A stable crystalline carbene, *J. Am. Chem. Soc.*, 113, 361, 1991.

71. Aggarwal, V. K., Emme, I., and Mereu, A., Unexpected side reactions of imidazolium-based ionic liquids in the base-catalysed Baylis-Hillman reaction, *Chem. Commun.*, 1612, 2002.

72. Yonker, C. R. and Linehan, J. C., A high-pressure NMR investigation of reaction chemistries in a simple salt hydrate, *J. Supercrit. Fluids*, 29, 257, 2004.

73. Mehnert, C. P., Supported ionic liquid catalysis, *Chem. Eur. J.*, 11, 50, 2005.

74. Giernoth, R. and Bankmann, D., Transition-metal free synthesis of perdeuterated imidazolium ionic liquidsby alkylation and H/D exchange, *Eur. J. Org. Chem.*, 2008 (in print). DOI: 10.1002/ejoc.200700784.

chapter fourteen

Ionic liquids and mass spectrometry

Andreas Tholey

Contents

14.1   Introduction

Mass spectrometry (MS) is one of the most universally used analytical techniques covering applications in almost all chemically oriented disciplines ranging from inorganic, organic, and physical chemistry to the environmental and life sciences. The most outstanding properties of this technique are its low sample consumption, the capability for automatization allowing a high sample throughput and the capability of coupling to separation techniques like gas or liquid chromatography. MS delivers information about the mass of an analyte as well as its isotope pattern both of which can be used for the identification of a compound. Further, detailed structural information can be obtained upon fragmentation of the analytes in MS/MS experiments. Thus, MS has become a valuable tool for quality control as well as for analysis of chemical processes. With the emerging importance of ionic liquids (ILs), this universal analytical method came into the focus of this field. The goal of this chapter is to show some basic principles and applications of MS in IL related research. After a brief introduction to the general principles of important MS techniques, the chapter is divided into three main parts. The first focus will be on the application of MS for the analysis and characterization of ILs, outlining the potential of MS for practical purposes. Important applications include the quality control and the monitoring of unwanted side reactions occurring in these solvents. Further, MS can be used for theoretical studies in order to elucidate physicochemical parameters which are important for a deeper understanding of the ILs. The second focus is on the application of MS for the analysis of compounds dissolved in IL. These methods have a high potential to be used for a straightforward monitoring of reactions performed in these solvents.

The third focus is the use of IL as supports for MS. The application of the IL enabling electrospray ionization (ESI) using nonpolar solvents will be demonstrated. A major emphasis will be put on the use of ILs as matrices for matrix-assisted laser desorption/ionization (MALDI) MS, a field gaining emerging interest in the last few years.

14.2   Basic principles of mass spectrometry

The basic principle of MS is the separation of analyte ions according to their m/z-ratio. For this purpose, prior to analysis, the analyte has to undergo two processes: first, the evaporation into the gas phase and second, unless it carries its charge intrinsically, the ionization. Depending on the ionization mechanism applied, the charge can be carried either by single electrons,

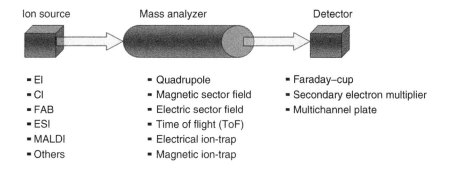

Figure 14.1 Schematic view of a mass spectrometer. Its basic parts are ion source, mass analyzer, and detector. Selected principles realized in modern mass spectrometers are assigned; EI—electron impact, CI—chemical ionization, FAB—fast atom bombardment, ESI—electrospray ionization, MALDI—matrix-assisted laser desorption/ionization. Different combinations of ion formation with mass separation can be realized.

by abstraction or removal of protons, or by cationization, for example, by sodium or potassium ions. Mass separation occurs under vacuum conditions in order to prevent loss of the analyte induced by collision with gas molecules; nevertheless, such collisions can be used to generate fragment ions under controlled conditions in MS/MS experiments.

The general assembly of any mass spectrometer contains three main parts: (i) a device for ion formation, the so-called ion source, (ii) a device for the separation of the ions (mass analyzer), and (iii) a detector (Figure 14.1).

The energy necessary for the ionization and evaporation of the analyte can be delivered either by use of thermal energy or by means of nonthermal energy sources. A number of techniques following different physical principles have been developed for the ion formation in MS, each having specific advantages for different classes of analytes.

A striking feature of the ILs is their low vapor pressure. This, on the other hand, is a factor hampering their investigation by MS. For example, a technique like electron impact (EI) MS, based on thermal evaporation of the sample prior to ionization of the vaporized analyte by collision with an electron beam, has only rarely been applied for the analysis of this class of compounds. In contrast, nonthermal ionization methods, like fast atom bombardment (FAB), secondary ion mass spectrometry (SIMS), atmospheric pressure chemical ionization (APCI), ESI, and MALDI suit better for this purpose. Measurement on the atomic level after burning the sample in a hot plasma (up to 8000°C), as realized in inductively coupled plasma (ICP) MS, has up to now only rarely been applied in the field of IL (characterization of gold particles dissolved in IL [1]). This method will potentially attract more interest in the future, especially, when the coupling of this method with chromatographic separations becomes a routine method.

In the following section, a brief outline of the basic principles of these methods is given.

For a more detailed description, a number of excellent textbooks are available.

14.2.1   Ionization techniques

14.2.1.1   Electrospray ionization

In ESI MS, a dissolved sample is sprayed through a capillary in an electric field which is situated in front of the vacuum inlet of the mass spectrometer [2]. Thus, in contrast to most other ionization techniques performed in high vacuum, the ionization process takes place at the atmospheric pressure. After leaving the capillary, the solvent forms a so-called Taylor-cone, which further forms a filament and finally, the spray of small droplets (Figure 14.2). These droplets carry charges on the surface; this is frequently supported by the acidification of the solvent. The droplets shrink is caused by the evaporation of the solvent. This leads to an increase of the charge-per-surface ratio, finally

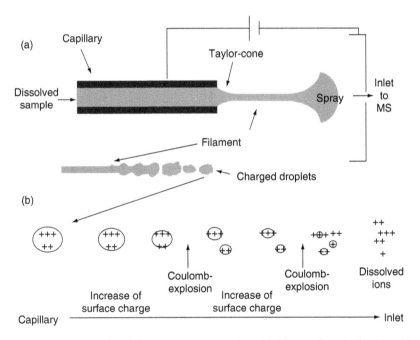

Figure 14.2   Principle of electrospray ionization. (a) The analyte is dissolved in an appropriate solvent and sprayed via a capillary into an electric field. Here, the liquid filament finally forms charged droplets. (b) The solvent of the charged droplets evaporates, resulting in an increase of the surface charge up to a critical boundary, at which a Coulomb explosion occurs. The newly formed droplets undergo the same process. The final products are the desolvated, naked ions, which are then entering the mass spectrometer.

leading to a coulomb explosion of the droplet caused by the repulsion of the charges. This process runs repeatedly until the desolvated analyte ions are formed. The ions can then be accelerated in the mass spectrometer and separated according to their mass to charge ratio. Several theories about the exact mechanisms of the electrospray process exist [3,4], but their discussion is far beyond the scope of this article. A major advantage of this method is that both ionization and evaporation occur under very smooth conditions, thus providing a soft ionization technique, which is especially appropriate for the analysis of polymeric biomolecules like peptides and proteins. Due to the straightforward coupling with liquid chromatography ESI MS is also a method of choice for the high-throughput analysis of small molecules. Because of the formation of multiple charged ions of the form $[M+nH]^{n+}$ (positive ion mode) or $[M-nH]^{n-}$ (negative ion mode), the mass range is nearly unlimited. The method is best-suited for the analysis of charged, polar compounds. Analytes incapable of acting as proton or metal ion donors/acceptors can hardly be measured by this method.

14.2.1.2 Atmospheric pressure chemical ionization

APCI is a method closely related to electrospray ionization. It uses ion–molecule reactions to produce ions from analyte molecules. The sample is electrohydrodynamically sprayed into the source (Figure 14.3). The evaporation of the solvent is often supported by a heated gas at temperatures between 80 and 400°C. Within the source, a plasma is created using a corona discharge needle that is placed close to the end of the metal capillary. In this plasma, proton transfer reactions occur, leading to the ionization of the analyte, mainly by the formation of $[M+H]^{+}$ ions. Compared to ESI MS, APCI MS is very well suited for the analysis of less-polar components and can therefore

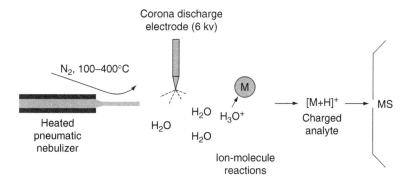

Figure 14.3 Principle of atmospheric pressure chemical ionization. The dissolved analyte is sprayed through a capillary. Evaporation of the solvent is supported by a heated gas stream. Within the source, a plasma is formed by a Corona discharge needle, which creates the charged reagent gas (here: H_3O^+). The ionization of the analyte (M) is performed by the transfer of the charge (proton) via ion–molecule reactions.

be regarded as a complementary technique of the latter. Further on, it can easily be coupled with liquid chromatography. Due to the generation of mainly singly charged ions, the mass range is low to moderate; most applications use this method for the analysis of compounds with masses <2000 Da.

14.2.1.3   Fast atom bombardment

In the fast atom bombardment (FAB) MS [5], the analyte is dissolved in a liquid matrix such as glycerol, thioglycerol, *m*-nitrobenzyl alcohol, or diethanolamine. A small amount (about 1 µL) of this mixture is placed on a target. The sample is then bombarded with a fast atom beam (Figure 14.4). Xenon atoms were most frequently used for this purpose, typical energies of the atom beam lie in the range of 6 keV. Upon collision of the atoms with the sample, both the matrix and the analyte are desorbed. Caused by the excess of the matrix over the analyte, most of the energy is taken up by the matrix, which leads to a reduced energy transfer onto the analyte. Thus, also molecules up to 6 kDa can be measured by this method without undergoing extensive fragmentation. The benefits of this method are its simple performance, tolerance for variations in sampling and ability to measure a large variety of compounds. However, a problem is the high chemical background induced by the presence of the matrix, which can hamper the analysis of

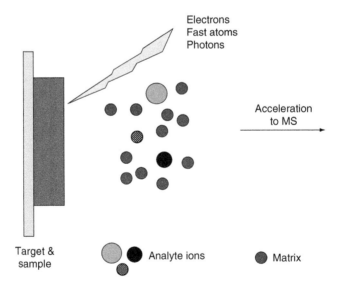

Figure 14.4 Generation of ions by desorption methods. The sample is placed on a target and then hit either by accelerated electrons (secondary ion mass spectrometry), accelerated atoms (fast atom bombardment) or laser light (laser desorption/ ionization, matrix-assisted laser desorption/ionization). In the case of FAB and MALDI, the analyte is additionally embedded in a matrix, which also is desorbed during these processes.

low-molecular weight compounds with masses < ~ about 300 Da. A further drawback is the need to find proper matrices, allowing for a homogeneous dissolving of the analyte; only a limited set of matrix substances are available for this purpose.

14.2.1.4   Secondary ion mass spectrometry

SIMS shares a number of properties with FAB MS. The ionization and desorption of the sample is achieved by the collision of the solid or liquid sample with an ion beam (Figure 14.4), but in contrast to FAB and MALDI MS, no additional matrices are added. Compared to the neutral atoms, ions can be focused and accelerated to higher kinetic energies, which allows for the ionization and desorption of higher mass analytes. Typically, masses up to 15,000 Da are measurable with SIMS. Due to the higher energy of the ion beam, fragmentation is more pronounced compared to other desorption methods (FAB, MALDI). Therefore, this method is mainly used for more stable analytes. As for FAB MS, the high chemical background in the low mass range may hamper the analysis of the compound in the lower mass range.

14.2.1.5   Matrix-assisted laser desorption/ionization

Since its introduction in the late 1980s [6], MALDI MS has become one of the most valuable tools for not only the investigation of polymeric biomolecules like peptides, proteins, and oligonucleotides but also for the analysis of technical polymers, small organic molecules, and low-molecular weight compounds of biological interest like amino acids, lipids, and carbohydrates [7].

The basic principle of MALDI MS (Figure 14.4) is based on the mixing of the analyte with a high excess (up to 100,000:1, mol:mol) of a low-molecular weight matrix consisting of small organic compounds (such as nicotinic acid), exhibiting a strong resonance absorption at the laser wavelength used. Most matrices are organic, aromatic compounds able to absorb the laser light applied in MALDI instruments. A more detailed description of the MALDI matrices is given in Chapter 5. The matrix–analyte mixture is then applied on a metal plate (target) where the solvent evaporates, leaving back a matrix–analyte cocrystallite. In the high vacuum of the mass spectrometer, pulsed laser shots with pulse lengths in the nanoseconds to picoseconds range are applied on the cocrystallite. This controllable energy transfer into the condensed phase of matrix and analyte results both to the desorption into the gas phase (plume) and the ionization of the analyte. The absorption of the majority of the laser energy by the matrix leads to a smooth ionization of the analyte, making MALDI a soft ionization technique. The mechanisms of the MALDI process, complex and by far not finally clarified, are far beyond the scope of this article. The fundamental processes are well-reviewed in Refs 8–10.

Ion formation occurs mainly by protonation (positive ion mode) or deprotonation (negative ion small cations (Na^+, K^+). In contrast to ESI, mainly singly charged ions are formed in MALDI MS. Because of the pulsed ion generation in MALDI, the combination with the time of flight (ToF) analyzer is the most widely used mass analyzer in combination with MALDI.

14.2.2   Mass analyzers

The ions formed can be separated and analyzed according to different physical properties like their behavior in magnetic (sector field mass analyzer, magnetic ion trap) or electric fields (quadrupoles, electric ion traps) or different flight times of ions of different sizes (tToF analyzers). An extensive discussion of these principles would go far beyond the scope of this chapter, but a number of excellent textbooks are available. An interesting feature, which can be realized by coupling of different analyzers in combination with collision chambers (e.g., quadrupole-ToF, triple quadrupole, ToF-ToF), is a possibility to perform MS/MS experiments. Here, the analyte can be fragmented, for example, by collision with gas molecules by collision-induced decomposition (CID); the second analyzer is then used to analyze the fragments formed during this process. With the help of such methods, valuable information about the structure of the analyte can be obtained.

14.3   Characterization of ionic liquids by mass spectrometry

MS delivers both information about the mass and the isotope pattern of a compound and can be used for the structural analysis upon performance of MS/MS experiments. Therefore, it is a valuable tool for the identification and characterization of an analyte as well as for the identification of impurities. Potential applications are the identification of IL in the quality control or in environmental studies. Unwanted by-products formed during the synthesis or by the hydrolysis of components of the ILs can be identified by this method. The analysis of the IL itself is also a prerequisite for the analysis of compounds dissolved in these media, as will be outlined in the section 14.4. Beside the identification of the ILs, a characterization of different properties like water miscibility and the formation of ion clusters, providing valuable information about the molecular structure of the IL, can be performed by means of MS techniques. The majority of studies reported up to now have dealt with ILs encompassing substituted imidazolium or pyridinium cations, therefore the following discussion concentrates on these compounds unless otherwise stated.

14.3.1   Analysis of anions and cations by mass spectrometry and mass spectrometry/mass spectrometry

Using SIMS, the structures of salts formed by mixing of 1-methyl-3-ethylimidazolium chloride with $AlCl_3$ could be determined [11] directly from the molten salts. The structure of the anions like $AlCl_4^-$ and $Al_2Cl_7^-$ could be confirmed. Additionally, unexpected anions like $[Al_2OCl_5]^-$ and $[Al_2Cl_6OH]^-$ were identified.

FAB MS has been applied in a number of studies to characterize ILs. Since most ILs are viscous liquids with negligible vapor pressure, the measurement of FAB MS is possible without the addition of a liquid matrix. In principle, ILs can therefore also be used as matrix substances for the FAB analysis of other analytes dissolved therein [12]. Spectra could be measured both in the positive and in the negative ion modes as has been demonstrated, for example, for butylpyridinium- chloroaluminates and gallates [12,13]. Beside the molecular ions, fragments mainly formed by the loss of the substituents of the central core of the cations, for example, butyl groups, were observed. Together with the isotope patterns, these fragments provide valuable information about the structure of newly composed compounds and help also to identify unexpected by-products like oxidized or hydrolyzed compounds in the ILs (see section 14.3.2).

The direct desorption of cations and anions of ILs by means of an interaction with laser light (laser desorption/ionization [LDI]) can also be used for the analysis of ILs [14]. *N*-substituted aromatic imidazole ring systems delivered strong signals in the positive ion mode when a nitrogen laser with a wavelength of 337 nm was used [14]. In case of 1-butyl-3-methylimidazolium salts, strong signals corresponding to the cation accompanied by an additional signal caused by the loss of the butyl group could be observed. In the case of 1,3-dimethyl-imidazolium salts, only the signal of the intact cation was observed. In both of the cations investigated no loss of methyl groups was observed. In the LDI-spectra measured in the negative ion mode, signals of the anions of the ILs could be measured. For example, anions like $[PF_6]^-$, $[BF_4]^-$, and $[(CH_3)_2PO_4]^-$ showed only signals of the unfragmented anions, whereas for $[C_8SO_4]^-$ ($[SO_4]^-$) and for $[Tf_2N]^-$ anions, less-intense signals of fragments ($[anion-CF_3]^-$ and $[anion-CF_3SO_2]^-$) could be observed in addition to the unfragmented anion [14].

Laser ablation ToF MS analysis, using IR lasers, has also been performed to characterize ILs [15]. It could be shown that intact ILs are ablated by this method. The neutral species ablated could be further investigated by vacuum UV postionization. Due to the faster heating rates achieved by this ablation process compared to the UV-LDI analysis, less fragmentation caused by the thermal decomposition of the analytes was observed.

In MALDI MS, the analyte is additionally embedded in higher molar excess of matrix molecules, which allows a soft desorption and ionization of the analyte. Typical molar ratios between the matrix and the analyte for measurement of low-molecular weight compounds (<500 Da) are in the range between 10:1 and 100:1, for higher masses (e.g., peptides and proteins) typical ratios are in the range of 1000:1 up to 100,000:1. Despite the capability to be measured by LDI MS, most classical ILs based on pyridinium or imidazolium cations are not capable of functioning as MALDI matrices [14,16]. The situation changes when classically used MALDI matrices (see Chapter 5) are added to IL:analyte mixtures. Under these conditions, the measurement of cations and anions of a number of ILs is possible, when appropriate amounts of known MALDI matrices are mixed with the ILs [14]. A drawback is the inherent occurrence of the matrix signals which are in the same mass range as the components of the IL. Therefore, MALDI MS is probably not the method of choice for the quality control of ILs. Nevertheless, the measurement of an IL by MALDI MS is a necessary prerequisite for the measurement of blank spectra, when low-molecular weight compounds dissolved in the IL are the analytes of interest. Generally, the same signals of the ILs as in the LDI analysis were found in MALDI MS both in positive and in negative ion modes, respectively, including the fragments described above together with a number of signals of the MALDI matrix itself [14].

ESI MS has gained increasing attention in the last few years both for the characterization of ILs and for detailed structural studies. ESI experiments can be performed by dissolving the IL in an organic solvent, for example, methanol but also in undiluted form [17]. The advantage of the latter approach is that unwanted side reactions between the solvent and the components of the IL are avoided. On the other hand, this method can lead to extensive pollution of the ion source, thus inducing the need for frequent cleaning of the source. Furthermore, the direct spraying of the undiluted IL is more amenable for low-viscosity ILs than for high viscosity molten salts. The use of elevated temperature, for example, applied via a heated curtain or drying gas, has been shown to support this process [17].

For the analysis of the ILs not accessible to solvolysis, the classical method using cosolvents is a viable alternative. ESI MS has been applied to characterize both the cations and the anions of IL [18–20]. Together with the detection of high-accuracy masses using Fourier-transform MS (FTMS) [21], the combination of ESI MS with MS/MS-experiments triggers this method to the most valuable method for the characterization of ILs. Fragmentation of IL components in MS experiments has been observed, in particular, for the cations, being dependent on the cone voltage applied [18]. A higher degree of fragmentation can be observed upon the increase of the cone voltage applied. A more pronounced and, caused by the possibility to select a certain precursor, a more controllable fragmentation can be reached by real MS/MS experiments, for example, by collision-induced decomposition (CID) [19,21]. A detailed study of the fragmentation behavior of N,N'-disubstituted imidazolium-based ILs was performed using sustained off-resonance ionization collision-induced decomposition (SORI CID) in combination with FT MS. The application of this method allows the study of detailed fragmentation mechanisms [21]. A number of unexpected fragmentations could be identified, which are potentially caused by the permanent presence of the positive charge in the imidazolium ring.

14.3.2   Analysis of impurities and by-products in ionic liquids

An important field of application of MS is the detection of impurities and products of unwanted side reactions, for example, products of oxidative processes and solvolysis, in ILs.

A number of hydrolysis products of chloroaluminate salts have been detected using FAB MS. The problem here is that under some circumstances FAB MS can provide only a restricted view on the surface of the sample that is more likely to be oxidized or hydrolyzed rather than the average sample [13]. Nevertheless, this problem is not directly related to MS itself, but rather to a problem of proper sample preparation. Inert conditions during sample preparation and construction of special sample inlet chambers for MS can help avoid these reactions. The addition of phosgene was suggested to reduce problems encountered with such side reactions [22], but here the problems encountered in working with an extremely toxic chemical have to be taken into account.

14.3.3   Determination of physicochemical properties of ionic liquids

MS can be used to study several intrinsic physicochemical properties of ILs, for example, the acidity of components of ILs or the determination of water miscibilities. Furthermore, MS delivers valuable information about noncovalent interactions, for example, responsible for the formation of quasi-molecular structures [23] formed by three-dimensional supramolecular polymeric networks within the IL.

14.3.3.1   Acidity

N,N-substituted imidazolium ions are known as precursors of relatively stable carbenes which are formed upon proton abstraction at C2. NMR experiments involving hydrogen-deuterium (H/D) exchange studies showed that the protons at C2 position of *N,N*-dialkylated imidazolium salts are slightly acidic having pKa values in the range of 20–23 [24]. In similar H/D-exchange experiments with ILs [C_4C_1Im][BF_4] and [C_1C_1Im][(C_1)$_2PO_4$] analyzed by MALDI MS, the LDI spectra of both cations showed clear evidence of an exchange of a proton by deuterium after the incubation of the IL with D_2O in basic solution [14]. Interestingly, the formation of the corresponding ions was increased, when the H/D-exchange experiment was performed under UV-irradiation.

14.3.3.2   Determination of water solubility of ionic liquids

Quantitative ESI MS has been applied to determine water solubility of several ILs [18]. The ion current (intensity) of the analyte signals was found to depend on the analyte concentration over a restricted range of concentrations. Nevertheless, relative quantifications performed by the use of internal standards deliver more reliable results. Isotopically labeled analogs of the analyte are best-suited for internal standards, but such compounds are only rarely available for ILs. Therefore, other ILs added in known amounts were used as internal standards [18]. Commonly, in MS, the amounts of the analyte and the standard should be in the same range to deliver reliable results. Using this method, the solubility of the ILs $[C_4C_1Im][PF_6]$ (18.6 $\pm$ 0.7 gL^{-1}), $[C_1(C_4)_3N][Tf_2N]$ (0.70 $\pm$ 0.08 gL^{-1}), and $[C_4py][Tf_2N]$ (6.0 $\pm$ 0.5 gL^{-1}) could be determined [18]. The advantage of this approach over methods based on the measurement of UV absorption is that both anions and cations of the ILs can, in principle, be determined simultaneously. Further, errors occurring from an increased solubility of hydrolysis products, that are known to have higher solubilities but cannot be discriminated by UV can be calculated by MS.

14.3.3.3   Analysis of supramolecular structures in ionic liquids

Supramolecular assemblies are the molecular base for some of the unique properties of ILs. Therefore, the knowledge of the nature, type, and strength of these structures [23] is a prerequisite for a deeper understanding of ILs as well as for the tailor-made design of new compounds. The most important noncovalent interactions responsible for the formation of such a structure are C–H hydrogen bonds [25]. Other interactions encompass the formation of clusters by ion pairing, which can be found, for example, in chloroaluminates [12].

Since the earliest applications of MS for the analysis of IL, higher-molecular ions assigned to molecular aggregates could be observed. For imidazolium-based ILs, clusters of the form $[Cat_xAn_{x-1}]^+$ can be observed in the positive ion mode in FAB [26], ESI [18,23], and APCI MSs [27], where x lies, depending on the ionization method applied, in the range between 2 and 12. The intensity of these clusters nearly exponentially decreases with n in electrospray MS [23], but for $n = 5$ a significantly increased signal intensity could be observed for different ILs.

To interpret such data it is important to have knowledge about the different mechanisms of the ionization methods used to study these complexes. The formation of ions, depending on the ionization technique applied, favors one or the other form of cluster ions. In FAB MS, preformed ions are desorbed from the solvent. In ESI MS, the ions evaporate from highly charged droplets to produce the gaseous ions. In APCI MS, the initial process is the evaporation of neutral clusters of the form A_nB_n, which is followed by ionization via proton transfer [27]. Due to these different mechanisms, the stoichiometries of the aggregates detected can vary. Clusters have also been observed in MALDI MS, but here the interpretation is difficult. MALDI is generally not the method chosen for the analysis of noncovalent interactions. Up to now it remains unclear whether the aggregates observed in MALDI experiments are desorbed directly from the sample or whether they are formed during the complex MALDI process.

In addition to the analysis of molecular masses of the clusters, important information about the structure of these aggregates can be obtained from

MS/MS experiments. For this purpose, different techniques can be applied. As outlined above, MS/MS information can be generated by the application of high cone voltages in ESI MS [18]. The extended energy of the ions upon acceleration in the electric field leads to a strong collision of the ions with gas molecules and residuals of the evaporating solvent. The second possibility is to perform CID experiments in collision chambers placed between two mass separation units (e.g., between two quadrupoles [triple quadrupole]; or between a quadrupole and a ToF mass analyzer), thus allowing real CID experiments to be performed under more controlled conditions. In such experimental setups it is possible to select a precursor, for example, the aggregate of interest, which can then be fragmented by collisions with gas molecules in the collision cell. The products (fragments) produced in this process can then be analyzed in the second mass spectrometer. Upon application of these experiments, the determination of the stoichiometries of clusters is straightforward.

An interesting application of such CID experiments is the determination of the bonding energies within the clusters. For this purpose, mixtures of different ILs were measured by ESI MS/MS [23]. The MS spectra showed clear evidence for the formation of mixed clusters, thus combinations of anions can be found if one IL with anions of the other IL is present (or vice versa combinations of anions and cations, depending on the ion polarity measured). When these mixed aggregates undergo collision-induced fragmentation, the preferential liberation of an anion/cation and the application of a method called Cooks kinetic method (CMK) based on the determination of rates of competitive dissociations [28] allows for the determination of the relative bonding energies of particular ions [23].

Noncovalent interactions between the IL and the ions dissolved therein are also of interest. For example, complex formation between ILs and metal ions could be studied by FAB MS [22]. In these experiments the formation of complexes of the form $[C_2C_1Im]_2–MCl_4$, with M = Ni^{2+}, Co^{2+}, or Mn^{2+} could be observed.

14.4   Analysis of compounds dissolved in ionic liquids

Two principal ways are amenable for the analysis of compounds dissolved in ILs: (i) the analytes can be separated from the ILs and investigated by suitable methods or (ii) the analytes can be investigated directly in the IL. The methods used most frequently are based on separation by high-pressure liquid chromatography (HPLC), the analysis of UV/VIS-absorption or fluorescence or NMR spectroscopy. These methods can be subject to several restrictions. The separation of mixtures involving ILs by HPLC is time-consuming and is in many cases still a challenge. Spectroscopic techniques can suffer from background absorption/fluorescence of the ILs and are difficult to perform in more complex reaction mixtures. Further, the presence of chromophoric or fluorophoric groups in the analyte is a prerequisite, which restricts the use of these methods. The use of MS can thus be a valuable alternative to these methods. Up to now, mainly LDI, MALDI, and ESI MS have been applied to analyze compounds dissolved in ILs.

An interesting approach using direct laser desorption MS was applied to study the products formed upon dissolution of divanadiumpentoxide in different ILs [29]. It showed again the potential of MS methods to serve as valuable tools both for the investigation of unwanted side reactions as well as for the generation of theoretical data allowing deeper insights to be acquired into complex interactions between the solvent and the analytes. In the same study, potential limitations of this method were described. These are caused by the formation of by-products under laser irradiation. For example, such species as $AlCl_3*VO_3$ and $AlCl_3*VO_2Cl_2^-$, which were not present in the IL without irradiation, were identified after laser irradiation.

In contrast to the analysis in pure ILs, amino acids, peptides, and proteins dissolved in ILs could be measured by MALDI MS after the addition of classical MALDI matrices.

Interestingly, a correlation between the molecular weight of the analyte and the ionizability in different ILs, dependent on the water miscibility of the latter, was observed [14]. As a rough rule, water-immiscible ILs seem to be more suitable for the detection of low-molecular weight analytes whereas biopolymers with higher masses can be analyzed more easily by MALDI MS when dissolved in water-miscible ILs. For example, signals for amino acid were observed only in water-immiscible ILs, for example, $[C_4C_1Im][PF_6]$ and $[C_4C_1Im][Tf_2N]$ but not in the presence of the water-miscible ILs like $[C_4C_1Im][C_8SO_4]$, $[C_4C_1Im][BF_4]$, and $[C_1C_1Im][(CH_3)_2PO_4]$. A similar behavior was observed for peptides, which were preferentially measurable in water-immiscible ILs $[C_4C_1Im][Tf_2N]$ and $[C_4C_1Im][PF_6]$. In contrast to these smaller analytes, protein signals were obtained also from water-immiscible ILs like $[C_4C_1Im][BF_4]$ and $[C_1C_1Im][(CH_3)_2PO_4]$ when the matrix sinapinic acid was added. For proteins a significant broadening of the signals could be observed, together with the formation of anion-protein adducts. For example, in the case of the IL $[C_4C_1Im][BF_4]$, $[BF_4]^-$ adducts could be observed. It has to be noted that successful measurements of analytes dissolved in ILs were only possible when a very restricted ratio between the IL, the analyte, and the matrix substance was applied [14]. The reasons for this limitation are yet unknown. Due to the need of an additional matrix, the investigation of analytes by MALDI MS, especially in the low mass range, can be hampered by a potential overlap between the analyte and the matrix signals.

The main prerequisite for the quantification of low-molecular weight compounds by MALDI MS is the use of internal standards and optimized analysis protocols. For example, the quantification of alanine using 1-^{13}C-alanine as labeled internal standard in the system $[C_4C_1Im][PF_6]$/CCA was possible with acceptable precision [14]. The method is therefore suited for screening processes, for example, for the comparison of enzyme activities in ILs. The detectability of peptides and proteins itself in this context also allows for an investigation of the enzyme itself and may be a useful tool for the detection of potential degradation products or modifications of the biocatalyst caused by side reactions with the components of ILs.

ESI MS has also been used for the measurement of the analytes dissolved in ILs. In contrast to MALDI MS, the measurement of low-molecular weight analytes is not hampered by the overlap of a signal with the matrix; nevertheless, the overlap of signals of the analyte with the components of IL or cosolvent used for the dissolution has to be taken into account in a particular experiment. Nevertheless, the analysis of ILs and analytes dissolved using ESI MS is straightforward. For example, a number of organometallo catalysts have been successfully analyzed by ESI MS. The major advantage of this approach is that these air- and water- sensitive catalysts do not have to be extracted from the reaction solution prior to analysis, thus also allowing postreaction analysis of the catalysts [30]. It has also been shown that direct measurement from undiluted IL solutions is possible when the analytes are solvent-sensitive [17]. A problem encountered with this approach is the pollution of the mass spectrometer with relatively high amounts of the nonvolatile ILs, thus leading to a need for frequent cleaning of the source. A potential alternative in this respect is the use of the direct probe method [31], using a modified electrospray source [20]. A fine stainless steel wire placed in front of the capillary helps in an effective suspension of the IL spray, thus reducing significantly the amount of IL entering the mass spectrometer (Figure 14.5).

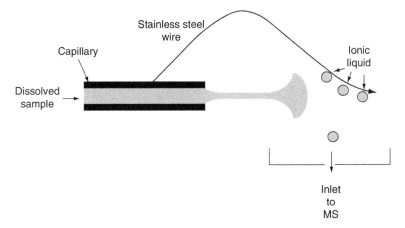

Figure 14.5 Modified-ESI source for the direct infusion of undiluted ILs. A stainless steel wire is placed in the spray, leading to the optimal vaporization of the IL. Additionally, an orthogonal ESI source is used. Only a part of the IL ions is transferred into the MS, thus minimizing pollution of the source. (Modified from Dyson, P. J. et al., Direct probe electrospray (and nanospray) ionization mass spectrometry of neat ionic liquids. *Chem. Commun.*, 2204, 2004. Reproduced by permission of the Royal Society of Chemistry.)

14.5   Ionic liquid as a tool for mass spectrometry

14.5.1   Ionic liquids as additive in electrospray ionization mass spectrometry

During the electrospray process, ions preformed in solution are transferred into the gas phase, mainly by the desolvation process described above. A critical parameter is the solvent used, that has to be volatile, capable of dissolving the analyte and has to have properties which allow the formation of a stable spray. The most commonly used solvents are water, methanol, or acetonitrile and mixtures thereof. For more apolar analytes, dichloromethane has been used as a solvent [32]. The use of apolar solvents like alkanes is restricted by the fact that they show a lack in conductivity [33]. In order to reduce this hampering factor, the addition of ILs was suggested. For example, the addition of ~10^{-5} M of phosphonium IL $[C_{14}(C_6)_3P][Tf_2N]$ to hexane leads to the formation of a stable spray. In the case of pentane, hexane, and cyclohexane, the relatively high concentration of the IL seems to be necessary for successful measurements. For solvents like benzene and toluene and more polar solvents (e.g., dichloromethane, diethyl ether), lower concentrations of the IL also lead to satisfactory results [33]. Beside the increase of conductivity, the addition of IL can also improve the solubility of certain analytes, for example, organometallic compounds, in the solvent, thus facilitating their analysis by ESI MS.

14.5.2   Ionic liquids as matrices for matrix-assisted laser desorption/ionization mass spectrometry

The matrix in MALDI MS fulfils several essential functions. First, the matrix absorbs the laser light via electronic (UV-MALDI) or vibrational (IR-MALDI) excitation and transfers this energy smoothly onto the analyte. Due to the high molar excess of the matrix over the analyte, the intermolecular interactions of analyte molecules are reduced, thus facilitating transfer into the gas phase. Last but not least, matrix–analyte interactions play an active role both in the ionization of the analyte as well as in its desorption [34].

Most of the commercially available MALDI mass spectrometers are equipped with UV-lasers, therefore the following discussion concentrates on MALDI matrices suited for the wavelength range between $\lambda = 266$ nm and $\lambda = 355$ nm. The majority of the UV-compatible matrices are small organic compounds absorbing in the range of $\lambda = 266$–355 nm. Many matrices possess hydroxyl- or amino groups in ortho- or para position and acidic groups or carbonyl functions (carboxylic groups, amides, ketones), but the presence of neither these carboxylic groups nor basic amino groups is a prerequisite. Further on, constitutional isomers of a compound can show

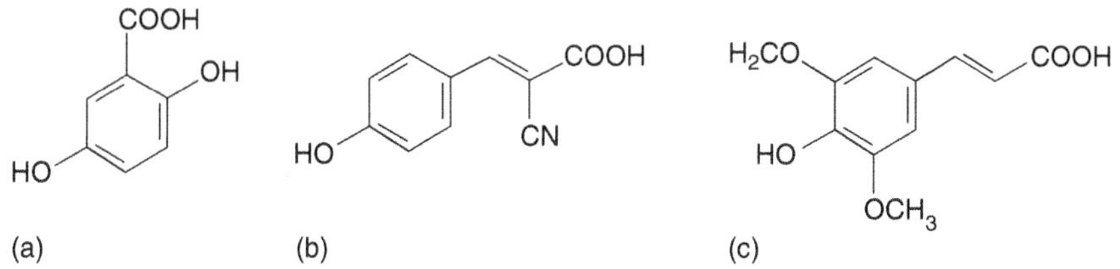

Figure 14.6 Structures of some commonly used MALDI matrices: (a) 2,5-dihydroxy benzoic acid (DHB), (b) α-cyano-4-hydroxy-cinnamic acid, and (c) sinapinic acid (SA). Combination of these acidic MALDI matrices with organic bases leads to the formation of ionic liquid matrices.

different properties in MALDI MS [35]. An important fact is that most of the matrix substances are not suited for the analysis of all classes of analytes.

Most commonly used matrix systems are derivatives of benzoic acid (e.g., 2,5-dihydroxybenzoic acid (DHB), derivatives of cinnamic acid (e.g., α-cyano-4-hydroxycinnamic acid (CHCA) or sinapinic acid (SA) (Figure 14.6) as well as heteroaromatic compounds containing nitrogen but numberless other substances and substance classes have been applied as matrices [36].

The choice of the matrix influences the ionization behavior, the formation of adducts, and the stability (or fragmentation) of the analytes and also has practical impact on the performance of the experiments. With the exception of the absorption behavior, there are no general rules for the prediction of a substance to be suited as a matrix, neither on the basis of chemical properties nor on that of physical properties. Therefore, in most cases, the search for new matrices applied an empirical approach taking into account a few practical rules, like the capability of solubility with the analyte, absorption behavior at the laser wavelength applied, inertness of the matrix, and vacuum stability. ILs show some of these properties and have therefore consequently been tested as potential MALDI matrices by Armstrong and coworkers [16], but as outlined above, the classical ILs based on substituted imidazolium or pyridinium cations were not suited for this purpose. In the same study, a new class of ILs capable of performing the MALDI process was introduced [16]. These new ILs, first named class II ILs [37], are organic salts formed by the equimolar mixtures of crystalline MALDI matrices, like CHCA, DHB or SA with organic bases, for example, tributylamine, pyridine, or 1-methylimidazole. This simple composition allows the design of numberless matrix combinations. In order to distinguish between these two classes of ILs, it was suggested to use the terms room-temperature ionic liquids (RTIL) for classically used ILs not suited as MALDI matrices and ionic liquid matrices (ILMs) [36,38] or ionic matrices [39] for ILs composed by a mixture of MALDI matrices with bases.

The synthesis of the ILM is straightforward and can be performed by simple combination by a (e.g., methanolic) solution of the acid component (the classically used MALDI matrix, e.g., DHB) with the organic base, followed by sonication and complete removal of the solvent [16]. The salt can then be redissolved in an appropriate solvent and used for MALDI sample preparation.

Many ILMs form smooth films with glycerol-like viscosities on the target [16,40], others form small crystals embedded in the viscous film [38] or are solids at room temperature [41].

ILMs have low vapor pressures [40] and are thus stable under vacuum conditions applied in MS ($1*10^{-7}$ to $1*10^{-9}$ bar) [16,38,40]. No obvious changes of sample shapes or spectra qualities were observed even after storage for more than 24 h in high vacuum. In some cases, a change in the shape of ILMs was observed after taking the target out of the vacuum, leading to a slight crystallization of former liquid samples. The formation of an ILM can also lead to a stabilization of matrices known to lack a long-time stability under high vacuum, as shown, for example, for the matrix 3-hydroxypicolinic acid [41].

A common problem using crystalline matrices in MALDI MS is the so-called hot-spot formation, which causes inhomogeneities on the sample, leading to significant variations of signal intensities at different positions of the sample spot. In contrast, the viscous liquid surface of ILMs is highly homogeneous and even partially crystallized matrices have shown significantly increased homogeneities compared to the solid matrices. Thus, the use of ILMs reduces the need for time-consuming search for hot-spots and additionally leads to a tremendous increase of spot-to-spot and shot-to-shot reproducibilities. These factors make ILMs valuable alternatives for applications in quantitative MALDI MS (see below) or for direct analysis of tissues by MALDI imaging [39].

A number of compound classes have been successfully analyzed using ILMs. Biopolymers like peptides [16,42–45], proteins [16], polynucleotides [41], as well phospholipids, technical polymers like polyethylene glycol [16] have been measured. One of the most interesting fields for the application of the ILM is the measurement of low-molecular weight compounds. The reason is a significant reduction of the number of matrix signals occurring in solid matrices (Figure 14.7), which bear the potential for overlap with analyte signals. The signals of the matrix are suppressed and in many cases even completely absent, which facilitates both measurement and interpretation of spectra.

It has been demonstrated that ILMs are suitable for qualitative and quantitative analyses of low-molecular weight compounds of biological interest, for example, carbohydrates, vitamins and amino acids [38], and glycolipids [40]. ILMs were further used for the direct analysis of alkaloids, anesthetics and antibiotics, separated by thin-layer chromatography (TLC) [46]. For this purpose, the ILM was spotted onto the fractions on the TLC-plates and the complete plate was measured in MALDI MS without the need for additional pretreatment of the TLC-samples. The mass deviation inherently caused by the inhomogeneous surface of the TLC-plate was balanced by using the

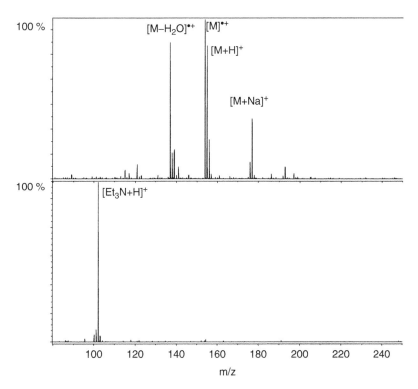

Figure 14.7 MALDI spectra of the MALDI matrix DHB (upper) and the ionic liquid matrix formed by equimolar mixture of DHB and triethylamine (lower). The triethylammonium-cation is the only signal in the spectrum of the ILM, the signal corresponding to DHB (as well as adducts and fragments thereof) are suppressed in the ILM. (From Tholey, A. and Heinzle, E., *Anal. Bioanal. Chem.*, 386, 24, 2006. With permission from Springer, Heidelberg, Germany.)

strong signal of the cation of the ILM (triethylamine) as internal standard. This fast and simple method can potentially be of great interest for studies in phytochemistry, organic synthesis, and process control. A further potential application in food and quality control was presented using the ILM for a screening for aflatoxines in chloroform extracts of peanuts [47].

Most ILMs are less acidic than the commonly used acidic matrices alone. This leads to the possibility to synthesize matrices with only weakly acidic or even neutral or basic pH values [48]. These matrices may be beneficial for the analysis of acid-labile compounds [40]. For example, these matrices were successfully used for the measurement of acid-labile compounds like sulphated oligosaccharides, which are a class of compounds with high biological relevance [49]. Using classical preparations, the detection of these challenging analytes was only possible after derivatization or in the form of noncovalent complexes formed with basic peptides. Upon use of the ILM, mainly DHB-butylamine and CHCA-1-methylimidazole, the measurement of sulphated oligosaccharides without prior complexation was possible for the first time.

Nevertheless, the addition of a small amount of TFA (0.1%) or 1% phosphoric acid [45], which is necessary in many cases to guarantee proper ionization, again makes the matrices acidic. Here, a tradeoff between the need for smooth conditions within the matrix on the one hand and on the ionization properties on the other hand has to be found for a particular analytical problem.

Despite these widespread applications, ILM is not equally well suited for all classes of analytes. Due to the need for increased laser energies/fluences for the ionization/desorption process, ILMs may only be of restricted suitability for some classes of analytes. For example for proteins, an extensive peak broadening caused potentially by the combination of extended neutral losses (e.g., of ammonia or water) and alkali-ion-adduct formation can be observed. On the other hand, the increased tendency of the ILM to favor sodium and potassium adduct formation makes it ideally suited for the measurement of carbohydrates [38,40], whereas in proteomics, this tendency of adduct formation is again an unwanted effect.

Among criteria and aspects speaking for or against the application of a particular MALDI, the most important are sensitivities, mass accuracies, and resolutions achievable and factors like the formation of adducts and the stability of the analytes both in terms of chemical, matrix-induced (on-target) modifications and the fragmentation behavior during ionization/desorption. These factors can be further influenced by the conditions applied for sample preparation or the addition of aiding substances like small amounts of acids.

The comparison of sensitivity and limits of detection achievable in ILMs and crystalline matrices have led to inconsistent results in different studies. For the measurement of peptides, a number of groups have reported comparable or even increased sensitivities in ILMs in positive [16,50] or negative ion mode [39], whereas others have reported decreased sensitivities [40]. For oligonucleotides [41], phospholipids [48], again increased sensitivities have been reported. For low-molecular weight analytes like sugars, decreased [40] or comparable limits of detection have been found [38].

Despite the use of balancing mechanisms of ion extraction, for example, delayed extraction, larger inhomogeneities frequently found in slowly grown crystalline matrix preparations are known to influence the mass accuracies. The highly homogeneous sample surfaces achievable in ILMs can drastically reduce these problems upon conversion of the crystalline matrices into ILMs [45].

Signal intensities in MALDI MS not only depend on the amount of the analyte but also on its chemical composition, for example, the sequence of amino acids in the case of peptides, and may underlie suppression by other components present in the sample [51,52]. Further, the hot-spot formation described above leads frequently to poor shot-to-shot and spot-to-spot reproducibilities. Both factors hamper the use of MALDI MS for quantitative measurements. Nevertheless, the potential of MALDI MS to perform quantitative measurements is now widely accepted [53–55]. The main steps toward quantitative MALDI MS are (i) the application of suited measurement protocols [56,57], (ii) the use of internal standards allowing for a relative quantification against a substance of known amount, and (iii) the improvement of sample homogeneity.

The enhanced homogeneity of samples resulting in the high reproducibility of signal intensities achievable in ILMs makes these matrices almost perfect alternatives for quantitative measurements. For example, both the accuracy achievable and the measurement time needed for quantitative measurements of low-molecular weight compounds of biological interest, for example, amino acids, sugars, and vitamins, was tremendously improved upon use of the ILM [38]. One application of this approach is the determination of enzyme activities and the screening for new enzyme variants [36]. For example, the screening of 10 genetically modified variants of the enzyme pyranose oxidase applying isotope labeled internal standards in combination with the ILM DHB-pyridine could be successfully performed with this method [58]. An interesting approach for the measurement of enzyme activities by direct screening from enzymatic reactions performed in the ILM was described [40]. An aqueous solution of the ILM DHB-butylamine was used both as a cosolvent for enzymatic reaction performed on a MALDI target as well as a matrix for direct monitoring of the microliter-scale enzyme reaction. Potential applications of this method lie in the field of monitoring of enzyme reactions taking place in organic solvents, for example, lipase catalyzed reactions.

Upon use of structurally modified variants as internal standards for the particular analytes, the relative quantification of oligonucleotides, peptides, and small proteins was demonstrated [44]. The potential of the ILM to allow quantitative analyses of peptides without the use of internal standards was presented recently [43]. Linear correlations between peptide amount and signal intensities could be found upon application of increased matrix-to-analyte ratios between 25,000 and 250,000 (mol:mol). The dynamic range of linearity thus spanned one order of magnitude. Unfortunately, the importance of the M/A ratio prevents the use of this method in samples with unknown orders of concentration, for example, in a proteomics environment. On the other hand, the method is applicable for the screening of enzyme-catalyzed reactions because the starting concentrations of the peptides are generally known in such assays.

14.6 Outlook

MS has a great potential to be established as a key method in almost all fields of IL research. In particular, soft ionization methods like ESI and APCI and to a lower extent MALDI MS are the methods of choice. Potential applications of MS lie, for example, in the analysis of reactions occurring in ILs, that is, for the study of biocatalytic reactions. Potential new applications can be the coupling of separation methods with MS and the application of MS for quantitative measurements with and in ILs.

ILs can also serve as tools for the MS itself. The introduction of the ILMs for MALDI MS has opened the way to a number of new applications for this method. A number of theoretical studies are necessary in order to fully understand the properties of the ILMs [36]. The basic processes of IL-MALDI are still only partially understood. Therefore, basic work remains to be done to explain theoretical aspects. The wide field of already indicated and other still unknown applications of the ILM seems to legitimate these efforts. Up to now, no consistent relationships have been found between the composition of an ILM and its ability to serve as a *good* matrix—a situation which is comparable to all other substances used as matrices. A deeper understanding of the theoretical background of the ILM is the prerequisite for a possible tailor-made creation of new matrices in the future.

References

1. Whitehead, J. A., Lawrance, G. A. and Mccluskey, A., Analysis of gold in solutions containing ionic liquids by inductively coupled plasma atomic emission spectrometry. *Australian J. Chem.*, 57, 151, 2004.
2. Fenn, J. B. et al., Electrospray ionization for mass spectrometry of large biomolecules. *Science*, 246, 64, 1989.
3. Dole, M. et al., Molecular beams of macro ions. *J. Chem. Phys.*, 49, 2240, 1968.
4. Iribarne, J. V., Dziedzic, P. J. and Thomson, B. A., Atmospheric pressure ion evaporation-mass spectrometry. *Int. J. Mass Spectrom Ion. Phys.*, 50, 331, 1983.
5. Barber, M. et al., Fast-atom-bombardment mass spectra of enkephalins. *Biochem. J.*, 197, 401, 1981.
6. Karas, M. and Hillenkamp, F., Laser desorption ionization of proteins with molecular masses exceeding 10,000 daltons. *Anal. Chem.*, 60, 2299, 1988.
7. Cohen, L. H. and Gusev, A. I., Small molecule analysis by MALDI mass spectrometry. *Anal. Bioanal. Chem.*, 373, 571, 2002.
8. Dreisewerd, K., The desorption process in MALDI. *Chem. Rev.*, 103, 395, 2003.
9. Karas, M. and Kruger, R., Ion formation in MALDI: The cluster ionization mechanism. *Chem. Rev.*, 103, 427, 2003.
10. Knochenmuss, R. and Zenobi, R., MALDI ionization: The role of in-plume processes. *Chem. Rev.*, 103, 441, 2003.
11. Franzen, G. et al., The anionic structure of room-temperature organic chloroaluminate melts from secondary ion mass-spectrometry. *Org. Mass Spec.*, 21, 443, 1986.
12. Ackermann, B. L., Tsarbopoulos, A. and Allison, J., Fast atom bombardment mass-spectrometric studies of the aluminum chloride N-butylpyridinium chloride molten-salt. *Anal. Chem.*, 57, 1766, 1985.
13. Wicelinski, S. P. et al., Fast atom bombardment mass-spectrometry of low-temperature chloroaluminate and chlorogallate melts. *Anal. Chem.*, 60, 2228, 1988.
14. Zabet-Moghaddam, M. et al., Matrix-assisted laser desorption/ionization mass spectrometry for the characterization of ionic liquids and the analysis of amino acids, peptides and proteins in ionic liquids. *J. Mass Spec.*, 39, 1494, 2004.

15. Dessiaterik, Y., Baer, T. and Miller, R. E., Laser ablation of imidazolium based ionic liquids. *J. Phys. Chem. A,* 110, 1500, 2006.
16. Armstrong, D. W. et al., Ionic liquids as matrixes for matrix-assisted laser desorption/ionization mass spectrometry. *Anal. Chem.,* 73, 3679, 2001.
17. Jackson, G. P. and Duckworth, D. C., Electrospray mass spectrometry of undiluted ionic liquids. *Chem. Commun.,* 522, 2004.
18. Alfassi, Z. B. et al., Electrospray ionization mass spectrometry of ionic liquids and determination of their solubility in water. *Anal. Bioanal. Chem.,* 377, 159, 2003.
19. Milman, B. L. and Alfassi, Z. B., Detection and identification of cations and anions of ionic liquids by means of electrospray mass spectrometry and tandem mass spectrometry. *Eur. J. Mass Spec.,* 11, 35, 2005.
20. Dyson, P. J. et al., Direct probe electrospray (and nanospray) ionization mass spectrometry of neat ionic liquids. *Chem. Commun.,* 2204, 2004.
21. Lesimple, A. et al., Electrospray mass spectral fragmentation study of N,N′-disubstituted imidazolium ionic liquids. *J. Am. Soc. Mass Spec.,* 17, 85, 2006.
22. Abdulsada, A. K. et al., A fast-atom-bombardment mass-spectrometric study of room-temperature 1-ethyl-3-methylimidazolium chloroaluminate(Iii) ionic liquids: Evidence for the existence of the decachlorotrialuminate(Iii) anion. *Org. Mass Spec.,* 28, 759, 1993.
23. Gozzo, F. C. et al., Gaseous supramolecules of imidazolium ionic liquids: Magic numbers and intrinsic strengths of hydrogen bonds. *Chem. Eur. J.,* 10, 6187, 2004.
24. Amyes, T. L. et al., Formation and stability of N-heterocyclic carbenes in water: The carbon acid pK(a) of imidazolium cations in aqueous solution. *J. Am. Chem. Soc.,* 126, 4366, 2004.
25. Dymek, C. J. et al., Evidence for the presence of hydrogen-bonded ion-ion interactions in the molten-salt precursor, 1-methyl-3-ethylimidazolium chloride. *J. Mol. Struc.,* 213, 25, 1989.
26. Abdulsada, A. K. et al., Evidence for the clustering of substituted imidazolium salts via hydrogen bonding under the conditions of fast atom bombardment mass spectrometry. *Eur. Mass Spec.,* 3, 245, 1997.
27. Dasilveira, B. A. et al., On the species involved in the vaporization of imidazolium ionic liquids in a steam-distillation-like process. *Angew. Chem. Int. Ed.,* 45, 7251, 2006.
28. Cooks, R. G. et al., Thermochemical determinations by the kinetic method. *Mass Spec. Rev.,* 13, 287, 1994.
29. Bell, R. C., Castleman, A. W. and Thorn, D. L., Vanadium oxide complexes in room-temperature chloroaluminate molten salts. *Inorg. Chem.,* 38, 5709, 1999.
30. Dyson, P. J., Mcindoe, J. S. and Zhao, D. B., Direct analysis of catalysts immobilised in ionic liquids using electrospray ionisation ion trap mass spectrometry. *Chem. Commun.,* 508, 2003.
31. Hong, C. M. et al., Generating electrospray from solutions predeposited on a copper wire. *Rapid Commun.Mass Spec.,* 13, 21, 1999.
32. Brayshaw, S. K. et al., Holding onto lots of hydrogen: A 12-hydride rhodium cluster that reversibly adds two molecules of H-2. *Angew. Chem. Int. Ed.,* 44, 6875, 2005.
33. Henderson, M. A. and Mcindoe, J. S., Ionic liquids enable electrospray ionisation mass spectrometry in hexane. *Chem. Commun.,* 2872, 2006.
34. Horneffer, V. et al., Matrix-analyte-interaction in MALDI MS: Pellet and nano-electrospray preparations. *Int. J. Mass Spectrom.,* 249/250, 426, 2006.

35. Ehring, H., Karas, M. and Hillenkamp, F., Role of photoionization and photochemistry in ionization processes of organic molecules and relevance for matrix-assisted laser desorption ionization mass spectrometry. *Org. Mass Spec.*, 27, 472, 1992.

36. Tholey, A. and Heinzle, E., Ionic (liquid) matrices for matrix-assisted laser desorption/ionization mass spectrometry-applications and perspectives. *Anal. Bioanal. Chem.*, 386, 24, 2006.

37. Anderson, J. L. et al., Characterizing ionic liquids on the basis of multiple solvation interactions. *J. Am. Chem. Soc.*, 124, 14247, 2002.

38. Zabet-Moghaddam, M., Heinzle, E. and Tholey, A., Qualitative and quantitative analysis of low molecular weight compounds by ultraviolet matrix-assisted laser desorption/ionization mass spectrometry using ionic liquid matrices. *Rapid Commun. Mass Spec.*, 18, 141, 2004.

39. Lemaire, R. et al., Solid ionic matrixes for direct tissue analysis and MALDI Imaging. *Anal. Chem.*, 78, 809, 2006.

40. Mank, M., Stahl, B. and Boehm, G., 2,5-Dihydroxybenzoic acid butylamine and other ionic liquid matrixes for enhanced MALDI-MS analysis of biomolecules. *Anal. Chem.*, 76, 2938, 2004.

41. Carda-Broch, S., Berthod, A. and Armstrong, D. W., Ionic matrices for matrix-assisted laser desorption/ionization time-of-flight detection of DNA oligomers. *Rapid Commun. Mass Spec.*, 17, 553, 2003.

42. Zabet-Moghaddam, M. et al., Pyridinium-based ionic liquid matrices can improve the identification of proteins by peptide mass-fingerprint analysis with matrix-assisted laser desorption/ionization mass spectrometry. *Anal. Bioanal. Chem.*, 384, 215, 2006.

43. Tholey, A., Zabet-Moghaddam, M. and Heinzle, E., Quantification of peptides for the monitoring of protease-catalyzed reactions by matrix-assisted laser desorption/ionization mass spectrometry using ionic liquid matrixes. *Anal. Chem.*, 78, 291, 2006.

44. Li, Y. L. and Gross, M. L., Ionic-liquid matrices for quantitative analysis by MALDI-TOF mass spectrometry. *J. Am. Soc. Mass Spec.*, 15, 1833, 2004.

45. Tholey, A., Ionic liquid matrices with phosphoric acid as matrix additive for the facilitated analysis of phosphopeptides by matrix-assisted laser desorption/ionization mass spectrometry. *Rapid Commun. Mass Spec.*, 20, 1761, 2006.

46. Santos, L. S. et al., Fast screening of low molecular weight compounds by thin-layer chromatography and "on-spot" MALDI-TOF mass spectrometry. *Anal. Chem.*, 76, 2144, 2004.

47. Catharino, R. R. et al., Aflatoxin screening by MALDI-TOF mass spectrometry. *Anal. Chem.*, 77, 8155, 2005.

48. Li, Y. L., Gross, M. L. and Hsu, F. F., Ionic-liquid matrices for improved analysis of phospholipids by MALDI-TOF mass spectrometry. *J. Am. Soc. Mass Spec.*, 16, 679, 2005.

49. Laremore, T. N. et al., Matrix-assisted laser desorption/ionization mass spectrometric analysis of uncomplexed highly sulfated oligosaccharides using ionic liquid matrices. *Anal. Chem.*, 78, 1774, 2006.

50. Jones, J. J. et al., Ionic liquid matrix-induced metastable decay of peptides and oligonucleotides and stabilization of phospholipids in MALDI FTMS analyses. *J. Am. Soc. Mass Spec.*, 16, 2000, 2005.

51. Cohen, S. L. and Chait, B. T., Influence of matrix solution conditions on the MALDI-MS analysis of peptides and proteins. *Anal. Chem.*, 68, 31, 1996.

52. Kussmann, M. et al., Matrix-assisted laser desorption/ionization mass spectrometric peptide mapping of the neural cell adhesion protein neurolin purified by sodium dodecyl sulfate polyacrylamide gel electrophoresis or acidic precipitation. *J. Mass Spec.,* 32, 483, 1997.

53. Duncan, M. W., Matanovic, G. and Cerpa-Poljak, A., Quantitative analysis of low molecular weight compounds of biological interest by matrix-assisted laser desorption ionization. *Rapid Commun. Mass Spec.,* 7, 1090, 1993.

54. Gusev, A. I. et al., Direct quantitative analysis of peptides using matrix assisted laser desorption ionization. *Anal. Bioanal. Chem.,* 354, 455, 1996.

55. Kang, M. J., Tholey, A. and Heinzle, E., Quantitation of low molecular mass substrates and products of enzyme catalyzed reactions using matrix-assisted laser desorption/ionization time-of-flight mass spectrometry. *Rapid Commun. Mass Spec.,* 14, 1972, 2000.

56. Kang, M. J., Tholey, A. and Heinzle, E., Application of automated matrix-assisted laser desorption/ionization time-of-flight mass spectrometry for the measurement of enzyme activities. *Rapid Commun. Mass Spec.,* 15, 1327, 2001.

57. Nicola, A. J. et al., Application of the fast-evaporation sample preparation method for improving quantification of angiotensin II by matrix-assisted laser desorption/ionization. *Rapid Commun. Mass Spec.,* 9, 1164, 1995.

58. Bungert, D. et al., Screening of sugar converting enzymes using quantitative MALDI-ToF mass spectrometry. *Biotechnol. Lett.,* 26, 1025, 2004.

chapter fifteen

Future prospects

Mikhel Koel

In this book, we did our best to introduce the trends in analytical chemistry to take in use the ILs. Our aim was to present more than just enlarged review next on the list of publications on ILs, which is growing almost exponentially already. We still can say that ILs are in their infancy, despite the enormous academic interest.

ILs have many fascinating properties, so different from conventional molecular solvents that make them of fundamental interest to all chemists and physicists. In principle it was confirmed that the use of ILs is opening new possibilities in different areas on analysis, and there are appearing new applications; among them are completely new approaches to conduct chemical and biological analysis using ILs. Even the use of ILs as additives like in liquid chromatography and electrophoresis is giving great advantage when compared to common materials.

Today the number of synthesized ILs is large, and among them a very many different cations and even a bigger number of anions are used. There is an almost limitless variety of liquids still to be discovered with a wide range of possible applications. The idea of *designer solvents* has grown in full extent and TSILs are a further direction in the developments. The basis for this activity is the easy preparation of salts with different ion constituents. This ability might best be described as the *chemical tunability* of ILs—class of solvents with members possessing similar physical properties but different chemical behavior. From this becomes evident that intensive study of physicochemical properties such as acidity, basicity, viscosity, solubility, and so on, and thermodynamic behavior when mixing with other solvents or even with other ILs must continue to fill gaps in our knowledge.

This tailoring for certain application must go hand in hand with theoretical work to develop acceptable thermodynamic models such as COSMO, or modified UNIFAC, also possibilities of QSPR approach must be used in full extent, giving the comprehensive approach for design of ILs. High-quality data on reference systems and the creation of the comprehensive database have to be strategized to make progress in the field of ILs. There are several proofs of principle for an advantageous use of some ILs, but the road to design and optimization of TSILs has a long way to go.

Before using the ILs, it must be remembered that they can be dramatically altered by the presence of impurities. Impurities will change the nature of these compounds. The influence of water and chloride anion on the viscosity and density of ILs has already been extensively discussed by many authors. Furthermore, the water content of ILs can affect the rates, directions, and selectivity of reactions and can be taken as cosolvent in extraction process.

Recent data and other scientific and engineering advances on ILs provide the potential for expanded opportunities in almost all of the separations technologies. Future separations needs are related to the pharmaceutical, biomedical, and other biotech industries, microelectronics, aerospace, and alternative fuels (i.e., hydrogen) segments of the economy. Exploration of the use of ILs for gas separation and gas storage, as well as a solvent media for reactions involving permanent gases is accompanied with intensive study of the solubility of gases in ILs. This is well related to use ILs as stationary phases in gas–liquid chromatography. ILs exhibit a unique dual nature selectivity that allows them to separate polar molecules like a polar stationary phase and nonpolar molecules like a nonpolar stationary phase. Further research on mixed IL stationary phases will serve to control and optimize selectivity for complicated analyte mixtures. In addition, the combination of cations and anions can be tuned to add further selectivity for more complex separations. The development of chiral stationary phases will likely mature as more chiral ILs are synthesized and evaluated.

In liquid chromatography and capillary electrophoresis where compounds usually called *ionic liquids* are working as additives in separation media can supply wider range of interaction modalities and their applications suggest they may be particularly advantageous in addressing the critical challenges especially of biomolecular separations. Surface-confined ILs are used as sorbents or capillary wall modifiers, and their specific features allow unique separation capability. Furthermore, the presence of a large ion with a delocalized charge can promote interactions with neutral aromatic species, a scaffold for introducing additional functionality and ion-exchange capability with potentially tunable hydrophobicity. Both gas and liquid chromatography are well-established techniques, and lots of standardized methods are also available. It makes this area very competitive and challenging for ILs.

This variability of ILs presents a challenge not only in instrumental separation methods like chromatography, but also in more common and used industry methods. As concerns extraction, it is associated with the possibility of different mechanisms of solute transfer, different mutual solubilities in biphasic systems, different solvation ability, and, probably, even different bulk phase structures. The success of the IL-based extraction systems, especially for metal ions lies in their ionic nature and possibility for ion exchange and nonvolatility. There is a need to develop more deep understanding of IL-based extraction systems taking the full advantage from the design of ILs; it will be undoubtedly beneficial for upcoming technological and analytical applications. The technique with great potential for larger applications is countercurrent chromatography which works with biphasic liquid systems. The CCC technique is mainly used to produce significant amounts of purified chemicals using as little solvent as possible. As ILs are able to form such biphasic systems with a number of solvents, they have great potential for use in CCC.

Major part of analytical chemistry is related to different spectroscopic techniques. In optical spectroscopy, ILs are already used as solvents for wide range of solutes to study their properties and behavior in conditions not available with organic solvents. ILs have their limits regarding the transparency, but knowing that many other limits are shifted far away. The use of ILs as solvents does not preclude the application of NMR techniques. After careful parameter adjustment, virtually all standard and advanced NMR techniques can be applied. The same can be said about mass spectrometry, which has a great potential to get a key method in almost all fields of IL research, including analytical applications with IL as necessary component to get good result.

A rapidly emerging field in analytical research involves the development of sensors and diagnostic devices centered on ILs as alternatives to molecular solvents and conventional materials. Microchips for analysis and combination with electrowetting phenomena of ILs could be the directions for more intensive studies. Most excitingly, novel applications may emerge in areas that were not even considered in the original concept. In addition, nanotechnology will impact the processes and separations in general with respect to scale and materials.

ILs have a special position in green chemistry also. Their nonvolatility makes them environmentally benign solvents, and in the beginning all studies on ILs were advertised as green chemistry. However, a closer look revealed that a more systematic approach to the assessment of greenness of chemistry and processes involving ILs is needed. Not all ILs are safe and nontoxic, and processes with ILs involved are not wasteless. Despite that, the great potential of ILs for green chemistry solutions is still not fully realized.

About the Contributors

Jared L. Anderson obtained his BS in 2000 at South Dakota State University and his PhD in analytical chemistry at Iowa State University in 2005. In August 2005, he joined the faculty at the University of Toledo (Ohio) as an assistant professor. In 2008, he was awarded a Faculty Early CAREER Award by the National Science Foundation. In the same year, he was awarded the Evangelos Theodosious Sigma Xi Young Faculty Research Award at the University of Toledo. His research interests include the use of ionic liquids in all aspects of separation science including analytical extractions, purification, and chromatography.

Gary A. Baker earned his PhD in chemistry from the University of New York at Buffalo in 2001 under the direction of Professor Frank V. Bright. Following postdoctoral training at Los Alamos National Laboratory, he joined the staff in the Chemical Sciences Division at Oak Ridge National Laboratory as a Eugene P. Wigner Fellow in 2005. His research interests currently include analytical applications of ionic liquids, protein–nanoparticle assembly-based biosensors, microcantilever technology, *green* nanoscience, and environmentally responsive materials.

Sheila N. Baker earned her PhD in chemistry from the University of New York at Buffalo in 2002 under the supervision of Professor Frank V. Bright. Following a postdoctoral appointment at Los Alamos National Laboratory, she worked as Senior Scientist at Protein Discovery, Inc. Recently she founded a small company, Science Catalyst, Inc. (Knoxville, Tennessee) which provides editing and proofreading services for scientific documents and consulting services in a range of areas. Now she is affiliated to Neutron Scattering Science Division, Oak Ridge National Laboratory.

Rolf W. Berg defended his dissertation (candidate of chemical engineering) in vibrational spectroscopy of coordination compounds at the Technical University of Denmark, Lyngby, Denmark in 1972. He has been a research fellow and after 1984 an associate professor at the Department of Chemistry for many years. He has had long-term scientific visits to Leicester University

(United Kingdom), Argonne National Laboratory (Illinois, United States), University of Trondheim (Norway), and Memorial University of St. Johns (New Foundland, Canada). Since 1992, he has been affiliated with Danish Academy of Natural Sciences. He is a member of the American Society for Applied Spectroscopy and of the editorial board of *Applied Spectroscopy Reviews*. His interests have included such areas as molten salt chemistry, crystallography, and material sciences.

Alain Berthod received his PhD in 1979 from the University of Lyon. He took an assistant professor's position at the French National Center for Scientific Research (CNRS) working in electrochemistry. In 1983 he was promoted as associate professor and in 1993 as research director. He focused on the developing and the use of micellar solutions and microemulsions in chromatography. His interests lie in the separation of chiral molecules and enantiorecognition mechanisms. He has contributed to the development of the countercurrent chromatography technique that uses a support-free liquid stationary phase. He was member of the editorial board of major analytical chemistry and chromatography journals. He is editor-in-chief of *Separation & Purification Reviews* (Taylor & Francis, Philadelphia, Pennsylvania).

Joan Frances Brennecke is the Keating-Crawford Professor of Chemical Engineering at the University of Notre Dame and Director of the Notre Dame Energy Center. She completed her PhD and MS (1989 and 1987) degrees at the University of Illinois at Urbana–Champaign and her BS at the University of Texas at Austin (1984). Her research interests are primarily in the development of less environmentally harmful solvents, including supercritical fluids and ionic liquids. Her most recent awards include the 2006 Professional Progress Award from the American Institute of Chemical Engineers and the J.M. Prausnitz Award, presented at the Eleventh International Conference on Properties and Phase Equilibria in 2007.

Frank V. Bright earned his PhD from Oklahoma State University in 1985 He was a postdoctoral fellow at Indiana University working with Gary M. Hieftje (1985–1987) before joining the faculty at the University at Buffalo, The State University of New York, in 1987. He is currently a UB Distinguished Professor, A. Conger Goodyear Chair, and Chemistry Department Chair.

Samuel Carda-Broch received his PhD in 2000 from the University of Valencia under the supervision of Dr. M.C. Garcia-Alvarez-Coque. He worked under a postdoctoral Marie Curie fellowship granted by the European Commission from 2001 to 2003 while working at the Laboratoire des Sciences Analytiques with Dr. A. Berthod. Since 2004, he has been a professor of analytical chemistry at Universidad Jaume I. Currently, his research focuses on drugs of abuse in biological fluids and formulations using micellar liquid chromatography.

Sheng Dai, leader of Nanomaterials Chemistry Group and senior research scientist at Chemical Sciences Division of Oak Ridge National Laboratory (ORNL) and adjunct professor at the University of Tennessee at Knoxville (UTK), received his PhD in chemistry from UTK in 1990. He has authored or coauthored more than 180 peer-reviewed journal or book publications. He currently holds five U.S. patents. His research interest includes chemical synthesis of novel materials, separation, catalysis, sensor development, and molecular recognition. Many of these publications are in the area of ionic liquids.

Urszula Domańska has been professor, Faculty of Chemistry, Warsaw University of Technology since February 1995. She has been the Head of the Physical Chemistry Division since September 1991 and vice director of the Institute of Fundamental Chemistry (1988–1990). She had long-term scientific visits as visiting professor: Laboratoire De Thermodynamique Et D'Analyse Chimique, University of Metz, France; University of Turku, Finland; Faculty of Science, Department of Chemistry, University of Natal, South Africa; Department of Chemical Engineering, Louisiana State University, United States. Her interests have included such areas of physical chemistry as thermodynamics, especially thermodynamics of phase equilibria, VLE, LLE, SLE, high-pressure SLE, separation science, calorimetry, correlation and prediction of physical–chemical properties, and ionic liquids. She is a member of the Polish Chemical Society; member of the Polish Association of Calorimetry and Thermal Analysis; member of IUPAC Commission on Solubility; member of International Association of Chemical Thermodynamics; and scientific advisor at the *Journal of Chemical Engineering Data*.

Vladimir M. Egorov was born in Odintsovo-10, Moscow Region, Russia, in 1982. He received his MS in chemistry from Moscow State University in 2005. The subject of his diploma thesis was to develop a method of analytical reagents immobilization on cellulose matrix by dissolution or reconstitution using ionic liquids. Currently, he is a postgraduate student at the MSU Chemistry Department. His research interests include (but not limited to) application and synthesis of novel ionic liquids and computational chemistry. He has been a prize-winner of numerous contests in chemistry, mathematics, and biology.

Ralf Giernoth studied chemistry at the University of Bonn (Germany) and finished his PhD (1999) with Joachim Bargon at the Institute of Physical and Theoretical Chemistry. From 1999 to 2001 he spent 2 years of postdoctoral research at the Dyson Perrins Laboratory, University of Oxford (United Kingdom) with John M. Brown. Since 2001, he has been doing independent research at the Institute of Organic Chemistry of the University of Cologne (Germany). His main fields of interest include *in situ* spectroscopy, catalysis, and chemistry in ionic liquids.

Christopher Hardacre read natural sciences at the University of Cambridge, where he also obtained his PhD in 1994. He was appointed to a lectureship in physical chemistry at the Queen's University of Belfast in 1995 and in 2003, he became a professor of physical chemistry. He is currently director of research in CenTACat, and his current interests include the understanding of gas and liquid phase catalytic processes for emission control, clean energy production, and fine chemical synthesis as well as the study and use of ionic liquids.

William T. Heller obtained his PhD in physics from Rice University in 1999. After postdoctoral training at Los Alamos National Laboratory, he moved to Oak Ridge National Laboratory for additional postdoctoral training before accepting a post as chemist/biophysicist in the Chemical Sciences Division of Oak Ridge National Laboratory. His work entails the application of neutron and x-ray scattering methods to the study of the structure and function of biological macromolecules and the characterization of materials.

Mihkel Kaljurand obtained his PhD degree in 1979 at the State University of Leningrad and DSc degree in analytical chemistry in 1990 at the Institute of Physical Chemistry of the USSR Academy of Sciences (Moscow). Since 1995 he has worked as a professor of analytical chemistry at Tallinn University of Technology.

In 1995 he was a research fellow at NASA, United States. His current research interests lie in environment-friendly analytical methods and their applicability bioactive compounds. He has received the National Science Prize twice (Estonia: 1991, 2006).

Zulema K. Lopez-Castillo received her BSc degree in chemical engineering from University of Sonora in 1995 and her MSc degree in chemical engineering from National Polytechnic Institute in 1998; both degrees in Mexico. She earned her PhD degree in chemical engineering from Texas A&M University in 2003 working in supercritical fluids. She is currently working as a postdoctoral research associate at the University of Notre Dame, working in the area of ionic liquids.

Huimin Luo received her PhD in organic chemistry from UTK in 1992. She became a staff scientist of Nuclear Science and Technology Division at ORNL in 1999 after a brief employment at a major pharmaceutical corporation. She was one of the ten national recipients of the Alexander Hollander Fellows in Life Sciences in 1993. Her major research interests and experience include organic synthesis of novel ionic liquids, macrocyclic ligands, perfluorinated cage compounds, nuclear medicine, and fluorine chemistry. She has published more than 30 papers in refereed journals in the aforementioned areas.

Taylor A. McCarty received her BA in biology and chemistry from SUNY Potsdam in 2003. She recently obtained her PhD in analytical chemistry from the University at Buffalo under the supervision of Dr. Frank Bright, where she is now the lab manager. Her research interests include the behavior of macromolecules such as polymers and proteins dissolved in environment-friendly solvent systems.

Berlyn Rose Mellein received her BS in chemical engineering at the University of California, Santa Barbara, in 2003, before joining the graduate program at the University of Notre Dame. She earned her MS in chemical engineering in 2007 and anticipates completion of her PhD degree in 2008.

Claire Lisa Mullan is currently a PhD student in the School of Chemistry and Chemical Engineering at Queen's University Belfast. Her project focuses on the determination of structural properties of ionic liquids and the nature of solute–solvent interactions, collaborating with both theoreticians and experimentalists.

Igor V. Pletnev was born in 1960 in Sheelit, Ekaterinburg region, Russia (curiously, this small Ural city is named after mineral *scheelite*, $CaWO_4$, and the great chemist *Carl Wilhelm Scheele*). In 1982 he graduated with honors from the Chemistry Department of Lomonosov Moscow State University (MSU). In 1986 he obtained his PhD in the two fields, analytical and inorganic chemistry, from the Vernadsky Institute of Geochemistry and Analytical Chemistry, Russian Academy of Sciences. Since then, he stayed at MSU Chemistry Department, where he now serves as Leading Research Scientist and head of the group at Analytical Chemistry division. In 2000 he was awarded the Russian State Prize in Science and Technology and in 2005 he received his doctorate (Doctor of Sciences, Russian post-PhD degree) from MSU. His research interests include separation science, coordination, and computational chemistry. He authored and coauthored more than 60 papers and chapters in 4 books. He has been supervisor of more than 10 PhD students, including the coauthors of this book.

Maria-Jose Ruiz-Angel received her PhD in 2003 under the supervision of Dr. M.C. Garcia-Alvarez-Coque at the University of Valencia. In 2004 she had a 2-year postdoctoral fellowship granted by the Spanish Ministry of

Education at the Laboratoire des Sciences Analytiques, Université Claude Bernard (France) under the supervision of Dr. A. Berthod. Since January 2007 she has a *Ramón y Cajal* research position at the University of Valencia.

Svetlana V. Smirnova was born in Yartsevo, Smolensk region, Russia. She graduated from MSU Chemistry Department, having specialization in analytical chemistry. She holds MS and PhD degrees from MSU. Her PhD thesis was devoted to the solvent extraction of amino acids. She works at the MSU Chemistry Department as an assistant professor (since 2002) and an associate professor (since 2005) giving the lectures and practice courses in analytical chemistry. Her scientific interests lie in the area of application of ionic liquids in extraction processes. She is the coauthor of more than 20 scientific publications.

Apryll M. Stalcup received her PhD in chemistry (1988) from Georgetown University in Washington, DC. After postdoctoral training at the University of Missouri–Rolla, she joined the Department of Chemistry at the University of Hawaii–Manoa in 1990 as an assistant professor. She moved to the Chemistry Department at the University of Cincinnati as an associate professor in 1996 and was promoted as a professor in 2002. Her research interests include liquid chromatography, capillary electrophoresis, chiral separations, and investigations of separation mechanisms.

Andreas Tholey studied chemistry at Saarland University in Saarbrücken, Germany. In his doctoral thesis at the German Cancer Research Centre (DKFZ) in Heidelberg, Germany, he studied the structural consequences of protein phosphorylation and the fragmentation behavior of phosphopeptides in nanoelectrospray mass spectrometry. During his postdoc in the department of Technische Biochemie at Saarland University (Professor E. Heinzle), he developed methods for the quantitative analysis of low-molecular weight compounds by MALDI mass spectrometry and applied these methods for the determination of enzyme activities in biochemistry and biotechnology. The application of ionic liquids as matrix substances for MALDI MS was an essential part of these developments. After the habilitation at Saarland University in April 2007, the focus of his work is now the development and application of mass spectrometric techniques for the analysis of proteomes and protein modifications.

Merike Vaher received her MSc degree in the field of polysaccharides in 1999 and PhD degree under the supervision of Professor Mihkel Kaljurand and Dr. Mihkel Koel in the field of ionic liquids in 2002 at Tallinn University of Technology, Estonia. Her current research interest is investigation of biologically active compounds by CE. She has received the National Science Prize twice (Estonia: 1987, 2006).

Tristan Gerard Alfred Youngs obtained his PhD in computational chemistry from the University of Reading in 2004, and moved to Queen's University Belfast to take up a research fellowship working on ionic liquids shortly afterward. His interests focus on the use of computational methodology to examine ionic liquids at the atomic and molecular levels, and works closely with experimentalists and theorists alike.

Index

D

E

Printed and bound by CPI Group (UK) Ltd, Croydon, CR0 4YY

24/10/2024

01778302-0014